T0202953

Universitext

Universitext

Universitext is a series of textbooks that presents material from a wide variety of mathematical disciplines at master's level and beyond. The books, often well class-tested by their author, may have an informal, personal even experimental approach to their subject matter. Some of the most successful and established books in the series have evolved through several editions, always following the evolution of teaching curricula, to very polished texts.

Thus as research topics trickle down into graduate-level teaching, first textbooks written for new, cutting-edge courses may make their way into *Universitext*.

More information about this series at http://www.springer.com/series/223

Hans Wilhelm Alt

Linear Functional Analysis

An Application-Oriented Introduction

Translated by Robert Nürnberg

 Springer

Hans Wilhelm Alt
Technische Universität München
Garching near Munich
Germany

Translation from German language edition:
Lineare Funktionalanalysis
by Hans Wilhelm Alt
Copyright © 2012, Springer Berlin Heidelberg
Springer Berlin Heidelberg is part of Springer Science + Business Media
All Rights Reserved

ISSN 0172-5939   ISSN 2191-6675   (electronic)
Universitext
ISBN 978-1-4471-7279-6   ISBN 978-1-4471-7280-2   (eBook)
DOI 10.1007/978-1-4471-7280-2

Library of Congress Control Number: 2016944464

Mathematics Subject Classification: 46N20, 46N40, 46F05, 47B06, 46G10

Printed on acid-free paper

This Springer imprint is published by Springer Nature
The registered company is Springer-Verlag London Ltd.

Preface

The present book is the English translation of a previous German edition, also published by Springer Verlag. The translation was carried out by Robert Nürnberg, who also did a marvellous job at detecting errors and mistakes in the original version. In addition, Andrei Iacob revised the English version.

The book originated in a series of lectures I gave for the first time at the University of Bochum in 1980, and since then it has been repeatedly used in many lectures by me and other mathematicians and during this time it has changed accordingly. I provide the reader with an introduction to Functional Analysis as a synthesis of Algebra, Topology, and Analysis, which is the source for basic definitions which are important for differential equations. The book includes a number of appendices in which special subjects are presented in more detail. Therefore its content is rich enough for a lecturer to find enough material to fill a course in functional analysis according to his special interests. The text can also be used as an additional source for lectures on partial differential equations or advanced numerical analysis.

It must be said that my strategy has been dictated by the desire to offer the reader an easy and fast access to the main theorems of linear functional analysis and, at the same time, to provide complete proofs. So there is a separate appendix where the Lebesgue integral is introduced in a complete functional analytic way, and an appendix whith details for Sobolev functions which complete the proofs of the embedding theorems. Therefore the text is self-contained and the reader will benefit from this fact.

Parallel to this edition, a revised German version has become available (Lineare Funktionalanalysis, 6. Edition, Springer 2012) with the same mathematical content. This is made possible by a common source text. Therefore one does not have to worry about the content in different versions. I am happy that this book is now accessible to a wider community.

If you find any errors or misprints in the text, please point them out to the author via email: "alt@ma.tum.de". This will help to improve the text of possible future editions.

I hope that this book is written in the good tradition of functional analysis and will serve its readers well. I thank Springer Verlag for making the publication of this edition possible and for their kind support over many years.

Technical University Munich, August 2015
H. W. Alt

Table of Contents

1 Introduction

Functional analysis deals with the structure of function spaces and the properties of continuous mappings between these spaces. Linear functional analysis, in particular, is confined to the analysis of linear mappings of this kind. Its development was based on the fundamental observation that the topological concepts of the Euclidean space $\mathbb{R}^n$ can be generalized to function spaces as well. To this end, functions are interpreted as points in a given space (see the cover page, where a part of the orthonormal system in 9.9 is shown). Given a set S, we consider the set of all maps $f : S \to \mathbb{R}$. Denoting this set by $\mathscr{F}(S; \mathbb{R})$ means that any point $f \in \mathscr{F}(S; \mathbb{R})$ defines a mapping $x \mapsto f(x)$ that assigns to each element $x \in S$ a unique $f(x) \in \mathbb{R}$. Then the set $\mathscr{F}(S; \mathbb{R})$ becomes a vector space if we define for all $f_1, f_2, f \in \mathscr{F}(S; \mathbb{R})$ and $\alpha \in \mathbb{R}$

$$(f_1 + f_2)(x) := f_1(x) + f_2(x) \,, \ (\alpha f)(x) := \alpha f(x) \quad \text{for } x \in S \,.$$

With the help of characteristic examples we now investigate similarities and differences between the Euclidean space $\mathbb{R}^n$ and some function spaces. The function spaces will be covered in more detail later on in the book.

First we consider the space $C^0(S)$ (see 3.2) of continuous functions $f : S \to \mathbb{R}$, where S is a bounded, closed set in $\mathbb{R}^n$. The supremum norm on $C^0(S)$ is defined by

$$\|f\|_{C^0} := \sup\{|f(x)| \,; \ x \in S\} \quad \text{for } f \in C^0(S) \,.$$

It satisfies the same norm axioms (see 2.4) as the Euclidean norm on $\mathbb{R}^n$,

$$\|x\|_{\mathbb{R}^n} := \left(\sum_{i=1}^{n} x_i^2 \right)^{\frac{1}{2}} \quad \text{for } x = (x_i)_{i=1,\ldots,n} = (x_1, \ldots, x_n) \in \mathbb{R}^n \,.$$

One difference between the two spaces is that $C^0(S)$, in contrast to $\mathbb{R}^n$, is an infinite-dimensional space, when S contains infinitely many points. This can be seen as follows. Let $x_i \in S$ for $i \in \mathbb{N}$ be pairwise distinct. Then for each $n \in \mathbb{N}$ we can find functions $\varphi_{n,i} \in C^0(S)$ for $i = 1, \ldots, n$, such that $\varphi_{n,i}(x_j) = \delta_{i,j}$ for $i, j = 1, \ldots, n$. Here

$$\delta_{i,j} := \begin{cases} 1 & \text{for } i = j, \\ 0 & \text{otherwise} \end{cases}$$

denotes the **Kronecker symbol**. Now if $\alpha_i \in \mathbb{R}$ for $i = 1, \ldots, n$ are such that

$$f := \sum_{i=1}^{n} \alpha_i \varphi_{n,i} = 0 \quad \text{in } C^0(S),$$

then it follows that $0 = f(x_j) = \alpha_j$ for $j = 1, \ldots, n$. Hence $\varphi_{n,1}, \ldots, \varphi_{n,n}$ are linearly independent and, since $n \in \mathbb{N}$ was chosen arbitrarily, the dimension of $C^0(S)$ cannot be finite. This changes the properties of the space significantly. For instance, while in $\mathbb{R}^n$ all bounded closed sets are compact (see the Heine-Borel theorem 4.7(7)), this is not the case in $C^0(S)$ (see the Arzelà-Ascoli theorem 4.12).

Also, the scalar product in $\mathbb{R}^n$,

$$(x, y)_{\mathbb{R}^n} := \sum_{i=1}^{n} x_i y_i \quad \text{for } x = (x_i)_{i=1,\ldots,n} , \ y = (y_i)_{i=1,\ldots,n} \in \mathbb{R}^n,$$

has an analogue for function spaces; indeed, define (cf. 3.16(3))

$$(f, g)_{L^2} := \int_S f(x)g(x) \, dx \quad \text{for } f, g \in C^0(S).$$

The corresponding norm $\|f\|_{L^2} := \sqrt{(f, f)_{L^2}}$ is bounded from above by the supremum norm, that is, there exists a constant $C < \infty$ such that

$$\|f\|_{L^2} \leq C \|f\|_{C^0} \quad \text{for all } f \in C^0(S)$$

(this follows from 3.18, if C denotes the square root of the Lebesgue measure of S). In general, a similar bound from below cannot be derived. To see this, consider the interval $S = [-1, 1] \subset \mathbb{R}$ and for $0 < \varepsilon < 1$ the functions f_ε defined by $f_\varepsilon(x) := \max\left(0, \frac{1}{\varepsilon}\left(1 - \frac{|x|}{\varepsilon}\right)\right)^{\frac{1}{2}}$, for which $\|f_\varepsilon\|_{C^0} = \varepsilon^{-\frac{1}{2}}$, but $\|f_\varepsilon\|_{L^2} = 1$. That means that the C^0-norm and the L^2-norm on $C^0(S)$ are not equivalent to each other (see 2.15); the C^0-norm is stronger than the L^2-norm. That is, the space $C^0(S)$, equipped with the L^2-norm, is not complete. For example, the functions g_k for $k \in \mathbb{N}$, $g_k(x) := (1-x)^k$ for $x \geq 0$, $g_k(x) := 1$ for $x \leq 0$, form a Cauchy sequence with respect to the L^2-norm, but there exists no function $g \in C^0(S)$ such that $\|g_k - g\|_{L^2} \to 0$ as $k \to \infty$.

In a situation like this we can apply a general principle in mathematics: completion (see 2.24). Similarly to defining the real numbers $\mathbb{R}$ as the completion of the rational numbers $\mathbb{Q}$, we can complete the space $C^0(S)$ with respect to the L^2-norm. Thus we obtain the complete space $L^2(S)$ of all square integrable, Lebesgue measurable functions on S (see 3.15 and 4.15(3)). In this space fundamental assertions hold, such as Lebesgue's convergence theorem (see 3.25).

We encounter a similar situation in a further generalization from the finite-dimensional case to the infinite-dimensional one. For the finite-dimensional case, let $E : S \to \mathbb{R}$ be a continuous function defined on a bounded closed set $S \subset \mathbb{R}^n$. We now look for a minimum of this function over S. The compactness of S and the continuity of E yield that such a minimum exists: E has an **absolute minimum** on S, that is, there exists an $x_0 \in S$ such that

$$E(x_0) = \inf_{x \in S} E(x).$$

The same holds true if we only assume that S is closed and if in addition we require that $E(x) \to \infty$ for $x \in S$ as $\|x\|_{\mathbb{R}^n} \to \infty$.

As an infinite-dimensional analogue we consider the following Dirichlet boundary value problem on an open, bounded set $\Omega \subset \mathbb{R}^n$. The given datum is a continuous function u_0 defined on the boundary $\partial\Omega$ of Ω, i.e. $u_0 \in C^0(\partial\Omega)$, and we want to find a continuous function $u : \overline{\Omega} \to \mathbb{R}$ that is twice continuously differentiable in Ω, such that

$$\Delta u(x) := \sum_{i=1}^{n} \frac{\partial^2}{\partial x_i^2} u(x) = 0 \quad \text{for } x \in \Omega,$$

$$u(x) = u_0(x) \quad \text{for } x \in \partial\Omega.$$

In applications, u is, for example, a stationary temperature distribution or the potential of a charge-free electric field. One approach to find a solution is to consider the corresponding **energy functional** (here identical to the **Dirichlet integral**)

$$E(u) := \frac{1}{2} \int_{\Omega} |\nabla u(x)|^2 \, dx,$$

where $\nabla u(x) := \left(\frac{\partial}{\partial x_1} u(x), \ldots, \frac{\partial}{\partial x_n} u(x) \right)$. Here we use the term **functional**, because E acts on functions, that is, E is a function defined on functions. In order to guarantee that $E(u) < \infty$, we initially define the domain of E to be

$$M := \left\{ v \in C^1(\overline{\Omega}) \, ; \, v = u_0 \text{ on } \partial\Omega \right\}, \quad \text{so} \quad E : M \to \mathbb{R},$$

where we assume that M is nonempty. If we now assume that $u \in M$ is an absolute minimum of E on M, then $E(u) \le E(u + \varepsilon\zeta)$ for all $\varepsilon \in \mathbb{R}$ and all $\zeta \in C^1(\overline{\Omega})$ such that $\zeta = 0$ in a neighbourhood of $\partial\Omega$. On noting that $\varepsilon \mapsto E(u + \varepsilon\zeta)$ is differentiable in ε (this function is quadratic in ε), it follows that

$$0 = \frac{d}{d\varepsilon} E(u + \varepsilon\zeta)\big|_{\varepsilon=0} = \int_{\Omega} \nabla\zeta(x) \bullet \nabla u(x) \, dx.$$

The fact that this identity holds for all functions ζ with the above mentioned properties contains all the information needed in order to derive a differential equation for u. That is why the functions ζ are also called **test functions**, and $u \in M$ is called a **weak solution** of the boundary value problem if the integral identity holds for all test functions. Introducing this solution concept allows the treatment of partial differential equations by means of functional analysis (see 6.5–6.8). We obtain the corresponding classical differential equation on assuming that $u \in C^2(\Omega)$, as integration by parts then yields that

$$0 = \int_\Omega \nabla \zeta(x) \bullet \nabla u(x) \, dx = - \int_\Omega \zeta(x) \Delta u(x) \, dx$$

for all test functions ζ. This implies $\Delta u = 0$ in Ω (cf. 4.22), and hence u is a solution of the original Dirichlet problem.

However, the existence of an absolute minimum $u \in M$ for a functional $E : M \to \mathbb{R}$ with $M \subset C^1(\overline{\Omega})$ is not established as easily as in the finite-dimensional case. For instance, if $\Omega = \,]0,1[$,

$$M_1 := \left\{ u \in C^1([0,1]) \; ; \; u(0) = 0, \; u'(1) = 1 \right\}$$

with $\|u\|_{C^1} := \|u\|_{C^0} + \|u'\|_{C^0}$ and

$$E_1(u) := \|u'\|_{C^0}^2 + \int_0^1 |u'(x)|^2 \, dx \,,$$

then M_1 is closed in $C^1([0,1])$ and E_1 is continuous with respect to the C^1-norm. Moreover, $E_1(u) \geq \|u'\|_{C^0}^2 \to \infty$ for $u \in M_1$ as $\|u\|_{C^1} \to \infty$, since for $u \in M_1$ and $x \in [0,1]$ we have

$$|u(x)| = \left| \int_0^x u'(y) \, dy \right| \leq \|u'\|_{C^0} \,,$$

and hence $\|u\|_{C^1}^2 \leq 4\|u'\|_{C^0}^2$. Consequently, all the assumptions are satisfied which lead in the above finite-dimensional case to the existence of an absolute minimum.

But E_1 does not have an absolute minimum on M_1. To see this, note that $E_1(u) \geq \|u'\|_{C^0}^2 \geq |u'(1)|^2 = 1$ for all $u \in M_1$. This lower bound also represents the infimum of E_1 over M_1, since the functions $u_\varrho(x) := \frac{1}{\varrho} x^\varrho$ for $\varrho > 1$ satisfy

$$\|u_\varrho'\|_{C^0}^2 = 1 \quad \text{and} \quad \int_0^1 |u_\varrho'(x)|^2 \, dx = \frac{1}{2\varrho - 1} \to 0 \text{ as } \varrho \to \infty.$$

Now, if $u \in M_1$ was an absolute minimum, i.e. $E_1(u) = 1$, then

$$\|u'\|_{C^0}^2 = 1 \quad \text{and} \quad \int_0^1 |u'(x)|^2 \, dx = 0 \,.$$

But the second equality implies $u' = 0$, which contradicts the first equality.

In conclusion, we note that the main difficulty in proving the existence of an absolute minimum lies in the fact that $C^1(\overline{\Omega})$ is equipped with a supremum norm, while the functional $E(u) = \frac{1}{2}\|\nabla u\|_{L^2}^2$ corresponds to an integral norm, which cannot be used to bound the C^1-norm (similarly to our first example). If, on the other hand, we equip $C^1(\overline{\Omega})$ with the integral norm

$$\|u\|_{W^{1,2}} := \|u\|_{L^2} + \|\nabla u\|_{L^2},$$

then (similarly to the first example) the space is no longer complete. But the completeness of the space under consideration is a crucial property in all existence proofs. Hence at times it becomes necessary to seek solutions to boundary value problems, or minima of functionals, in a larger class of functions. For instance, on completing the space $C^1(\overline{\Omega})$ with respect to the above $W^{1,2}$-norm (see 3.27), and thus obtaining the **Sobolev space** $W^{1,2}(\Omega)$, we can consider the functional E to be defined on $W^{1,2}(\Omega)$ rather than on $C^1(\overline{\Omega})$. In this new space, the above variational problem admits a solution (see 8.17).

As a third example we consider the infinite-dimensional analogue of matrices. The set of all sequences with only finitely many nonzero terms is defined by

$$c_* := \left\{ x = (x_k)_{k \in \mathbb{N}} ; \ x_k \in \mathbb{R} \text{ for } k \in \mathbb{N}, \text{ and there exists an } n \in \mathbb{N}, \right.$$
$$\left. \text{such that } x_k = 0 \text{ for all } k > n \right\}.$$

A linear map $T : c_* \to c_*$ is characterized by the values T_{ij}, the i-th coordinate of $T(\mathbf{e}_j)$. Here $\mathbf{e}_j$ corresponds to the j-th unit vector of the Euclidean space, that is, $\mathbf{e}_j := (\delta_{j,k})_{k \in \mathbb{N}} \in c_*$. In other words

$$Tx = \sum_{i \in \mathbb{N}} \left(\sum_{j \in \mathbb{N}} T_{ij} x_j \right) \mathbf{e}_i,$$

where in each sum only finitely many terms are nonzero, with their number depending on x. Hence T can be represented by a matrix $(T_{ij})_{i,j \in \mathbb{N}}$ with infinitely many rows and columns.

For finite matrices, i.e. in the finite-dimensional case, a linear map $T : \mathbb{R}^n \to \mathbb{R}^n$ is injective if and only if it is surjective. However, if we consider the **shift operator** $T : c_* \to c_*$, defined by

$$T(x_1, x_2, x_3, \ldots) := (0, x_1, x_2, x_3, \ldots),$$

then T is injective, but not surjective. Nevertheless, later on we will see that the above property of finite matrices carries over to certain maps, namely to compact perturbations of the identity (see the Fredholm alternative 11.11). Chapters 11 and 12 are devoted to the spectral theory of such operators. There we will generalize results from linear algebra that provide normal forms for finite-dimensional matrices. For instance, the Jordan normal form of matrices corresponds to the spectral theorem for compact operators (see 11.9

and 11.13), while the fact that every symmetric matrix is diagonalizable corresponds to the spectral theorem for compact normal operators (see 12.11 and 12.12). In function spaces such operators occur in the analysis of differential and integral equations.

As a final example we consider a **Sturm-Liouville problem**. A solution to the Sturm-Liouville problem is given by a function $u \in C^2([0,1])$ satisfying the differential equation

$$Tu := -(pu')' + qu = f$$

and, for instance, satisfying the boundary conditions

$$u(0) = 0, \quad u'(1) = 0.$$

We assume that the right-hand side of the differential equation satisfies $f \in C^0([0,1])$, while for the coefficients we assume e.g. $q \in C^0([0,1])$ and $p \in C^1([0,1])$, with p being a strictly positive function, i.e. there exists a number $c > 0$, such that $p(x) \geq c$ for all $x \in [0,1]$.

The Sturm-Liouville problem can be formulated as an integral equation. Then one looks for a function $u \in C^0([0,1])$ such that $u = K_f u$, where

$$(K_f u)(x) := \int_0^x \frac{1}{p(y)} \int_y^1 (f - qu)(z)\, dz\, dy.$$

If $u \in C^0([0,1])$ is a solution to this integral equation, i.e. $u = K_f u$, then the integral representation and the assumptions on p, q, f yield that $u \in C^2([0,1])$, and that both the differential equation and the boundary conditions are satisfied.

It follows from the Banach fixed point theorem that the integral equation admits a unique solution. This is true whenever K_f is a **contraction mapping**, i.e. if there exists a number $\theta < 1$, such that

$$\|K_0 u\| \leq \theta \|u\| \quad \text{for all } u \in C^0([0,1]),$$

where $\|\cdot\|$ denotes the supremum norm (it is also possible to use other, equivalent norms, which can lead to improved contraction factors). For instance, for $p = 1$ we have

$$|K_0 u(x)| = \left| \int_0^x \int_y^1 (qu)(z)\, dz\, dy \right| = \left| \int_0^1 (qu)(z) \min(z, x)\, dz \right|$$

$$\leq \|u\| \int_0^1 |zq(z)|\, dz,$$

and hence the boundary value problem has a unique solution if

$$p = 1 \quad \text{and} \quad \int_0^1 z|q(z)|\, dz < 1.$$

However, this unduly restricts the class of admissible functions q. In order to be able to treat more general q, we reformulate the problem and attempt to solve an infinite-dimensional system of linear equations. To this end, let $\{e_i \, ; \; i \in \mathbb{N}\}$ be a linearly independent set in the function space

$$V := \left\{ v \in C^2([0,1]) \; ; \; v(0) = 0, \; v'(1) = 0 \right\}$$

and define

$$a_{ij} := \int_0^1 e_i(x)(Te_j)(x)\,\mathrm{d}x \quad \text{and} \quad f_i := \int_0^1 e_i(x)f(x)\,\mathrm{d}x\,.$$

Using the formal ansatz $u = \sum_{j \in \mathbb{N}} u_j e_j$ it then follows from $Tu = f$ that formally

$$\sum_{j \in \mathbb{N}} a_{ij} u_j = f_i \quad \text{for all } i \in \mathbb{N}\,.$$

If the e_i form a Schauder basis (see 9.3) with respect to the L^2-norm, then this infinite-dimensional system of equations is even formally equivalent to the differential equation. For, with an arbitrary function $\zeta = \sum_{i \in \mathbb{N}} \alpha_i e_i \in V$ and since $Tu = \sum_{j \in \mathbb{N}} u_j Te_j$, it follows from the system of equations that

$$0 = \sum_{i \in \mathbb{N}} \alpha_i \left(\sum_{j \in \mathbb{N}} a_{ij} u_j - f_i \right)$$

$$= \sum_{i \in \mathbb{N}} \alpha_i \left(\int_0^1 \sum_{j \in \mathbb{N}} u_j e_i(x)(Te_j)(x)\,\mathrm{d}x - \int_0^1 e_i(x)f(x)\,\mathrm{d}x \right)$$

$$= \int_0^1 \left(\sum_{i \in \mathbb{N}} \alpha_i e_i(x) \right) \left(\sum_{j \in \mathbb{N}} u_j (Te_j)(x) - f(x) \right) \mathrm{d}x$$

$$= \int_0^1 \zeta(x)\big((Tu)(x) - f(x)\big)\,\mathrm{d}x\,,$$

and hence (similarly to the Dirichlet problem above) that the differential equation is fulfilled. Remember, that this conclusion was formal.

We now assume that we can choose for each i the e_i as normalized eigenvector of T corresponding to the eigenvalue λ_i, i.e.

$$Te_i = \lambda_i e_i\,, \quad \int_0^1 e_i(x)^2\,\mathrm{d}x = 1\,.$$

It follows for $i, j \in \mathbb{N}$ that

$$(\lambda_i - \lambda_j) \int_0^1 e_i(x)e_j(x)\,\mathrm{d}x$$

$$= \int_0^1 (Te_i)(x)e_j(x)\,\mathrm{d}x - \int_0^1 e_i(x)(Te_j)(x)\,\mathrm{d}x = 0\,.$$

For the last identity we have used the fact that T is a self-adjoint operator. To see this, note that, for $u, v \in V$,

$$\int_0^1 v(x)(Tu)(x)\,\mathrm{d}x$$

$$= -\int_0^1 v(x)(pu')'(x)\,\mathrm{d}x + \int_0^1 q(x)v(x)u(x)\,\mathrm{d}x$$

$$= -\underbrace{\left[\, v(x)p(x)u'(x) \,\right]_{x=0}^{x=1}}_{=0} + \int_0^1 \big(p(x)v'(x)u'(x) + q(x)v(x)u(x)\big)\,\mathrm{d}x$$

is symmetric in v and u. Hence for $\lambda_i \neq \lambda_j$ it follows that

$$a_{ij} = \int_0^1 e_i(x)(Te_j)(x)\,\mathrm{d}x = \lambda_j \int_0^1 e_i(x)e_j(x)\,\mathrm{d}x = 0\,.$$

Moreover, setting $N := \{i \in \mathbb{N}\,;\ \lambda_i = 0\}$ and assuming that all eigenvalues λ_i with $i \notin N$ are pairwise distinct, yields that

$$a_{ij} = \lambda_i \delta_{i,j} \quad \text{for all } i, j \in \mathbb{N}.$$

Hence the (formal) infinite-dimensional system of linear equations is reduced to diagonal form and reads

$$\lambda_i u_i = f_i \quad \text{for all } i \in \mathbb{N}.$$

We obtain the solvability condition

$$f_i = 0 \quad \text{for } i \in N,$$

and, formally, the solution

$$u = \sum_{i \notin N} \frac{1}{\lambda_i}\left(\int_0^1 e_i(x)f(x)\,\mathrm{d}x\right)e_i + \sum_{i \in N} \alpha_i e_i\,,$$

where the α_i, $i \in N$, can be chosen arbitrarily. Moreover we see that, analogously to linear algebra, the number of linearly independent functions corresponding to the eigenvalue 0, i.e. the number of degrees of freedom for the solution u, agrees with the number of side constraints for the datum f (cf. 11.6 and 12.8).

Thus we have reduced the Sturm-Liouville problem to an eigenvalue problem for the operator T. Here we note that we employed arguments which are analogous to matrix calculus, but which are merely formal for infinite matrices. Of course, these need to be justified and this will be the subject of Chapters 11 and 12.

2 Preliminaries

In this chapter we introduce a number of fundamental structures in general spaces: topology, metric, norm, and scalar product. They are the natural generalizations of the corresponding concepts in the Euclidean space $\mathbb{R}^n$.

The most detailed structure is given by a scalar product in a $\mathbb{K}$-vector space, where here and throughout we take either $\mathbb{K} = \mathbb{R}$, i.e. $\mathbb{K}$ is the set of **real numbers**, or $\mathbb{K} = \mathbb{C}$, i.e. $\mathbb{K}$ is the set of **complex numbers**. For $\alpha \in \mathbb{K}$ we use the notation

$$|\alpha| := \sqrt{\alpha\overline{\alpha}} \quad \text{with} \quad \overline{\alpha} := \begin{cases} \operatorname{Re}\alpha - \mathrm{i}\operatorname{Im}\alpha & \text{for } \mathbb{K} = \mathbb{C}, \\ \alpha & \text{for } \mathbb{K} = \mathbb{R}, \end{cases}$$

and if $\alpha \in \mathbb{C}$ and for example

$$\alpha > 0, \quad \text{we implicitly assume that } \alpha \in \mathbb{R} \subset \mathbb{C}.$$

2.1 Scalar product. Let X be a $\mathbb{K}$-vector space. We call a map $(x_1, x_2) \mapsto (x_1, x_2)_X$ from $X \times X$ to $\mathbb{K}$ a **sesquilinear form** if for all $\alpha \in \mathbb{K}$ and for all $x, x_1, x_2, y, y_1, y_2 \in X$ one has

(S1) $(\alpha x, y)_X = \alpha(x, y)_X$,
$(x, \alpha y)_X = \overline{\alpha}(x, y)_X$,

(S2) $(x_1 + x_2, y)_X = (x_1, y)_X + (x_2, y)_X$,
$(x, y_1 + y_2)_X = (x, y_1)_X + (x, y_2)_X$.

This means that $(\cdot_1, \cdot_2)_X$ is **linear** in the first argument and **conjugate linear** in the second argument. Where no ambiguities arise, one can also write (x_1, x_2) in place of $(x_1, x_2)_X$. The sesquilinear form is called **symmetric** (also called a **Hermitian form**) if for all $x, y \in X$ one has

(S3) $(x, y)_X = \overline{(y, x)_X}$ (**Symmetry**).

A sesquilinear form is called **positive semidefinite** if for all $x \in X$

(S4') $(x, x)_X \geq 0$ (and then $(x, x)_X \in \mathbb{R}$)   (**Positivity**)

and **positive definite** if for all $x \in X$

(S4) $(x, x)_X \geq 0$ and in addition: $(x, x)_X = 0 \iff x = 0$.

9

For Hermitian forms $(x, x)_X = \overline{(x, x)_X}$ is real-valued, and for positive semidefinite Hermitian forms it is always nonnegative, in which case we define

$$\|x\|_X := \sqrt{(x, x)_X} .$$

A positive definite Hermitian form is also called a **scalar product** or **inner product**, and then the pair $(X, (\cdot_1, \cdot_2)_X)$ is called a **pre-Hilbert space**. If this scalar product in the vector space X is fixed, then we also say that X is a **pre-Hilbert space**.

The following lemma contains the fundamental properties of a scalar product.

2.2 Lemma. Let $(x_1, x_2) \mapsto (x_1, x_2)_X$ from $X \times X$ to $\mathbb{K}$ be a positive semidefinite Hermitian form and $\|x\| := \sqrt{(x, x)_X}$ for $x \in X$. Then it holds for all $x, y \in X$ and all $\alpha \in \mathbb{K}$ that

(1) $\|\alpha x\| = |\alpha| \cdot \|x\|$ (**Homogeneity**),

(2) $|(x, y)_X| \le \|x\| \cdot \|y\|$ (**Cauchy-Schwarz inequality**),

(3) $\|x + y\| \le \|x\| + \|y\|$ (**Triangle inequality**),

(4) $\|x + y\|^2 + \|x - y\|^2 = 2(\|x\|^2 + \|y\|^2)$ (**Parallelogram law**).

Proof (1). $\|\alpha x\|^2 = (\alpha x, \alpha x)_X = \alpha (x, \alpha x)_X = \alpha \overline{\alpha} (x, x)_X = |\alpha|^2 \|x\|^2$. $\square$

Proof (2). Let $x, y \in X$. It holds for $\alpha, \beta \in \mathbb{K} \setminus \{0\}$ (we want to set $\alpha = \|x\|$ and $\beta = \|y\|$) that

$$0 \le \left\| \frac{x}{\alpha} - \frac{y}{\beta} \right\|^2 = \frac{\|x\|^2}{|\alpha|^2} + \frac{\|y\|^2}{|\beta|^2} - 2\mathrm{Re}\left(\frac{(x, y)_X}{\alpha \overline{\beta}} \right), \qquad (2\text{-}1)$$

and hence for $\alpha > 0$ and $\beta > 0$, upon multiplying the inequality by $\alpha \beta > 0$, that

$$2\mathrm{Re}(x, y)_X \le \frac{\beta}{\alpha} \|x\|^2 + \frac{\alpha}{\beta} \|y\|^2 .$$

Setting $\alpha = \|x\| + \varepsilon$, $\beta = \|y\| + \varepsilon$ with $\varepsilon > 0$ yields that

$$2\mathrm{Re}(x, y)_X \le (\|y\| + \varepsilon) \cdot \frac{\|x\|^2}{\|x\| + \varepsilon} + (\|x\| + \varepsilon) \cdot \frac{\|y\|^2}{\|y\| + \varepsilon}$$

$$\le (\|y\| + \varepsilon) \cdot \|x\| + (\|x\| + \varepsilon) \cdot \|y\| .$$

As this holds for all $\varepsilon > 0$, it follows that

$$\mathrm{Re}(x, y)_X \le \|x\| \cdot \|y\| .$$

On replacing x with $\overline{(x, y)_X} x$ we obtain

$$|(x, y)_X|^2 \le |(x, y)_X| \cdot \|x\| \cdot \|y\|$$

and then cancelling in the case $(x, y)_X \ne 0$ gives the desired result. If $(x, y)_X = 0$, then the claim is trivial. $\square$

Proof (3). On recalling (2) we have

$$\|x+y\|^2 = \|x\|^2 + \|y\|^2 + 2\mathrm{Re}\,(x\,,y)_X$$
$$\leq \|x\|^2 + \|y\|^2 + 2\|x\| \cdot \|y\| = (\|x\| + \|y\|)^2\,.$$

$\square$

Proof (4). The first identity in the proof of (3) was

$$\|x+y\|^2 = \|x\|^2 + \|y\|^2 + 2\mathrm{Re}\,(x\,,y)_X\,.$$

Replacing y by $-y$ yields, since $(x\,,-y)_X = -\,(x\,,y)_X$, that

$$\|x-y\|^2 = \|x\|^2 + \|y\|^2 - 2\mathrm{Re}\,(x\,,y)_X\,. \tag{2-2}$$

Adding the two identities gives the result. $\square$

2.3 Orthogonality. Let X be a pre-Hilbert space over $\mathbb{K}$ and for $x \in X$ let $\|x\|_X := \sqrt{(x\,,x)_X}$ as in 2.1.

(1) Let $x,y \in X$. If $(x\,,y)_X = 0$, we say that x and y are **perpendicular**, or that they are **orthogonal vectors**. Then

$$\|x-y\|_X^2 = \|x\|_X^2 + \|y\|_X^2 \qquad \textbf{(\textit{Pythagoras' theorem})}.$$

(2) If Y and Z are two subspaces (see 4.4(2)) of a vector space X, then the sum

$$Y + Z := \{y + z \in X\,;\; y \in Y \text{ and } z \in Z\}$$

is again a subspace. The sum is called a **direct sum**, and we write $Y \oplus Z = Y + Z$, if $Y \cap Z = \{0\}$. If X is a pre-Hilbert space, then the subspaces are called **orthogonal** if $(y\,,z)_X = 0$ for all $y \in Y$ and $z \in Z$. Clearly it then holds that $Y \cap Z = \{0\}$ and we denote the subspace $Y \oplus Z$ also by $Y \perp Z$. The **orthogonal complement** of a subspace Y is defined by

$$Y^\perp := \{x \in X\,;\; (y\,,x)_X = 0 \text{ for all } y \in Y\} \text{ (see also 9.17)}\,.$$

It holds that $Y \cap Y^\perp = \{0\}$.

(3) For $x,y \in X \setminus \{0\}$ the Cauchy-Schwarz inequality 2.2(2) then reads

$$|\gamma| \leq 1 \quad \text{with} \quad \gamma := \left(\frac{x}{\|x\|_X}\,,\frac{y}{\|y\|_X}\right)_X\,.$$

Here equality holds if and only if x and y are linearly dependent.

(4) If $\mathbb{K} = \mathbb{R}$, then in (3) there exists a unique

$$\theta \in [0,\pi] \quad \text{such that} \quad \gamma = \cos(\theta).$$

We call θ the **angle** between x and y. It follows from (3) that x and y are linearly dependent if and only if $\theta = 0$ or $\theta = \pi$, and they are orthogonal if and only if $\theta = \frac{\pi}{2}$.

Proof (1). The theorem of Pythagoras follows from (2-2). □

Proof (2). This essentially contains only definitions. □

Proof (3). If x and y are linearly dependent, it is obvious that $|\gamma| = 1$. If $|\gamma| = 1$, then on setting $\alpha = \|x\|_X$, $\beta = \overline{\gamma}\|y\|_X$, equation (2-1) becomes

$$0 \leq \left\| \frac{x}{\alpha} - \frac{y}{\beta} \right\|^2 = 2 - 2\mathrm{Re}\left(\frac{(x,\,y)_X}{\|x\|_X \cdot \gamma\|y\|_X} \right) = 0\,,$$

which implies

$$\frac{x}{\alpha} = \frac{y}{\beta}\,,$$

hence x and y are linearly dependent. □

Proof (4). By (3), the vectors x and y are linearly dependent if and only if $1 = |\gamma| = |\cos(\theta)|$, which means $\theta = 0$ or $\theta = \pi$. By (1), the vectors x and y are orthogonal if and only if $(x,\,y)_X = 0$, which means $\cos(\theta) = 0$, that is, $\theta = \frac{\pi}{2}$. □

The standard example is the n-dimensional **Euclidean space** $\mathbb{R}^n$. The **Euclidean scalar product** and the **Euclidean norm** (for clarity these will be denoted by special symbols) are defined by

$$x \bullet y := \sum_{i=1}^n x_i y_i \quad \text{and} \quad |x| := \sqrt{x \bullet x} = \left(\sum_{i=1}^n x_i^2 \right)^{\frac{1}{2}}$$

for $x = (x_i)_{i=1,\dots,n} \in \mathbb{R}^n$, $y = (y_i)_{i=1,\dots,n} \in \mathbb{R}^n$. In the complex space $\mathbb{C}^n$ we define correspondingly

$$z \bullet w := \sum_{i=1}^n z_i \overline{w_i} \in \mathbb{C} \quad \text{and} \quad |z| := \sqrt{z \bullet z} = \left(\sum_{i=1}^n z_i \overline{z_i} \right)^{\frac{1}{2}} \in \mathbb{R}$$

for $z = (z_i)_{i=1,\dots,n} \in \mathbb{C}^n$, $w = (w_i)_{i=1,\dots,n} \in \mathbb{C}^n$. The infinite-dimensional analogue of Euclidean space is the sequence space (see 2.23).

A fundamental step in the development of functional analysis was the introduction of norms $x \mapsto \|x\|_X$ that are not induced by a scalar product as in 2.1, but are instead only characterized by the homogeneity and the triangle inequality in 2.2.

2.4 Norm. Let X be a $\mathbb{K}$-vector space. The pair $(X, \|\cdot\|)$ is called a *normed space* if $\|\cdot\| : X \to \mathbb{R}$ satisfies the following conditions for $x, y \in X$ and $\alpha \in \mathbb{K}$:

(**N1**) $\|x\| \geq 0$ (*Positivity*),
          and: $\|x\| = 0 \iff x = 0$,
(**N2**) $\|\alpha x\| = |\alpha| \cdot \|x\|$ (*Homogeneity*),
(**N3**) $\|x + y\| \leq \|x\| + \|y\|$ (*Triangle inequality*).

We then say that the map $\|\cdot\| : X \to \mathbb{R}$ is a *norm* on X. If a norm $\|\cdot\|_X : X \to \mathbb{R}$ is fixed on the vector space X, then we also call X a *normed space*.

Note that the property $(x = 0 \implies \|x\| = 0)$ in (N1) follows independently from (N2) on setting $\alpha = 0$ there. We call $\|\cdot\|$ a *seminorm* if we take (N1) without the property $(\|x\| = 0 \implies x = 0)$. It then follows from (N2) and (N3) that the set $Z := \{z \in X; \ \|z\| = 0\}$ is a subspace of X, and hence

$$x \sim y \quad :\iff \quad x - y \in Z$$

defines an equivalence relation "$\sim$" on X. Now let $\widetilde{X}$ be the set X together with the equivalence relation

$$x = y \text{ in } \widetilde{X} \quad :\iff \quad x \sim y \quad \iff \quad x - y \in Z.$$

Then all the vector space properties carry over from X to $\widetilde{X}$, and $(\widetilde{X}, \|\cdot\|)$ is a normed space (see remark). A common notation for the *factor space* or *quotient space* $\widetilde{X}$ is X/Z.

Remark: Let X be an arbitrary set, with "$\sim$" an arbitrary equivalence relation on X, and then let $\widetilde{X}$ be the set X with this equivalence relation, that is,

$$x = y \text{ in } \widetilde{X} \quad :\iff \quad x \sim y \text{ in } X.$$

A map $f : \widetilde{X} \to S$ to another set S is said to be *well defined* if

$$x = y \text{ in } \widetilde{X} \quad \implies \quad f(x) = f(y) \text{ in } S. \tag{2-3}$$

Hence, when defining a map on $\widetilde{X}$, condition (2-3) always needs to be verified.

Similarly, given a map $f : X \to S$, then this also defines a map from $\widetilde{X}$ to S, if (2-3) is satisfied for f. Analogous results hold for maps defined on e.g. $X \times X$.

In the case of a seminorm as discussed above, it can be easily shown that this is satisfied for the maps $(x, y) \mapsto x + y$ from $X \times X$ to X and $(\alpha, x) \mapsto \alpha x$ from $\mathbb{K} \times X$ to X, as well as for the map $x \mapsto \|x\|$ from X to $\mathbb{R}$.

In Section 3 we will introduce the most important norms in spaces of continuous and integrable functions. These norms are derived from the following norms in $\mathbb{K}^n$.

2.5 Example. For $1 \leq p \leq \infty$ the ***p-norm*** on $\mathbb{K}^n$ is defined by

$$|x|_p := \begin{cases} \left(\displaystyle\sum_{i=1}^{n} |x_i|^p \right)^{\frac{1}{p}} & \text{for } 1 \leq p < \infty, \\[2ex] \displaystyle\max_{i=1,\dots,n} |x_i| & \text{for } p = \infty, \end{cases}$$

where $x = (x_i)_{i=1,\dots,n} \in \mathbb{K}^n$. For $p = 2$ the ***Euclidean norm*** of x is $|x|_2 = |x|$. Alternative notations for the ***maximum norm*** $|x|_\infty$ are $|x|_{\max}$ and $|x|_{\sup}$, while the ***sum norm*** $|x|_1$ is also denoted by $|x|_{\text{sum}}$.

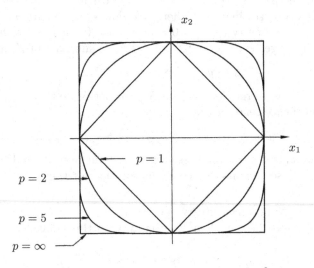

Fig. 2.1. *Unit spheres for p-norms in* $\mathbb{R}^2$

Proof. All of the norm axioms are easily verified, apart from the triangle inequality in the case $1 < p < \infty$ for $n \geq 2$. However, this follows from the ***Hölder inequality*** (proof to follow)

$$|x \bullet y| \leq |x|_p \cdot |y|_{p'} \tag{2-4}$$

for $x = (x_1, \dots, x_n)$, $y = (y_1, \dots, y_n)$, where p' is the ***dual exponent*** to p, i.e. it is defined by $\frac{1}{p} + \frac{1}{p'} = 1$.

Note: This inequality is a special case of the general Hölder inequality in 3.18 for the counting measure on $\{1, \dots, n\}$. Here we give a different proof.

The inequality (2-4) can, for instance, be shown by induction, on employing the inequality for $n = 2$. To this end, let $x' := (x_1, \dots, x_{n-1})$, $y' := (y_1, \dots, y_{n-1})$ and observe that

$$|x \bullet y| \leq |x' \bullet y'| + |x_n| \cdot |y_n|$$
$$\leq |x'|_p \cdot |y'|_{p'} + |x_n| \cdot |y_n| \quad \text{(induction hypothesis)}$$
$$\leq \left| (|x'|_p , |x_n|) \right|_p \cdot \left| (|y'|_{p'} , |y_n|) \right|_{p'} \quad \text{(inequality for } n = 2)$$
$$= |x|_p \cdot |y|_{p'} .$$

The inequality for $n = 2$ follows immediately from the elementary inequality

$$a_1 b_1 + a_2 b_2 \leq (a_1^p + a_2^p)^{\frac{1}{p}} \cdot (b_1^{p'} + b_2^{p'})^{\frac{1}{p'}} \quad \text{for} \quad a_1, a_2, b_1, b_2 \geq 0. \quad (2\text{-}5)$$

This holds trivially if one of the numbers is equal to 0. Otherwise, dividing by $a_1 b_1$ and setting $\alpha := a_2^p a_1^{-p}$, $\beta := b_2^{p'} b_1^{-p'}$ yields the equivalent inequality

$$1 + \alpha^{\frac{1}{p}} \beta^{\frac{1}{p'}} \leq (1 + \alpha)^{\frac{1}{p}} \cdot (1 + \beta)^{\frac{1}{p'}} \quad \text{for } \alpha, \beta > 0, \quad (2\text{-}6)$$

which we will prove now. For fixed $r := \alpha^{\frac{1}{p}} \cdot \beta^{\frac{1}{p'}}$ we have that

$$\alpha = \left(r \beta^{-\frac{1}{p'}} \right)^p = r^p \beta^{-\frac{p}{p'}} = r^p \beta^{1-p} =: \psi(\beta),$$

since $\frac{p}{p'} = p - 1$. Then the inequality reads

$$1 + r \leq \varphi(\beta) := (1 + \psi(\beta))^{\frac{1}{p}} \cdot (1 + \beta)^{\frac{1}{p'}},$$

and the right-hand side is minimal, if $\varphi'(\beta) = 0$. Now

$$\varphi'(\beta) = \varphi(\beta) \cdot \left(\frac{\psi'(\beta)}{p(1 + \psi(\beta))} + \frac{1}{p'(1 + \beta)} \right) = \frac{\varphi(\beta)}{p'\beta} \cdot \left(\frac{\beta}{1 + \beta} - \frac{\psi(\beta)}{1 + \psi(\beta)} \right),$$

since $\psi'(\beta) = -\psi(\beta) \cdot \frac{p-1}{\beta}$. Hence $\varphi'(\beta) = 0$ means $\beta = \psi(\beta)$, and so $\beta = r$, $\alpha = r$. This proves (2-6), and therefore the Hölder inequality (2-5) follows.

On letting $z_i := |x_i + y_i|^{p-1}$, $z = (z_1, \ldots, z_n)$, we have that

$$|x_i + y_i|^p \leq |x_i| \, z_i + |y_i| \, z_i .$$

The Hölder inequality then implies, since $p' \cdot (p - 1) = p$, that

$$|x + y|_p^p \leq (|x_i|)_{i=1,\ldots,n} \bullet z + (|y_i|)_{i=1,\ldots,n} \bullet z$$
$$\leq |x|_p \cdot |z|_{p'} + |y|_p \cdot |z|_{p'} = (|x|_p + |y|_p) \cdot |x + y|_p^{p-1},$$

which yields $|x + y|_p \leq |x|_p + |y|_p$. $\qquad\qquad\square$

We now interpret the norm $\|x\|$ of x as the distance of the point x from the origin 0 and replace $\|x\|$ with a value $d(x, 0)$, where $d : X \times X \to \mathbb{R}$ is a map for which only the triangle inequality has to hold. This notion of a distance can be defined in arbitrary sets.

2.6 Metric. A *metric space* is a pair (X, d), where X is a set and

$$d : X \times X \to \mathbb{R} \text{ for all } x, y, z \in X$$

has the following properties:

(M1) $d(x, y) \geq 0$ (*Positivity*),
    and: $d(x, y) = 0 \iff x = y$,
(M2) $d(x, y) = d(y, x)$ (*Symmetry*),
(M3) $d(x, y) \leq d(x, z) + d(z, y)$ (*Triangle inequality*).

We then call $d(x, y)$ the *distance* between the points x and y. The map $d : X \times X \to \mathbb{R}$ is called a *metric* on X. If a metric $d_X : X \times X \to \mathbb{R}$ is fixed on the set X, then we also call X a *metric space*. If (X, d) is a metric space and $A \subset X$, then (A, d) is also a metric space, with d restricted to $A \times A$.

Without the property $(d(x, y) = 0 \implies x = y)$ in (M1) we call d a *semimetric*. Then the *factor space* of X with respect to d is given as follows: The properties of the semimetric imply that

$$x \sim y \quad :\iff \quad d(x, y) = 0$$

defines an equivalence relation "$\sim$" on X. Now let $\widetilde{X}$ be the set X equipped with the equivalence relation

$$x = y \text{ in } \widetilde{X} \quad :\iff \quad x \sim y \quad \iff \quad d(x, y) = 0.$$

Then (M3) implies that d is also well defined on $\widetilde{X} \times \widetilde{X}$, and that $(\widetilde{X}, d)$ is a metric space (see the remark in 2.4).

2.7 Fréchet metric. In vector spaces X, metrics d are often defined by

$$d(x, y) = \varrho(x - y) \quad \text{for } x, y \in X,$$

where $\varrho : X \to \mathbb{R}$ satisfies the following properties for all $x, y \in X$:

(F1) $\varrho(x) \geq 0$ (*Positivity*),
    and: $\varrho(x) = 0 \iff x = 0$,
(F2) $\varrho(x) = \varrho(-x)$ (*Symmetry*),
(F3) $\varrho(x + y) \leq \varrho(x) + \varrho(y)$ (*Triangle inequality*).

A map $\varrho : X \to \mathbb{R}$ satisfying (F1)–(F3) is called a *Fréchet metric*. Any norm $x \mapsto \|x\|$ on X is a Fréchet metric and hence defines the *induced metric* $d(x, y) := \|x - y\|$.

We begin with some elementary examples.

2.8 Examples of metrics.

(1) A bounded Fréchet metric on $\mathbb{K}^n$ that is not a norm is given by

$$\varrho(x) := \frac{|x|}{1 + |x|} \quad \text{for } x \in \mathbb{K}^n.$$

(2) Let $-\infty$, $+\infty$ be two distinct elements that do not belong to $\mathbb{R}$. One can then define a metric on $\mathbb{R} \cup \{\pm\infty\}$ by

$$d(x,y) := |g(x) - g(y)| \quad \text{for } x, y \in \mathbb{R} \cup \{\pm\infty\},$$

where

$$g(x) := \begin{cases} -1 & \text{for } x = -\infty, \\ \dfrac{x}{1+|x|} & \text{for } x \in \mathbb{R}, \\ +1 & \text{for } x = +\infty. \end{cases}$$

(3) Let ∞ be an element that does not belong to $\mathbb{R}^n$. One can then define a metric on $\mathbb{R}^n \cup \{\infty\}$ by

$$d(x,y) := |\tau_{\text{stereo}}(x) - \tau_{\text{stereo}}(y)| .$$

Here

$$\tau_{\text{stereo}} : \mathbb{R}^n \cup \{\infty\} \longrightarrow \{ y \in \mathbb{R}^n \times \mathbb{R} = \mathbb{R}^{n+1} ; \; |y - (0, \tfrac{1}{2})| = \tfrac{1}{2} \} ,$$

where the image is the ball $B_{\frac{1}{2}}^{\mathbb{R}^{n+1}}((0, \tfrac{1}{2}))$ with respect to the Euclidean metric, is defined by

$$\tau_{\text{stereo}}(x) := \begin{cases} \dfrac{(x, |x|^2)}{1+|x|^2} & \text{for } x \in \mathbb{R}^n, \\ (0, 1) & \text{for } x = \infty. \end{cases}$$

Remark: The inverse $\tau_{\text{stereo}}^{-1}$ is the **stereographic projection**, i.e. $y = \tau_{\text{stereo}}(x)$ with $|y - (0, \tfrac{1}{2})| = \tfrac{1}{2}$ and $y \neq (0,1)$ is given by

$$(1-a)(0,1) + ay = (x,0) \quad \text{for an } a \in \mathbb{R}.$$

Proof (1). The function $\varphi(s) := \frac{s}{1+s}$ for $s \geq 0$ satisfies

$$\varphi(s) \leq \varphi(\tilde{s}) \quad \text{for } 0 \leq s \leq \tilde{s},$$

$$\varphi(s_1 + s_2) = \frac{s_1}{1 + s_1 + s_2} + \frac{s_2}{1 + s_1 + s_2} \leq \varphi(s_1) + \varphi(s_2) \quad \text{for } s_1, s_2 \geq 0.$$

Apply the above for $s = |x + y| \leq |x| + |y| = \tilde{s}$, $s_1 = |x|$, $s_2 = |y|$. $\qquad \square$

Proof (2) *and* (3). Use that g, resp. τ_{stereo}, is injective and employ the triangle inequality in $\mathbb{R}$, resp. $\mathbb{R}^{n+1}$. $\qquad \square$

With the help of the distance between two points we now define the distance between two sets. As a special case we obtain the definition of balls with respect to a given metric.

2.9 Balls and distance between sets. Let (X, d) be a metric space. For two sets $A_1, A_2 \subset X$ the **distance** between A_1 and A_2 is defined by

$$\operatorname{dist}(A_1, A_2) := \inf \left\{ d(x, y) \; ; \; x \in A_1, \; y \in A_2 \right\},$$

where $\inf \emptyset := \infty$ (so that $\operatorname{dist}(A, \emptyset) = \infty$). For $x, y \in X$ it holds that $d(x, y) = \operatorname{dist}(\{x\}, \{y\})$. For $x \in X$ the **distance** from x to $A \subset X$ is defined by

$$\operatorname{dist}(x, A) := \operatorname{dist}(\{x\}, A) = \inf \left\{ d(x, y) \; ; \; y \in A \right\}.$$

For $r > 0$ the **r-neighbourhood of the set A** is defined by

$$\mathrm{B}_r(A) := \left\{ x \in X \; ; \; \operatorname{dist}(x, A) < r \right\},$$

and $\mathrm{B}_r(x) := \mathrm{B}_r(\{x\})$ is called the **ball around x with radius r** or, alternatively, the **r-neighbourhood of the point x**. We have

$$\mathrm{B}_r(x) = \left\{ y \in X \; ; \; d(y, x) < r \right\}.$$

The **diameter** of a subset $A \subset X$ is defined by

$$\operatorname{diam}(A) := \sup \left\{ d(x, y) \; ; \; x, y \in A \right\},$$

if $A \neq \emptyset$, and $\operatorname{diam}(\emptyset) := 0$ (or make the convention that $\sup \emptyset := 0$). A set $A \subset X$ is called **bounded** if $\operatorname{diam}(A) < \infty$.

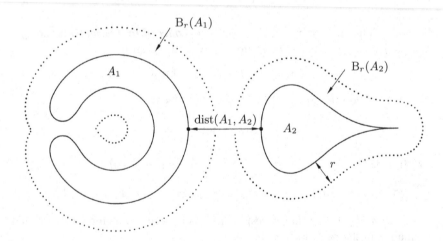

Fig. 2.2. *Metric definitions in $\mathbb{R}^2$ with respect to $x \mapsto |x|_2$*

The concept of a ball plays an important role in definitions and proofs for metric spaces. It can be used, for instance, to introduce the following notion of an "open subset" (see 2.10). In functional analysis the concept of open sets is applied to function spaces. Depending on the chosen distance the notion of open sets is different, therefore one obtains different results for the considered class of functions. In 3.2, 3.3 and 3.7 this applies to function spaces with respect to supremum norms, in 3.15 to function spaces with respect to integral norms, and in 3.13 to function spaces equipped with distances that are induced by a measure.

2.10 Open and closed sets. Let (X, d) be a metric space. For $A \subset X$ the *interior* of A (notation: $\text{intr}_X(A)$ or $\text{intr}(A)$ or $\mathring{A}$) is defined by

$$\text{intr}(A) := \{ x \in X \; ; \; B_\varepsilon(x) \subset A \text{ for an } \varepsilon > 0 \} \subset A,$$

and the *closure* of A (or the *closed hull*, notation: $\text{clos}_X(A)$ or $\text{clos}(A)$ or $\overline{A}$) is defined by

$$\text{clos}(A) := \{ x \in X \; ; \; B_\varepsilon(x) \cap A \neq \emptyset \text{ for all } \varepsilon > 0 \} \supset A.$$

It holds that $x \in \text{clos}(A)$ if and only if $\text{dist}(x, A) = 0$. Using quantifiers, the above definitions can be written as

$$x \in \text{intr}(A) \quad \Longleftrightarrow \quad \exists \, \varepsilon > 0 : B_\varepsilon(x) \setminus A = \emptyset,$$
$$x \in \text{clos}(A) \quad \Longleftrightarrow \quad \forall \, \varepsilon > 0 : B_\varepsilon(x) \cap A \neq \emptyset,$$

or

$$x \in \text{intr}(A) \quad \Longleftrightarrow \quad \exists \, \varepsilon > 0 : \forall \, y \in B_\varepsilon(x) : y \in A,$$
$$x \in \text{clos}(A) \quad \Longleftrightarrow \quad \forall \, \varepsilon > 0 : \exists \, y \in B_\varepsilon(x) : y \in A.$$

A subset $A \subset X$ is called *open* if $\text{intr}(A) = A$, and $A \subset X$ is called *closed* if $\text{clos}(A) = A$. The complement of a closed set is open and the complement of an open set is closed. The *boundary* of $A \subset X$ (notation: $\text{bdry}_X(A)$ or $\text{bdry}(A)$ or ∂A) is defined by

$$\text{bdry}(A) := \text{clos}(A) \setminus \text{intr}(A)$$
$$= \text{clos}(A) \cap \text{clos}(X \setminus A) = \text{bdry}(X \setminus A)$$

and, being an intersection of closed sets, is a closed set. We have

$$X = \text{intr}(A) \cup \text{bdry}(A) \cup \text{intr}(X \setminus A),$$

where the union is disjoint.

We now consider on X only the class of open sets. This class is characterized by the fact that arbitrary unions of open sets and finite intersections of open sets are still open sets.

2.11 Topology. A *topological space* is a pair $(X, \mathcal{T})$, where X is a set and $\mathcal{T}$ is a system of subsets of X (the elements of $\mathcal{T}$ are called *open sets*), with the following properties:

(T1) $\emptyset \in \mathcal{T}$, $X \in \mathcal{T}$,
(T2) $\mathcal{T}' \subset \mathcal{T}$ $\implies$ $\bigcup_{U \in \mathcal{T}'} U \in \mathcal{T}$,
(T3) $U_1, U_2 \in \mathcal{T}$ $\implies$ $U_1 \cap U_2 \in \mathcal{T}$.

A topological space is called a **Hausdorff space** if in addition the following *separation axiom* is satisfied:

(T4) For $x_1, x_2 \in X$ with $x_1 \neq x_2$ there exist $U_1, U_2 \in \mathcal{T}$ such that $x_1 \in U_1$, $x_2 \in U_2$, and $U_1 \cap U_2 = \emptyset$, the same with quantifiers:
$\forall \, x_1, x_2 \in X, \, x_1 \neq x_2 \, : \, \exists \, U_1, U_2 \in \mathcal{T} \, : \, x_1 \in U_1, \, x_2 \in U_2, \, U_1 \cap U_2 = \emptyset$.

A subset $A \subset X$ is called **closed** with respect to $\mathcal{T}$ if $X \setminus A \in \mathcal{T}$, that is, with respect to $\mathcal{T}$, the complement of an open set is closed, and the complement of a closed set is open. We define for $A \subset X$ (note the remark in 2.12 below)

$$\mathrm{intr}_{(X,\mathcal{T})} (A) := \big\{ x \in X \, ; \, U \subset A \text{ for some } U \in \mathcal{T} \text{ with } x \in U \big\} \subset A,$$
$$\mathrm{clos}_{(X,\mathcal{T})} (A) := \big\{ x \in X \, ; \, U \cap A \neq \emptyset \text{ for all } U \in \mathcal{T} \text{ with } x \in U \big\} \supset A.$$

Alternative notations are $\mathrm{intr}\,(A) := \mathrm{intr}_{(X,\mathcal{T})} (A)$ or $\mathring{A} := \mathrm{intr}_{(X,\mathcal{T})} (A)$ and $\mathrm{clos}\,(A) := \mathrm{clos}_{(X,\mathcal{T})} (A)$ or $\overline{A} := \mathrm{clos}_{(X,\mathcal{T})} (A)$. It holds that

$$A = \mathrm{intr}_{(X,\mathcal{T})} (A) \quad \Longleftrightarrow \quad A \in \mathcal{T},$$
$$A = \mathrm{clos}_{(X,\mathcal{T})} (A) \quad \Longleftrightarrow \quad X \setminus A \in \mathcal{T}.$$

If $A \subset X$, then $(A, \mathcal{T}_A)$ is a topological space with the **relative topology**

$$\mathcal{T}_A := \{ U \cap A ; \; U \in \mathcal{T} \}.$$

The following is the standard construction of a topology and it shows that for a metric space the definitions of interior and closure in 2.11 (with respect to a topology) and in 2.10 (with respect to a metric) are the same.

2.12 Proposition. Let (X, d) be a metric space and, on recalling the definition of the interior of a set in 2.10 (we write $\mathrm{intr}_{(X,d)} (A)$ instead of $\mathrm{intr}_X (A)$), let

$$\mathcal{T} := \{ A \subset X \, ; \; \mathrm{intr}_{(X,d)} (A) = A \}.$$

Then $(X, \mathcal{T})$ is a topological space and, in particular, a Hausdorff space. We call $\mathcal{T}$ the topology **induced** by the metric d.
Remark: For all subsets $A \subset X$ it holds $\mathrm{intr}_{(X,d)} (A) = \mathrm{intr}_{(X,\mathcal{T})} (A)$ and $\mathrm{clos}_{(X,d)} (A) = \mathrm{clos}_{(X,\mathcal{T})} (A)$.

Proof of the proposition. In order to show axiom (T3), let $A_1, A_2 \in \mathcal{T}$ and $x \in A_1 \cap A_2$. Then $\mathrm{intr}_{(X,d)} (A_1) = A_1$ and $\mathrm{intr}_{(X,d)} (A_2) = A_2$ with the definition as in 2.10. Hence there exist $\varepsilon_1, \varepsilon_2 > 0$ such that $B_{\varepsilon_1} (x) \subset A_1$ and $B_{\varepsilon_2} (x) \subset A_2$. Setting $\varepsilon := \min(\varepsilon_1, \varepsilon_2) > 0$ yields $B_\varepsilon (x) \subset A_1 \cap A_2$, and hence $A_1 \cap A_2 \in \mathcal{T}$. For the proof of (T4) let $x \neq y$. Then the triangle inequality yields that

$$B_r(x) \cap B_r(y) = \emptyset \text{ for } r := \tfrac{1}{2}d(x,y) > 0 \,,$$

and $B_r(x), B_r(y) \in \mathcal{T}$ (see E2.2(2)).   □

2.13 Definition. Let $(X,\mathcal{T})$ be a topological space. A subset $A \subset X$ is called **dense** in X if $\mathrm{clos}\,(A) = X$, and X is called **separable** if X contains a countable dense subset. A subset $A \subset X$ is called separable if the relative topological space $(A, \mathcal{T}_A)$ is separable. Hence, if (X,d) is a metric space, a subset $A \subset X$ is separable if the metric space (A, d) is separable.

2.14 Comparison of topologies. Let $\mathcal{T}_1, \mathcal{T}_2$ be two topologies on a set X. We say that $\mathcal{T}_2$ is **stronger** (or **finer**) than $\mathcal{T}_1$, or equivalently that $\mathcal{T}_1$ is **weaker** (or **coarser**) than $\mathcal{T}_2$, if

$$\mathcal{T}_1 \subset \mathcal{T}_2 \,.$$

Let d_1, d_2 be two metrics on X and $\mathcal{T}_1, \mathcal{T}_2$ the corresponding induced topologies (see 2.11). Then the metric d_2 is said to be **stronger** (**weaker**) than d_1 if $\mathcal{T}_2$ is stronger (weaker) than $\mathcal{T}_1$. The metrics d_1 and d_2 are called **equivalent**, if $\mathcal{T}_1 = \mathcal{T}_2$. Similarly, a norm is said to be **stronger** (**weaker**) than another norm, and two norms are called **equivalent** if this holds for the induced metrics, respectively.

2.15 Comparison of norms. Let $\|\cdot\|_1$ and $\|\cdot\|_2$ be two norms on a $\mathbb{K}$-vector space X. Then

(1) $\|\cdot\|_2$ is stronger than $\|\cdot\|_1$ if and only if there exists a positive number C such that

$$\|x\|_1 \le C\|x\|_2 \quad \text{for all } x \in X \,.$$

(2) The two norms are equivalent if and only if there exist positive numbers c and C such that

$$c\|x\|_2 \le \|x\|_1 \le C\|x\|_2 \quad \text{for all } x \in X \,.$$

Proof (1). Let $B_r^i(x)$ denote the balls and $\mathcal{T}_i$ the topologies with respect to the norms $\|\cdot\|_i$. Let $\mathcal{T}_1 \subset \mathcal{T}_2$. Since $B_1^1(0) \in \mathcal{T}_1$ (see E2.2(2)), $B_1^1(0)$ is open with respect to $\|\cdot\|_2$ and, in particular, 0 lies in the interior (with respect to $\|\cdot\|_2$) of $B_1^1(0)$, i.e.

$$B_\varepsilon^2(0) \subset B_1^1(0) \quad \text{for some } \varepsilon > 0.$$

This means, for $x \in X$, $x \ne 0$, that

$$\left\| \frac{\varepsilon x}{2\|x\|_2} \right\|_2 = \frac{\varepsilon}{2} < \varepsilon, \quad \text{therefore} \quad \left\| \frac{\varepsilon x}{2\|x\|_2} \right\|_1 < 1,$$

$$\text{that is } \|x\|_1 \le \frac{2}{\varepsilon}\|x\|_2 \,.$$

Conversely, if the inequality in assertion (1) holds, then

$$B_r^2(x) \subset B_{Cr}^1(x) \qquad \text{for all } x \in X \text{ and } r > 0.$$

Let $A \in \mathcal{T}_1$. Then $A = \text{intr}_{d_1}(A)$ with respect to $\mathcal{T}_1$, and for $x \in A$ there is an $\varepsilon > 0$ such that

$$B_\varepsilon^1(x) \subset A, \qquad \text{therefore} \qquad B_{\frac{\varepsilon}{C}}^2(x) \subset A.$$

This proves that $A \in \mathcal{T}_2$. □

Proof (2). Apply (1) twice. □

2.16 Examples.

(1) The p-norms on $\mathbb{K}^n$ defined in 2.5 are pairwise equivalent, since for $1 \leq p < \infty$

$$|x|_\infty \leq |x|_p \leq n^{\frac{1}{p}} |x|_\infty.$$

(2) The Euclidean norm and the Fréchet metric in 2.8(1) induce the same topology on $\mathbb{K}^n$, since for $y \in \mathbb{K}^n$

$$|y| \leq 2\varrho(y) \text{ if } \varrho(y) \leq \tfrac{1}{2}, \qquad \varrho(y) \leq |y|.$$

Hence, $B_{\frac{r}{2}}^{\text{metric}}(x) \subset B_r^{\text{norm}}(x) \subset B_r^{\text{metric}}(x)$ for $0 < r \leq 1$.

(3) For open sets $U \subset \mathbb{R} \cup \{\pm\infty\}$ with respect to the metric in 2.8(2) it holds that

$$
\begin{aligned}
x \in U \cap \mathbb{R} &\iff \left]x - \varepsilon, x + \varepsilon\right[\subset U && \text{for an } \varepsilon > 0, \\
+\infty \in U &\iff \left]\tfrac{1}{\varepsilon}, +\infty\right] \subset U && \text{for an } \varepsilon > 0, \\
-\infty \in U &\iff \left[-\infty, -\tfrac{1}{\varepsilon}\right[\subset U && \text{for an } \varepsilon > 0.
\end{aligned}
$$

(4) For open sets $U \subset \mathbb{K}^n \cup \{\infty\}$ with respect to the metric in 2.8(3) it holds that

$$
\begin{aligned}
x \in U \cap \mathbb{K}^n &\iff \{y \in \mathbb{K}^n ; |y - x| < \varepsilon\} \subset U && \text{for an } \varepsilon > 0, \\
\infty \in U &\iff \{y \in \mathbb{K}^n ; |y| > \tfrac{1}{\varepsilon}\} \subset U && \text{for an } \varepsilon > 0.
\end{aligned}
$$

One of the most important concepts in analysis is the notion of a limit and the resulting concept of continuity. Given a mapping $f : X \to Y$ between Hausdorff spaces X and Y, then f is **continuous** at $x_0 \in X$ (see 2.17(4) below) if

$$f(x_0) = \lim_{x \to x_0} f(x) \quad \text{in } Y.$$

This is the well-known notion of continuity in the analysis of Euclidean spaces. We now generalize this concept as follows: Given sets S, X, Y and mappings $\varphi : S \to X, f : S \to Y$, we consider two points $x_0 \in X, y_0 \in Y$ and the question is whether the function values $f(s)$ are "close to" y_0 if $\varphi(s)$ is "close to" x_0. In metric spaces we can define the notion of closeness with the help of balls around x_0 and y_0, and similarly in topological spaces with the help of open sets that contain x_0 and y_0, respectively.

Usually we have that $S \subset X$ and $\varphi(s) = s$ for $s \in S$. But often this is not the case. A nontrivial example is given in A3.17 (there S is a system of sets with $f(E) := |\nu(E)|$ and $\varphi(E) := \mu(E)$ for $E \in S$, hence $X = \mathbb{R}$ and $Y = \mathbb{R}$).

2.17 Convergence and continuity. Let S be a set, $(X, \mathcal{T}_X)$ and $(Y, \mathcal{T}_Y)$ Hausdorff spaces, and

$$\varphi : S \to X, \quad x_0 \in X, \quad f : S \to Y, \quad y_0 \in Y.$$

We say that

$$f(s) \text{ \textbf{\textit{converges to}} } y_0 \text{ in } Y \text{ (with respect to } \mathcal{T}_Y)$$
$$\text{\textbf{\textit{as}} } \varphi(s) \text{ \textbf{\textit{goes to}} } x_0 \text{ in } X \text{ (with respect to } \mathcal{T}_X),$$

and use the notation

$$f(s) \to y_0 \text{ in } Y \quad \text{as} \quad \varphi(s) \to x_0 \text{ in } X,$$

if the following holds for all $U_0 \subset X$, $V_0 \subset Y$:

$$x_0 \in U_0 \in \mathcal{T}_X, \qquad \text{There exists a } U \in \mathcal{T}_X \text{ such that } x_0 \in U \subset U_0,$$
$$y_0 \in V_0 \in \mathcal{T}_Y \quad \Longrightarrow \quad \varphi^{-1}(U) \neq \emptyset \text{ and } f(\varphi^{-1}(U)) \subset V_0.$$

The conclusion states that for a $U \in \mathcal{T}_X$ with $x_0 \in U \subset U_0$ it holds that

$$s \in S, \ \varphi(s) \in U \quad \Longrightarrow \quad f(s) \in V_0,$$

and that $\varphi(s) \in U$ for at least one $s \in S$. We have (see E2.4):

(1) Given x_0, f, φ, there exists at most one such $y_0 \in Y$. Hence we write

$$y_0 = \lim_{\varphi(s) \to x_0} f(s),$$

and call y_0 the **limit** of $f(s)$ as $\varphi(s)$ goes to x_0.

(2) $x_0 \in \text{clos}(\varphi(S))$ and $y_0 \in \text{clos}(f(S))$.

(3) The most important special case is: $S \subset X$ and $\varphi(s) = s$ for $s \in S$. Then, for points $x_0 \in \text{clos}(S)$ and $y_0 \in Y$, the definition

$$f(x) \to y_0 \text{ in } Y \text{ as } x \to x_0 \text{ in } X, \text{ i.e. } y_0 = \lim_{x \to x_0} f(x),$$

is equivalent to

$$V \in \mathcal{T}_Y, \ y_0 \in V \quad \Longrightarrow \quad \begin{array}{l} \text{There exists a } U \in \mathcal{T}_X \text{ such that } x_0 \in U \\ \text{and } f(U \cap S) \subset V, \end{array}$$

in words: For every open set V containing y_0 there exists an open set U containing x_0 such that $f(U \cap S)$ is contained in V.

(4) If in (3) in addition $x_0 \in S$, then it follows that $y_0 = f(x_0)$, i.e.

$$f(x_0) = \lim_{x \to x_0} f(x).$$

In this case f is called **continuous at the point** x_0.

(5) If $S = X$, then $f : X \to Y$ is called a **continuous map** if f is continuous at all points $x_0 \in X$. This is equivalent to

$$V \in \mathcal{T}_Y \quad \Longrightarrow \quad f^{-1}(V) \in \mathcal{T}_X,$$

in words: The mapping f has the property that the inverse image of each open set in Y is open in X.

2.18 Convergence in metric spaces. Let (X, d_X) and (Y, d_Y) be metric spaces and let $A \subset X$ and $f : A \to Y$.

(1) Let $x_0 \in \operatorname{clos}(A)$ and $y_0 \in Y$. Then

$$f(x) \to y_0 \text{ in } Y \text{ as } x \to x_0 \text{ in } X$$

if and only if:

> For all $\varepsilon > 0$ there exists a $\delta > 0$, such that
> $$x \in A, \ d_X(x, x_0) < \delta \Longrightarrow d_Y(f(x), y_0) < \varepsilon,$$

i.e. if and only if

$$d_Y(f(x), y_0) \to 0 \text{ as } d_X(x, x_0) \to 0 \quad (\text{in } \mathbb{R}).$$

Using quantifiers this property can be written as:

$$\forall \, \varepsilon > 0 : \exists \, \delta > 0 : \forall \, x \in A : \ d_X(x, x_0) < \delta \Longrightarrow d_Y(f(x), y_0) < \varepsilon.$$

(2) Let $X = \mathbb{K}^n \cup \{\infty\}$ (equipped with the metric in 2.8(3)) and let $A \subset \mathbb{K}^n$ be unbounded. Then $\infty \in \operatorname{clos}(A)$, and $x \to \infty$ in $\mathbb{K}^n \cup \{\infty\}$ means that $|x| \to +\infty$ in $\mathbb{R} \cup \{\pm\infty\}$ (equipped with the metric in 2.8(2)). Let $y \in Y$. Then

$$f(x) \to y \text{ in } Y \text{ as } |x| \to +\infty$$

if and only if:

> For all $\varepsilon > 0$ there exists a $\delta > 0$, such that
> $$x \in A, \ |x| > \tfrac{1}{\delta} \Longrightarrow d_Y(f(x), y) < \varepsilon.$$

Using quantifiers this property can be written as:

$$\forall \, \varepsilon > 0 : \exists \, \delta > 0 : \forall \, x \in A : \ |x| > \tfrac{1}{\delta} \Longrightarrow d_Y(f(x), y) < \varepsilon.$$

(3) Let $X = \mathbb{R} \cup \{\pm\infty\}$ (equipped with the metric in 2.8(2)) and let $A = \mathbb{N}$, i.e. $(y_j)_{j \in \mathbb{N}}$ with $y_j := f(j)$ is a sequence in Y. It then holds for $y \in Y$ that

$$y_j \to y \text{ in } Y \text{ as } j \to +\infty$$

if and only if:

> For all $\varepsilon > 0$ there exists a $k \in \mathbb{N}$ such that
> $$j \in \mathbb{N}, \; j > k \Longrightarrow d_Y(y_j, y) < \varepsilon.$$

Using quantifiers this property can be written as:

$$\forall \, \varepsilon > 0 : \exists \, k \in \mathbb{N} : \forall \, j \in \mathbb{N} : j > k \Longrightarrow d_Y(y_j, y) < \varepsilon.$$

(4) In metric spaces convergence is equivalent to *sequential convergence*, that is, the convergence in (1) holds if and only if for all sequences $(x_j)_{j \in \mathbb{N}}$ in A:

$$x_j \to x_0 \text{ as } j \to \infty \quad \Longrightarrow \quad d_Y\big(f(x_j), y_0\big) \to 0 \text{ as } j \to \infty. \tag{2-7}$$

Proof (1). Use the fact that balls $B_\varepsilon(y_0)$ belong to the topology $\mathcal{T}_Y$ induced by d_Y and that

$$y_0 \in V \in \mathcal{T}_Y \quad \Longrightarrow \quad B_\varepsilon(y_0) \subset V \text{ for an } \varepsilon > 0.$$

Likewise in X, every $B_\delta(x_0) \in \mathcal{T}_X$, and if $x_0 \in U \in \mathcal{T}_X$, then $B_\delta(x_0) \subset U$ for some $\delta > 0$. $\qquad\square$

Proof (2). Follows from (1), on noting that for $0 < \delta' < 1$ for the ball $B_{\delta'}(\infty)$ with respect to the stereographic projection the following is true:

$$x \in B_{\delta'}(\infty) \quad \Longleftrightarrow \quad |x| > \sqrt{\delta'^{-2} - 1} =: \delta^{-1}.$$

$\qquad\square$

Proof (3). Similarly to (2), by choosing $\frac{1}{\delta} \le k < \frac{1}{\delta} + 1$. $\qquad\square$

Proof (4). Assume that (1) holds and that $x_j \to x_0$ in X as $j \to \infty$. Then given $\varepsilon > 0$, there exists a $\delta > 0$ such that $d_Y(f(x), y_0) < \varepsilon$ for $x \in A$ with $d_X(x, x_0) < \delta$. Then (3) yields the existence of a $k \in \mathbb{N}$ such that $d_X(x_j, x_0) < \delta$ for $j > k$. Consequently $d_Y(f(x_j), y_0) < \varepsilon$. This proves the claim in (4).

Conversely, assume that the convergence statement in (1) is not true. Then we have to negate the assertion

$$\forall \, \varepsilon > 0 : \exists \, \delta > 0 : \forall \, x \in A : d_X(x, x_0) < \delta \Longrightarrow d_Y\big(f(x), y_0\big) < \varepsilon.$$

The negation is:

$$\exists \, \varepsilon > 0 : \forall \, \delta > 0 : \exists \, x \in A : d_X(x, x_0) < \delta \text{ and } d_Y\big(f(x), y_0\big) \ge \varepsilon.$$

Consequently there exist an $\varepsilon > 0$ and, for $\delta_j := \frac{1}{j}$, $j \in \mathbb{N}$, an $x_j \in A$ such that

$$d_X(x_j, x_0) < \delta_j \text{ and } d_Y\big(f(x_j), y_0\big) \ge \varepsilon.$$

In particular, $x_j \to x_0$ in X as $j \to \infty$, but $d_Y(f(x_j), y_0) \ge \varepsilon$ for all $j \in \mathbb{N}$. This contradicts (2-7). $\qquad\square$

2.19 Note. In 2.18(3) we identified sequences in Y with maps from $\mathbb{N}$ to Y. This can be generalized to arbitrary sets I and A. Here the notation $(a_i)_{i \in I}$, with $a_i \in A$ for $i \in I$, defines a map $i \mapsto a_i$ from I to A. The set of all of these maps is denoted by A^I and I is also called *index set*,

$$A^I := \{(a_i)_{i \in I} \; ; \; \forall \, i \in I \, : \, a_i \in A\}.$$

In this book, I is usually a subset of $\mathbb{N}$. Examples are the sequence space $\mathbb{K}^{\mathbb{N}}$ in 2.23 and the set $X^{\mathbb{N}}$ in 2.24. In addition one can identify $\mathbb{K}^n$ with $\mathbb{K}^{\{1,\dots,n\}}$. In general it is important to note that $(a_i)_{i \in I}$ is well distinguished from the subset $\{a_i \in A \, ; \; i \in I\} \subset A$ (relevant in e.g. 9.3).

The analysis of limits in metric spaces is often based on inequalities, which we also call "estimates" or "bounds"; this is especially true in function spaces. Usually performing the limit is not trivial and consists of a "nested limit".

2.20 Note (Nested limits). We make the following remark on convergence proofs. By a *nested limit* for sequences defined on $\mathbb{N}$ we understand the following. Let $a_i \geq 0$, $b_{k,i} \geq 0$, $c_k \geq 0$ for $i, k \in \mathbb{N}$ with the property

$$a_i \leq \underbrace{b_{k,i}}_{\substack{\to \, 0 \text{ as } i \to \infty \\ \text{for a fixed } k}} + \underbrace{c_k}_{\to \, 0 \text{ as } k \to \infty} .$$

From this we deduce that $(a_i)_{i \in \mathbb{N}}$ is a *null sequence*, i.e.

$$a_i \to 0 \quad \text{as } i \to \infty .$$

To see this, assume that the inequality $a_i \leq b_{k,i} + c_k$ holds for $i, k \in \mathbb{N}$, that $c_k \to 0$ as $k \to \infty$ and that for each $k \in \mathbb{N}$ we have that $b_{k,i} \to 0$ as $i \to \infty$. For an arbitrary $\varepsilon > 0$ we can then choose a $k_\varepsilon \in \mathbb{N}$ such that $c_{k_\varepsilon} < \varepsilon$. Moreover, for this k_ε there exists an i_ε such that $b_{k_\varepsilon,i} < \varepsilon$ for all $i > i_\varepsilon$. Hence we have that

$$a_i \leq b_{k_\varepsilon,i} + c_{k_\varepsilon} < 2\varepsilon \quad \text{for all } i > i_\varepsilon .$$

This proves that $(a_i)_{i \in \mathbb{N}}$ is a null sequence.

This book contains many such limit considerations. A first example you can find in the proof of 2.23(2). In these cases the detailed argumentation will either be omitted, or dramatically shortened to something like:

First choose k large, then choose i large.

Also nested limits with more than two indices are used.

One of the most important concepts in metric spaces is the

2.21 Completeness. Let (X, d) be a metric space.

(1) A sequence $(x_k)_{k \in \mathbb{N}}$ in X is called a **Cauchy sequence** if

$$d(x_k, x_l) \to 0 \quad \text{as } (k, l) \to (\infty, \infty).$$

One usually writes $k, l \to \infty$ in place of $(k, l) \to (\infty, \infty)$.

Remark: Here convergence of $(k, l) \in \mathbb{N}^2 \subset (\mathbb{R} \cup \{\pm\infty\})^2$ is understood with respect to the product metric $d_2(a, b) := d_1(a_1, b_1) + d_1(a_2, b_2)$ for $a = (a_1, a_2)$ and $b = (b_1, b_2)$ in $(\mathbb{R} \cup \{\pm\infty\})^2$, where d_1 is the metric on $\mathbb{R} \cup \{\pm\infty\}$ as defined in 2.8(2).

(2) If $(x_k)_{k \in \mathbb{N}}$ is a sequence in X, then a point $x \in X$ is called a **cluster point** of this sequence if there exists a **subsequence** $(x_{k_i})_{i \in \mathbb{N}}$ (i.e. a sequence $(k_i)_{i \in \mathbb{N}}$ in $\mathbb{N}$ with $k_i \to \infty$ as $i \to \infty$) such that $x = \lim_{i \to \infty} x_{k_i}$.

Remark: The set of all cluster points of a sequence $(x_k)_{k \in \mathbb{N}}$ in X is identical to the closed set

$$\bigcap_{m \in \mathbb{N}} \mathrm{clos}_X \left(\{ x_k \in X \,;\; k \geq m \} \right). \tag{2-8}$$

(3) The metric space (X, d) is called **complete** if every Cauchy sequence in X has a cluster point in X.

Remark: Because every Cauchy sequence can have at most one cluster point, this means that every Cauchy sequence in X has a limit in X.

2.22 Banach spaces and Hilbert spaces.

(1) A normed $\mathbb{K}$-vector space X is called a **Banach space** if it is complete with respect to the induced metric.

(2) A Banach space X is called a **Banach algebra** if it is an algebra satisfying

$$\|xy\|_X \leq \|x\|_X \cdot \|y\|_X \quad \text{for all } x, y \in X. \tag{2-9}$$

Here X is an **algebra** if a product $(x, y) \mapsto xy \in X$ is defined on X which satisfies the associative law, the distributive law and $\alpha(xy) = (\alpha x)y = x(\alpha y)$ for all $\alpha \in \mathbb{K}$ and all $x, y \in X$. The algebra is called **commutative** if $xy = yx$ for all $x, y \in X$.

(3) A pre-Hilbert space that is complete with respect to the induced metric is called **Hilbert space**.

The basic example of a complete space is the space of real numbers $\mathbb{R}$, where the axiom of completeness in $\mathbb{R}$ is precisely the additional axiom compared to the space of rational numbers $\mathbb{Q}$. From the completeness of $\mathbb{R}$ one can then deduce (see E2.6) that $\mathbb{R}^n$ and $\mathbb{C}^n$ are complete (with respect to any of the metrics introduced in 2.5 and 2.8). As the simplest infinite-dimensional example we now consider

2.23 Sequence spaces. We denote by $\mathbb{K}^{\mathbb{N}}$ the set of all sequences (defined on $\mathbb{N}$) with values in $\mathbb{K}$:

$$\mathbb{K}^{\mathbb{N}} := \left\{\, x = (x_i)_{i\in\mathbb{N}} \; ; \; x_i \in \mathbb{K} \text{ for } i \in \mathbb{N} \,\right\}.$$

The canonical unit vectors in $\mathbb{K}^{\mathbb{N}}$ are given by

$$\mathbf{e}_i := (0,\dots,0,\,1\,,0,\dots) \quad \text{for } i \in \mathbb{N}.$$

$$\uparrow$$

$$i\text{-th component}$$

Then:

(1) The set $\mathbb{K}^{\mathbb{N}}$ becomes a metric space with the Fréchet metric

$$\varrho(x) := \sum_{i\in\mathbb{N}} 2^{-i}\frac{|x_i|}{1+|x_i|} \quad \text{for } x = (x_i)_{i\in\mathbb{N}} \in \mathbb{K}^{\mathbb{N}}.$$

(2) Let $x^k = \left(x_i^k\right)_{i\in\mathbb{N}} \in \mathbb{K}^{\mathbb{N}}$ and $x = (x_i)_{i\in\mathbb{N}} \in \mathbb{K}^{\mathbb{N}}$. Then

$$\varrho(x^k - x) \to 0 \text{ as } k \to \infty$$
$$\Longleftrightarrow \text{ For every } i\colon (\, x_i^k \to x_i \text{ as } k \to \infty \,).$$

(3) The set $\mathbb{K}^{\mathbb{N}}$ equipped with this metric is complete.

(4) For $x = (x_i)_{i\in\mathbb{N}} \in \mathbb{K}^{\mathbb{N}}$ we define

$$\|x\|_{\ell^p} := \left(\sum_{i\in\mathbb{N}} |x_i|^p\right)^{\frac{1}{p}} \in [0,\infty]\,, \quad \text{if } 1 \le p < \infty,$$

$$\|x\|_{\ell^\infty} := \sup_{i\in\mathbb{N}} |x_i| \in [0,\infty]\,,$$

and consider for $1 \le p \le \infty$ the set (for the case $0 < p < 1$ see E4.11)

$$\ell^p(\mathbb{K}) := \left\{\, x \in \mathbb{K}^{\mathbb{N}} \; ; \; \|x\|_{\ell^p} < \infty \,\right\}.$$

Then the set $\ell^p(\mathbb{K})$ with the norm $x \mapsto \|x\|_{\ell^p}$ is a Banach space.

(5) If $p = 2$, then $\ell^2(\mathbb{K})$ becomes a Hilbert space with the scalar product

$$(x\,,\,y)_{\ell^2} := \sum_{i\in\mathbb{N}} x_i\,\overline{y_i} \quad \text{for } x, y \in \ell^2(\mathbb{K}).$$

Proof (1). Let $\varrho_0(s) := \frac{|s|}{1+|s|}$ for $s \in \mathbb{K}$. Then

$$\varrho(x) = \sum_{i=1}^{\infty} 2^{-i}\varrho_0(x_i) \le \sum_{i=1}^{\infty} 2^{-i} = 1\,,$$

and hence $\varrho(x)$ is always finite. The triangle inequality for ϱ follows as in 2.8(1). $\qquad\square$

Proof (2). Let x^k, $x \in \mathbb{K}^{\mathbb{N}}$ with $\varrho(x^k - x) \to 0$ as $k \to \infty$. Then $\varrho_0(x_i^k - x_i) \le 2^i \varrho(x^k - x) \to 0$ for all i, and hence $\left| x_i^k - x_i \right| \to 0$ as $k \to \infty$. Conversely, assuming that $x_i^k \to x_i$ as $k \to \infty$ for all i yields that

$$\varrho(x^k - x) \le \underbrace{\sum_{i=1}^{j} 2^{-i} \varrho_0(x_i^k - x_i)}_{\to\, 0 \text{ as } k \to \infty \text{ for any } j} + \underbrace{2^{-j}}_{\to\, 0 \text{ as } j \to \infty}.$$

Consequently, $\varrho(x^k - x) \to 0$ as $k \to \infty$. □

Proof (3). If $\left(x^k \right)_{k \in \mathbb{N}}$ is a Cauchy sequence in $\mathbb{K}^{\mathbb{N}}$, then, similarly to the above, it follows that $\left(x_i^k \right)_{k \in \mathbb{N}}$ is a Cauchy sequence in $\mathbb{K}$ for any i. Hence there exist the limit

$$x_i := \lim_{k \to \infty} x_i^k \quad \text{in } \mathbb{K}.$$

On setting $x := (x_i)_{i \in \mathbb{N}}$ it follows from (2) that $\varrho(x^k - x) \to 0$ as $k \to \infty$. □

Proof (4). This is a special case of the more general result in 3.16 for the counting measure on $\mathbb{N}$. Here we give a separate proof.

Let $x = (x_i)_{i \in \mathbb{N}}$ and $y = (y_i)_{i \in \mathbb{N}}$ be in $\ell^p(\mathbb{K})$ and for $n \in \mathbb{N}$ define $x^n := (x_1, \ldots, x_n)$, $y^n := (y_1, \ldots, y_n)$. It follows from 2.5 that

$$\left| x^n + y^n \right|_p \le \left| x^n \right|_p + \left| y^n \right|_p \le \| x \|_{\ell^p} + \| y \|_{\ell^p} < \infty.$$

Letting $n \to \infty$ this implies that $x + y \in \ell^p(\mathbb{K})$, with

$$\| x + y \|_{\ell^p} \le \| x \|_{\ell^p} + \| y \|_{\ell^p}.$$

Hence $\ell^p(\mathbb{K})$ is a normed space. In order to show completeness, let $\left(x^k \right)_{k \in \mathbb{N}}$, with $x^k = \left(x_i^k \right)_{i \in \mathbb{N}} \in \ell^p(\mathbb{K})$, be a Cauchy sequence in $\ell^p(\mathbb{K})$. As $\left| x_i^k - x_i^l \right| \le \left\| x^k - x^l \right\|_{\ell^p}$ we have that $\left(x_i^k \right)_{k \in \mathbb{N}}$ are Cauchy sequences in $\mathbb{K}$, and hence there exist $x_i := \lim_{k \to \infty} x_i^k \in \mathbb{K}$. This implies for $n \in \mathbb{N}$ in the case $p < \infty$ that as $l \to \infty$

$$\sum_{i=1}^{n} \left| x_i^k - x_i \right|^p \longleftarrow \sum_{i=1}^{n} \left| x_i^k - x_i^l \right|^p \le \left\| x^k - x^l \right\|_{\ell^p}^p,$$

and so

$$\left(\sum_{i=1}^{n} \left| x_i^k - x_i \right|^p \right)^{\frac{1}{p}} \le \limsup_{l \to \infty} \left\| x^k - x^l \right\|_{\ell^p} =: \varepsilon_k < \infty$$

for all n. Hence $x^k - x \in \ell^p(\mathbb{K})$, and consequently $x \in \ell^p(\mathbb{K})$, and it holds that $\left\| x^k - x \right\|_{\ell^p} \le \varepsilon_k \to 0$ as $k \to \infty$. In the case $p = \infty$ we can argue analogously. □

The set of real numbers $\mathbb{R}$ may be defined as the completion of the rational numbers $\mathbb{Q}$. This procedure can be generalized to arbitrary metric spaces.

2.24 Completion. Let (X, d) be a (not necessarily complete) metric space. Consider the set $X^{\mathbb{N}}$ of all sequences in X and define

$$\widetilde{X} := \big\{ \widetilde{x} = (x_j)_{j \in \mathbb{N}} \in X^{\mathbb{N}} \ ; \ (x_j)_{j \in \mathbb{N}} \text{ is a Cauchy sequence in } X \big\}$$

with the equivalence relation

$$(x_j)_{j \in \mathbb{N}} = (y_j)_{j \in \mathbb{N}} \text{ in } \widetilde{X} \quad :\Longleftrightarrow \quad (d(x_j, y_j))_{j \in \mathbb{N}} \text{ is a null sequence.}$$

Then $(\widetilde{X}, \widetilde{d})$ is a complete metric space, where $\widetilde{d}$ is defined by

$$\widetilde{d}\big((x_j)_{j \in \mathbb{N}}, (y_j)_{j \in \mathbb{N}}\big) := \lim_{j \to \infty} d(x_j, y_j).$$

Moreover, the rule $J(x) := (x)_{j \in \mathbb{N}}$ defines an injective map $J : X \to \widetilde{X}$ which is *isometric*, i.e.

$$\widetilde{d}\big(J(x), J(y)\big) = d(x, y) \quad \text{for all } x, y \in X.$$

For $(x_j)_{j \in \mathbb{N}} \in \widetilde{X}$ it holds that $\widetilde{d}\big((x_j)_{j \in \mathbb{N}}, J(x_i)\big) \to 0$ as $i \to \infty$, and so $J(X)$ is dense in $\widetilde{X}$.

Conclusion: The above shows that for any metric space (X, d) there exist a complete metric space $(\widetilde{X}, \widetilde{d})$ and an injective isometric map $J : X \to \widetilde{X}$ such that $J(X)$ is dense in $\widetilde{X}$. It is then natural to identify elements $x \in X$ with $J(x) \in \widetilde{X}$.

Proof. For $\widetilde{x} = (x_i)_{i \in \mathbb{N}}$ and $\widetilde{y} = (y_i)_{i \in \mathbb{N}}$ in $\widetilde{X}$ we have

$$|d(x_j, y_j) - d(x_i, y_i)| \leq |d(x_j, y_j) - d(x_i, y_j)| + |d(x_i, y_j) - d(x_i, y_i)|$$
$$\leq d(x_j, x_i) + d(y_j, y_i) \quad \text{(triangle inequality)}$$
$$\to 0 \quad \text{as } i, j \to \infty,$$

and hence there exists

$$\widetilde{d}(\widetilde{x}, \widetilde{y}) := \lim_{i \to \infty} d(x_i, y_i).$$

Similarly, it follows for $\widetilde{x}^1 = \widetilde{x}^2$ in $\widetilde{X}$ and $\widetilde{y}^1 = \widetilde{y}^2$ in $\widetilde{X}$ that

$$\big|d(x_i^2, y_i^2) - d(x_i^1, y_i^1)\big| \to 0 \quad \text{as } i \to \infty.$$

This shows that $\widetilde{d} : \widetilde{X} \times \widetilde{X} \to \mathbb{R}$ is well defined (see the remark in 2.4). Furthermore, it follows that $\widetilde{d}(\widetilde{x}, \widetilde{y}) = 0$ if $\widetilde{x} = \widetilde{y}$ in $\widetilde{X}$, and the triangle inequality carries over from d to $\widetilde{d}$. Hence $\widetilde{d}$ is a metric on $\widetilde{X}$.

In order to show completeness, let $(x^k)_{k \in \mathbb{N}}$ be a Cauchy sequence in $\widetilde{X}$, where $x^k = (x_j^k)_{j \in \mathbb{N}}$ for $k \in \mathbb{N}$. Given $k \in \mathbb{N}$ choose j_k such that

$$d(x_i^k, x_j^k) \leq \tfrac{1}{k} \quad \text{for } i,j \geq j_k \,.$$

Then

$$
\begin{aligned}
d(x_{j_k}^k, x_{j_l}^l) &\leq d(x_{j_k}^k, x_j^k) + d(x_j^k, x_j^l) + d(x_j^l, x_{j_l}^l) \\
&\leq \frac{1}{k} + d(x_j^k, x_j^l) + \frac{1}{l} \quad \text{for } j \geq j_k, j_l \\
&\to \frac{1}{k} + \widetilde{d}(x^k, x^l) + \frac{1}{l} \quad \text{as } j \to \infty \\
&\to 0 \quad \text{as } k, l \to \infty.
\end{aligned}
\tag{2-10}
$$

Hence we have that $x^\infty := \left(x_{j_l}^l\right)_{l \in \mathbb{N}} \in \widetilde{X}$ and

$$
\begin{aligned}
\widetilde{d}(x^l, x^\infty) &\leftarrow d(x_k^l, x_k^\infty) \quad \text{as } k \to \infty \\
&\leq d(x_k^l, x_{j_l}^l) + d(x_{j_l}^l, x_{j_k}^k) \leq \frac{1}{l} + d(x_{j_l}^l, x_{j_k}^k) \quad \text{for } k \geq j_l \\
&\to 0 \quad \text{as } k, l \to \infty \quad \text{(recall (2-10))}.
\end{aligned}
$$

The assertions on J are easily verified. $\qquad\square$

This means that every metric space that is not complete can be extended to a complete space. Examples of completions are the space of Lebesgue integrable functions in Appendix A3 and the Sobolev spaces in 3.27.

E2 Exercises

E2.1 Open and closed sets. If $(X, \mathcal{T})$ is a topological space, then it holds for $A \subset X$ that:

(1) $X \setminus \text{clos}(A) = \text{intr}(X \setminus A)$.

(2) $\text{intr}(A)$ is open, and $\text{clos}(A)$ is closed.

(3) $A \in \mathcal{T} \iff A = \text{intr}(A)$.

(4) $X \setminus A \in \mathcal{T} \iff A = \text{clos}(A)$.

Solution (1). From the negation of the definition of a closure in 2.11 it follows for $x \in X \setminus \text{clos}(A)$ that there exists an $U \in \mathcal{T}$ with $x \in U$ and $U \cap A = \emptyset$. This means $U \subset X \setminus A$ and $U \in \mathcal{T}$ with $x \in U$, and this is the definition of a point $x \in \text{intr}(X \setminus A)$. $\qquad\square$

Solution (2). Let $\mathcal{T}' := \{U \in \mathcal{T}; \ U \subset A, \ U \cap \text{intr}(A) \neq \emptyset\}$. On recalling the definition of the interior of A we then have that

$$\text{intr}(A) \subset V := \bigcup_{U \in \mathcal{T}'} U \in \mathcal{T}.$$

Moreover, $x \in U \in \mathcal{T}'$ implies that $U \in \mathcal{T}$ and $x \in U \subset A$, and so $x \in \text{intr}(A)$. Hence, $\text{intr}(A) = V \in \mathcal{T}$. The second claim now follows from (1). $\qquad\square$

Solution (3). If $A \in \mathcal{T}$ and $x \in A$, then $x \in U := A$ with $U \in \mathcal{T}$, and so $A \subset \operatorname{intr}(A) \subset A$. Conversely, $A = \operatorname{intr}(A) \in \mathcal{T}$ by (2). □

Solution (4). Follows from (3) on noting (1). □

E2.2 Distance and neighbourhoods. Let (X, d) be a metric space and $A \subset X$. Then:

(1) $\operatorname{dist}(\cdot, A)$ is a Lipschitz continuous function with Lipschitz constant ≤ 1, where equality holds if $X \setminus \overline{A}$ is nonempty.

(2) The neighbourhoods $B_r(A)$ for $r > 0$ are open sets. In particular, all balls $B_r(x)$ for $x \in X$ and $r > 0$ are open.

(3) For $r_1, r_2 > 0$, one has $B_{r_1}(B_{r_2}(A)) \subset B_{r_1+r_2}(A)$, and equality holds if X is a normed space.

Solution (1). Let $x, y \in X$. Given $\varepsilon > 0$ choose $a \in A$ such that $d(x, a) \leq \operatorname{dist}(x, A) + \varepsilon$. On employing the triangle inequality it then follows that

$$\operatorname{dist}(y, A) - \operatorname{dist}(x, A) \leq d(y, a) - d(x, a) + \varepsilon \leq d(y, x) + \varepsilon.$$

A symmetry argument then yields that

$$|\operatorname{dist}(y, A) - \operatorname{dist}(x, A)| \leq d(x, y).$$

This corresponds to the definition of Lipschitz continuity in 3.7 with Lipschitz constant ≤ 1. If $x \in X \setminus \overline{A}$, then $B_\varepsilon(x) \cap A = \emptyset$ for an $\varepsilon > 0$, and hence $\operatorname{dist}(x, A)$ is positive. Now choose for every $\varepsilon > 0$ a $y \in A$ such that $d(x, y) \leq (1 + \varepsilon)\operatorname{dist}(x, A)$. It follows that

$$|\operatorname{dist}(y, A) - \operatorname{dist}(x, A)| = \operatorname{dist}(x, A) \geq \frac{1}{1 + \varepsilon} d(x, y),$$

which shows that the Lipschitz constant is equal to 1. □

Solution (2). Let $x \in B_r(A)$ and $\delta := r - \operatorname{dist}(x, A) > 0$. If $y \in B_\delta(x)$, then, by (1),

$$\operatorname{dist}(y, A) \leq \operatorname{dist}(x, A) + d(x, y) < \operatorname{dist}(x, A) + \delta = r,$$

and so $B_\delta(x) \subset B_r(A)$. □

Solution (3). Let $x \in B_{r_1}(B_{r_2}(A))$, i.e. $\operatorname{dist}(x, B_{r_2}(A)) < r_1$. Then there exists a $y \in B_{r_2}(A)$ with $d(x, y) < r_1$. It follows from (1) that

$$\operatorname{dist}(x, A) \leq \operatorname{dist}(y, A) + d(x, y) < r_2 + r_1.$$

Now let X be a normed space and $x \in B_{r_1+r_2}(A)$. Then there exists a $y \in A$ with $\|x - y\| < r_1 + r_2$. It follows for

$$z := (1 - s)x + sy, \quad s := \frac{r_2}{r_1 + r_2},$$

that

$$\|z - y\| = (1 - s)\|x - y\| < r_1 \quad \text{and} \quad \|x - z\| = s\|x - y\| < r_2\,,$$

and so $x \in B_{r_2}(B_{r_1}(A))$. □

E2.3 Construction of metrics. Let $\psi : [0, \infty[\to [0, \infty[$ be a continuously differentiable strictly monotone function with $\psi(0) = 0$ and nonincreasing derivative ψ'. Then

$$d \text{ is a metric on } X \quad \Longrightarrow \quad \psi \circ d \text{ is a metric on } X\,.$$

Example:

$$\psi(t) := \frac{t}{1 + t}\,.$$

Solution. We have to verify the metric axioms in 2.6 for $\psi \circ d$. The axiom (M1) is satisfied, since

$$\psi(d(x, y)) = 0 \quad \Longleftrightarrow \quad d(x, y) = 0 \quad \Longleftrightarrow \quad x = y\,.$$

The axiom (M3) follows from

$$\psi(d(x, y)) \le \psi(d(x, z) + d(z, y)) = \psi(d(x, z)) + \int_0^{d(z,y)} \psi'(d(x, z) + t)\, dt$$

$$\le \psi(d(x, z)) + \int_0^{d(z,y)} \psi'(t)\, dt = \psi(d(x, z)) + \psi(d(z, y))\,.$$

□

E2.4 Convergence. Prove the assertions on convergence in 2.17.

Proof 2.17(1). Assume that $f(s) \to y_1$ and $f(s) \to y_2$ in Y as $\varphi(s) \to x_0$ with $y_1 \neq y_2$. As Y is a Hausdorff space, there exist $y_1 \in V_1 \in \mathcal{T}_Y$ and $y_2 \in V_2 \in \mathcal{T}_Y$ such that $V_1 \cap V_2 = \emptyset$. However, the definition of convergence yields that there exists a $U_1 \in \mathcal{T}_X$ such that $x_0 \in U_1$ and $f(\varphi^{-1}(U_1)) \subset V_1$, and then a $U_2 \in \mathcal{T}_X$ such that $x_0 \in U_2 \subset U_1, \varphi^{-1}(U_2) \neq \emptyset$ and $f(\varphi^{-1}(U_2)) \subset V_2$. As $U_2 \subset U_1$ it follows that $f(\varphi^{-1}(U_2)) \subset V_2 \cap V_1 = \emptyset$, and so $\varphi^{-1}(U_2) = \emptyset$, which is a contradiction. □

Proof 2.17(2). For $x_0 \in U_0 \in \mathcal{T}_X$ the definition of convergence gives that $\varphi^{-1}(U_0) \neq \emptyset$, i.e. $\varphi(S) \cap U_0 \neq \emptyset$, and so $x_0 \in \text{clos}(\varphi(S))$. In addition it follows from the definition of convergence that for $y_0 \in V_0 \in \mathcal{T}_Y$ there exists an $s \in S$ with $f(s) \in V_0$, and so $y_0 \in \text{clos}(f(S))$. □

Proof 2.17(3). Choosing $U_0 = X$ and $V_0 = V$ yields convergence in 2.17(3). Conversely, set $V = V_0$. Then if $x_0 \in U \in \mathcal{T}_X$ with $f(U \cap S) \subset V$ as in 2.17(3), it holds for $\widetilde{U} = U \cap U_0$ that

$$\widetilde{U} \cap S \neq \emptyset \text{ (since } x_0 \in \text{clos}(S)) \text{ and } f(\widetilde{U} \cap S) \subset V_0\,.$$

□

Proof 2.17(4). Let $y_0 \in V \in \mathcal{T}_Y$ and then U as in 2.17(3). It follows from $x_0 \in U \cap S$ that $f(x_0) \in V$. As Y is a Hausdorff space, this implies that $f(x_0) = y_0$. $\qquad\qquad$ □

Proof 2.17(5). Let f be continuous, and $V \in \mathcal{T}_Y$ with $x_0 \in f^{-1}(V)$. Since f is continuous at x_0, there exists a $U \in \mathcal{T}_X$ such that $x_0 \in U$ and $f(U) \subset V$, i.e. $x_0 \in U \subset f^{-1}(V)$. Hence $f^{-1}(V) \in \mathcal{T}_X$. Conversely, let $x_0 \in X$ and $f(x_0) \in V \in \mathcal{T}_Y$. Then $x_0 \in U := f^{-1}(V) \in \mathcal{T}_X$, which proves the continuity of f in x_0. $\qquad\qquad$ □

E2.5 Examples of continuous maps.

(1) Let $\mathcal{T}_1$, $\mathcal{T}_2$ be two topologies on X. Then the *identity* Id $: X \to X$, defined by $\mathrm{Id}(x) := x$, is a continuous map from $(X, \mathcal{T}_2)$ to $(X, \mathcal{T}_1)$ if and only if $\mathcal{T}_2$ is stronger than $\mathcal{T}_1$.

(2) If (X, d) is a metric space, then $d : X \times X \to \mathbb{R}$ is continuous.

(3) If $(X, \|\cdot\|)$ is a normed space, then the norm is a continuous map from X to $\mathbb{R}$.

(4) Let $(\cdot_1, \cdot_2)$ be a scalar product on the $\mathbb{K}$-vector space X, let $\|\cdot\|$ be the corresponding induced norm and consider the normed space $(X, \|\cdot\|)$. Then the scalar product is a continuous map from $X \times X$ to $\mathbb{K}$.

Solution (2). Use E2.2(1). $\qquad\qquad$ □

Solution (3). This follows from (2) and the definition of the induced metric in 2.6. $\qquad\qquad$ □

Solution (4). Employ the Cauchy-Schwarz inequality 2.2(2). $\qquad\qquad$ □

E2.6 Completeness of Euclidean space. The set $\mathbb{K}^n$ is complete with respect to all of the metrics given in 2.5 and 2.8.

Solution. First show the completeness with respect to the ∞-norm in 2.5: If $\left(x^k\right)_{k \in \mathbb{N}}$ is a Cauchy sequence with respect to this norm, $x^k = \left(x_i^k\right)_{i=1,\ldots,n}$, then $\left|x_i^k - x_i^l\right| \leq \left\|x^k - x^l\right\|_\infty$, and so $\left(x_i^k\right)_{k \in \mathbb{N}}$ are Cauchy sequences in $\mathbb{K}$, which means that there exist $x_i = \lim_{k \to \infty} x_i^k$ in $\mathbb{K}$ (because $\mathbb{R}$ and $\mathbb{C}$ are complete, with the completeness of the latter following from that of $\mathbb{R}^2$, which is shown here). Hence $\left|x_i^k - x_i\right| \to 0$ as $k \to \infty$ for every $i \in \{1, \ldots, n\}$, which implies that $\left\|x^k - x\right\|_\infty \to 0$ as $k \to \infty$.

The completeness with respect to the other metrics then follows from the results in 2.16. $\qquad\qquad$ □

E2.7 Incomplete function space. Let $I := [a, b] \subset \mathbb{R}$ be an interval with $a < b$, and for $n \in \mathbb{N}$ let

$$\mathcal{P}_n := \{f : I \to \mathbb{R}; \; f \text{ is a polynomial of degree } \leq n\}.$$

Then $\mathcal{P} := \bigcup_{n \in \mathbb{N}} \mathcal{P}_n$ equipped with

$$\|f\|_\infty := \sup_{x \in I} |f(x)| \quad \text{for } f \in \mathcal{P}$$

is a normed space that is not complete.

Solution. The norm axioms are easily verified. Setting

$$f(x) := e^x = \sum_{i=1}^\infty \frac{1}{i!} x^i, \quad f_n(x) := \sum_{i=1}^n \frac{1}{i!} x^i$$

we have that

$$\sup_{x \in I} |f_n(x) - f(x)| \to 0 \quad \text{as } n \to \infty.$$

Hence $(f_n)_{n \in \mathbb{N}}$ is a Cauchy sequence in $\mathcal{P}$. If $g = \lim_{n \to \infty} f_n$ existed in $\mathcal{P}$, it would follow that $|f_n(x) - g(x)| \le \|f_n - g\|_\infty \to 0$ as $n \to \infty$ for all $x \in I$, and so $g = f \notin \mathcal{P}$, which is a contradiction. □

E2.8 On completeness. Let (X, d) be a metric space. Then:

(1) If (X, d) is complete and $Y \subset X$ is closed, then (Y, d) is also a complete metric space.

(2) If $Y \subset X$ and (Y, d) is complete, then Y is closed in X (as a subset of the metric space (X, d)).

Solution (1). If $(x^k)_{k \in \mathbb{N}}$ is a Cauchy sequence in Y, then it is also a Cauchy sequence in X. The completeness of X yields that it has a limit $x \in X$. As Y is closed it follows that $x \in Y$. □

Solution (2). Let $(x^k)_{k \in \mathbb{N}}$ be a sequence in Y converging in X to $x \in X$. Since Y is equipped with the metric d, it is a Cauchy sequence in Y. The completeness of Y yields that it has a limit $y \in Y$. Now y must also be the limit of the sequence in X, and so $x = y \in Y$. □

E2.9 Hausdorff distance between sets. Let (X, d) be a metric space and

$$\mathcal{A} := \{A \subset X ; \ A \text{ is nonempty, bounded and closed}\}.$$

The **_Hausdorff distance_** between $A_1 \in \mathcal{A}$ and $A_2 \in \mathcal{A}$ is defined by

$$d_H(A_1, A_2) := \inf\{\varepsilon > 0 ; \ A_1 \subset B_\varepsilon(A_2) \text{ and } A_2 \subset B_\varepsilon(A_1)\}.$$

Then d_H is a metric on $\mathcal{A}$, and for $A, B \in \mathcal{A}$ we have

$$d_H(A, B) = \max \left(\sup_{a \in A} \text{dist}(a, B), \ \sup_{b \in B} \text{dist}(b, A) \right)$$

$$= \sup_{x \in M} |\text{dist}(x, A) - \text{dist}(x, B)|$$

for any set M with $A \cup B \subset M \subset X$.

Solution. If $d_H(A_1, A_2) = 0$, then

$$A_1 \subset \bigcap_{\varepsilon > 0} B_\varepsilon(A_2) = \overline{A_2} = A_2 \,,$$

and similarly $A_2 \subset A_1$. Moreover, d_H is symmetric by definition. Given $A_1, A_2, A_3 \in \mathcal{A}$ and $\delta > 0$, there exist numbers $\varepsilon_1 > 0$, $\varepsilon_2 > 0$ such that

$$\varepsilon_1 \leq d_H(A_1, A_2) + \delta, \ A_1 \subset B_{\varepsilon_1}(A_2), \ A_2 \subset B_{\varepsilon_1}(A_1),$$
$$\varepsilon_2 \leq d_H(A_2, A_3) + \delta, \ A_2 \subset B_{\varepsilon_2}(A_3), \ A_3 \subset B_{\varepsilon_2}(A_2).$$

By E2.2(3),

$$A_1 \subset B_{\varepsilon_1}(B_{\varepsilon_2}(A_3)) \subset B_{\varepsilon_1 + \varepsilon_2}(A_3) \,,$$
$$A_3 \subset B_{\varepsilon_2}(B_{\varepsilon_1}(A_1)) \subset B_{\varepsilon_1 + \varepsilon_2}(A_1) \,,$$

and hence

$$d_H(A_1, A_3) \leq \varepsilon_1 + \varepsilon_2 \leq d_H(A_1, A_2) + d_H(A_2, A_3) + 2\delta \,.$$

This shows that d_H defines a metric.

Now let $A, B \in \mathcal{A}$, $d := d_H(A, B)$ and

$$d_{\max} := \max \left(\sup_{a \in A} \mathrm{dist}(a, B), \ \sup_{b \in B} \mathrm{dist}(b, A) \right),$$
$$d_{\sup} := \sup_{x \in M} |\mathrm{dist}(x, A) - \mathrm{dist}(x, B)|.$$

Then $d_{\sup} \geq d_{\max}$, on noting that

$$d_{\sup} \geq \sup_{x \in B} |\mathrm{dist}(x, A) - 0|,$$

and applying a symmetry argument. Moreover, $d_{\max} \geq d$, as for $\delta > 0$ we have that

$$B \subset B_{d_{\max} + \delta}(A) \,,$$

and hence, by a symmetry argument, that $d_{\max} + \delta \geq d$. Furthermore, $d \geq d_{\max}$, since $B \subset B_\varepsilon(A)$ and $A \subset B_\varepsilon(B)$ implies that

$$\mathrm{dist}(b, A) < \varepsilon \quad \text{for } b \in B \quad \text{and} \quad \mathrm{dist}(a, B) < \varepsilon \quad \text{for } a \in A \,,$$

and so $d_{\max} \leq \varepsilon$. Finally, $d_{\max} \geq d_{\sup}$, because for $x \in X$ and $\delta > 0$, there exists a $b \in B$ such that

$$\mathrm{dist}(x, B) \geq d(x, b) - \delta \,.$$

Thanks to E2.2(1),

$$\mathrm{dist}(x, A) - \mathrm{dist}(x, B) \leq \mathrm{dist}(x, A) - d(x, b) + \delta \leq \mathrm{dist}(b, A) + \delta \,,$$

and hence, by a symmetry argument, $d_{\sup} \leq d_{\max} + \delta$.   □

3 Function spaces

In this chapter we introduce the most important function spaces occurring in analysis. They are the spaces of continuous and differentiable functions, also called classical function spaces (see 3.2–3.7), the spaces of integrable functions, also called Lebesgue spaces (see 3.15–3.21), and the Sobolev spaces (see 3.27–3.29). Sobolev spaces combine properties regarding differentiability with those concerning integrability, and they play a fundamental role in the treatment of differential equations. Although in this chapter we almost exclusively consider functions with values in Banach spaces, for an understanding of the basic results it is sufficient to replace the Banach space Y with a Euclidean space $\mathbb{R}^k$. However, for more advanced topics, including applications to parabolic differential equations, it will be crucial to consider the case where Y is itself a function space.

3.1 Bounded functions. Let S be a set and let Y be a Banach space over $\mathbb{K}$ with norm $y \mapsto |y|$. We define the set of **bounded functions** (or **bounded maps**) on S with values in Y by

$$B(S;Y) := \left\{ f : S \to Y \; ; \; f(S) \text{ is a bounded subset of } Y \right\}.$$

This is a subset of the set $\mathscr{F}(S;Y)$ of all functions from S to Y (see the Introduction). On defining

$$
\begin{aligned}
(f_1 + f_2)(x) &:= f_1(x) + f_2(x) \quad \text{for } x \in S, \\
(\alpha f)(x) &:= \alpha f(x) \quad \text{for } x \in S
\end{aligned}
\tag{3-1}
$$

for functions f_1, f_2, f and $\alpha \in \mathbb{K}$, the set $B(S;Y)$ becomes a $\mathbb{K}$-vector space, and with the **supremum norm**

$$\|f\|_{B(S)} \left(\text{or } \|f\|_{\mathrm{sup}} \right) := \sup_{x \in S} |f(x)|$$

a Banach space. We use the abbreviation $B(S)$ for $B(S;\mathbb{K})$ and use similar abbreviations for all the function spaces below.

Remark: Strictly speaking, the norm $\|f\|_{B(S)}$ would need to be written $\|f\|_{B(S;Y)}$. But since then $\|f\|_{B(S;Y)} = \|\,|f|\,\|_{B(S;\mathbb{R})}$, we drop the image space Y in the subscript for notational convenience, and we will proceed similarly for all the remaining function spaces.

Proof. In order to prove completeness, let $(f_k)_{k\in\mathbb{N}}$ be a Cauchy sequence in $B(S;Y)$. Then

$$|f_k(x) - f_l(x)| \leq \|f_k - f_l\|_{\sup} \to 0 \quad \text{as } k, l \to \infty,$$

for all $x \in S$ and hence $(f_k(x))_{k\in\mathbb{N}}$ is a Cauchy sequence in Y. As Y is complete, there exists

$$f(x) := \lim_{k\to\infty} f_k(x) \quad \text{in } Y.$$

It follows for $x \in S$ that

$$|f(x) - f_k(x)| = \lim_{l\to\infty} |f_l(x) - f_k(x)| \leq \liminf_{l\to\infty} \|f_l - f_k\|_{\sup} < \infty.$$

That means $f - f_k \in B(S;Y)$ and hence also $f \in B(S;Y)$, with

$$\|f - f_k\|_{\sup} \leq \liminf_{l\to\infty} \|f_l - f_k\|_{\sup} \to 0 \quad \text{as } k \to \infty.$$

$\square$

A special class of bounded functions are continuous functions on bounded closed subsets of $\mathbb{R}^n$.

3.2 Continuous functions on compact sets. If $S \subset \mathbb{R}^n$ is closed and bounded and Y is a Banach space over $\mathbb{K}$ with norm $y \mapsto |y|$, then

$$C^0(S;Y) \ \big(\text{or } C(S;Y)\big) := \big\{\, f : S \to Y \ ; \ f \text{ is continuous on } S \,\big\}$$

is the set of **continuous functions** on S with values in Y. Then $C^0(S;Y)$ is a closed subspace of $B(S;Y)$, and so if it is equipped with the supremum norm

$$\|f\|_{C^0(S)} := \|f\|_{B(S)} = \sup_{x\in S} |f(x)|$$

it becomes a Banach space. Where no confusion can arise, we will also write $\|f\|_{C^0}$ instead of $\|f\|_{C^0(S)}$. It is easily seen that $C^0(S) := C^0(S;\mathbb{K})$ is a commutative Banach algebra with the product $(fg)(x) := f(x) \cdot g(x)$ for $f, g \in C^0(S)$ and $x \in S$.

Remark: The Heine-Borel theorem (see 4.7(7)) states that the closed and bounded subsets of $\mathbb{R}^n$ that we consider here are precisely the compact subsets of $\mathbb{R}^n$. This property will play a crucial role in the following considerations.

Note: The space $C^0(S;Y)$ can be defined more generally for compact topological spaces S.

Proof. Every $f \in C^0(S; Y)$ is bounded, and so an element of $B(S; Y)$: Given $x \in S$, since f is continuous at x, there exists a $\delta_x > 0$ such that $f(B_{\delta_x}(x)) \subset B_1(f(x))$. The balls $B_{\delta_x}(x)$ for $x \in S$ form an open cover of S. Since S is compact (see 4.7(7)), there exist finitely many points $x_1, \ldots, x_m \in S$ such that

$$S \subset \bigcup_{i=1}^m B_{\delta_{x_i}}(x_i) , \quad \text{and so} \quad f(S) \subset \bigcup_{i=1}^m B_1(f(x_i)) ,$$

which is a bounded set in Y.

For $f_1, f_2 \in C^0(S; Y)$ we have that $f_1 + f_2$ and αf_1, for $\alpha \in \mathbb{K}$, are continuous, and hence $C^0(S; Y)$ is a subspace of $B(S; Y)$. Moreover, $C^0(S; Y)$ is closed in $B(S; Y)$: Let $(f_i)_{i \in \mathbb{N}}$ be a Cauchy sequence in $B(S; Y)$ with functions $f_i \in C^0(S; Y)$. The completeness of $B(S; Y)$ yields that the sequence has a limit $f \in B(S; Y)$. For $x, y \in S$ it then holds that

$$|f(y) - f(x)| \leq \underbrace{|f_i(y) - f_i(x)|}_{\to\, 0 \text{ as } y \to x \text{ for any } i} + 2 \cdot \underbrace{\|f - f_i\|_{B(S)}}_{\to\, 0 \text{ as } i \to \infty} ,$$

which shows that $f \in C^0(S; Y)$. Hence $C^0(S; Y)$ is closed and, equipped with the $B(S)$-norm, complete (see E2.8(1)). □

Proof Note. The boundedness of continuous functions $f : S \to Y$ can be seen as follows: For $x \in S$ we have that $U_x := f^{-1}(B_1(f(x)))$ is an open set in S. The compactness of S yields that S is covering compact (see 4.6(1)). Hence the cover $(U_x)_{x \in S}$ has a finite subcover, i.e. there exist $x_1, \ldots, x_m \in S$ such that $S \subset \cup_{i=1}^m U_{x_i}$. This yields the boundedness of $f(S)$ as before. □

We now consider the space of continuous functions on general subsets of $\mathbb{R}^n$ and we equip this space with a metric, similarly to the construction for sequence spaces (see 2.23(1)).

3.3 Continuous functions. Let $S \subset \mathbb{R}^n$, so that there exists an ***exhaustion*** $(K_i)_{i \in \mathbb{N}}$ with bounded closed sets $K_i \subset \mathbb{R}^n$ (i.e. compact sets, see 4.7(7)) such that

$$S = \bigcup_{i \in \mathbb{N}} K_i \text{ and } \emptyset \neq K_i \subset K_{i+1} \subset S \text{ for } i \in \mathbb{N} ,$$
$$x \in S \implies B_\delta(x) \cap S \subset K_i \text{ for a } \delta > 0 \text{ and an } i \in \mathbb{N} . \tag{3-2}$$

Moreover, let Y be a Banach space over $\mathbb{K}$ and let

$$C^0(S; Y) := \{ f : S \to Y ;\ f \text{ is continuous on } S \}$$

be the set of ***continuous functions*** on S with values in Y. Then it holds that:

(1) $C^0(S; Y)$ with the operations (3-1) is a $\mathbb{K}$-vector space.

(2) Equipped with the Fréchet metric

$$\varrho(f) := \sum_{i \in \mathbb{N}} 2^{-i} \frac{\|f\|_{C^0(K_i)}}{1 + \|f\|_{C^0(K_i)}} \quad \text{for } f \in C^0(S; Y)$$

this is a complete metric space.

(3) The topology induced by this metric is independent of the choice of exhaustion.

(4) If $S \subset \mathbb{R}^n$ is bounded and closed (i.e. compact), then this topology coincides with that induced by the norm in 3.2.

Example: The property (3-2) is satisfied for open sets $S \subset \mathbb{R}^n$ and for closed sets $S \subset \mathbb{R}^n$ (see E3.2).

Proof (2). The properties of a metric can be shown as in 2.23(1). If $(f_k)_{k \in \mathbb{N}}$ is a Cauchy sequence in $C^0(S; Y)$, then $\|f_k - f_l\|_{C^0(K_i)} \to 0$ as $k, l \to \infty$ for all $i \in \mathbb{N}$ (see E3.1). Since, by 3.2, $C^0(K_i; Y)$ is complete, there exist functions $g_i \in C^0(K_i; Y)$ such that $\|f_k - g_i\|_{C^0(K_i)} \to 0$ as $k \to \infty$, and in particular $f_k(x) \to g_i(x)$ as $k \to \infty$ for all $x \in K_i$. That means that $g_{i_1} = g_{i_2}$ on $K_{i_1} \cap K_{i_2}$ for all $i_1, i_2 \in \mathbb{N}$, and so there exists a function $f : S \to Y$ such that $f = g_i$ on K_i for all i.

Now f is continuous, since for $x \in S$ we have that $S \cap B_\delta(x) \subset K_i$ for a $\delta > 0$ and an $i \in \mathbb{N}$. Then $f = g_i$ on $S \cap B_\delta(x)$, and so f is continuous at x. Finally, because $\|f_k - f\|_{C^0(K_i)} \to 0$ as $k \to \infty$ for all i, it follows (see E3.1) that $\varrho(f_k - f) \to 0$ as $k \to \infty$. □

Proof (3). Let $\left(\tilde{K}_j\right)_{j \in \mathbb{N}}$ be another exhaustion with bounded closed sets and let $\tilde{\varrho}$ be the corresponding Fréchet metric. From the second property in (3-2) for the sequence $(K_i)_{i \in \mathbb{N}}$ it follows that:

$$K \subset S \text{ compact} \quad \Longrightarrow \quad K \subset K_i \text{ for an } i \in \mathbb{N}$$

(use (3-2) for $x \in K$ and the covering compactness of K from 4.6(1)). For $j \in \mathbb{N}$ we have that $\tilde{K}_j \subset S$ is compact, and so there exists an $i_j \in \mathbb{N}$ such that $\tilde{K}_j \subset K_{i_j}$. By using induction on j, we can choose i_j such that $i_{j_1} > i_{j_2}$ for $j_1 > j_2$. Now let $\delta > 0$ and $\varrho(f) < \delta$. Then for every $l \in \mathbb{N}$

$$\tilde{\varrho}(f) \leq \sum_{j > l} 2^{-j} + \sum_{j \leq l} 2^{-j} \frac{\|f\|_{C^0(K_{i_j})}}{1 + \|f\|_{C^0(K_{i_j})}}$$

$$\leq 2^{-l} + c_l \cdot \delta \quad \text{with } c_l := \max_{j \leq l} 2^{i_j - j}.$$

Given $\varepsilon > 0$, choose l such that $2^{-l} \leq \frac{\varepsilon}{2}$ and then $\delta > 0$ sufficiently small, so that $c_l \cdot \delta < \frac{\varepsilon}{2}$, which implies that

$$\varrho(f) < \delta \quad \Longrightarrow \quad \widetilde{\varrho}(f) < \varepsilon.$$

This proves that the topology induced by ϱ is stronger than the topology induced by $\widetilde{\varrho}$. A symmetry argument yields the converse. □

Proof (4). If S itself is bounded and closed, we can choose $K_i = S$ for all $i \in \mathbb{N}$, in which case

$$\varrho(f) = \frac{\|f\|_{C^0(S)}}{1 + \|f\|_{C^0(S)}}.$$

This metric is equivalent to the metric induced by the $C^0(S)$-norm (as in 2.16(2)). □

Observe that for open sets $\Omega \subset \mathbb{R}^n$, functions in $C^0(\Omega)$ may grow arbitrarily towards the boundary of Ω, e.g. $x \mapsto e^{\frac{1}{x}}$ is in $C^0(]0, 1[)$. It is possible to show that there exists no norm on $C^0(\Omega)$ that induces the same topology as the metric introduced in 3.3.

On the other hand, functions that vanish outside a compact subset of Ω play an important role in functional analysis as so-called **test functions** (see the account on distributions in 5.17 and beyond). With this in mind, we introduce the following definition:

3.4 Support of a function. Let $S \subset \mathbb{R}^n$ and let Y be a Banach space. For a map $f : S \to Y$ we call

$$\operatorname{supp}(f) := \operatorname{clos}(\{x \in S;\ f(x) \neq 0\}) \subset \operatorname{clos}(S)$$

the **support** of f. For $S \subset \mathbb{R}^n$ as in 3.3 we then define

$$C_0^0(S; Y) \ (\text{or } C_c^0(S; Y))$$
$$:= \{ f \in C^0(S; Y) \ ;\ \operatorname{supp}(f) \text{ is a compact subset of } S \}.$$

Remark: If $S \subset \mathbb{R}^n$ is open, then for $f \in C_0^0(S; Y)$ there exists a neighbourhood of the boundary $\mathrm{B}_\varepsilon(\partial S)$, $\varepsilon > 0$, such that $f = 0$ in $S \cap \mathrm{B}_\varepsilon(\partial S)$.

3.5 Differentiable functions. Let $\Omega \subset \mathbb{R}^n$ be open and let Y be a Banach space. We consider maps $f : \Omega \to Y$ and start by introducing the usual notations for derivatives, where in the following $\mathbf{e}_1, \ldots, \mathbf{e}_n$ denote the **canonical unit vectors** in $\mathbb{R}^n$. Let $x \in \Omega$. If the limit

$$\partial_i f(x) \quad \left(\text{or } \frac{\partial}{\partial x_i} f(x),\ \frac{\partial f}{\partial x_i}(x) \right)$$
$$:= \lim_{h \to 0} \frac{1}{h} \big(f(x + h\mathbf{e}_i) - f(x)\big) \quad \text{in } Y$$

exists, then $\partial_i f(x)$ is called the **i-th partial derivative** of f at the point x.

The map $f : \Omega \to Y$ is called **continuously differentiable** if all partial derivatives $\partial_i f$, $i = 1, \ldots, n$, exist at all points and if they are continuous maps from Ω to Y. Then

$$Df(x)(v) := \sum_{i=1}^{n} v \bullet \mathbf{e}_i \cdot \partial_i f(x) \quad \text{for } v \in \mathbb{R}^n$$

defines a linear map $Df(x) : \mathbb{R}^n \to Y$, the **derivative** of f at the point x. Moreover, denoting the norm in Y by $|\cdot|$, it holds that

$$f(y) = f(x) + Df(x)(y - x) + |y - x| \cdot \varepsilon_x(y)$$
$$\text{with } \varepsilon_x(y) \to 0 \text{ as } y \to x.$$

For $v \in \mathbb{R}^n$ we call $\partial_v f(x) := Df(x)(v)$ the **directional derivative of f in the direction of v** at the point x.

For $m \geq 2$ we call $f : \Omega \to Y$ **m times continuously differentiable** if all the iterated partial derivatives

$$\partial_{i_1} \partial_{i_2} \cdots \partial_{i_k} f \quad \text{with } i_1, \ldots, i_k \in \{1, \ldots, n\}, \ k \leq m,$$

exist and are continuous maps from Ω to Y. Then these iterated derivatives do not depend on the order in which the individual partial derivatives are applied. That is why higher order partial derivatives are indexed as follows: We call s an n-dimensional **multi-index of order k** if

$$s = (s_1, \ldots, s_n) \in \mathbb{Z}^n \quad \text{with } s_i \geq 0 \text{ for } i = 1, \ldots, n,$$
$$k = |s| := s_1 + \ldots + s_n,$$

i.e. $|s| = |s|_1$ corresponds to the sum norm in $\mathbb{R}^n$. For multi-indices s we then define **partial derivatives of higher order** by

$$\partial^s f(x) := \partial_1^{s_1} \cdots \partial_n^{s_n} f(x), \text{ where inductively}$$
$$\partial_i^k f(x) := \partial_i \big(\partial_i^{k-1} f\big)(x) \text{ for } k > 0, \quad \partial_i^0 f(x) := f(x).$$

The number $|s|$ is the **order** of the partial derivative $\partial^s f$. Further notations for multi-indices are

$$r \leq s \quad :\Longleftrightarrow \quad r_i \leq s_i \text{ for } i = 1, \ldots, n,$$
$$\binom{s}{r} := \prod_{i=1}^{n} \binom{s_i}{r_i} \quad (\textbf{binomial coefficient}),$$
$$s! := \prod_{i=1}^{n} s_i!,$$
$$x^s := \prod_{i=1}^{n} x_i^{s_i} \quad \text{for } x \in \mathbb{R}^n.$$

Additional notations for partial derivatives are

$$\nabla f(x) := (\partial_1 f(x), \ldots, \partial_n f(x)), \quad D^l f(x) := (\partial^s f(x))_{|s|=l}.$$

We now consider spaces of differentiable functions.

3.6 Space of differentiable functions. Let $\Omega \subset \mathbb{R}^n$ be open and bounded, and let $m \geq 0$ be an integer. Then we define

$$C^m(\overline{\Omega}; Y) := \{ \; f : \Omega \to Y \; ; \; f \text{ is } m \text{ times continuously differentiable in } \Omega$$
$$\text{and, for } |s| \leq m, \; \partial^s f \text{ can be continuously extended to } \overline{\Omega} \; \} \; .$$

Assertion: The set $C^m(\overline{\Omega}; Y)$ is a vector space and with the norm

$$\|f\|_{C^m(\overline{\Omega})} := \sum_{|s| \leq m} \|\partial^s f\|_{C^0(\overline{\Omega})}$$

it becomes a Banach space.

Further definitions: Similarly to 3.3, for $S \subset \mathbb{R}^n$ as in 3.3 with compact sets $K_i = \overline{\Omega}_i$, Ω_i open, we can define the complete metric space $C^m(S; Y)$ by using the norms of $C^m(\overline{\Omega}_i; Y)$. For $S = \Omega \subset \mathbb{R}^n$ open we can define, similarly to 3.4, the set $C_0^m(\Omega; Y)$.

Proof. We now prove the completeness of $C^m(\overline{\Omega}; Y)$ for the case $m = 1$; the general case then follows by induction on m. Let $(f_k)_{k \in \mathbb{N}}$ be a Cauchy sequence in $C^1(\overline{\Omega}; Y)$. Then $(f_k)_{k \in \mathbb{N}}$ and $(\partial_i f_k)_{k \in \mathbb{N}}$ for $i = 1, \ldots, n$ are Cauchy sequences in $C^0(\overline{\Omega}; Y)$. Hence, on recalling 3.2, there exist $f, g_i \in C^0(\overline{\Omega}; Y)$ such that $f_k \to f$ and $\partial_i f_k \to g_i$ uniformly on $\overline{\Omega}$. For $x \in \Omega$ and y close to x the fundamental theorem of calculus for Y-valued functions yields, on defining $x_t := (1 - t)x + ty$ for $0 \leq t \leq 1$, that

$$f_k(x_1) - f_k(x_0) = \int_0^1 \frac{d}{dt} f_k(x_t) \, dt = \int_0^1 (y - x) \bullet \nabla f_k(x_t) \, dt \, ,$$

where $(y - x) \bullet \nabla f_k(x_t) = \sum_{i=1}^n (y - x)_i \partial_i f_k(x_t)$. It follows that

$$|f_k(y) - f_k(x) - (y - x) \bullet \nabla f_k(x)|$$

$$= \left| \int_0^1 (y - x) \bullet (\nabla f_k(x_t) - \nabla f_k(x)) \, dt \right|$$

$$\leq \int_0^1 |\nabla f_k(x_t) - \nabla f_k(x)| \, dt \cdot |y - x|$$

$$\leq \left(2\|\nabla f_k - g\|_{C^0(\overline{\Omega})} + \sup_{0 \leq t \leq 1} |g(x_t) - g(x)| \right) \cdot |y - x| \, ,$$

where $g := (g_1, \ldots, g_n)$. On letting $k \to \infty$ this yields that

$$|f(y) - f(x) - (y - x) \bullet g(x)| \leq \underbrace{\sup_{0 \leq t \leq 1} |g(x_t) - g(x)|}_{\to 0 \text{ as } y \to x} \cdot |y - x| \, .$$

That means that f is differentiable at x with $\nabla f(x) = g(x)$. Hence $f \in C^1(\overline{\Omega}; Y)$ and in this space it holds that $f_k \to f$ as $k \to \infty$. $\qquad \square$

Subspaces of continuous and differentiable functions can be obtained by considering functions f with a given **modulus of continuity** σ, i.e.

$$|f(x) - f(y)| \leq C_f \cdot \sigma(|x - y|) \quad \text{for all } x, y,$$

where $\sigma : [0, \infty[\rightarrow [0, \infty[$ is continuous and strictly monotonically increasing with $\sigma(0) = 0$. The most important special case is $\sigma(s) = s^\alpha$ with $0 < \alpha \leq 1$:

3.7 Hölder continuous functions. Let $S \subset \mathbb{R}^n$ and let Y be a Banach space with norm $y \mapsto |y|$ as before. For $0 < \alpha \leq 1$ and $f : S \rightarrow Y$ we call

$$\text{Höl}_\alpha(f, S) := \sup \left\{ \frac{|f(x) - f(y)|}{|x - y|^\alpha} \; ; \; x, y \in S, \; x \neq y \right\} \in [0, \infty]$$

the **Hölder constant** of f on S to the exponent α, and in the special case $\alpha = 1$ we call $\text{Lip}(f, S) := \text{Höl}_1(f, S)$ the **Lipschitz constant**. If $\Omega \subset \mathbb{R}^n$ is open and bounded and $m \geq 0$, then the corresponding **Hölder spaces** are defined by

$$C^{m,\alpha}(\overline{\Omega}; Y) := \left\{ f \in C^m(\overline{\Omega}; Y) \; ; \; \text{Höl}_\alpha(\partial^s f, \overline{\Omega}) < \infty \text{ for } |s| = m \right\}.$$

These are Banach spaces with the norm

$$\|f\|_{C^{m,\alpha}(\overline{\Omega})} := \sum_{|s| \leq m} \|\partial^s f\|_{C^0(\overline{\Omega})} + \sum_{|s| = m} \text{Höl}_\alpha(\partial^s f, \overline{\Omega}).$$

Functions in $C^{0,\alpha}(\overline{\Omega}; Y)$ are called **Hölder continuous** on $\overline{\Omega}$, and **Lipschitz continuous** in the special case $\alpha = 1$.

Remark: The space $C^{0,\alpha}(S; Y)$ can be defined for any bounded closed set $S \subset \mathbb{R}^n$.

Remark: Similarly to 3.3 and 3.6 one can also define the metric spaces $C^{m,\alpha}(\Omega; Y)$.

Proof. We now prove the completeness of the space $C^{0,\alpha}(S; Y)$, where $S \subset \mathbb{R}^n$ is bounded and closed. For $C^{m,\alpha}(S; Y)$ with $m \geq 1$ apply the argument given below to the derivatives of order m in addition to the completeness of the space $C^m(S; Y)$.

Let $(f_k)_{k \in \mathbb{N}}$ be a Cauchy sequence in $C^{0,\alpha}(S; Y)$. Then it is also a Cauchy sequence in $C^0(S; Y)$, and due to the completeness of $C^0(S; Y)$ there exists an $f \in C^0(S; Y)$ such that $\|f - f_k\|_{C^0(S)} \rightarrow 0$ as $k \rightarrow \infty$. Now it holds for $x, y \in S$, $x \neq y$, and as $l \rightarrow \infty$ that

$$\frac{|(f - f_k)(x) - (f - f_k)(y)|}{|x - y|^\alpha} \quad \longleftarrow \quad \frac{|(f_l - f_k)(x) - (f_l - f_k)(y)|}{|x - y|^\alpha}$$

$$\leq \quad \text{Höl}_\alpha(f_l - f_k, S).$$

Hence,

$$\text{Höl}_\alpha(f - f_k, S) \le \liminf_{l \to \infty} \text{Höl}_\alpha(f_l - f_k, S) \to 0 \quad \text{as } k \to \infty.$$

That means that $f \in C^{0,\alpha}(S; Y)$ and $f_k \to f$ in $C^{0,\alpha}(S; Y)$ as $k \to \infty$.   □

3.8 Infinitely differentiable functions. The vector space of *infinitely differentiable functions* on an open set $\Omega \subset \mathbb{R}^n$ with values in a Banach space Y is defined by

$$C^\infty(\Omega; Y) := \bigcap_{m \in \mathbb{N}} C^m(\Omega; Y). \tag{3-3}$$

Initially this defines $C^\infty(\Omega; Y)$ only as a vector space.

Similarly to 3.4, one can also define $C_0^m(\Omega; Y)$ and $C_0^\infty(\Omega; Y)$. An example of a function in $C_0^\infty(\mathbb{R}^n; \mathbb{R})$ is given in E3.3. One way to define a topology for the space $C_0^\infty(\Omega; \mathbb{R})$ will be given in 5.20.

Measures and Integrals

In the second part of this chapter we will introduce spaces of measurable and integrable functions. To this end, we will first give the definition of a general measure and describe some examples of commonly used measures.

3.9 Measures. Let S be an arbitrary set and let $\mathcal{B}$ be a nonempty system of subsets of S. If $\mathcal{B}$ is a Boolean ring or a Boolean algebra (see A3.1), then $\mathcal{B}$ is called a *σ-ring* or a *σ-algebra*, respectively, if additionally

$$E_i \in \mathcal{B} \text{ for } i \in \mathbb{N} \implies \bigcup_{i \in \mathbb{N}} E_i \in \mathcal{B}.$$

We call $(S, \mathcal{B}, \mu)$ a *measure space*, and then μ a *measure* on $\mathcal{B}$, if the following holds:

(1) $\mathcal{B}$ is a σ-algebra, which according to the above definition means that:

$$\emptyset \in \mathcal{B},$$
$$E \in \mathcal{B} \implies S \setminus E \in \mathcal{B},$$
$$E_i \in \mathcal{B} \text{ for } i \in \mathbb{N} \implies \bigcup_{i \in \mathbb{N}} E_i \in \mathcal{B}.$$

(2) $\mu : \mathcal{B} \to [0, \infty]$ with $\mu(\emptyset) = 0$ is *σ-additive*, i.e.

$$\begin{matrix} E_i \in \mathcal{B} \text{ for } i \in \mathbb{N} \\ \text{pairwise disjoint} \end{matrix} \implies \mu\left(\bigcup_{i \in \mathbb{N}} E_i\right) = \sum_{i \in \mathbb{N}} \mu(E_i).$$

(3) If $N \in \mathcal{B}$ with $\mu(N) = 0$ and $E \subset N$, then $E \in \mathcal{B}$.

Condition (3) is called the **completeness** of the measure space. Sets $N \in \mathcal{B}$ with $\mu(N) = 0$ are called **μ-null sets** and we say that a statement holds **μ-almost everywhere** if it holds outside of a μ-null set. Sets $E \in \mathcal{B}$ are called **μ-measurable**.

A measure μ is called **σ-finite measure** and $(S, \mathcal{B}, \mu)$ is called a **σ-finite measure space** if in addition

(4) There exist $S_m \in \mathcal{B}$, $m \in \mathbb{N}$, such that $\mu(S_m) < \infty$ and $\bigcup_{m \in \mathbb{N}} S_m = S$.

3.10 Examples of measures. Let $\mathcal{B}$ be a Boolean ring of subsets of S.

(1) For the **discrete measure** (or **counting measure**) μ on $S = \mathbb{N}$ the system $\mathcal{B}$ consists of all subsets of $\mathbb{N}$ and let for $E \subset \mathbb{N}$

$$\mu(E) \in \{0\} \cup \mathbb{N} \cup \{\infty\} \text{ be the number of elements in } E.$$

(2) For the **Lebesgue measure** let $S = \mathbb{R}^n$ and let $\mathcal{B}_0$ be the set of all finite unions of disjoint, semi-open cuboids (see A3.3) in $\mathbb{R}^n$ and let

$$L^n \left(\underset{i=1}{\overset{n}{\times}} \, [a_i, b_i[\right) := \prod_{i=1}^{n} (b_i - a_i).$$

The set $\mathcal{B}_0$ is a Boolean algebra (not a σ-algebra) and L^n can be extended to an additive function $L^n : \mathcal{B}_0 \to \mathbb{R}$. Moreover, L^n is then also σ-additive on $\mathcal{B}_0$ (see A3.3). Then an extension principle (see A3.15) yields that:

Lemma: There exists a smallest σ-finite measure space $(\mathbb{R}^n, \mathcal{B}, \mu)$ with $\mathcal{B}_0 \subset \mathcal{B}$ and $\mu = L^n$ on $\mathcal{B}_0$.

We call μ the **Lebesgue measure** on $\mathbb{R}^n$ and denote it by $L^n := \mu$. The system $\mathcal{B}$ of all Lebesgue measurable sets "consists of" Lebesgue null sets and Borel sets, that is, for $E \in \mathcal{B}$ there exist Borel sets E_1, E_2 with $E_1 \subset E \subset E_2$ and $L^n(E_2 \setminus E_1) = 0$. The system of all **Borel sets** is the smallest σ-algebra that contains $\mathcal{B}_0$ (or, alternatively, all open sets or all closed sets), i.e. it is given by

$$\bigcap_{\substack{\mathcal{B}_0 \subset \tilde{\mathcal{B}}, \\ \tilde{\mathcal{B}} \ \sigma\text{-algebra}}} \tilde{\mathcal{B}}.$$

(3) For the **Dirac measure** let S be an arbitrary set, and let $\mathcal{B}$ be the system of all subsets of S. For a given $x \in S$ we define a measure δ_x by

$$\delta_x(E) := \begin{cases} 1, & \text{if } x \in E, \\ 0, & \text{otherwise.} \end{cases}$$

(4) We consider the **surface measure** on a given hypersurface. Let S be a smooth surface patch in $\mathbb{R}^n$, $n \geq 1$, which is parameterized over $\mathbb{R}^{n-1}$, i.e.

$$S = \left\{ (y, g(y)) \in \mathbb{R}^n \; ; \; y \in D \right\},$$

where D is an open bounded subset of $\mathbb{R}^{n-1}$ and $g \in C^1(\overline{D}; \mathbb{R})$. Let $\mathcal{B}_0$ be the Borel sets of S, and for any $E \in \mathcal{B}_0$ define

$$\widetilde{E} := \left\{ y \in D \; ; \; (y, g(y)) \in E \right\},$$

which again is a Borel set, and hence L^{n-1}-measurable, and define

$$\mathrm{H}^{n-1}(E) := \int_{\widetilde{E}} \sqrt{1 + |\nabla g(y)|^2} \, \mathrm{dL}^{n-1}(y).$$

In this way $E \mapsto \mathrm{H}^{n-1}(E)$ becomes a measure $(S, \mathcal{B}, \mathrm{H}^{n-1})$ on the set S, which is called the **surface measure** on S or the $(n-1)$-dimensional **Hausdorff measure** on S. In A8.5 we will generalize this definition to the case of Lipschitz continuous functions g. Similarly, one can introduce m-dimensional Hausdorff measures H^m on m-dimensional surface patches. The Hausdorff measure H^0 is the counting measure. It is also possible to define the Hausdorff measure, without using a parameterization, on all Borel sets in $\mathbb{R}^n$, which yields a generalization of the above definition. This generalized measure is the basis of geometric measure theory (see e.g. [Simon]).

3.11 Measurable functions. Let $(S, \mathcal{B}, \mu)$ be a measure space and (Y, d) a metric space. A map $f : S \to Y$ is called **μ-measurable** if

(1) $U \subset Y$ open $\implies$ $f^{-1}(U) \in \mathcal{B}$.

(2) There exists a μ-null set N such that $f(S \setminus N)$ is separable.

Remark: If the space Y is itself separable (e.g. $Y = \mathbb{R}^n$), then condition (2) is trivially satisfied (use 4.18(1) and 4.17(2)).

3.12 Lemma. The following hold:

(1) If $f_1 : S \to Y_1$ and $f_2 : S \to Y_2$ are measurable, then also $(f_1, f_2) : S \to Y_1 \times Y_2$ is measurable.

(2) If $f : S \to Y$ is measurable, Z is a Banach space, and $\varphi : Y \to Z$ is continuous, then also $\varphi \circ f$ is measurable.

(3) If $f_j : S \to Y$ is measurable for $j \in \mathbb{N}$ and the limit $f := \lim_{j \to \infty} f_j$ exists almost everywhere, then also f is measurable.

(4) If $f_j : S \to \mathbb{R}$ are measurable, then also $\inf_{j \in \mathbb{N}} f_j$ and also $\liminf_{j \to \infty} f_j$ are measurable, if they are finite almost everywhere. (On the null set, on which the limits are $-\infty$, we can define the function values arbitrarily.)

Proof (3). Let $f(x) = \lim_{j \to \infty} f_j(x)$ for $x \in S \setminus N$ with $\mu(N) = 0$, such that $f_j(S \setminus N)$ is separable for all j. If $U \subset Y$ is open, let

$$U_i := \{ y \in U ;\ \mathrm{B}_{\frac{1}{i}}(y) \subset U \} .$$

Then the following statement is true

$$\forall\, x \in S \setminus N :\ \big(f(x) \in U \iff \exists\, i, k \in \mathbb{N} :\ \forall\, j \geq k :\ f_j(x) \in U_i \big),$$

that is, for $x \in S \setminus N$ it holds that $f(x) \in U$ if and only if there exist i, k with $f_j(x) \in U_i$ for all $j \geq k$, or in set notation,

$$f^{-1}(U) \setminus N = \bigcup_i \bigcup_k \bigcap_{j \geq k} f_j^{-1}(U_i) \setminus N \in \mathcal{B}.$$

In addition, $f(S \setminus N) \subset \overline{\bigcup_j f_j(S \setminus N)}$ is separable (see 4.17(1)).   □

Proof (4). Let $g(x) := \inf_{j \in \mathbb{N}} f_j(x) > -\infty$ for $x \in S \setminus N$ with $\mu(N) = 0$. Then for all $a \in \mathbb{R}$

$$g^{-1}\big([a, \infty[\,\big) \setminus N = \bigcap_j f_j^{-1}\big([a, \infty[\,\big) \setminus N \in \mathcal{B},$$

which implies that g is measurable. Similarly, we have that $g_k := \inf_{j \geq k} f_j$ are measurable, and hence, on noting (3), also $\liminf_{j \to \infty} f_j = \lim_{k \to \infty} g_k$.   □

A further example of a metric space is the

3.13 Space of measurable functions. With the notation as in 3.11 let

$$M(\mu; Y) := \big\{ f : S \to Y ;\ f \text{ is } \mu\text{-measurable} \big\}$$

with the equivalence relation

$$f = g \text{ in } M(\mu; Y) \quad :\iff \quad f = g \ \mu\text{-almost everywhere.}$$

If $\mu(S) < \infty$ then this space becomes a metric space with the metric

$$d_\mu(f, g) := \inf\{ r \geq 0 ;\ \mu(\{ d(f, g) > r \}) \leq r \},$$

where we use the abbreviation

$$\{ d(f, g) > r \} := \{ x \in S ;\ d(f(x), g(x)) > r \} .$$

We say that a sequence $(f_k)_{k \in \mathbb{N}}$ is *convergent in measure* μ to f if $d_\mu(f_k, f) \to 0$ as $k \to \infty$. This is equivalent to

$$\mu(\{ d(f_k, f) > \varepsilon \}) \to 0 \quad \text{as } k \to \infty \tag{3-4}$$

for all $\varepsilon > 0$.

Proof. It follows from $\mu(S) < \infty$ that d_μ is a bounded function on $M(\mu; Y)$. If $d_\mu(f, g) = 0$, then

$$\mu(\{ d(f, g) > \varepsilon \}) \leq r \quad \text{for all } 0 < r \leq \varepsilon,$$

i.e. $\mu(\{d(f,g) > \varepsilon\}) = 0$ for $\varepsilon > 0$, and hence $\mu(\{d(f,g) > 0\}) = 0$, that is $f = g$ in $M(\mu; Y)$, which proves axiom (M1) in 2.6. For $f, g, h \in M(\mu; Y)$ and $r > d_\mu(f, h)$, $s > d_\mu(h, g)$ we have

$$\{d(f,g) > r + s\} \subset \{d(f,h) + d(h,g) > r + s\}$$
$$\subset \{d(f,h) > r\} \cup \{d(h,g) > s\}.$$

Consequently

$$\mu(\{d(f,g) > r + s\}) \leq \mu(\{d(f,h) > r\}) + \mu(\{d(h,g) > s\})$$
$$\leq r + s,$$

and so $d_\mu(f, g) \leq r + s$, which proves the triangle inequality for d_μ.   □

The following theorem holds.

3.14 Theorem. If Y is complete, then $M(\mu; Y)$ with d_μ is a complete metric space.

Proof. Indeed, let $(f_k)_{k \in \mathbb{N}}$ be a Cauchy sequence in $M(\mu; Y)$. Then there exists a monotone subsequence $(k_i)_{i \in \mathbb{N}}$ with

$$\mu(\{d(f_l, f_{k_i}) > 2^{-i}\}) \leq 2^{-i} \quad \text{for } l \geq k_i.$$

On setting

$$\tilde{E}_j := \bigcup_{i \geq j} \{d(f_{k_{i+1}}, f_{k_i}) > 2^{-i}\}$$

we have that $\mu(\tilde{E}_j) \leq 2^{1-j}$ and for $x \notin \tilde{E}_j$ and $i_2 \geq i_1 \geq j$ it holds that

$$d(f_{k_{i_2}}(x), f_{k_{i_1}}(x)) \leq \sum_{i \geq j} d(f_{k_{i+1}}(x), f_{k_i}(x)) \leq 2^{1-j}. \tag{3-5}$$

With $E_m := \bigcup_{j \geq m} \tilde{E}_j$ it follows that $(f_{k_i}(x))_{i \in \mathbb{N}}$ for $x \notin E_m$ is a Cauchy sequence in Y, and $\mu(E_m) \leq 2^{2-m}$. Hence there exists

$$f(x) := \lim_{i \to \infty} f_{k_i}(x) \quad \text{in } Y \text{ for } x \notin E := \bigcap_{m \in \mathbb{N}} E_m,$$

where $\mu(E) = 0$. On noting 3.12(3) this implies that f is a measurable function. Moreover, by (3-5),

$$d(f(x), f_{k_m}(x)) \leq 2^{1-m} \quad \text{for } x \notin E_m,$$

and so

$$\mu(\{d(f, f_{k_m}) > 2^{1-m}\}) \leq \mu(E_m) \leq 2^{2-m}.$$

This implies that $d_\mu(f, f_{k_m}) \leq 2^{2-m} \to 0$ as $m \to \infty$. Hence $(f_k)_{k \in \mathbb{N}}$ has a cluster point.   □

We now introduce the standard spaces of integrable functions. They are based on the definition of the Lebesgue integral (see Appendix A3).

3.15 Lebesgue spaces. Let $(S, \mathcal{B}, \mu)$ be a measure space and let Y be a Banach space over $\mathbb{K}$ with norm $y \mapsto |y|$. For a real number p we define in the case $1 \leq p < \infty$

$$L^p(\mu; Y) := \{ f : S \to Y ; \ f \text{ is } \mu\text{-measurable and } |f|^p \in L(\mu; \mathbb{R}) \}$$

and in the case $p = \infty$

$$L^\infty(\mu; Y) := \{ f : S \to Y ; \ f \text{ is } \mu\text{-measurable and } \mu\text{-essentially bounded} \} ,$$

in each case with the equivalence relation

$$f = g \text{ in } L^p(\mu; Y) \quad :\Longleftrightarrow \quad f = g \ \mu\text{-almost everywhere.}$$

Here $|f|$ denotes the function $x \mapsto |f(x)|$, which according to 3.12(2) is μ-measurable. For $p = \infty$ we call f **essentially bounded** with respect to μ if

$$\sup_{x \in S \setminus N} |f(x)| < \infty \quad \text{for a } \mu\text{-null set } N \in \mathcal{B}.$$

Hence, for $f \in L^p(\mu; Y)$ the quantities

$$\|f\|_{L^p} := \left(\int_S |f|^p \, d\mu \right)^{\frac{1}{p}} \quad \text{for } 1 \leq p < \infty,$$

$$\|f\|_{L^\infty} := \inf_{N : N \subset S, \ \mu(N) = 0} \left(\sup_{x \in S \setminus N} |f(x)| \right)$$

are well defined and lie in $[0, \infty[$. (For the case $0 < p < 1$ see E4.11.)

3.16 Theorem. Under the assumptions in 3.15 the following hold:

(1) If $1 \leq p \leq \infty$, then $L^p(\mu; Y)$ with $f \mapsto \|f\|_{L^p}$ is a Banach space.

(2) If $p = 1$, then $L^p(\mu; Y)$ coincides with the space of Lebesgue integrable functions in Appendix A3, i.e. $L^1(\mu; Y) = L(\mu; Y)$.

(3) If $p = 2$ and if Y is a Hilbert space with scalar product $(y_1, y_2) \mapsto (y_1, y_2)_Y$ (e.g. $Y = \mathbb{K}^l$ with the Euclidean scalar product $(y_1, y_2) \mapsto y_1 \bullet y_2$), then the space $L^2(\mu; Y)$ with

$$(f, g)_{L^2} := \int_S (f(x), g(x))_Y \, d\mu(x) \tag{3-6}$$

becomes a Hilbert space.

(4) If $p = \infty$, then for $f \in L^\infty(\mu; Y)$ there exists a μ-null set $N \in \mathcal{B}$ such that

$$\|f\|_{L^\infty} = \sup_{x \in S \setminus N} |f(x)| \quad (N \text{ depends on } f \text{ (!)}). \tag{3-7}$$

Moreover,

$$\|f\|_{L^\infty} = \operatorname*{ess\,sup}_{S} |f|,$$

where the **essential supremum** for a μ-measurable essentially bounded function $g : S \to \mathbb{R}$ is defined by

$$\operatorname*{ess\,sup}_{S} g := \inf \left\{ \sup_{x \in S \setminus N} g(x) \; ; \; N \subset S \text{ with } \mu(N) = 0 \right\}.$$

Proof (1). (See 3.17–3.21). The Minkowski inequality 3.20, which is shown with the help of the Hölder inequality 3.19, yields that $L^p(\mu; Y)$ is a vector space. The completeness for $p = \infty$ is shown in 3.17 and for $p < \infty$ in 3.21. $\qquad\square$

Proof (2). Follows from Bochner's criterion in A3.19(1). $\qquad\square$

Proof (3). Let $f, g \in L^2(\mu; Y)$. Since $|(f(x), g(x))_Y| \leq |f(x)| \cdot |g(x)|$ for $x \in S$, it follows from the Hölder inequality 3.18 and the majorant criterion A3.19(2) that $(f, g)_{L^2}$ is well defined. Clearly $(f, g) \mapsto (f, g)_{L^2}$ is sesquilinear and $(f, f)_{L^2} = \|f\|_{L^2}^2$. $\qquad\square$

Proof (4). The fact that countable unions of null sets are null sets immediately yields (3-7). $\qquad\square$

We now introduce some special notations. We write $L^p(\mu)$ in place of $L^p(\mu; \mathbb{K})$. Then $f \in L^p(\mu; Y)$ implies that $|f| \in L^p(\mu)$.

If $\mu = L^n \llcorner E$ is the Lebesgue measure on a Lebesgue measurable set $E \subset \mathbb{R}^n$, i.e. $\mu(A) := L^n(E \cap A)$, then we also write $L^p(E; Y)$ in place of $L^p(\mu; Y)$. In addition, in integrals we often replace $dL^n(x)$ by dx, where x is the integration variable:

$$\int_E f(x)\, dx := \int_E f(x)\, dL^n(x) = \int_{\mathbb{R}^n} \mathcal{X}_E f\, dL^n.$$

We note that for the counting measure μ in 3.10(1) it holds that $L^p(\mu) = \ell^p(\mathbb{K})$, which is easy to show.

Two important theorems that characterize the convergence of a sequence in $L^p(\mu; Y)$ are Lebesgue's convergence theorem (see 3.25) and Vitali's convergence theorem (see 3.23). But first we address the completeness of $L^p(\mu; Y)$.

3.17 Lemma. $L^\infty(\mu; Y)$ is complete.

Proof. Let $(f_k)_{k\in\mathbb{N}}$ be a Cauchy sequence in $L^\infty(\mu; Y)$. Then there exist a constant C and a μ-null set N such that for $x \in S \setminus N$ the following hold

$$|f_k(x)| \le \|f_k\|_{L^\infty} \le C < \infty \quad \text{for all } k,$$
$$|f_k(x) - f_l(x)| \le \|f_k - f_l\|_{L^\infty} \to 0 \quad \text{as } k, l \to \infty.$$

Hence for $x \in S$ there exists

$$f(x) := \begin{cases} \lim_{k\to\infty} f_k(x) & \text{in } Y \text{ for } x \in S \setminus N, \\ 0 & \text{for } x \in N. \end{cases}$$

The function f is measurable (see 3.12(3)) and bounded, and for $x \in S \setminus N$

$$|f(x) - f_k(x)| = \lim_{l\to\infty} |f_l(x) - f_k(x)| \le \liminf_{l\to\infty} \|f_l - f_k\|_{L^\infty},$$

and hence

$$\|f - f_k\|_{L^\infty} \le \liminf_{l\to\infty} \|f_l - f_k\|_{L^\infty} \longrightarrow 0 \quad \text{as } k \to \infty.$$

$\square$

3.18 Lemma (Hölder's inequality). Let $m \in \mathbb{N}$ and $f_i \in L^{p_i}(\mu)$ for $i = 1, \ldots, m$ with $p_i \in [1, \infty]$, and let $q \in [1, \infty]$ such that

$$\sum_{i=1}^{m} \frac{1}{p_i} = \frac{1}{q}. \tag{3-8}$$

Then the product $f_1 \cdots f_m \in L^q(\mu)$ and

$$\left\| \prod_{i=1}^{m} f_i \right\|_{L^q} \le \prod_{i=1}^{m} \|f_i\|_{L^{p_i}}. \tag{3-9}$$

Observe: Here we set $\frac{1}{p_i} = 0$ if $p_i = \infty$ and similarly $\frac{1}{q} = 0$ if $q = \infty$.

Standard case: Let $p, p' \in [1, \infty]$ such that $\frac{1}{p} + \frac{1}{p'} = 1$. Then if $f \in L^p(\mu)$ and $g \in L^{p'}(\mu)$, it holds that $f \cdot g \in L^1(\mu)$ with

$$\|f \cdot g\|_{L^1} \le \|f\|_{L^p} \cdot \|g\|_{L^{p'}}. \tag{3-10}$$

Remark: When forming products of functions, for simplicity we considered only scalar-valued functions, i.e. the case $Y = \mathbb{K}$. The Hölder inequality for the standard case can be generalized to vector-valued functions, i.e. $Y = \mathbb{K}^l$, and then reads as

$$\|f \bullet g\|_{L^1} \leq \|f\|_{L^p} \cdot \|g\|_{L^{p'}} \quad \text{for } f \in L^p(\mu; \mathbb{K}^l), \ g \in L^{p'}(\mu; \mathbb{K}^l).$$

We mention that the Hölder inequality also holds for functions with values in certain Banach spaces Y, for example a Banach algebra Y or a Hilbert space Y. It also generalizes to the case of multiple products.

Proof of the standard case. (This case corresponds to $m = 2$ and $q = 1$ in the general case.) It holds that fg is measurable (by 3.12(1) and 3.12(2)). For the limiting case $p = 1$ we have that $p' = \infty$ and note that $|(fg)(x)| \leq \|g\|_{L^\infty} |f(x)|$ for almost all x. The majorant criterion (see A3.19(2)) then yields that $fg \in L^1(\mu)$. The limiting case $p = \infty$ follows due to symmetry. Moreover, the claim holds trivially if $\|f\|_{L^p} = 0$ or $\|g\|_{L^{p'}} = 0$, since then $fg = 0$ almost everywhere.

Hence let $1 < p < \infty$ and let $\|f\|_{L^p} > 0$ and $\|g\|_{L^{p'}} > 0$. For $a, b \geq 0$ we have that

$$ab \leq \frac{1}{p} a^p + \frac{1}{p'} b^{p'} \quad (\textbf{Young's inequality}). \tag{3-11}$$

This elementary inequality can be shown as follows: For $a > 0$ and $b > 0$ take the logarithm on both sides and obtain, due to the concavity of the logarithm, that

$$\log(ab) = \log a + \log b = \tfrac{1}{p}\log a^p + \tfrac{1}{p'}\log b^{p'} \leq \log\left(\tfrac{1}{p} a^p + \tfrac{1}{p'} b^{p'}\right).$$

Now setting

$$a = \frac{|f(x)|}{\|f\|_{L^p}} \quad \text{and} \quad b = \frac{|g(x)|}{\|g\|_{L^{p'}}}$$

yields

$$\frac{|f(x)g(x)|}{\|f\|_{L^p}\|g\|_{L^{p'}}} \leq \frac{|f(x)|^p}{p\|f\|_{L^p}^p} + \frac{|g(x)|^{p'}}{p'\|g\|_{L^{p'}}^{p'}}.$$

As the right-hand side is integrable with respect to x, it follows that $fg \in L^1(\mu)$ on recalling the majorant criterion A3.19(2). Integration over x then yields that

$$\frac{\int_S |fg|\,d\mu}{\|f\|_{L^p}\|g\|_{L^{p'}}} \leq \frac{1}{p}\frac{\int_S |f|^p\,d\mu}{\|f\|_{L^p}^p} + \frac{1}{p'}\frac{\int_S |g|^{p'}\,d\mu}{\|g\|_{L^{p'}}^{p'}} = \frac{1}{p} + \frac{1}{p'} = 1.$$

$\square$

Proof of the general case. Clearly, by (3-8), $p_i \in [q, \infty]$. If $p_i = q$ for some i, then $p_j = \infty$ for all $j \neq i$. On the other hand, if $p_j = \infty$ for some j, then

$$\left|\prod_{i=1}^m f_i(x)\right| \leq \|f_j\|_{L^\infty} \cdot \left|\prod_{i:i\neq j} f_i(x)\right| \quad \text{and} \quad \sum_{i:i\neq j} \frac{1}{p_i} = \frac{1}{q},$$

that is, the claim is inductively reduced to $m - 1$ functions. Thus, let $q < p_i < \infty$ for $i = 1, \ldots, m$. We give two different proofs.

1st possibility. Use the generalized **Young's inequality**

$$\frac{1}{q} \prod_{i=1}^{m} a_i^q \leq \sum_{i=1}^{m} \frac{1}{p_i} a_i^{p_i} \quad \text{for } a_i \geq 0, \ i = 1, \ldots, m, \tag{3-12}$$

for exponents as in (3-8). On rewriting $\tilde{p}_i := \frac{p_i}{q}$, this elementary inequality follows, as in the proof of (3-11) (now for m terms), from the concavity of the logarithm. Then continue analogously to the proof of the standard case.

2nd possibility. This proceeds by induction on m and uses the standard case. We have

$$\sum_{i=1}^{m-1} \frac{1}{p_i} = \frac{1}{q} - \frac{1}{p_m} =: \frac{1}{r} \quad \text{with } 1 < r < \infty,$$

and so the induction hypothesis yields for $g := f_1 \cdots f_{m-1}$ that

$$\|g\|_{L^r} \leq \prod_{i=1}^{m-1} \|f_i\|_{L^{p_i}}.$$

Now apply the standard case for $\widetilde{f} := |f_m|^q \in L^{\frac{p_m}{q}}(\mu)$ and $\widetilde{g} := |g|^q \in L^{\frac{r}{q}}(\mu)$ and obtain, since $\frac{q}{p_m} + \frac{q}{r} = 1$, that

$$\left\| \prod_{i=1}^{m} f_i \right\|_{L^q} = \|f_m g\|_{L^q} = \left\| \widetilde{f} \widetilde{g} \right\|_{L^1}^{\frac{1}{q}} \leq \left(\left\| \widetilde{f} \right\|_{L^{\frac{p_m}{q}}} \cdot \left\| \widetilde{g} \right\|_{L^{\frac{r}{q}}} \right)^{\frac{1}{q}}$$

$$= \|f_m\|_{L^{p_m}} \cdot \|g\|_{L^r} \leq \prod_{i=1}^{m} \|f_i\|_{L^{p_i}}.$$

$\square$

Proof of remark. Since $|f(x) \bullet g(x)| \leq |f(x)| \cdot |g(x)|$ due to the Cauchy-Schwarz inequality in $\mathbb{R}^l$, it follows from the scalar Hölder inequality that

$$\|f \bullet g\|_{L^1} \leq \||f| \cdot |g|\|_{L^1} \leq \||f|\|_{L^p} \cdot \||g|\|_{L^{p'}} = \|f\|_{L^p} \cdot \|g\|_{L^{p'}},$$

where the integrability of $f \bullet g$ follows from that of $|f| \cdot |g|$, thanks to the majorant criterion A3.19(2).

In a Banach algebra Y the product, which we denote by $(y_1, y_2) \mapsto y_1 y_2$, is continuous. For the standard case it then follows, by 3.12(1) and 3.12(2), that $x \mapsto f(x)g(x)$ is measurable. In addition we have the pointwise inequality $|f(x)g(x)| \leq |f(x)| \cdot |g(x)|$, where $y \mapsto |y|$ denotes the norm in Y. This yields the claim as in the scalar case. In an analogous fashion this carries over to multiple products.

$\square$

The generalization of the majorant criterion A3.19(2) to L^p-spaces is given by the following

3.19 Lemma (Majorant criterion). Let $f : S \to Y$ be a μ-measurable function, $1 \le p < \infty$ and $g \in L^1(\mu; \mathbb{R})$, $g \ge 0$, with

$$|f|^p \le g \quad \mu\text{-almost everywhere.}$$

Then $f \in L^p(\mu; Y)$.

Proof. As $|f|^p$ is measurable on recalling 3.12(2), it follows from A3.19(2) that $|f|^p \in L^1(\mu)$. As f is measurable, we have that $f \in L^p(\mu; Y)$ by definition. $\square$

3.20 Lemma (Minkowski inequality). If $f, g \in L^p(\mu; Y)$, then $f + g \in L^p(\mu; Y)$ and

$$\|f + g\|_{L^p} \le \|f\|_{L^p} + \|g\|_{L^p}.$$

Proof. For $p = 1$ and $p = \infty$ this follows from the pointwise triangle inequality. Furthermore, we have the elementary inequality

$$|a + b|^p \le 2^{p-1}(|a|^p + |b|^p) \quad \text{for } a, b \in \mathbb{R} \text{ and } 1 \le p < \infty. \tag{3-13}$$

Hence for $1 < p < \infty$

$$|f + g|^p \le (|f| + |g|)^p \le 2^{p-1}(|f|^p + |g|^p)$$

pointwise and so $f + g \in L^p(\mu; Y)$ on noting 3.19, where the measurability of $f + g$ follows from 3.12(1) and 3.12(2). A more convenient inequality is

$$|f + g|^p \le |f| \cdot |f + g|^{p-1} + |g| \cdot |f + g|^{p-1}.$$

We have that $|f|, |g|, |f + g| \in L^p(\mu)$ and, with p' as in 3.18, it then holds that $|f + g|^{p-1} \in L^{p'}(\mu)$ since $p'(p - 1) = p$. Hence the Hölder inequality 3.18 implies that

$$\int_S |f + g|^p \, d\mu \le \int_S |f| \cdot |f + g|^{p-1} \, d\mu + \int_S |g| \cdot |f + g|^{p-1} \, d\mu$$
$$\le \|f\|_{L^p} \cdot \left\||f + g|^{p-1}\right\|_{L^{p'}} + \|g\|_{L^p} \cdot \left\||f + g|^{p-1}\right\|_{L^{p'}}$$
$$= \left(\|f\|_{L^p} + \|g\|_{L^p}\right) \cdot \left(\int_S |f + g|^p \, d\mu\right)^{1 - \frac{1}{p}}.$$

If $\int_S |f + g|^p \, d\mu = 0$, the claim holds trivially. Otherwise the desired result follows from cancellation. $\square$

3.21 Fischer-Riesz theorem. $L^p(\mu; Y)$ is complete for $1 \le p < \infty$.

Proof. Let $(f_k)_{k\in\mathbb{N}}$ be a Cauchy sequence in $L^p(\mu;Y)$. Since every Cauchy sequence has at most one cluster point, it is sufficient to show convergence of a subsequence. To this end we choose a monotone subsequence $(k_i)_{i\in\mathbb{N}}$ such that

$$\|f_k - f_l\|_{L^p} \leq 2^{-i} \quad \text{for } k, l \geq k_i.$$

In the following we denote the subsequence $(f_{k_i})_{i\in\mathbb{N}}$ again by $(f_k)_{k\in\mathbb{N}}$, as the remainder of the sequence is no longer needed. This convention for the transition to subsequences will be used repeatedly in this book. It then holds that

$$\sum_{k\in\mathbb{N}} \|f_{k+1} - f_k\|_{L^p} \leq \sum_{k\in\mathbb{N}} 2^{-k} < \infty.$$

Let

$$g_l := \sum_{k=1}^{l} |f_{k+1} - f_k|. \tag{3-14}$$

By Fatou's lemma (see A3.20) and the Minkowski inequality

$$\int_S \left(\lim_{l\to\infty} g_l^p\right) d\mu \leq \liminf_{l\to\infty} \int_S g_l^p \, d\mu = \left(\liminf_{l\to\infty} \|g_l\|_{L^p}\right)^p$$

$$\leq \left(\sum_{k\in\mathbb{N}} \|f_{k+1} - f_k\|_{L^p}\right)^p < \infty.$$

Therefore,

$$\lim_{l\to\infty} g_l(x) < \infty \quad \text{for } \mu\text{-almost all } x.$$

From the definition of g_l in (3-14) it follows that $(f_k(x))_{k\in\mathbb{N}}$ is a Cauchy sequence in Y for μ-almost all x, hence the limit

$$f(x) := \lim_{k\to\infty} f_k(x) \quad \text{in } Y \text{ for } \mu\text{-almost all } x$$

exists. It follows from 3.12(3) that f is measurable and appealing once more to Fatou's lemma yields that

$$\int_S |f - f_l|^p \, d\mu \leq \liminf_{k\to\infty} \int_S |f_k - f_l|^p \, d\mu$$

$$= \left(\liminf_{k\to\infty} \|f_k - f_l\|_{L^p}\right)^p \longrightarrow 0 \quad \text{as } l \to \infty.$$

$\square$

Essential for applications is the characterization of convergent sequences in $L^p(\mu;Y)$ via the pointwise convergence of the function values in Y. To this end we first establish the proposition 3.22(1), which will be used frequently in this book. For $p = 1$ Lebesgue's convergence theorem A3.21 gives a convergence criterion for almost everywhere convergent sequences. Here we will establish its generalization to L^p-spaces (see 3.25). This result is readily reduced to the very general theorem 3.23 due to Vitali, or, alternatively, it can be shown with the help of Fatou's lemma similarly to the proof of A3.21.

3.22 Lemma. Let $f_k \in L^p(\mu; Y)$ for $k \in \mathbb{N}$ with $1 \le p < \infty$.

(1) If $f \in L^p(\mu; Y)$, then:

$$\begin{array}{l} \|f - f_k\|_{L^p} \to 0 \\ \text{as } k \to \infty \end{array} \implies \begin{array}{l} \text{There exists a subsequence } (k_i)_{i \in \mathbb{N}}, \text{ such that} \\ f_{k_i} \to f \text{ as } i \to \infty \text{ } \mu\text{-almost everywhere.} \end{array}$$

(2) If $(f_k)_{k \in \mathbb{N}}$ is a Cauchy sequence in $L^p(\mu; Y)$ and $f : S \to Y$, then:

$$\begin{array}{l} f_k \to f \text{ as } k \to \infty \\ \mu\text{-almost everywhere} \end{array} \implies \begin{array}{l} f \in L^p(\mu; Y) \text{ and} \\ \|f - f_k\|_{L^p} \to 0 \text{ as } k \to \infty. \end{array}$$

Remark: For (1) see E3.5.

Proof (1). We have $|f - f_k|^p \to 0$ in $L^1(\mu; Y)$. Now apply A3.11. □

Proof (2). It follows from theorem 3.21 that there exists an $\widetilde{f} \in L^p(\mu; Y)$ such that $\left\| \widetilde{f} - f_k \right\|_{L^p} \to 0$ as $k \to \infty$. By (1), there exists a subsequence $(f_{k_i})_{i \in \mathbb{N}}$, with $f_{k_i} \to \widetilde{f}$ μ-almost everywhere. The assumption in (2) then gives that $\widetilde{f} = f$ μ-almost everywhere. □

3.23 Vitali's convergence theorem. Let $f_k \in L^p(\mu; Y)$ with $1 \le p < \infty$, and let $f_k \to f$ μ-almost everywhere as $k \to \infty$. Then the following are equivalent:

(1) $f \in L^p(\mu; Y)$ and $\|f_k - f\|_{L^p} \to 0$ as $k \to \infty$.
(2) It holds that

$$\sup_k \int_E |f_k|^p \, d\mu \longrightarrow 0 \quad \text{as } \mu(E) \to 0,$$

and for any $\varepsilon > 0$ there exists a μ-measurable set E_ε such that $\mu(E_\varepsilon) < \infty$ and

$$\sup_k \int_{S \setminus E_\varepsilon} |f_k|^p \, d\mu \le \varepsilon.$$

Proof (1)$\Rightarrow$(2). It follows from the Minkowski inequality 3.20 that

$$\left(\int_E |f_k|^p \, d\mu \right)^{\frac{1}{p}} \le \left(\int_E |f|^p \, d\mu \right)^{\frac{1}{p}} + \underbrace{\left(\int_S |f_k - f|^p \, d\mu \right)^{\frac{1}{p}}}_{\to 0 \text{ as } k \to \infty}.$$

Hence for $\gamma > 0$ there exists a $k_\gamma \in \mathbb{N}$ such that for $k > k_\gamma$ and all μ-measurable sets E

$$\left(\int_E |f_k|^p \, d\mu\right)^{\frac{1}{p}} \leq \left(\int_E |f|^p \, d\mu\right)^{\frac{1}{p}} + \gamma.$$

For every function $g \in L^1(\mu; \mathbb{R})$ with $g \geq 0$ it follows from A3.17(2) that for $\gamma > 0$ there exists a $\delta(g, \gamma) > 0$ such that

$$\int_E g \, d\mu \leq \gamma^p \quad \text{for } \mu(E) \leq \delta(g, \gamma).$$

Employing this for $g = |f|^p$ and $g = |f_k|^p$ for $k \leq k_\gamma$ yields that

$$\text{for all } k \in \mathbb{N}: \quad \left(\int_E |f_k|^p \, d\mu\right)^{\frac{1}{p}} \leq 2\gamma,$$

$$\text{if } \mu(E) \leq \min(\delta(|f|^p, \gamma), \delta(|f_1|^p, \gamma), \ldots, \delta(|f_{k_\gamma}|^p, \gamma)).$$

This proves the first claim in (2).

The second claim follows correspondingly, upon substituting E in place of $S \setminus E$ in the above argument and using the fact that for every function $g \in L^1(\mu; \mathbb{R})$ with $g \geq 0$, for $\gamma > 0$ there exists a μ-measurable set $A(g, \gamma)$ with $\mu(A(g, \gamma)) < \infty$ such that (see below)

$$\int_{S \setminus A(g, \gamma)} g \, d\mu \leq \gamma^p. \tag{3-15}$$

Then we obtain in much the same way as before that

$$\text{for all } k \in \mathbb{N}: \quad \left(\int_{S \setminus E} |f_k|^p \, d\mu\right)^{\frac{1}{p}} \leq 2\gamma$$

$$\text{for } E := A(|f|^p, \gamma) \cup A(|f_1|^p, \gamma) \cup \ldots \cup A(|f_{k_\gamma}|^p, \gamma).$$

For the proof of (3-15) consider $A_\varepsilon := \{g \geq \varepsilon\} := \{x \in S; \ g(x) \geq \varepsilon\}$. Since

$$\int_S g \, d\mu \geq \int_{A_\varepsilon} g \, d\mu \geq \varepsilon \mu(A_\varepsilon),$$

we have $\mu(A_\varepsilon) < \infty$. Moreover, $A_{\varepsilon_1} \subset A_{\varepsilon_2}$ for $\varepsilon_1 > \varepsilon_2$ and

$$A := \bigcup_{\varepsilon > 0} A_\varepsilon = \{g > 0\}.$$

It then follows from A3.17(2), by choosing a decreasing null sequence for ε, that

$$\int_{S \setminus A_\varepsilon} g \, d\mu \longrightarrow \int_{S \setminus A} g \, d\mu = \int_{\{g=0\}} g \, d\mu = 0.$$

$\square$

Proof (2)⇒(1). Let E_ε be as in (2). Then Egorov's theorem A3.18 yields that $f_k \to f$ as $k \to \infty$ μ-uniformly on E_ε, i.e. there exists an $A_\varepsilon \subset E_\varepsilon$ with $\mu(E_\varepsilon \setminus A_\varepsilon) \le \varepsilon$ such that $f_k \to f$ uniformly on A_ε as $k \to \infty$. It follows that

$$\int_S |f_k - f_l|^p \, d\mu \le \int_{A_\varepsilon} |f_k - f_l|^p \, d\mu + \int_{S \setminus A_\varepsilon} |f_k - f_l|^p \, d\mu$$

$$\le \underbrace{\mu(A_\varepsilon) \cdot \operatorname*{ess\,sup}_{A_\varepsilon} |f_k - f_l|^p}_{\substack{\to\, 0 \text{ as } k,l \to \infty \\ \text{for all } \varepsilon}}$$

$$+ \underbrace{2^{p-1} \sup_{m \in \mathbb{N}} \int_{S \setminus E_\varepsilon} |f_m|^p \, d\mu}_{\le\, \varepsilon} + \underbrace{2^{p-1} \sup_{m \in \mathbb{N}} \int_{E_\varepsilon \setminus A_\varepsilon} |f_m|^p \, d\mu}_{\to\, 0 \text{ as } \varepsilon \to 0},$$

where the last term converges to 0 as $\varepsilon \to 0$ by (2), since $\mu(E_\varepsilon \setminus A_\varepsilon) \le \varepsilon \to 0$. Therefore, $(f_k)_{k \in \mathbb{N}}$ is a Cauchy sequence in $L^p(\mu; Y)$. Since $f_k \to f$ μ-almost everywhere, the claim follows from 3.22(2). □

As a consequence we obtain:

3.24 Corollary. Let $1 \le p < \infty$. Then

$$f_k \to f \text{ in } L^p(\mu; Y) \text{ as } k \to \infty \quad \Longrightarrow \quad |f_k|^p \to |f|^p \text{ in } L^1(\mu) \text{ as } k \to \infty.$$

Proof (with Vitali's convergence theorem). We first assume that $f_k \to f$ μ-almost everywhere. Now apply the conclusion from 3.23(1) to 3.23(2) for $f_k, f \in L^p(\mu; Y)$, and then conversely apply the conclusion from 3.23(2) to 3.23(1) for $g_k := |f_k|^p, g := |f|^p \in L^1(\mu; \mathbb{R})$. This yields the claim. For general f_k, f we use an indirect proof. Assume that there exist an $\varepsilon_0 > 0$ and a subsequence $(|f_{k_i}|^p)_{i \in \mathbb{N}}$ such that $\| |f_{k_i}|^p - |f|^p \|_{L^1} \ge \varepsilon_0$ for $i \in \mathbb{N}$. Since $f_{k_i} \to f$ in $L^p(\mu; Y)$ as $i \to \infty$, it follows from 3.22(1) that there exists a subsequence $(f_{k_{i_m}})_{m \in \mathbb{N}}$ such that $f_{k_{i_m}} \to f$ μ-almost everywhere as $m \to \infty$. Now applying the above conclusion to this subsequence leads to a contradiction. □

Proof (without Vitali's convergence theorem). An alternative proof, which does not use Vitali's convergence theorem, is as follows: For $p = 1$ we immediately obtain the desired result on noting that $||f_k| - |f|| \le |f_k - f|$. Hence let $p > 1$. We employ an elementary inequality: For $M > 1$ let $\delta_M > 0$ be the unique number such that, for all $a > 0$,

$$(1 + a)^p \le 1 + M \cdot a^p \quad \Longleftrightarrow \quad a \ge \delta_M.$$

(The existence of δ_M is easily established on noting that $a \mapsto (1 + M \cdot a^p)^{\frac{1}{p}}$ is a strictly convex function.) Clearly we have that $\delta_M \searrow 0$ as $M \nearrow \infty$. Hence for $y_0, y_1 \in Y$ with $0 < |y_0| \le |y_1|$ we have

$$0 \le |y_1|^p - |y_0|^p \le |y_0|^p \left(\left(1 + \frac{|y_1 - y_0|}{|y_0|} \right)^p - 1 \right)$$

$$\le \begin{cases} M \cdot |y_1 - y_0|^p & \text{if } \dfrac{|y_1 - y_0|}{|y_0|} \ge \delta_M, \\[2ex] ((1 + \delta_M)^p - 1)|y_0|^p & \text{if } \dfrac{|y_1 - y_0|}{|y_0|} \le \delta_M. \end{cases}$$

This yields the following inequality:

$$\int_S \big||f_k|^p - |f|^p\big|\, d\mu$$

$$\le M \underbrace{\int_S |f_k - f|^p\, d\mu}_{\to\, 0 \text{ as } k \to \infty} + ((1 + \delta_M)^p - 1)\underbrace{\int_S (|f|^p + |f_k|^p)\, d\mu}_{\text{bounded in } k}\,.$$

This gives the desired result, on choosing M sufficiently large (so that δ_M becomes small), and then k sufficiently large depending on M. □

3.25 Lebesgue's general convergence theorem. Let $f_k, f : S \to Y$ be μ-measurable, let $g_k \to g$ in $L^1(\mu; \mathbb{R})$ as $k \to \infty$ and let $1 \le p < \infty$. Suppose that

$$f_k \to f \quad \mu\text{-almost everywhere as } k \to \infty,$$
$$|f_k|^p \le g_k \quad \mu\text{-almost everywhere for all } k \in \mathbb{N}.$$

Then it follows that $f_k, f \in L^p(\mu; Y)$ and $f_k \to f$ in $L^p(\mu; Y)$ as $k \to \infty$.

Proof (without Vitali's convergence theorem). This is a generalization of the proof of Lebesgue's convergence theorem A3.21 (first presented in [Alt3]). We begin with the case where $g_k \to g$ μ-almost everywhere as $k \to \infty$. Let

$$h_k := \tfrac{1}{2}(g_k + g) - \tfrac{1}{2^p}|f_k - f|^p\,.$$

The elementary inequality (3-13) implies that

$$h_k \ge \tfrac{1}{2}(g_k + g) - \tfrac{1}{2}(|f_k|^p + |f|^p) \ge 0 \quad \mu\text{-almost everywhere.}$$

The assumptions yield that $h_k \to g$ μ-almost everywhere and

$$\int_S h_k\, d\mu \le \int_S \frac{1}{2}(g_k + g)\, d\mu \longrightarrow \int_S g\, d\mu \quad \text{as } k \to \infty.$$

It follows from Fatou's lemma that

$$\int_S g\, d\mu \le \liminf_{k \to \infty} \int_S h_k\, d\mu$$

$$= \underbrace{\lim_{k \to \infty} \int_S \frac{1}{2}(g_k + g)\, d\mu}_{= \int_S g\, d\mu} - \frac{1}{2^p}\limsup_{k \to \infty} \int_S |f_k - f|^p\, d\mu\,,$$

and hence we obtain the desired result

$$\limsup_{k\to\infty} \int_S |f_k - f|^p \, d\mu = 0 \, .$$

For general g_k, g we use an indirect proof, similarly to the proof of 3.24. To this end, we first need to show that $f \in L^p(\mu; Y)$. It follows from 3.22(1) that there exists a subsequence $(g_{k_j})_{j \in \mathbb{N}}$ such that $g_{k_j} \to g$ μ-almost everywhere as $j \to \infty$. Since by assumption $|f_{k_j}|^p \leq g_{k_j}$ μ-almost everywhere, we have that $|f|^p \leq g$ μ-almost everywhere. The majorant criterion 3.19 then implies that $f \in L^p(\mu; Y)$. Now assume that there exist an $\varepsilon_0 > 0$ and a subsequence $(f_{k_i})_{i \in \mathbb{N}}$ such that $\|f_{k_i} - f\|_{L^p} \geq \varepsilon_0$ for $i \in \mathbb{N}$. Since $g_{k_i} \to g$ in $L^1(\mu; \mathbb{R})$ as $i \to \infty$, there exists a subsequence $(g_{k_{i_m}})_{m \in \mathbb{N}}$ such that $g_{k_{i_m}} \to g$ μ-almost everywhere as $m \to \infty$. Applying the above conclusion to this subsequence leads to a contradiction. □

Proof (with Vitali's convergence theorem). We now provide an alternative proof which uses Vitali's convergence theorem. On noting the bounds on f_k we have that $f_k \in L^p(\mu; Y)$, recall 3.19. Again we begin by assuming that $g_k \to g$ μ-almost everywhere. It follows from Vitali's convergence theorem (implication 3.23(1)$\Rightarrow$3.23(2)) for the functions g_k, g that

$$\sup_k \int_E |f_k|^p \, d\mu \leq \sup_k \int_E g_k \, d\mu \longrightarrow 0 \quad \text{as } \mu(E) \to 0 \, ,$$

and similarly for the result corresponding to the second claim in 3.23(2). Conversely, applying Vitali's convergence theorem (implication 3.23(2)$\Rightarrow$3.23(1)) now for the functions f_k yields the desired result.

For general g_k, g use an indirect proof as in the first proof above. □

In proofs it is often convenient to approximate L^p functions by smooth functions. The following result shows that for the Lebesgue measure this is possible by using continuous functions.

3.26 Lemma. As before, let $(S, \mathcal{B}, \mu)$ be a measure space and let $f \in L^p(\mu; Y)$ with $1 \leq p < \infty$. Then:

(1) There exists a sequence $(f_k)_{k \in \mathbb{N}}$ of step functions with steps in $\mathcal{B}$ such that $\|f - f_k\|_{L^p} \to 0$ as $k \to \infty$.
Note: If $\mu(S) < \infty$ and $Y = \mathbb{K}^m$, then this also holds for $p = \infty$.
(2) If $S = \mathbb{R}^n$ and μ is the Lebesgue measure, then there exists a sequence $(f_k)_{k \in \mathbb{N}}$ of functions $f_k \in C_0^0(\mathbb{R}^n; Y)$ such that $\|f - f_k\|_{L^p} \to 0$ as $k \to \infty$.
Observe: This does not hold for $p = \infty$, since the uniform limit of continuous functions is again a continuous function.

Proof (1). Let $\varepsilon > 0$ and set $E_\varepsilon := \{\varepsilon \leq |f| \leq \frac{1}{\varepsilon}\}$. Then $E_\varepsilon \in \mathcal{B}$ and

$$\int_S |f|^p \, d\mu \geq \varepsilon^p \mu(E_\varepsilon) \, ,$$

and so $\mu(E_\varepsilon) < \infty$. On noting that $\mathcal{X}_{E_\varepsilon} f$ is measurable and that $|\mathcal{X}_{E_\varepsilon} f| \le \frac{1}{\varepsilon}\mathcal{X}_{E_\varepsilon} \in L^1(\mu; \mathbb{R})$, it follows that $\mathcal{X}_{E_\varepsilon} f \in L^1(\mu; Y)$, recall 3.19. The construction of the Lebesgue integral (see axiom (L5) in A3.16) yields the existence of step functions $g_{\varepsilon k}$ (with steps in $\mathcal{B}_0$) such that $g_{\varepsilon k} \to \mathcal{X}_{E_\varepsilon} f$ in $L^1(\mu; Y)$ as $k \to \infty$. Then

$$f_{\varepsilon k}(x) := \begin{cases} g_{\varepsilon k}(x) & \text{if } x \in E_\varepsilon \text{ and } |g_{\varepsilon k}(x)| \le \dfrac{2}{\varepsilon}, \\[2mm] \dfrac{2g_{\varepsilon k}(x)}{\varepsilon |g_{\varepsilon k}(x)|} & \text{if } x \in E_\varepsilon \text{ and } |g_{\varepsilon k}(x)| > \dfrac{2}{\varepsilon}, \\[2mm] 0 & \text{if } x \in S \setminus E_\varepsilon, \end{cases}$$

defines step functions $f_{\varepsilon k}$ (with steps in $\mathcal{B}$), and for $x \in E_\varepsilon$ with $|g_{\varepsilon k}(x)| > \frac{2}{\varepsilon}$ it holds that

$$|f_{\varepsilon k}(x) - f(x)| \le \frac{3}{\varepsilon} \le 3(|g_{\varepsilon k}(x)| - |f(x)|) \le 3|g_{\varepsilon k}(x) - f(x)|.$$

Hence we also have that $f_{\varepsilon k} \to \mathcal{X}_{E_\varepsilon} f$ in $L^1(\mu; Y)$ as $k \to \infty$ and

$$\int_S |f - f_{\varepsilon k}|^p \, d\mu \le \underbrace{\int_{S \setminus E_\varepsilon} |f|^p \, d\mu}_{\substack{\to\, 0 \text{ as } \varepsilon \to 0, \\ \text{recall A3.17(2)}}} + \Big(\frac{3}{\varepsilon}\Big)^{p-1} \underbrace{\int_S |\mathcal{X}_{E_\varepsilon} f - f_{\varepsilon k}| \, d\mu}_{\substack{\to\, 0 \text{ as } k \to \infty \\ \text{for all } \varepsilon}}.$$

$\square$

Proof (1) *Note.* Let $R := \|f\|_{L^\infty} > 0$. Now $\overline{B_R(0)} \subset \mathbb{K}^m$ is compact. Hence for $k \in \mathbb{N}$ there exists a partition of $\overline{B_R(0)}$ into finitely many disjoint Borel sets A_j, $1 \le j \le n_k$, such that $\text{diam}(A_j) \le \frac{1}{k}$ and $A_j \ne \emptyset$. Choose $a_j \in A_j$. Then

$$\left\| f - \sum_{j=1}^{n_k} \mathcal{X}_{f^{-1}(A_j)} \, a_j \right\|_{L^\infty} \le \frac{1}{k}.$$

$\square$

Proof (2). It follows from (1) that f can be approximated in the L^p-norm by step functions with steps in $\mathcal{B}$. Hence the claim is reduced to the case $Y = \mathbb{R}$ and $f = \mathcal{X}_E$ with $E \in \mathcal{B}$ and $L^n(E) < \infty$. But then $f \in L^1(\mathbb{R}^n)$ and, by definition of the Lebesgue integral (see axiom (L5) in A3.16), there exist step functions g_k which approximate f in the L^1-norm. In addition, $f_k := \max(0, \min(1, g_k))$ are also such step functions and it follows from

$$|f - f_k|^p \le |f - f_k| \le |f - g_k|$$

that $f_k \to f$ in $L^p(\mathbb{R}^n)$. As f_k has steps in $\mathcal{B}_0$, the claim is further reduced to the case that $f = \mathcal{X}_Q$, where $Q = [a, b[$ with $a, b \in \mathbb{R}^n$. But then $f_\varepsilon \to f$ in $L^p(\mathbb{R}^n)$ as $\varepsilon \searrow 0$, if, for example,

$$f_\varepsilon(x) := \prod_{i=1}^{n} \max\left(0, \min\left(1, \frac{g_i(x_i)}{\varepsilon}\right)\right),$$

$$g_i(\xi) := \frac{b_i - a_i}{2} - \left|\xi - \frac{b_i + a_i}{2}\right|.$$

$\square$

Sobolev spaces

In the Introduction it was illustrated that in the calculus of variations one encounters norms with respect to which the initially chosen function spaces are not complete. The reason is that in their definition a combination of derivatives and integrals occur. For example, take the set $X := C^1(\overline{\Omega})$ and define the norm by

$$\|f\|_X := \sqrt{\int_\Omega \left(|f(x)|^2 + |\nabla f(x)|^2\right) \mathrm{d}x}.$$

We note that $f_\varepsilon(x) := \sqrt{|x - x_0|^2 + \varepsilon^2}$, with $x_0 \in \Omega$, for $\varepsilon \searrow 0$ is a Cauchy sequence with respect to this norm, but the function $\lim_{\varepsilon \to 0} f_\varepsilon(x) = |x - x_0|$ does not belong to $C^1(\overline{\Omega})$. Hence X is not a complete space with respect to the norm $f \mapsto \|f\|_X$.

The functional analysis approach to solve such variational problems consists in the completion of X to a space $\widetilde{X}$. This is in order to show the existence of a "weak solution", i.e. a solution in $\widetilde{X}$ (see e.g. 6.5–6.8 and 8.16–8.18). If X is as in the above example, then the completion $\widetilde{X}$ will be the completion of a classical function space with respect to a norm containing integrals. The spaces obtained in this manner are called Sobolev spaces.

3.27 Sobolev spaces. Let $m \geq 0$ be an integer and $1 \leq p \leq \infty$. If $\Omega \subset \mathbb{R}^n$ is open, then let $\widetilde{X}$ be the completion (see 2.24) of the normed vector space

$$X := \left\{ f \in C^\infty(\Omega) \,;\, \|f\|_X < \infty \right\} \quad \text{with} \quad \|f\|_X := \sum_{|s| \leq m} \|\partial^s f\|_{L^p(\Omega)}.$$

We now want to characterize $\widetilde{X}$. If $(f_j)_{j \in \mathbb{N}} \in \widetilde{X}$, then $(\partial^s f_j)_{j \in \mathbb{N}}$ are Cauchy sequences in $L^p(\Omega)$, and hence it follows from 3.17 and 3.21 that there exist uniquely defined functions $f^{(s)} \in L^p(\Omega)$ such that

$$\partial^s f_j \to f^{(s)} \quad \text{in } L^p(\Omega) \text{ as } j \to \infty. \tag{3-16}$$

The relation between the functions $f^{(s)}$ arises from the rule of integration by parts for the functions f_j, which yields that

$$\int_\Omega \partial^s \zeta \cdot f_j \, \mathrm{dL}^n = (-1)^{|s|} \int_\Omega \zeta \cdot \partial^s f_j \, \mathrm{dL}^n \quad \text{for all } \zeta \in C_0^\infty(\Omega).$$

Hence, since $f_j = \partial^0 f_j \to f^{(0)}$ and $\partial^s f_j \to f^{(s)}$ for $j \to \infty$ (see (3-16)), using the Hölder inequality we have that

$$\int_\Omega \partial^s \zeta \cdot f^{(0)} \, \mathrm{dL}^n = (-1)^{|s|} \int_\Omega \zeta \cdot f^{(s)} \, \mathrm{dL}^n \quad \text{for all } \zeta \in C_0^\infty(\Omega). \tag{3-17}$$

Therefore, we define the following **Sobolev space** of order $m \in \mathbb{N}$ with exponent p, $1 \le p \le \infty$, by

$$W^{m,p}(\Omega) := \{\, f \in L^p(\Omega) \; ; \; \text{for } |s| \le m \text{ there exist } f^{(s)} \in L^p(\Omega)$$
$$\text{such that } f^{(0)} = f \text{ and (3-17) hold} \,\}$$

and equip this Sobolev space with the norm

$$\|f\|_{W^{m,p}(\Omega)} := \sum_{|s| \le m} \left\| f^{(s)} \right\|_{L^p(\Omega)}.$$

We will show that for $p < \infty$ the space $\widetilde{X}$ is completely characterized by $W^{m,p}(\Omega)$.

But first a few remarks on this definition. Other commonly used notations for $W^{m,p}(\Omega)$ are $H^{m,p}(\Omega)$, $H_p^m(\Omega)$, and for the special case $p = 2$ also $H^m(\Omega)$ for $H^{m,2}(\Omega)$. Sometimes these Sobolev spaces are defined as the completion of functions in $C^m(\overline{\Omega})$ (i.e. $\overline{\Omega}$ instead of Ω) for an open and bounded set Ω. If Ω has a smooth boundary, then the two definitions coincide (see A8.7 in connection with 3.28).

Given f, the functions $f^{(s)}$ in the above definition of $W^{m,p}(\Omega)$ are uniquely defined. To see this let $\widetilde{f}^{(s)}$ have the same properties. Then

$$\int_\Omega \zeta \bigl(\widetilde{f}^{(s)} - f^{(s)}\bigr) \, \mathrm{dL}^n = 0 \quad \text{for all } \zeta \in C_0^\infty(\Omega),$$

and hence $\widetilde{f}^{(s)} = f^{(s)}$ almost everywhere in Ω (this follows from 4.22, as the Hölder inquality yields that $g := \widetilde{f}^{(s)} - f^{(s)} \in L^p(\Omega') \subset L^1(\Omega')$ for bounded open sets $\Omega' \subset\subset \Omega$).

For smooth functions $f \in C^m(\Omega)$ with $\partial^s f \in L^p(\Omega)$ for $|s| \le m$ it follows from the rule of integration by parts that $f \in W^{m,p}(\Omega)$, and in particular $f^{(s)} = \partial^s f$. Therefore, for $f \in W^{m,p}(\Omega)$ we call

$$\partial^s f := f^{(s)}$$

the **weak derivatives** of f.

We now return to the space $\widetilde{X}$. Setting

$$J\bigl((f_k)_{k\in\mathbb{N}}\bigr) := \lim_{k\to\infty} f_k \quad \text{(limit in } L^p(\Omega), \text{ see (3-16) for } s = 0)$$

we define a linear map

$$J : \widetilde{X} \to W^{m,p}(\Omega),$$

which, on recalling the equivalence relation in $\widetilde{X}$ (see 2.24), is injective and, by definition of the $W^{m,p}$-norm, preserves the norm. To see the latter, observe that

$$\left\| (f_k)_{k \in \mathbb{N}} \right\|_{\widetilde{X}} = \lim_{k \to \infty} \| f_k \|_X = \lim_{k \to \infty} \sum_{|s| \leq m} \| \partial^s f_k \|_{L^p}$$

$$= \sum_{|s| \leq m} \| \partial^s f \|_{L^p} = \| f \|_{W^{m,p}}.$$

It remains to investigate whether J is surjective. But this can only hold for $p < \infty$, as for $p = \infty$ the image $J(\widetilde{X}) = W^{m,\infty}(\Omega) \cap C^m(\Omega)$ is a proper subspace of $W^{m,\infty}(\Omega)$. For example, the function $f(x) := |x_n|$ belongs to $W^{1,\infty}(\Omega)$ (this follows similarly to E3.7, where the one-dimensional case is considered). However, in the case $0 \in \Omega$ this function does not belong to the space $C^1(\Omega)$. For $p < \infty$ the surjectivity of J is formulated in theorem 3.28 below.

The fact that J is surjective and norm preserving yields that $W^{m,p}(\Omega)$ for $p < \infty$ is a Banach space, on noting that $\widetilde{X}$ is complete. But this can also be shown independently of this observation, and then also for $p = \infty$. To this end, let $(f_k)_{k \in \mathbb{N}}$ be a Cauchy sequence in $W^{m,p}(\Omega)$. Then $(\partial^s f_k)_{k \in \mathbb{N}}$ for $|s| \leq m$ are Cauchy sequences in $L^p(\Omega)$. By the Fischer-Riesz theorem (theorem 3.21 in the case $p < \infty$ and lemma 3.17 in the case $p = \infty$), there exist $f^{(s)} \in L^p(\Omega)$ such that

$$\partial^s f_k \to f^{(s)} \quad \text{in } L^p(\Omega) \text{ as } k \to \infty.$$

Moreover, as before the rule of integration by parts (3-17) carries over from $\partial^s f_k$ to $f^{(s)}$, and hence $f := f^{(0)} \in W^{m,p}(\Omega)$ with $\partial^s f = f^{(s)}$ for $|s| \leq m$.

3.28 Theorem. If $f \in W^{m,p}(\Omega)$ with $1 \leq p < \infty$, then there exist $f_j \in W^{m,p}(\Omega) \cap C^\infty(\Omega)$ such that $\| f - f_j \|_{W^{m,p}(\Omega)} \to 0$ as $j \to \infty$.

Note: We will give the proof of this theorem in Chapter 4 (see 4.24). For the proof we will need the approximation of functions by means of convolutions, a fundamental technique in analysis that is not yet available to us.

We will see later on (see 10.13) that functions in $W^{m,p}(\Omega)$ can be identified with classical (continuous and continuously differentiable) functions, but only if m and p are sufficiently large, or more precisely if $m - \frac{n}{p}$ is greater than zero. In the case $n = 1$, Sobolev functions always admit a continuous representative (see the remark in E3.6). For $n \geq 2$, Sobolev functions are in general not continuous functions. Noncontinuous examples are given in 10.7 and E10.7. What is the consequence? The motivation for introducing Sobolev spaces was to give solutions to differential equations. However, in this way we obtain only weak solutions in the Sobolev space. Hence a regularity theory is needed, which guarantees that the solutions of certain variational problems are smooth (continuous, differentiable, etc.) functions (see e.g. Appendix A12). Moreover, many variational problems also include side conditions, for instance in the form of boundary conditions. Hence these boundary conditions need to be formulated in Sobolev spaces as well. In fact, one can show that, in a weak sense, Sobolev functions have boundary values (see A8.6). Sobolev functions on Ω with weak boundary values 0 on the boundary of Ω can also be described in a simpler way, that is, as the limit of smooth functions with compact support in Ω (see also A8.10):

3.29 $W_0^{m,p}(\Omega)$-spaces. Let $\Omega \subset \mathbb{R}^n$ be open, let $m \geq 0$ be an integer and $1 \leq p < \infty$. Then the **Sobolev space with zero boundary values** of order m with exponent p is defined by

$$W_0^{m,p}(\Omega) := \{ f \in W^{m,p}(\Omega) ; \text{ there exist } f_k \in C_0^\infty(\Omega) \text{ such that}$$
$$\|f - f_k\|_{W^{m,p}} \to 0 \text{ as } k \to \infty \}.$$

Other commonly used notations are $H_0^{m,p}$, $\mathring{H}^{m,p}$, H_{p0}^m, $\mathring{W}^{m,p}$. The above defined space $W_0^{m,p}(\Omega)$ is a closed subspace of $W^{m,p}(\Omega)$.

Remark: If $\Omega \subset \widetilde{\Omega}$ and $f \in W_0^{m,p}(\Omega)$, then the function defined by $\widetilde{f} := f$ in Ω and $\widetilde{f} := 0$ in $\widetilde{\Omega} \setminus \Omega$ belongs to $W_0^{m,p}(\widetilde{\Omega})$.

Proof of Remark. Let $f_k \in C_0^\infty(\Omega)$ be as in the definition. Then the analogously extended functions $\widetilde{f_k}$ belong to $C_0^\infty(\widetilde{\Omega}) \subset W^{m,p}(\widetilde{\Omega})$ and converge in the $W^{m,p}$-norm to $\widetilde{f}$.   □

The space $W_0^{1,2}(\Omega)$ is used to solve a boundary value problem in 6.8.

E3 Exercises

E3.1 On uniform convergence. Let $S \subset \mathbb{R}^n$ be as in 3.3 and let Y be a Banach space. Then the following are equivalent for $f, f_k \in C^0(S;Y)$, $k \in \mathbb{N}$:

(1) $f_k \longrightarrow f$ in $C^0(S;Y)$ as $k \to \infty$.

(2) $\|f_k - f\|_{C^0(K_m)} \longrightarrow 0$ as $k \to \infty$ for all $m \in \mathbb{N}$.

(3) $\|f_k - f\|_{C^0(K)} \longrightarrow 0$ as $k \to \infty$ for all bounded and closed sets $K \subset S$.

E3.2 Exhaustion property. Which of the following sets S satisfies the exhaustion property in 3.3 ?

(1) $S \subset \mathbb{R}^n$ closed.

(2) $S \subset \mathbb{R}^n$ open.

(3) $S = S_0 := \{(x_1, x_2) \in \mathbb{R}^2 ; x_1 > 0, x_2 \geq 0\}$.

(4) $S = S_0 \cup \{(0,0)\}$ with S_0 as in (3).

Solution (1). Let $x_0 \in S$ and choose $K_m := S \cap \overline{B_m(x_0)}$.   □

Solution (2). For $\delta > 0$,

$$S_\delta := \{x \in \mathbb{R}^n \,;\, \text{dist}(x, \mathbb{R}^n \setminus S) \geq \delta\}$$

is closed. Choose $K_m := S_{\frac{1}{m}} \cap \overline{B_m(0)}$ for m sufficiently large. $\square$

Solution (3). Choose $K_m := \{(x_1, x_2) \in \mathbb{R}^2 \,;\, \frac{1}{m} \leq x_1 \leq m,\ 0 \leq x_2 \leq m\}$. $\square$

Solution (4). The property is not satisfied, since $S \cap B_\delta(0) \subset K$, K closed, implies that $K \not\subset S$. $\square$

E3.3 A test function. Letting

$$f(x) := \begin{cases} \exp\left(-\dfrac{1}{1 - |x|^2}\right) & \text{for } |x| < 1, \\ 0 & \text{otherwise,} \end{cases}$$

defines a function $f \in C_0^\infty(\mathbb{R}^n; \mathbb{R})$.

E3.4 L^p-norm as $p \to \infty$. Let $(S, \mathcal{B}, \mu)$ be a *bounded measure space*,
i.e. $\mu(S) < \infty$, and in addition nontrivial, i.e. $\mu(S) > 0$. For μ-measurable functions $f : S \to Y$ and $1 \leq p < \infty$ let

$$\Phi_p(f) := \begin{cases} \left(\dfrac{1}{\mu(S)} \displaystyle\int_S |f(x)|^p \, d\mu(x)\right)^{\frac{1}{p}} & \text{if } f \in L^p(\mu; Y), \\ \infty & \text{otherwise.} \end{cases}$$

Then $p \mapsto \Phi_p(f)$ is monotonically nondecreasing and for $f \in L^\infty(\mu; Y)$ we have that

$$\|f\|_{L^\infty} = \lim_{p \to \infty} \Phi_p(f) = \lim_{p \to \infty} \|f\|_{L^p}.$$

Solution. For $1 \leq p < q$ with $\Phi_q(f) < \infty$ let $r := \frac{q}{p}$ and let r' be the dual exponent, i.e. $\frac{1}{r} + \frac{1}{r'} = 1$. The Hölder inequality then yields that $\Phi_p(f) < \infty$ and

$$\Phi_p(f) = \left(\frac{1}{\mu(S)} \|1 \cdot |f|^p\|_{L^1}\right)^{\frac{1}{p}}$$

$$\leq \left(\frac{1}{\mu(S)} \|1\|_{L^{r'}} \cdot \||f|^p\|_{L^r}\right)^{\frac{1}{p}} = \left(\mu(S)^{\frac{1}{r'} - 1} \|f\|_{L^q}^{\frac{q}{r}}\right)^{\frac{1}{p}} = \Phi_q(f).$$

In addition, for $f \in L^\infty(\mu; Y)$ we have

$$\int_S |f|^p \, d\mu \leq \mu(S) \|f\|_{L^\infty}^p, \quad \text{and so} \quad \Phi_p(f) \leq \|f\|_{L^\infty}.$$

Moreover, for all $\kappa > 0$,

$$\int_S |f|^p \, d\mu \geq \mu(\{|f| \geq \kappa\}) \cdot \kappa^p \,,$$

and hence

$$\Phi_p(f) \geq \left(\frac{\mu(\{|f| \geq \kappa\})}{\mu(S)}\right)^{\frac{1}{p}} \cdot \kappa \quad \longrightarrow \kappa$$

as $p \to \infty$, if $\mu(\{|f| \geq \kappa\}) > 0$, which is satisfied for $\kappa < \|f\|_{L^\infty}$. This proves the desired result. $\qquad\qquad\square$

E3.5 Subsequences. Show that in 3.22(1) choosing a subsequence is necessary in general.

Solution. For $l, k \in \mathbb{N} \cup \{0\}$, $0 \leq k < 2^l$, let $n := 2^l + k$ and

$$f_n := \mathcal{X}_{[k2^{-l}, (k+1)2^{-l}]} \,.$$

It then follows for every $p \in [1, \infty[$ that $f_n \to 0$ in $L^p([0,1])$ as $n \to \infty$, but for every $x \in [0,1]$ the sequence $(f_n(x))_{n \in \mathbb{N}}$ has two distinct cluster points 0 and 1, i.e. the function sequence $(f_n)_{n \in \mathbb{N}}$ does not converge at any given point (and in particular it does not converge almost everywhere). However, $(f_{2^l}(x))_{l \in \mathbb{N}}$ converges for all $x \in [0,1]$ (to 1 for $x = 0$ and to 0 for $x > 0$), and hence the subsequence $(f_{2^l})_{l \in \mathbb{N}}$ converges at all points (and in particular it converges almost everywhere). $\qquad\qquad\square$

E3.6 Fundamental theorem of calculus. Let $I \subset \mathbb{R}$ be an open interval.

(1) If $f \in W^{1,1}(I)$, then for almost all $x_1, x_2 \in I$ (with respect to the one-dimensional Lebesgue measure)

$$f(x_2) - f(x_1) = \int_{x_1}^{x_2} f'(x) \, dx \,.$$

(2) Conversely, if $f, g \in L^1(I)$ and

$$f(x_2) - f(x_1) = \int_{x_1}^{x_2} g(x) \, dx$$

for almost all $x_1, x_2 \in I$, then $f \in W^{1,1}(I)$ with $f' = g$.

Remark: This is related to the concept of **absolutely continuous functions** on I, defined by

$$AC(I) := \left\{ \varphi \in C^0(\overline{I}) \; ; \; \text{there exists a } g \in L^1(I) \text{ such that for } x_1, x_2 \in I \right.$$
$$\left. \varphi(x_2) - \varphi(x_1) = \int_{x_1}^{x_2} g(x) \, dx \right\} \,.$$

Proposition (2) states that $AC(I) \subset W^{1,1}(I)$. If $f \in W^{1,1}(I)$ and $x_1 \in I$ is fixed such that the identity in (1) holds for almost all $x_2 \in I$, then $f = \varphi$ almost everywhere, where

$$\varphi(y) := f(x_1) + \int_{x_1}^{y} f'(x)\, dx \quad \text{for } y \in I.$$

As $\varphi \in AC(I) \subset C^0(\overline{I})$, we see that f agrees almost everywhere with a continuous function, or in other words, φ is the unique continuous representative of f.

Solution (1). It follows from 3.28 that there exist $f_k \in W^{1,1}(I) \cap C^\infty(I)$ such that $f_k \to f$ in $W^{1,1}(I)$. It holds for all $x_1, x_2 \in I$ that

$$f_k(x_2) - f_k(x_1) = \int_{x_1}^{x_2} f_k{}'(x)\, dx.$$

Since $f_k{}' \to f'$ in $L^1(I)$, the right-hand side converges to the desired expression. Moreover, it follows from 3.22(1) that there exists a subsequence $(f_{k_i})_{i \in \mathbb{N}}$ such that $f_{k_i}(x) \to f(x)$ for almost all $x \in I$. □

Solution (2). Let $\zeta \in C_0^\infty(I)$. Choose $x_\pm \in I$ such that $\zeta(x) = 0$ for all x outside of $[x_-, x_+]$ and such that the assumption holds for $x_1 = x_-$ and almost all x_2. Then

$$\int_I f(x)\zeta'(x)\, dx = \int_{x_-}^{x_+} (f(x) - f(x_-))\, \zeta'(x)\, dx = \int_{x_-}^{x_+} \int_{x_-}^{x} g(y)\, \zeta'(x)\, dy\, dx$$

$$= \int_{x_-}^{x_+} \left(\int_{y}^{x_+} \zeta'(x)\, dx \right) g(y)\, dy = -\int_I \zeta(y)\, g(y)\, dy.$$

□

E3.7 Left- and right-hand limit. Let $I \subset \mathbb{R}$ be an open interval, $x_0 \in I$ and $f \in W^{1,1}(I \setminus \{x_0\})$. Then there exist

$$f_-(x_0) := \operatorname*{ess\,lim}_{x \nearrow x_0} f(x) \quad \text{and} \quad f_+(x_0) := \operatorname*{ess\,lim}_{x \searrow x_0} f(x).$$

In addition,

$$f \in W^{1,1}(I) \quad \Longleftrightarrow \quad f_-(x_0) = f_+(x_0).$$

Note: For the multidimensional case see A8.9.

Definition: We call "ess lim" the **essential limit**. In general "ess lim" is the limit for points outside an appropriately chosen null set. In the concrete case above consider the continuous representative of f from the remark in E3.6.

Solution $\Rightarrow$. For almost all $x_1, x_2 \in I$ with $x_1, x_2 > x_0$ or $x_1, x_2 < x_0$ it follows from E3.6 that

$$|f(x_2) - f(x_1)| = \left| \int_{x_1}^{x_2} f'(x)\, dx \right| \leq \int_{x_1}^{x_2} |f'(x)|\, dx \longrightarrow 0 \qquad \text{as } x_1, x_2 \to x_0$$

(see A3.17(2)). Hence the limits $f_-(x_0)$ and $f_+(x_0)$ exist. If $f \in W^{1,1}(I)$, then it follows similarly for almost all $x_1, x_2 \in I$ with $x_1 < x_0 < x_2$ that

$$|f(x_2) - f(x_1)| \leq \int_{x_1}^{x_2} |f'(x)|\, dx \longrightarrow 0 \qquad \text{as } x_1, x_2 \to x_0,$$

and so $f_-(x_0) = f_+(x_0)$. $\qquad\square$

Solution $\Leftarrow$. For almost all $x_1 < y_1 < x_0 < y_2 < x_2$ in I it follows once again from E3.6 that

$$(f(x_2) - f(x_1)) - (f(y_2) - f(y_1)) = f(y_1) - f(x_1) + f(x_2) - f(y_2)$$
$$= \int_{x_1}^{y_1} f'(x)\, dx + \int_{y_2}^{x_2} f'(x)\, dx = \int_{x_1}^{x_2} f'(x)\, dx - \int_{y_1}^{y_2} f'(x)\, dx .$$

The right-hand side converges to

$$\int_{x_1}^{x_2} f'(x)\, dx$$

as $y_1, y_2 \to x_0$, while $f(y_2) - f(y_1) \to f_+(x_0) - f_-(x_0) = 0$. Hence $f \in W^{1,1}(I)$ on recalling E3.6. $\qquad\square$

E3.8 Estimating the Hölder norm by the $W^{1,p}$-norm. Let $1 < p \leq \infty$, $\alpha := 1 - \frac{1}{p}$ and $I := [a, b] \subset \mathbb{R}$. Then there exists a constant $C < \infty$ such that for all $f \in C^1(I)$ and all $x_0 \in I$,

$$\|f\|_{C^{0,\alpha}(I)} \leq |f(x_0)| + C \cdot \|f'\|_{L^p(I)} .$$

Note: See also theorem 10.13.

Solution. For $a \leq x_1 < x_2 \leq b$ the Hölder inequality (with $\frac{1}{p'} = 1 - \frac{1}{p} = \alpha$) yields

$$|f(x_2) - f(x_1)| = \left| \int_{x_1}^{x_2} f'(x)\, dx \right| \leq \int_{x_1}^{x_2} 1 \cdot |f'(x)|\, dx$$
$$\leq \left(\int_{x_1}^{x_2} 1\, dx \right)^{\frac{1}{p'}} \left(\int_{x_1}^{x_2} |f'(x)|^p\, dx \right)^{\frac{1}{p}} \leq (x_2 - x_1)^\alpha \|f'\|_{L^p(I)} .$$

This implies that

$$|f(x_2)| \leq |f(x_1)| + (b-a)^\alpha \|f'\|_{L^p(I)}$$

and

$$\frac{|f(x_2) - f(x_1)|}{|x_2 - x_1|^\alpha} \leq \|f'\|_{L^p(I)},$$

and hence the desired result with $C = 1 + (b-a)^\alpha$.   □

A3 Lebesgue's integral

Here we give the construction of Lebesgue's integral for σ-subadditive measures. In the context of functional analysis it appeared to be appropriate to carry out this construction based on a completion principle. Moreover, it is adequate for functional analysis to consider functions with values in Banach spaces Y. Here $Y = \mathbb{R}$ or $Y = \mathbb{R}^m$ are the standard cases, and for the understanding of the construction it suffices to consider for Y this Euclidean case.

The construction of Lebesgue's integral can be found in A3.1–A3.16. These give rise to the fundamental properties of the space $L(\mu)$ of integrable functions in A3.16, the so-called "axioms of Lebesgue's theory". From these axioms we will deduce the most important properties of Lebesgue integrable functions in A3.11–A3.21. In particular, in A3.15 we will show that the set of characteristic functions in $L(\mu)$ provides a σ-additive extension of the original measure.

A3.1 Assumptions. Let S be a set and let $\mathcal{B}_0$ be a nonempty system of subsets of S that forms a *(Boolean) ring*, i.e.

$$E_1, E_2 \in \mathcal{B}_0 \quad \Longrightarrow \quad E_1 \setminus E_2 \in \mathcal{B}_0 \text{ and } E_1 \cup E_2 \in \mathcal{B}_0.$$

Then it also holds that $\emptyset \in \mathcal{B}_0$ and $E_1 \cap E_2 = E_1 \setminus (E_1 \setminus E_2) \in \mathcal{B}_0$. The system of sets $\mathcal{B}_0$ is called a *(Boolean) algebra* if in addition $S \in \mathcal{B}_0$. Then $\mathcal{B}_0$ is characterized by the properties

(1) $\emptyset \in \mathcal{B}_0$,

(2) $E \in \mathcal{B}_0 \quad \Longrightarrow \quad S \setminus E \in \mathcal{B}_0$,

(3) $E_1, E_2 \in \mathcal{B}_0 \quad \Longrightarrow \quad E_1 \cup E_2 \in \mathcal{B}_0 \text{ and } E_1 \cap E_2 \in \mathcal{B}_0$.

In the following, let $(S, \mathcal{B}_0, \mu)$ be a *pre-measure space*, i.e. $\mathcal{B}_0$ is a Boolean algebra of subsets of S and

(4) $\mu : \mathcal{B}_0 \to [0, \infty]$ with $\mu(\emptyset) = 0$

is an *additive measure*, i.e.

(5) $E_1, \ldots, E_m \in \mathcal{B}_0$ pairwise disjoint $\quad \Longrightarrow \quad \mu\left(\bigcup_{i=1}^m E_i\right) = \sum_{i=1}^m \mu(E_i)$,

and, in addition, *σ-subadditive*, i.e.

(6) $E, E_i \in \mathcal{B}_0$ for $i \in \mathbb{N}$, $E \subset \bigcup_{i \in \mathbb{N}} E_i$ $\implies$ $\mu(E) \leq \sum_{i \in \mathbb{N}} \mu(E_i)$.

A3.2 Consequences. It follows from A3.1(4) and the additivity A3.1(5) that μ is **monotone** on $\mathcal{B}_0$, i.e.

(1) $E_1, E_2 \in \mathcal{B}_0$, $E_1 \subset E_2$ $\implies$ $\mu(E_1) \leq \mu(E_2)$,

and that μ is **subadditive** on $\mathcal{B}_0$, i.e.

(2) $E, E_i \in \mathcal{B}_0$, $i = 1, \ldots, m$, $E \subset \bigcup_{i=1}^m E_i$ $\implies$ $\mu(E) \leq \sum_{i=1}^m \mu(E_i)$.

Together with the σ-subadditivity A3.1(6) this yields that μ is **σ-additive** on $\mathcal{B}_0$, i.e.

(3) If $E_i \in \mathcal{B}_0$ for $i \in \mathbb{N}$ are pairwise disjoint, then

$$\bigcup_{i \in \mathbb{N}} E_i \in \mathcal{B}_0 \quad \implies \quad \mu\left(\bigcup_{i \in \mathbb{N}} E_i\right) = \sum_{i \in \mathbb{N}} \mu(E_i).$$

Proof (1). It holds that $\mu(E_2) = \mu(E_2 \setminus E_1) + \mu(E_1) \geq \mu(E_1)$.   □

Proof (2). Define inductively

$$A_0 := \emptyset \quad \text{and} \quad A_i := E_i \setminus \bigcup_{j < i} A_j \text{ for } i = 1, \ldots, m.$$

Then the A_i are pairwise disjoint with $E \subset \bigcup_{i=1}^m A_i$. It follows from (1) and A3.1(5) that

$$\mu(E) \leq \mu\left(\bigcup_{i=1}^m A_i\right) = \sum_{i=1}^m \mu(A_i) \leq \sum_{i=1}^m \mu(E_i).$$

□

Proof (3). The inequality "$\leq$" follows from A3.1(6). In addition, combining (1) and A3.1(5) yields that

$$\mu\left(\bigcup_{i \in \mathbb{N}} E_i\right) \geq \mu\left(\bigcup_{i=1}^m E_i\right) = \sum_{i=1}^m \mu(E_i) \quad \text{for all } m.$$

□

A3.3 Example (Elementary Lebesgue measure). As an example let $S = \mathbb{R}^n$ and let $\mathcal{B}_0$ consist of all finite unions of disjoint semi-open cuboids, where **semi-open cuboids** are sets of the form

$$[a, b[:= \{x \in \mathbb{R}^n \, ; \; a_i \leq x_i < b_i \text{ for } i = 1, \ldots, n\}$$

with $-\infty \leq a_i < b_i \leq +\infty$. Let

$$\mu([a, b[) := \prod_{i=1}^n (b_i - a_i),$$

with the value being ∞ if there is a $b_i = \infty$ or an $a_i = -\infty$. Now, if a semi-open cuboid E is the disjoint union of semi-open cuboids $E_1, \ldots, E_m$, then it is straightforward to show that

$$\mu(E) = \sum_{i=1}^{m} \mu(E_i) \ .$$

Hence it is possible to uniquely extend μ additively to $\mathcal{B}_0$. It holds that μ is σ-subadditive on $\mathcal{B}_0$, and so the assumptions in A3.1, in particular A3.1(6), are satisfied.

Definition: We denote this measure by $L^n := \mu$.

Proof of σ-subadditivity. Let E and E_i be semi-open cuboids as in A3.1(6). Without loss of generality, let $E_i = [a^i, b^i[$ with $a^i, b^i \in \mathbb{R}^n$, and let $E \in \mathcal{B}_0$ be given as a disjoint union

$$E = \bigcup_{i=1}^{l} [\alpha^i, \beta^i[\quad \text{with } \alpha^i, \beta^i \in \mathbb{R}^n, \ \alpha_j^i < \beta_j^i.$$

For small $\delta > 0$ consider the set

$$E^\delta := \bigcup_{i=1}^{l} [\alpha^{i\delta}, \beta^{i\delta}[\quad \text{with } \alpha_j^{i\delta} := \alpha_j^i + \delta \ < \ \beta_j^{i\delta} := \beta_j^i - \delta \ ,$$

which belongs to $\mathcal{B}_0$. As $\overline{E^\delta} \subset E \subset \bigcup_{i \in \mathbb{N}} E_i$, for $\varepsilon > 0$ the sets

$$]a^{i\varepsilon}, b^{i\varepsilon}[\quad \text{with } a_j^{i\varepsilon} := a_j^i - \varepsilon(b_j^i - a_j^i) \ , \ b_j^{i\varepsilon} := b_j^i + \varepsilon(b_j^i - a_j^i),$$

form an open cover of the closed bounded set $\overline{E^\delta}$. It follows that $\overline{E^\delta}$ (see 4.7(7) and 4.6) is already covered by finitely many open cuboids $]a^{i_1\varepsilon}, b^{i_1\varepsilon}[, \ldots,]a^{i_m\varepsilon}, b^{i_m\varepsilon}[$. Consequently

$$\mu(E^\delta) \le \sum_{k=1}^{m} \mu\big([a^{i_k\varepsilon}, b^{i_k\varepsilon}[\big) \quad \text{(recall A3.2(2))}$$

$$= \sum_{k=1}^{m} (1 + 2\varepsilon)^n \mu(E_{i_k}) \le (1 + 2\varepsilon)^n \sum_{i \in \mathbb{N}} \mu(E_i) \ .$$

Additivity and the definition of μ yield that, as $\delta \to 0$,

$$\mu(E^\delta) = \sum_{i=1}^{l} \mu\big([\alpha^{i\delta}, \beta^{i\delta}[\big) \longrightarrow \sum_{i=1}^{l} \mu\big([\alpha^i, \beta^i[\big) = \mu(E) \ .$$

On letting $\varepsilon \to 0$ we obtain the desired result. $\qquad\qquad\square$

A3.4 Definition (Outer measure and null sets). Let μ be as in A3.1.

(1) The *outer measure* μ^* corresponding to μ is defined by

$$\mu^*(A) := \inf \Big\{ \sum_{i \in \mathbb{N}} \mu(E_i) \ ; \ A \subset \bigcup_{i \in \mathbb{N}} E_i \ , \ E_i \in \mathcal{B}_0 \Big\}$$

for $A \subset S$. As it is possible to restrict this definition to disjoint sets E_i, we have

$$\mu^*(A) = \inf \left\{ \lim_{i \to \infty} \mu(A_i) \; ; \; A \subset \bigcup_{i \in \mathbb{N}} A_i \; , \; A_i \in \mathcal{B}_0 \; , \; A_i \subset A_{i+1} \right\}.$$

It follows that μ^* is σ-subadditive, and A3.1(6) yields that

$$\mu = \mu^* \text{ on } \mathcal{B}_0. \tag{A3-1}$$

(2) We say that

$$N \subset S \text{ is a } \boldsymbol{\mu\text{-null set}} \quad :\Longleftrightarrow \quad \mu^*(N) = 0.$$

Any subset of a μ-null set is a μ-null set. Countable unions of μ-null sets are μ-null sets. We say that a statement holds $\boldsymbol{\mu\text{-almost everywhere}}$ if it holds outside of a μ-null set.

Note: If $(S, \mathcal{B}_0, \mu)$ is a measure space (see 3.9), then N is a μ-null set if and only if $N \in \mathcal{B}_0$ with $\mu(N) = 0$.

Proof (1). It is $\mu \geq \mu^*$ by the definition of the outer measure and A3.1(6) implies that $\mu \leq \mu^*$. □

A3.5 Step functions. In the following let $(S, \mathcal{B}_0, \mu)$ be a pre-measure space as in A3.1 and let Y be a Banach space with norm $y \mapsto |y|$. The set of **step functions** with respect to $(S, \mathcal{B}_0, \mu)$ with values in Y is defined by

$$T(\mu; Y) := \{ f : S \to Y \; ; \; f(S) \text{ is finite,}$$
$$f^{-1}(\{y\}) \in \mathcal{B}_0 \text{ for } y \in Y,$$
$$\mu(f^{-1}(\{y\})) < \infty \text{ for } y \neq 0 \}$$

with the equivalence relation

$$f = g \text{ in } T(\mu; Y) \quad :\Longleftrightarrow \quad f = g \; \mu\text{-almost everywhere}.$$

Since step functions have steps in $\mathcal{B}_0$, it follows that

$$f = g \text{ in } T(\mu; Y) \quad \Longleftrightarrow \quad \mu(\{x \in S; \; f(x) \neq g(x)\}) = 0.$$

It turns out that step functions are precisely those functions $f : S \to Y$ that can be written as

$$f = \sum_{i=1}^{m} \mathcal{X}_{E_i} \alpha_i \quad \text{with } m \in \mathbb{N}, \; \alpha_i \in Y, \; E_i \in \mathcal{B}_0, \; \mu(E_i) < \infty.$$

(Observe that this representation is not unique.) This implies that $T(\mu; Y)$ is a vector space.

Definition: We denote by $\mathcal{X}_E$ the **characteristic function** of the set E, which is given by

$$\mathcal{X}_E(x) := \begin{cases} 1 & \text{for } x \in E, \\ 0 & \text{for } x \notin E. \end{cases}$$

A3.6 Elementary integral. For $f \in T(\mu; Y)$ we define the elementary μ-*integral* of f over S by

$$\int_S f \, d\mu := \sum_{y \in Y \setminus \{0\}} \mu(f^{-1}(\{y\})) \, y \, .$$

Then:

(1) It holds that

$$\int_S f \, d\mu = \sum_{i=1}^m \mu(E_i) \alpha_i \, , \quad \text{if } f = \sum_{i=1}^m \mathcal{X}_{E_i} \alpha_i \text{ as in A3.5.}$$

(2) The elementary integral is a linear map from $T(\mu; Y)$ to Y.

(3) For every $f \in T(\mu; Y)$, the function $x \mapsto |f(x)| \in \mathbb{R}$, denoted by $|f|$, belongs to $T(\mu; \mathbb{R})$ and

$$\left| \int_S f \, d\mu \right| \le \int_S |f| \, d\mu \, .$$

(4) The set $T(\mu; Y)$ with

$$\|f\|_{T(\mu)} := \int_S |f| \, d\mu$$

is a normed space.

Construction of Lebesgue's integral

The aim now is to describe the completion (see 2.24)

$$\widetilde{T}(\mu; Y) := \widetilde{T(\mu; Y)} \text{ of the normed space } T(\mu; Y) \, ,$$

that is, to construct an isomorphism between $\widetilde{T}(\mu; Y)$ and a function space. In this way, the completion $\widetilde{T}(\mu; Y)$, i.e. the set of Cauchy sequences in $T(\mu; Y)$, serves as a model for the set of functions that are Lebesgue integrable with respect to μ. Before we start, let us introduce the following notations: For $f \in T(\mu; Y)$ and $E \in \mathcal{B}_0$ we have that $\mathcal{X}_E f \in T(\mu; Y)$ and we define

$$\int_E f \, d\mu := \int_S \mathcal{X}_E f \, d\mu \, .$$

If $f \in T(\mu; \mathbb{R})$ and $a \in \mathbb{R}$, we set e.g.

$$\{f > a\} := \{x \in S; \; f(x) > a\} \in \mathcal{B}_0 \, . \tag{A3-2}$$

The crucial observations for the construction of Lebesgue's integral are:

A3.7 Lemma. Let $(f_k)_{k\in\mathbb{N}} \in \tilde{T}(\mu; Y)$. Then:

(1) There exist a μ-null set N and a subsequence $(f_{k_i})_{i\in\mathbb{N}}$ such that there exists

$$f(x) := \lim_{i\to\infty} f_{k_i}(x) \quad \text{in } Y \text{ for all } x \in S \setminus N.$$

(2) For the function f in (1)

$$(f_k)_{k\in\mathbb{N}} = 0 \text{ in } \tilde{T}(\mu; Y) \quad \Longleftrightarrow \quad f = 0 \; \mu\text{-almost everywhere.}$$

Proof (1). Choose a subsequence $(f_{k_i})_{i\in\mathbb{N}}$ such that $k_i < k_{i+1}$ for $i \in \mathbb{N}$ and

$$\|f_k - f_l\|_{T(\mu)} = \int_S |f_k - f_l|\,\mathrm{d}\mu \le 2^{-i} \quad \text{for } k, l \ge k_i.$$

Let

$$g_j := \sum_{i=1}^{j} |f_{k_i} - f_{k_{i+1}}| \in T(\mu; Y).$$

As $(g_j)_{j\in\mathbb{N}}$ is a monotonically increasing sequence of functions, there exists

$$g(x) := \lim_{j\to\infty} g_j(x) \in [0, \infty[\quad \text{for all } x \in S.$$

Now

$$\int_S g_j\,\mathrm{d}\mu = \sum_{i=1}^{j} \int_S |f_{k_i} - f_{k_{i+1}}|\,\mathrm{d}\mu \le \sum_{i=1}^{\infty} 2^{-i} = 1.$$

For every $\varepsilon > 0$ we then have that $A_j := \{g_j > \frac{1}{\varepsilon}\} \in \mathcal{B}_0$, with

$$1 \ge \int_S g_j\,\mathrm{d}\mu \ge \frac{1}{\varepsilon}\mu(A_j).$$

In addition, $A_j \subset A_{j+1}$ and

$$N := \{g = \infty\} \subset \{g > \tfrac{1}{\varepsilon}\} \subset \bigcup_{j\in\mathbb{N}} A_j.$$

It follows from A3.4 that

$$\mu^*(N) \le \lim_{j\to\infty} \mu(A_j) \le \varepsilon.$$

Hence N is a μ-null set and $g(x) < \infty$ for $x \in S \setminus N$, and so $(f_{k_i}(x))_{i\in\mathbb{N}}$ is a Cauchy sequence in Y for $x \in S \setminus N$. $\square$

Proof (2)$\Rightarrow$. We assume without loss of generality (drop the sequence elements that do not belong to the subsequence in (1)) that

$$f(x) = \lim_{k\to\infty} f_k(x) \quad \text{for } x \in S \setminus N.$$

We need to show that $\{f \neq 0\} \setminus N$ is a μ-null set. Since $\|f_k\|_{T(\mu)} \to 0$ as $k \to \infty$, there exists a subsequence $(f_{k_i})_{i \in \mathbb{N}}$ such that

$$\int_S |f_{k_i}| \, \mathrm{d}\mu \leq 2^{-i}.$$

For $\varepsilon > 0$ define

$$N_{\varepsilon,i} := \{|f_{k_i}| > \varepsilon\} \in \mathcal{B}_0, \quad N_\varepsilon := \{|f| > \varepsilon\} \setminus N.$$

Then it holds for every $i_0 \in \mathbb{N}$ that

$$N_\varepsilon \subset \bigcup_{i > i_0} N_{\varepsilon,i},$$

and so

$$\mu^*(N_\varepsilon) \leq \sum_{i > i_0} \mu(N_{\varepsilon,i})$$

$$\leq \sum_{i > i_0} \frac{1}{\varepsilon} \int_S |f_{k_i}| \, \mathrm{d}\mu \leq \frac{1}{\varepsilon} 2^{-i_0} \longrightarrow 0 \quad \text{as } i_0 \to \infty.$$

Hence N_ε is a μ-null set for all $\varepsilon > 0$, which yields that $\{|f| > 0\} \setminus N$ is a μ-null set. $\qquad\square$

Proof (2)$\Leftarrow$. We need to show that $\|f_k\|_{T(\mu;Y)} \to 0$ for a subsequence $k \to \infty$. We may again assume that $0 = f(x) = \lim_{k \to \infty} f_k(x)$ for $x \in S \setminus N$. Moreover we can assume (by choosing a subsequence) that

$$\int_S |f_k - f_{k+1}| \, \mathrm{d}\mu \leq 2^{-k}.$$

We note that $E_l := \{f_l \neq 0\} \in \mathcal{B}_0$ with $\mu(E_l) < \infty$, and hence it holds for $l < k$ and $\varepsilon > 0$ that

$$\int_S |f_k| \, \mathrm{d}\mu \leq \int_{E_l} |f_k| \, \mathrm{d}\mu + \int_{S \setminus E_l} |f_k| \, \mathrm{d}\mu$$

$$\leq \varepsilon\mu(E_l) + \int_{\{|f_k| > \varepsilon\}} |f_k| \, \mathrm{d}\mu + \int_{S \setminus E_l} |f_k - f_l| \, \mathrm{d}\mu$$

$$\leq \underbrace{\varepsilon\mu(E_l)}_{\substack{\to 0 \text{ as } \varepsilon \to 0 \\ \text{for any } l}} + \int_{\{|f_k| > \varepsilon\}} |f_l| \, \mathrm{d}\mu + 2\underbrace{\int_S |f_k - f_l| \, \mathrm{d}\mu}_{\to 0 \text{ as } k,l \to \infty},$$

with the second term on the right-hand side being

$$\leq \|f_l\|_{\sup} \mu(\{|f_k| > \varepsilon\}).$$

If we can show that for every $\varepsilon > 0$

$$\mu(\{|f_k| > \varepsilon\}) \to 0 \quad \text{as } k \to \infty$$

(i.e. $(f_k)_{k \in \mathbb{N}}$ is μ-measure convergent to 0, see (3-4)), then the desired result follows. For the proof define for $l \geq k$

$$E_{k,l} := \left\{ \sum_{i=k}^{l} |f_i - f_{i+1}| > \varepsilon \right\} \in \mathcal{B}_0 \,.$$

Then

$$\varepsilon \mu(E_{k,l}) \leq \int_S \sum_{i=k}^{l} |f_i - f_{i+1}| \, d\mu \leq \sum_{i=k}^{l} 2^{-i} \leq 2^{-k+1} \,,$$

and in addition $E_{k,l} \subset E_{k,l+1}$ for $l \geq k$. For points $x \in S \setminus N \setminus \bigcup_{l \geq k} E_{k,l}$ it holds that

$$|f_k(x)| \leq \underbrace{|f_{l+1}(x)|}_{\to \, 0 \text{ as } l \to \infty} + \underbrace{\sum_{i=k}^{l} |f_i(x) - f_{i+1}(x)|}_{\leq \, \varepsilon \text{ for all } l} \,,$$

and so

$$\{|f_k| > \varepsilon\} \subset N \cup \bigcup_{l \geq k} E_{k,l} \,,$$

which implies that

$$\mu(\{|f_k| > \varepsilon\}) = \mu^*(\{|f_k| > \varepsilon\}) \leq \mu^*(N) + \mu^* \left(\bigcup_{l \geq k} E_{k,l} \right)$$

$$\leq \lim_{l \to \infty} \mu(E_{k,l}) \leq \frac{1}{\varepsilon} 2^{-k+1} \to 0 \quad \text{as } k \to \infty.$$

$\square$

A3.8 Lebesgue integrable functions. Let $\widetilde{T}(\mu; Y)$ be the completion of $T(\mu; Y)$ (see 2.24). Recalling lemma A3.7, we consider the set

$$L(\mu; Y) := \big\{ f : S \to Y \; ; \; \text{there exists a sequence}$$
$$(f_k)_{k \in \mathbb{N}} \in \widetilde{T}(\mu; Y) \text{ such that} \qquad \text{(A3-3)}$$
$$f = \lim_{k \to \infty} f_k \; \mu\text{-almost everywhere} \big\}$$

with the equivalence relation

$$f = g \text{ in } L(\mu; Y) \quad :\Longleftrightarrow \quad f = g \; \mu\text{-almost everywhere}.$$

In the following we also write $L(\mu)$ instead of $L(\mu; \mathbb{K})$. Functions in $L(\mu; Y)$ are called **μ-integrable**. Clearly, $L(\mu; Y)$ is a vector space which contains $T(\mu; Y)$, where the equivalence relation in $L(\mu; Y)$ restricted to $T(\mu; Y)$ is the same as in A3.5.

Proposition: Defining

$$J\big((f_k)_{k\in\mathbb{N}}\big) := f \quad \text{with } f \text{ as in (A3-3)}$$

yields a vector space isomorphism between $\widetilde{T}(\mu;Y)$ and $L(\mu;Y)$.

Proof. It follows from A3.7 that for $(f_k)_{k\in\mathbb{N}} \in \widetilde{T}(\mu;Y)$ there exists a unique f in $L(\mu;Y)$ such that for a subsequence $(f_{k_i})_{i\in\mathbb{N}}$ it holds that $f = \lim_{i\to\infty} f_{k_i}$ almost everywhere. To see this, let $\left(f_{\widetilde{k}_i}\right)_{i\in\mathbb{N}}$ be another subsequence and $\widetilde{f} = \lim_{i\to\infty} f_{\widetilde{k}_i}$ almost everywhere. Then $\left(f_{k_i} - f_{\widetilde{k}_i}\right)_{i\in\mathbb{N}} = 0$ in $\widetilde{T}(\mu;Y)$, and so $f - \widetilde{f} = 0$ almost everywhere by A3.7(2), i.e. $f = \widetilde{f}$ in $L(\mu;Y)$. This defines a map

$$J : \widetilde{T}(\mu;Y) \to L(\mu;Y),$$

which is obviously linear. Moreover, it follows from A3.7(2) that it is injective and the definition of $L(\mu;Y)$ in (A3-3) yields that it is surjective. □

A3.9 Lebesgue integral. Let f, $(f_k)_{k\in\mathbb{N}}$ be as in the definition of $L(\mu;Y)$ in (A3-3). Then it follows from A3.6(3) and A3.6(4) that

$$\left| \int_S f_k \, d\mu - \int_S f_l \, d\mu \right| \le \int_S |f_k - f_l| \, d\mu \longrightarrow 0 \quad \text{as } k, l \to \infty.$$

As Y is complete, there exists

$$\lim_{k\to\infty} \int_S f_k \, d\mu \quad \text{in } Y.$$

The integral of f with respect to the measure μ is defined by

$$\int_S f \, d\mu \ \left(\text{or } \int_S f(x) \, d\mu(x) \right) := \lim_{k\to\infty} \int_S f_k \, d\mu,$$

which is independent of the choice of the sequence $(f_k)_{k\in\mathbb{N}}$ in the definition (A3-3) (this follows as in the proof of A3.8).

The space $L(\mu;Y)$ together with the above defined integral has the following properties (L1)–(L5), which we call the axioms of the Lebesgue integration theory.

A3.10 Theorem (Axioms of the Lebesgue integral). For the integral defined in A3.9 it holds that:

(L1) $T(\mu;Y) \subset L(\mu;Y)$ and the integral is linear on $L(\mu;Y)$, with

$$\int_S \mathcal{X}_E \alpha \, d\mu = \mu(E)\alpha \quad \text{for } E \in \mathcal{B}_0 \text{ with } \mu(E) < \infty \text{ and } \alpha \in Y.$$

(L2) If $f \in L(\mu; Y)$, then $|f| \in L(\mu; \mathbb{R})$ and

$$\left| \int_S f \, d\mu \right| \le \int_S |f| \, d\mu \, .$$

(L3) For $f \in L(\mu; Y)$ and $\varepsilon > 0$,

$$\int_S |f| \, d\mu \ge \varepsilon \mu^*(\{|f| > \varepsilon\}) \, .$$

(L4) $L(\mu; Y)$ is a Banach space with the norm

$$\|f\|_{L(\mu)} := \int_S |f| \, d\mu \, .$$

(L5) $T(\mu; Y)$ is dense in $L(\mu; Y)$.

Proof (L1). If $f \in T(\mu; Y)$, then $J((f)_{k \in \mathbb{N}}) = f$, where J is the isomorphism from A3.8. In addition we note that the integral of f in A3.9 coincides with the elementary integral in A3.6.   □

Proof (L2). If $J((f_k)_{k \in \mathbb{N}}) = f$ with $(f_k)_{k \in \mathbb{N}} \in \widetilde{T}(\mu; Y)$, then it follows from the triangle inequality in Y that

$$\int_S ||f_k| - |f_l|| \, d\mu \le \int_S |f_k - f_l| \, d\mu \longrightarrow 0 \quad \text{as } k, l \to \infty,$$

and hence $(|f_k|)_{k \in \mathbb{N}} \in \widetilde{T}(\mu; \mathbb{R})$. Since there exists a subsequence $(k_i)_{i \in \mathbb{N}}$ such that $f_{k_i} \to f$ μ-almost everywhere, we have that $|f_{k_i}| \to |f|$ μ-almost everywhere as $i \to \infty$, and so $J((|f_k|)_{k \in \mathbb{N}}) = |f|$ (where here J is the isomorphism associated with $T(\mu; \mathbb{R})$). Noting that

$$\left| \int_S f_k \, d\mu \right| \le \int_S |f_k| \, d\mu$$

yields the inequality in (L2).   □

Proof (L3). Let $J((f_k)_{k \in \mathbb{N}}) = f$, that is, for a μ-null set N and a subsequence $k \to \infty$ (again denoted by $(f_k)_{k \in \mathbb{N}}$) we have that

$$f_k(x) \to f(x) \quad \text{for } x \in S \setminus N \, .$$

On choosing a further subsequence (and retaining the notation as above) we can assume without loss of generality that

$$\int_S |f_{k+1} - f_k| \, d\mu \le 2^{-k} \, .$$

Let $x \in S \setminus N$. For $0 < \delta < \varepsilon$ we have that

$$|f(x)| > \varepsilon \quad \Longrightarrow \quad |f_k(x)| > \delta \text{ or } |f(x) - f_k(x)| > \varepsilon - \delta.$$

In the latter case,

$$\begin{aligned}|f(x) - f_k(x)| &= \lim_{i \to \infty} |f_i(x) - f_k(x)| \\ &\leq \sum_{i \geq k} |f_{i+1}(x) - f_i(x)|,\end{aligned}$$

so that there exists an $l > k$, depending on x, such that

$$\sum_{k \leq i \leq l} |f_{i+1}(x) - f_i(x)| > \varepsilon - \delta.$$

Hence

$$\{|f| > \varepsilon\} \subset N \cup \{|f_k| > \delta\} \cup \bigcup_{l \geq k} \Big\{ \sum_{k \leq i \leq l} |f_{i+1} - f_i| > \varepsilon - \delta \Big\}.$$

As the set in the rightmost union is monotonically increasing in l, it follows from the definition of the outer measure that

$$\begin{aligned}\mu^*(\{|f| > \varepsilon\}) &\leq \mu(\{|f_k| > \delta\}) + \lim_{l \to \infty} \mu\Big(\Big\{ \sum_{k \leq i \leq l} |f_{i+1} - f_i| > \varepsilon - \delta \Big\}\Big) \\ &\leq \frac{1}{\delta} \int_S |f_k| \, d\mu + \frac{1}{\varepsilon - \delta} \sum_{k \leq i \leq l} \int_S |f_{i+1} - f_i| \, d\mu \\ &\leq \frac{1}{\delta} \int_S |f_k| \, d\mu + \frac{1}{\varepsilon - \delta} 2^{1-k}.\end{aligned}$$

Recalling the proof of (L2) we have that

$$\int_S |f| \, d\mu = \lim_{k \to \infty} \int_S |f_k| \, d\mu,$$

and so letting $k \to \infty$ and then $\delta \to \varepsilon$ we obtain the desired result. $\qquad \square$

Proof (L4) *and* (L5). If $J\big((f_k)_{k \in \mathbb{N}}\big) = f$, then it follows from the proof of (L2) that

$$\|f\|_{L(\mu)} := \int_S |f| \, d\mu = \lim_{k \to \infty} \int_S |f_k| \, d\mu = \big\|(f_k)_{k \in \mathbb{N}}\big\|_{\tilde{T}(\mu;Y)},$$

i.e. J is isometric. As $\tilde{T}(\mu;Y)$ is complete, it follows that $L(\mu;Y)$ with the above norm is a Banach space. If $J\big((f_k)_{k \in \mathbb{N}}\big) = f$, then $J\big((f_k - f_j)_{k \in \mathbb{N}}\big) = f - f_j$ for all j and

$$\|f - f_j\|_{L(\mu)} = \lim_{k \to \infty} \int_S |f_k - f_j| \, d\mu \longrightarrow 0 \quad \text{as } j \to \infty,$$

i.e. every $f \in L(\mu;Y)$ can be approximated in the $L(\mu;Y)$-norm by step functions. $\qquad \square$

Extension of measures

This concludes the construction of the Lebesgue integral. We now derive the most important properties of this integral. We note that to this end, we will only (!) make use of the properties (L1)–(L5). This means: All the results of Lebesgue's integration theory can be derived from the properties (L1)–(L5), and that is why these properties are called "axioms".

A3.11 Lemma. If $f_k \to f$ in $L(\mu; Y)$ as $k \to \infty$, then there exists a subsequence $(k_i)_{i \in \mathbb{N}}$ such that $f_{k_i} \to f$ μ-almost everywhere as $i \to \infty$.

Proof. Choose a subsequence $(k_i)_{i \in \mathbb{N}}$ with

$$\int_S |f - f_{k_i}| \, \mathrm{d}\mu \le 2^{-i} \,.$$

For $\varepsilon > 0$ we have that

$$N_\varepsilon := \left\{ \limsup_{i \to \infty} |f - f_{k_i}| > \varepsilon \right\} \subset \bigcup_{i \ge j} \{ |f - f_{k_i}| > \varepsilon \}$$

for all j, and hence

$$\mu^*(N_\varepsilon) \le \sum_{i \ge j} \mu^*(\{ |f - f_{k_i}| > \varepsilon \}) \,,$$

which, on recalling (L3), is

$$\le \frac{1}{\varepsilon} \sum_{i \ge j} \int_S |f - f_{k_i}| \, \mathrm{d}\mu \le \frac{1}{\varepsilon} 2^{1-j} \longrightarrow 0 \quad \text{as } j \to \infty.$$

This yields that N_ε is a μ-null set, and consequently so is

$$\left\{ \limsup_{i \to \infty} |f - f_{k_i}| > 0 \right\}.$$

□

A3.12 Conclusions.

(1) *Monotonicity of the integral.* For $f, g \in L(\mu; \mathbb{R})$ it holds that:

$$g \ge f \ \mu\text{-almost everywhere} \quad \Longrightarrow \quad \int_S g \, \mathrm{d}\mu \ge \int_S f \, \mathrm{d}\mu \,.$$

(2) *Convergence criterion.* If $(f_k)_{k \in \mathbb{N}}$ is a Cauchy sequence in $L(\mu; Y)$ and $f_k \to f$ μ-almost everywhere as $k \to \infty$, then $f \in L(\mu; Y)$ and $\|f - f_k\|_{L(\mu)} \to 0$ as $k \to \infty$.

(3) *Monotone convergence theorem.* Let $f_k \in L(\mu; \mathbb{R})$ for $k \in \mathbb{N}$ and let $f : S \to \mathbb{R}$. Moreover, let $0 \le f_k \nearrow f$ μ-almost everywhere as $k \to \infty$ (that is, f_k converges from below monotonically to f) and

$$\limsup_{k \to \infty} \int_S f_k \, d\mu < \infty.$$

Then $f \in L(\mu; \mathbb{R})$ and $f_k \to f$ in $L(\mu; \mathbb{R})$ as $k \to \infty$.

Note: In particular,

$$\int_S f \, d\mu = \lim_{k \to \infty} \int_S f_k \, d\mu.$$

Proof (1). By (L2),

$$\int_S g \, d\mu - \int_S f \, d\mu = \int_S |g - f| \, d\mu \ge \left| \int_S (g - f) \, d\mu \right| \ge 0.$$

$\square$

Proof (2). We have from (L4) that $L(\mu; Y)$ is complete, and hence there exists a $g \in L(\mu; Y)$ such that $f_k \to g$ in $L(\mu; Y)$ as $k \to \infty$. It follows from A3.11 that there exists a subsequence $(k_i)_{i \in \mathbb{N}}$ such that

$$f_{k_i} \to g \quad \mu\text{-almost everywhere as } i \to \infty.$$

Hence, $f = g$ μ-almost everywhere, i.e. $f = g \in L(\mu; Y)$. $\square$

Proof (3). It follows from (1) that the integrals of f_k form a monotone sequence in $\mathbb{R}$, and the assumptions state that this sequence is bounded. Hence there exists

$$\lim_{k \to \infty} \int_S f_k \, d\mu.$$

It follows for $l \ge k$ that

$$\int_S |f_l - f_k| \, d\mu = \int_S f_l \, d\mu - \int_S f_k \, d\mu \longrightarrow 0 \quad \text{as } k, l \to \infty.$$

Now the desired result follows from (2). $\square$

Proof (3) *Note.* It follows from (L2), which implies the continuity of the integral with respect to the $L(\mu)$-norm, that

$$\left| \int_S f \, d\mu - \int_S f_k \, d\mu \right| \le \int_S |f - f_k| \, d\mu \longrightarrow 0 \quad \text{as } k \to \infty,$$

since $f_k \to f$ in $L(\mu; \mathbb{R})$. By the way this also follows from the convergence of the integrals, since due to the fact that $f \ge f_k$ we have

$$\|f - f_k\|_{L(\mu)} = \int_S |f - f_k| \, d\mu = \int_S f \, d\mu - \int_S f_k \, d\mu,$$

which converges to 0 as $k \to \infty$. Here we use the linearity of the integral. $\square$

For the Lebesgue measure in A3.3 we have the following additional approximation property:

A3.13 Remark. For $\mu = L^n$ as in A3.3 it holds that: Every function in $L(\mu; Y)$ can be approximated by functions in $C_0^0(\mathbb{R}^n; Y)$.

Proof. Functions from $C_0^0(\mathbb{R}^n; Y)$ belong to $L(\mu; Y)$, because they can be uniformly approximated by step functions on semi-open cuboids (cf. the definition of the Riemann integral in 6.22). We know from (L5) that functions in $L(\mu; Y)$ can be approximated by step functions in $T(\mu; Y)$, and hence it is sufficient to consider functions $\mathcal{X}_E$ for semi-open cuboids $E = [a, b[, \, a, b \in \mathbb{R}^n$. For $\varepsilon > 0$ let

$$E_\varepsilon := \{x \in \mathbb{R}^n \,; \, a_i - \varepsilon \le x_i < b_i + \varepsilon \text{ for } i = 1, \dots, n\},$$
$$f_\varepsilon(x) := \max\left(0 \,, \, 1 - \tfrac{1}{\varepsilon}\mathrm{dist}(x, E)\right).$$

Then $f_\varepsilon \in C_0^0(\mathbb{R}^n)$ and $\mathcal{X}_E \le f_\varepsilon \le \mathcal{X}_{E_\varepsilon}$. It follows from A3.12(1) that

$$\int_S |f_\varepsilon - \mathcal{X}_E| \, d\mu \le \int_S (\mathcal{X}_{E_\varepsilon} - \mathcal{X}_E) \, d\mu = \mu(E_\varepsilon \setminus E) \longrightarrow 0$$

as $\varepsilon \to 0$. □

We started with a measure, see A3.1,

$$(S, \mathcal{B}_0, \mu) \,, \quad \mu : \mathcal{B}_0 \to [0, \infty] \, \sigma\text{-additive}, \quad \mathcal{B}_0 \text{ an algebra.}$$

Next, we will construct a **σ-additive extension**

$$(S, \mathcal{B}, \bar{\mu}) \,, \quad \bar{\mu} : \mathcal{B} \to [0, \infty] \, \sigma\text{-additive}, \quad \mathcal{B} \text{ a } \sigma\text{-algebra.}$$

Here extension means that

$$\mathcal{B}_0 \subset \mathcal{B} \,, \quad \bar{\mu} = \mu \text{ on } \mathcal{B}_0 \,.$$

The construction is carried out by considering integrable sets based on the Lebesgue integral.

A3.14 Integrable sets. Let $\mathcal{B}_1$ be the smallest σ-algebra (for the definition see 3.9) that contains $\mathcal{B}_0$. It holds for **μ-integrable sets** E, i.e. sets $E \subset S$ with $\mathcal{X}_E \in L(\mu; \mathbb{R})$, that:

(1) There exist $E_k \in \mathcal{B}_0$ such that $\mathcal{X}_{E_k} \to \mathcal{X}_E$ in $L(\mu; \mathbb{R})$ as $k \to \infty$.
(2) There exists an $E' \in \mathcal{B}_1$ such that $\mathcal{X}_E = \mathcal{X}_{E'}$ μ-almost everywhere.
(3) It is $\int_S \mathcal{X}_E \, d\mu = \mu^*(E)$.
(4) For all $A \in \mathcal{B}_1$, it is $\mathcal{X}_{E \cap A} \in L(\mu; \mathbb{R})$.

Proof (1). It follows from (L5) that there exist $f_k \in T(\mu; \mathbb{R})$ with $f_k \to \mathcal{X}_E$ in $L(\mu; \mathbb{R})$ as $k \to \infty$. On defining

$$E_k := \{f_k > \tfrac{1}{2}\}$$

we have that $|\mathcal{X}_{E_k} - \mathcal{X}_E| \leq 2|f_k - \mathcal{X}_E|$. Now A3.12(1) yields that $\mathcal{X}_{E_k} \to \mathcal{X}_E$ in $L(\mu; \mathbb{R})$. □

Proof (2). Let $\mathcal{X}_{E_k}$ be functions with the property in (1), where on choosing a subsequence we can assume that

$$\|\mathcal{X}_{E_k} - \mathcal{X}_E\|_{L(\mu)} \leq 2^{-k}, \tag{A3-4}$$

which will be needed in the proof of (3) below. In addition, it follows from A3.11 that there exists a μ-null set N such that for a further subsequence $k \to \infty$ (the assumption (A3-4) will then still hold)

$$\mathcal{X}_{E_k}(x) \to \mathcal{X}_E(x) \quad \text{as } k \to \infty \text{ for all } x \in S \setminus N.$$

This pointwise convergence implies in set notation that

$$E \setminus N = \bigcap_j \bigcup_{i \geq j} (E_i \setminus N) = E' \setminus N, \tag{A3-5}$$

where

$$E' := \bigcap_j \bigcup_{i \geq j} E_i \in \mathcal{B}_1.$$

Hence, $\mathcal{X}_E = \mathcal{X}_{E'}$ μ-almost everywhere. □

Proof (3). Let E_k, $k \in \mathbb{N}$, be as above (with the properties (A3-4) and (A3-5)). It follows from (A3-5) for all $j \in \mathbb{N}$ that

$$E \setminus N \subset \bigcup_{i \geq j} E_i \setminus N \subset E_j \cup \bigcup_{i > j} (E_i \setminus E_{i-1})$$

and hence

$$\mu^*(E \setminus N) \leq \mu(E_j) + \sum_{i > j} \mu(E_i \setminus E_{i-1})$$

$$= \int_S \mathcal{X}_{E_j} \, d\mu + \sum_{i > j} \int_S \mathcal{X}_{E_i \setminus E_{i-1}} \, d\mu$$

$$\leq \int_S \mathcal{X}_{E_j} \, d\mu + \underbrace{\sum_{i > j} (2^{-i} + 2^{-(i-1)})}_{= 3 \cdot 2^{-j}},$$

where we have used

$$\int_S \mathcal{X}_{E_i \setminus E_{i-1}} \, d\mu \leq \int_S |\mathcal{X}_{E_i} - \mathcal{X}_{E_{i-1}}| \, d\mu$$

$$\leq \|\mathcal{X}_{E_i} - \mathcal{X}_E\|_{L(\mu)} + \|\mathcal{X}_{E_{i-1}} - \mathcal{X}_E\|_{L(\mu)}$$

and (A3-4). Consequently,

$$\mu^*(E) \le \mu^*(E \setminus N) + \mu^*(N) = \mu^*(E \setminus N)$$
$$\le \int_S \mathcal{X}_{E_j} \, d\mu + 3 \cdot 2^{-j} \longrightarrow \int_S \mathcal{X}_E \, d\mu \quad \text{as } j \to \infty. \tag{A3-6}$$

In particular, the outer measure $\mu^*(E)$ is finite. Hence it follows from the definition of the outer measure in A3.4 that for every $\varepsilon > 0$ there exist sets $A_i \in \mathcal{B}_0$, $i \in \mathbb{N}$ such that

$$E \subset A := \bigcup_{i \in \mathbb{N}} A_i, \quad A_i \subset A_{i+1}, \quad \mu^*(E) + \varepsilon \ge \lim_{i \to \infty} \mu(A_i).$$

On noting that $\mathcal{X}_{A_i}(x) \nearrow \mathcal{X}_A(x)$ as $i \to \infty$ for all $x \in S$ and that $\mu(A_i)$ is bounded, it follows from the monotone convergence theorem A3.12(3) that $\mathcal{X}_A \in L(\mu; \mathbb{R})$ and

$$\mu(A_i) = \int_S \mathcal{X}_{A_i} \, d\mu \nearrow \int_S \mathcal{X}_A \, d\mu \quad \text{as } i \to \infty,$$

and hence, by A3.12(1),

$$\mu^*(E) + \varepsilon \ge \int_S \mathcal{X}_A \, d\mu \ge \int_S \mathcal{X}_E \, d\mu.$$

Letting $\varepsilon \to 0$ and recalling (A3-6) yields the desired result. □

Proof (4). Let

$$\mathcal{M} := \{ A \subset S \,;\, \mathcal{X}_{E \cap A} \in L(\mu; \mathbb{R}) \}.$$

It holds that $\mathcal{B}_0 \subset \mathcal{M}$. To see this, let $A \in \mathcal{B}_0$ and let $\mathcal{X}_{E_k}$ be as in (1). Then

$$|\mathcal{X}_{E_k \cap A} - \mathcal{X}_{E_l \cap A}| \le |\mathcal{X}_{E_k} - \mathcal{X}_{E_l}|,$$

and hence, on recalling A3.12(1), $(\mathcal{X}_{E_k \cap A})_{k \in \mathbb{N}}$ is a Cauchy sequence in $L(\mu; \mathbb{R})$. We have from A3.11 that $\mathcal{X}_{E_k} \to \mathcal{X}_E$ μ-almost everywhere for a subsequence $k \to \infty$, and hence also $\mathcal{X}_{E_k \cap A} \to \mathcal{X}_{E \cap A}$ μ-almost everywhere. Now A3.12(2) implies that $\mathcal{X}_{E \cap A} \in L(\mu; \mathbb{R})$, i.e. $A \in \mathcal{M}$. Moreover,

$$A_1, A_2 \in \mathcal{M} \quad \Longrightarrow \quad A_1 \cap A_2 \in \mathcal{M}.$$

To see this, let $E_{ik} \in \mathcal{B}_0$ be the corresponding sets to $E \cap A_i$ from (1), i.e. $\mathcal{X}_{E_{ik}} \to \mathcal{X}_{E \cap A_i}$ in $L(\mu; \mathbb{R})$ as $k \to \infty$. Then

$$|\mathcal{X}_{E_{1k} \cap E_{2k}} - \mathcal{X}_{E_{1l} \cap E_{2l}}| \le |\mathcal{X}_{E_{1k}} - \mathcal{X}_{E_{1l}}| + |\mathcal{X}_{E_{2k}} - \mathcal{X}_{E_{2l}}|,$$

and hence $(\mathcal{X}_{E_{1k} \cap E_{2k}})_{k \in \mathbb{N}}$ is a Cauchy sequence in $L(\mu; \mathbb{R})$. It follows from A3.11 that there exists a subsequence $k \to \infty$ such that $\mathcal{X}_{E_{ik}} \to \mathcal{X}_{E \cap A_i}$ μ-almost everywhere for $i = 1, 2$. Then we also have that $\mathcal{X}_{E_{1k} \cap E_{2k}} \to \mathcal{X}_{E \cap A_1 \cap A_2}$ μ-almost everywhere. Hence it follows from A3.12(2) that $\mathcal{X}_{E \cap A_1 \cap A_2}$ belongs to $L(\mu; \mathbb{R})$, i.e. $A_1 \cap A_2 \in \mathcal{M}$. In addition,

$$A \in \mathcal{M} \quad \Longrightarrow \quad S \setminus A \in \mathcal{M},$$

since $\mathcal{X}_E, \mathcal{X}_{E \cap A} \in L(\mu; \mathbb{R})$ implies that

$$\mathcal{X}_{E \setminus A} = \mathcal{X}_E - \mathcal{X}_{E \cap A} \in L(\mu; \mathbb{R}).$$

Finally,

$$A_i \in \mathcal{M} \text{ with } A_i \subset A_{i+1} \text{ for } i \in \mathbb{N} \quad \Longrightarrow \quad A := \bigcup_{i \in \mathbb{N}} A_i \in \mathcal{M},$$

because as $i \to \infty$ we have that $\mathcal{X}_{E \cap A_i} \nearrow \mathcal{X}_{E \cap A} \leq \mathcal{X}_E \in L(\mu; \mathbb{R})$, which, on recalling A3.12(3), implies $\mathcal{X}_{E \cap A} \in L(\mu; \mathbb{R})$.

It follows from the above established properties of $\mathcal{M}$ that $\mathcal{M}$ is a σ-algebra that contains $\mathcal{B}_0$. Hence $\mathcal{B}_1 \subset \mathcal{M}$. $\qquad \square$

A3.15 Measure extension. Let $(S, \mathcal{B}_0, \mu)$ be a pre-measure space as in A3.1 and let $\mathcal{B}_1$ be the σ-algebra induced by $\mathcal{B}_0$ from A3.14. Let

$$\mathcal{B} := \left\{ E \subset S \ ; \ \mathcal{X}_E = \mathcal{X}_{E'} \ \mu\text{-almost everywhere for an } E' \in \mathcal{B}_1 \right\}$$

and define $\bar{\mu} : \mathcal{B} \to [0, \infty]$ by

$$\bar{\mu}(E) := \begin{cases} \displaystyle\int_S \mathcal{X}_E \, d\mu & \text{if } \mathcal{X}_E \in L(\mu; \mathbb{R}), \\ \infty & \text{otherwise.} \end{cases}$$

Then:

(1) $\mathcal{X}_E \in L(\mu; \mathbb{R}) \quad \Longrightarrow \quad E \in \mathcal{B}$ and $\bar{\mu}(E) = \mu^*(E)$.

(2) $\mathcal{B}$ is a σ-algebra and $\bar{\mu} : \mathcal{B} \to [0, \infty]$ is σ-additive.

(3) N is a μ-null set (i.e. $\mu^*(N) = 0$) $\quad \Longleftrightarrow \quad N \in \mathcal{B}$ with $\bar{\mu}(N) = 0$.

(4) $\bar{\mu}$ is an extension of μ from $\mathcal{B}_0$ to $\mathcal{B}$ and $(S, \mathcal{B}, \bar{\mu})$ is a measure space.

Interpretation: This shows that there exists a measure extension $(\mathcal{B}, \bar{\mu})$ of $(\mathcal{B}_0, \mu)$ which is given by the outer measure μ^* of μ. In the following we will always write μ instead of $\bar{\mu}$.

Proof (1). Follows immediately from A3.14(2) and A3.14(3). $\qquad \square$

Proof (2). On noting that $\mathcal{B}_1$ is a σ-algebra and that countable unions of μ-null sets are again μ-null sets (see A3.4), we have that $\mathcal{B}$ is a σ-algebra.

In order to show that $\bar{\mu}$ is additive, consider two disjoint sets $E_1, E_2 \in \mathcal{B}$. If $\bar{\mu}(E_1 \cup E_2) < \infty$, i.e. $\mathcal{X}_{E_1 \cup E_2} \in L(\mu; \mathbb{R})$, choose $E_1', E_2' \in \mathcal{B}_1$ for E_1 and E_2 as in the definition of $\mathcal{B}$. Then $\mathcal{X}_{E_1' \cup E_2'} = \mathcal{X}_{E_1 \cup E_2}$ in $L(\mu; \mathbb{R})$, and so $E_1' \cup E_2'$ is an integrable set. Now A3.14(4) yields that E_1' and E_2' are integrable sets, and hence so are E_1 and E_2. On noting that

$$\mathcal{X}_{E_1 \cup E_2} = \mathcal{X}_{E_1} + \mathcal{X}_{E_2}, \tag{A3-7}$$

the additivity of the integral yields that

$$\bar{\mu}(E_1 \cup E_2) = \bar{\mu}(E_1) + \bar{\mu}(E_2).$$ (A3-8)

If, on the other hand, $\bar{\mu}(E_1 \cup E_2) = \infty$, then $E_1 \cup E_2$ is not an integrable set, and so by (A3-7) the sets E_1 and E_2 cannot both be integrable, which again implies (A3-8).

It remains to show that $\bar{\mu}$ is σ-additive on $\mathcal{B}$. To this end, let $E_i \in \mathcal{B}$, $i \in \mathbb{N}$, be pairwise disjoint with $E := \bigcup_i E_i \in \mathcal{B}$. We first consider the case when all the sets E_i are integrable. The above established additivity of $\bar{\mu}$ yields for $k \in \mathbb{N}$ that

$$C_k := \sum_{i \le k} \bar{\mu}(E_i) = \bar{\mu}(\widetilde{E}_k) = \int_S \mathcal{X}_{\widetilde{E}_k} \, \mathrm{d}\mu \quad \text{with } \widetilde{E}_k := \bigcup_{i \le k} E_i,$$

and $\mathcal{X}_{\widetilde{E}_k} \nearrow \mathcal{X}_E$ as $k \to \infty$. If the C_k are bounded, then the monotone convergence theorem A3.12(3) yields that $\mathcal{X}_E \in L(\mu; \mathbb{R})$ and

$$\bar{\mu}(E) = \int_S \mathcal{X}_E \, \mathrm{d}\mu = \lim_{k \to \infty} \int_S \mathcal{X}_{\widetilde{E}_k} \, \mathrm{d}\mu = \sum_{i \in \mathbb{N}} \bar{\mu}(E_i).$$

If $C_k \to \infty$ as $k \to \infty$, then it follows from $\widetilde{E}_k \subset E$ and the monotonicity of the integral A3.12(1) that E is not integrable, and so

$$\bar{\mu}(E) = \infty = \sum_i \bar{\mu}(E_i).$$ (A3-9)

It remains to consider the case when one of the sets E_i is not integrable. As before we obtain that then E cannot be integrable and so (A3-9) holds trivially.   □

Proof (3). If N is a μ-null set, then $\mathcal{X}_N = \mathcal{X}_\emptyset = 0$ μ-almost everywhere, and so $\mathcal{X}_N = 0$ in $L(\mu; \mathbb{R})$, whence $N \in \mathcal{B}$ with $\bar{\mu}(N) = 0$. Conversely, if $N \in \mathcal{B}$ with $\bar{\mu}(N) = 0$, then

$$\|\mathcal{X}_N\|_{L(\mu)} = \int_S \mathcal{X}_N \, \mathrm{d}\mu = 0,$$

and hence the norm property (L4) yields that $\mathcal{X}_N = 0$ in $L(\mu; \mathbb{R})$, i.e. $\mathcal{X}_N = 0$ μ-almost everywhere, which means that N is a μ-null set.   □

Proof (4). Let $E \in \mathcal{B}_0$. If $\bar{\mu}(E) < \infty$, then it follows from (L3) that

$$\mu(E) = \mu^*(E) \le \int_S \mathcal{X}_E \, \mathrm{d}\mu = \bar{\mu}(E) < \infty.$$

If $\mu(E) < \infty$, then $\mathcal{X}_E \in T(\mu; \mathbb{R})$, and hence $\bar{\mu}(E) = \mu(E)$, recall (L1). This shows that $\bar{\mu}$ is an extension of μ. It follows from (2) that $\mathcal{B}$ is a σ-algebra and that $\bar{\mu}$ is σ-additive.

On recalling the definition of a measure space in 3.9, it remains to show that $(S, \mathcal{B}, \bar{\mu})$ is complete. But this follows from (3), since subsets of μ-null sets are again μ-null sets.   □

Properties of Lebesgue's integral

A3.16 Measurable functions (see 3.11). Let $(S, \mathcal{B}, \mu)$ be a measure space. A map $f : S \to Y$ is called μ-**measurable** if

(1) $U \subset Y$ open $\implies f^{-1}(U) \in \mathcal{B}$.
(2) There exists a μ-null set N such that $f(S \setminus N)$ is separable.

In the following we want to show that integrable functions $f : S \to Y$ are precisely those measurable functions for which $|f|$ is integrable. This result follows from A3.19 below, on setting $g = |f|$ in A3.19(2).

A3.17 Lemma. Let $f \in L(\mu; Y)$. Then the following is true:

(1) For $E \in \mathcal{B}$ it holds that $\mathcal{X}_E f \in L(\mu; Y)$.
(2) Define

$$\nu(E) := \int_E f \, d\mu := \int_S \mathcal{X}_E f \, d\mu \in Y \quad \text{for } E \in \mathcal{B}.$$

Then $\nu : \mathcal{B} \to Y$ is σ-additive and

$$|\nu(E)| \to 0 \quad \text{as } \mu(E) \to 0. \tag{A3-10}$$

Definition: In particular, if $Y = \mathbb{R}$ and $f = \mathcal{X}_A$ then

$$\nu(E) = \int_E \mathcal{X}_A \, d\mu = \mu(A \cap E) =: (\mu \llcorner A)(E)$$

is the measure μ restricted to A.

Proof (1). By (L5), we can choose step functions $f_k \in T(\mu; Y)$ such that $f_k \to f$ in $L(\mu; Y)$ as $k \to \infty$. For a given $E \in \mathcal{B}$, choose $E' \in \mathcal{B}_1$ as in A3.14(2). Then A3.14(4) yields that $\mathcal{X}_{E'} f_k \in L(\mu; Y)$, and hence also $\mathcal{X}_E f_k \in L(\mu; Y)$. It follows from

$$|\mathcal{X}_E f_k - \mathcal{X}_E f_l| \le |f_k - f_l|$$

that $(\mathcal{X}_E f_k)_{k \in \mathbb{N}}$ is a Cauchy sequence in $L(\mu; Y)$. Moreover, it follows from A3.11 that $f_k \to f$ μ-almost everywhere for a subsequence $k \to \infty$, and hence also $\mathcal{X}_E f_k \to \mathcal{X}_E f$. Then A3.12(2) yields that $\mathcal{X}_E f \in L(\mu; Y)$.   □

Proof (2). The additivity of ν follows from the additivity of the integral. Now let $E = \bigcup_{i \in \mathbb{N}} E_i$ with $E_i \subset E_{i+1}$. Then $\mathcal{X}_{E_i} f \to \mathcal{X}_E f$ pointwise as $i \to \infty$. Moreover,

$$\int_{E_i} |f| \, \mathrm{d}\mu \le \int_E |f| \, \mathrm{d}\mu < \infty$$

and the sequence of integrals is nondecreasing in i. Hence it holds for $i < j$ that

$$\int_S \left| \mathcal{X}_{E_j} f - \mathcal{X}_{E_i} f \right| \mathrm{d}\mu = \int_{E_j \setminus E_i} |f| \, \mathrm{d}\mu$$

$$= \int_{E_j} |f| \, \mathrm{d}\mu - \int_{E_i} |f| \, \mathrm{d}\mu \longrightarrow 0 \quad \text{as } i, j \to \infty,$$

i.e. $(\mathcal{X}_{E_i} f)_{i \in \mathbb{N}}$ is a Cauchy sequence in $L(\mu; Y)$. It then follows from A3.12(2) that $\mathcal{X}_{E_i} f \to \mathcal{X}_E f$ in $L(\mu; Y)$ and hence also $\nu(E_i) \to \nu(E)$ as $i \to \infty$. This shows that ν is σ-additive. For the proof of (A3-10) choose step functions

$$f_k = \sum_{i=1}^{n_k} \mathcal{X}_{E_{ki}} \alpha_{ki}$$

such that $\| f - f_k \|_{L(\mu)} \to 0$ as $k \to \infty$, recall (L5). Then we have that

$$|\nu(E)| \le \int_S |f - f_k| \, \mathrm{d}\mu + \int_S |f_k| \, \mathrm{d}\mu$$

$$\le \underbrace{\int_S |f - f_k| \, \mathrm{d}\mu}_{\to 0 \text{ as } k \to \infty} + \sum_{i=1}^{n_k} |\alpha_{ki}| \underbrace{\mu(E \cap E_{ki})}_{\substack{\to 0 \text{ as } \mu(E) \to 0 \\ \text{for every } k \text{ and } i}}.$$

$\square$

A3.18 Egorov's theorem. Let $\mu(S) < \infty$ and let $f_j, f : S \to Y$ be μ-measurable. Then the following are equivalent:

(1) $f_j \to f$ μ-almost everywhere as $j \to \infty$.
(2) $f_j \to f$ **μ-uniformly**, i.e. for $\varepsilon > 0$ there exists an $E_\varepsilon \in \mathcal{B}$ such that $\mu(S \setminus E_\varepsilon) \le \varepsilon$ and

$$f_j \to f \quad \text{uniformly on } E_\varepsilon \text{ as } j \to \infty.$$

Proof (2)$\Rightarrow$(1). Let $E := \bigcup_{i \in \mathbb{N}} E_{\frac{1}{i}}$. Then we have that $\mu(S \setminus E) = 0$ and $f_j(x) \to f(x)$ for $x \in E$. $\square$

Proof (1)$\Rightarrow$(2). Let $E \in \mathcal{B}$ with $\mu(S \setminus E) = 0$ such that for all $x \in E$ it holds that: $f_j(x) \to f(x)$ as $j \to \infty$. Consider the sets

$$E_{k,i} := \left\{ x \in E \, ; \, |f_j(x) - f(x)| < \tfrac{1}{i} \text{ for all } j \ge k \right\} \in \mathcal{B}.$$

On noting that, for every i,

$$E \subset \bigcup_k E_{k,i} \quad \text{with } E_{k,i} \subset E_{k+1,i} \, ,$$

we see that $\mu(E \setminus E_{k,i}) \to 0$ as $k \to \infty$. Hence, for a given $\varepsilon > 0$ there exists for all i a k_i such that

$$\mu(E \setminus E_{k_i,i}) \leq \varepsilon \cdot 2^{-i} \, .$$

Letting

$$E_\varepsilon := \bigcap_i E_{k_i,i} \, ,$$

we have $\mu(S \setminus E_\varepsilon) \leq \varepsilon$ and

$$\sup_{x \in E_\varepsilon} |f_j(x) - f(x)| < \tfrac{1}{i} \quad \text{for all } i \text{ and all } j \geq k_i,$$

i.e. f_j converges uniformly on E_ε to f as $j \to \infty$.   □

A3.19 Theorem.

(1) **Bochner's criterion.** If $f : S \to Y$, then

$$f \in L(\mu; Y) \quad \Longleftrightarrow \quad f \text{ is } \mu\text{-measurable and } |f| \in L(\mu; \mathbb{R}) \, .$$

(2) **Majorant criterion.** If $f : S \to Y$ and $g \in L(\mu; \mathbb{R})$, then

$$\begin{array}{l} f \text{ is } \mu\text{-measurable and} \\ |f| \leq g \ \mu\text{-almost everywhere} \end{array} \quad \Longrightarrow \quad f \in L(\mu; Y) \, .$$

Proof (1)$\Rightarrow$. It follows from (L5) that there exist $f_k \in T(\mu; Y)$ such that $\|f - f_k\|_{L(\mu)} \to 0$ as $k \to \infty$, and then A3.11 yields that $f_k \to f$ almost everywhere for a subsequence $k \to \infty$. As the f_k are measurable, the measurability of f follows from 3.12(3). Moreover, since $||f_k| - |f_l|| \leq |f_k - f_l|$ we have that $(|f_k|)_{k \in \mathbb{N}}$ is a Cauchy sequence in $L(\mu; \mathbb{R})$. As in addition $|f_k| \to |f|$ almost everywhere (for the above subsequence), it follows that $|f| \in L(\mu; \mathbb{R})$, recall A3.12(2).   □

Proof (1)$\Leftarrow$. This is the special case $g = |f|$ in (2).   □

Proof (2). We begin with the special case of a Euclidean image space $Y = \mathbb{R}^m$. For $k \in \mathbb{N}$ choose a cover of $\partial B_1(0) \subset \mathbb{R}^m$ consisting of disjoint nonempty Borel sets E_j, $j = 1, \ldots, j_k$, with diameter less than $\frac{1}{k}$ (use 4.7(7) and 4.6(3)). Choose $\alpha_j \in E_j$ and a null set N such that $|f(x)| \leq g(x)$ for $x \in S \setminus N$. Then, for $i = 1, \ldots, k$ and $j = 1, \ldots, j_k$, consider the disjoint sets

$$E_{i,j} := \left\{ x \in S \setminus N \, ; \, f(x) \neq 0 \, , \, \frac{i-1}{k} < \frac{|f(x)|}{g(x)} \leq \frac{i}{k} \, , \, \frac{f(x)}{|f(x)|} \in E_j \right\} \, .$$

We have from the assumptions and (1), respectively, that f and g are measurable, and so 3.12(2) yields that $E_{i,j} \in \mathcal{B}$. Let

$$f_k := \sum_{i,j} \frac{i-1}{k} \mathcal{X}_{E_{i,j}} g\alpha_j .$$

As $g \in L(\mu; \mathbb{R})$ it holds (trivially) that $g\alpha_j \in L(\mu; Y)$ and hence $f_k \in L(\mu; Y)$, thanks to A3.17(1). For $x \in E_{i,j}$ we have that

$$|f_k(x) - f(x)| \leq \left| \frac{i-1}{k} g(x) \left(\alpha_j - \frac{f(x)}{|f(x)|} \right) \right|$$

$$+ \left| \left(\frac{i-1}{k} - \frac{|f(x)|}{g(x)} \right) g(x) \frac{f(x)}{|f(x)|} \right| \leq g(x) \cdot \frac{1}{k} + \frac{1}{k} \cdot g(x) .$$

It follows that $|f_k - f| \leq \frac{2}{k} g$ almost everywhere. This yields that

$$\int_S |f_k - f_l| \, d\mu \leq \int_S \left(\frac{2}{k} + \frac{2}{l} \right) g \, d\mu \longrightarrow 0 \qquad \text{as } k, l \to \infty.$$

On noting that $f_k \to f$ almost everywhere, the desired result follows from A3.12(2).

Now we consider the case of an arbitrary Banach space Y, where we can assume that Y is separable. (Otherwise replace Y, on recalling 3.11, with $\text{clos}\left(\text{span}(f(S \setminus N))\right)$, where N is a null set such that $f(S \setminus N)$ is separable, and then set $f = 0$ on N.) Now if $\{\alpha_j; \, j \in \mathbb{N}\}$ is a dense subset of $\partial B_1(0) \subset Y$ (use 4.17(2)), then for every $k \in \mathbb{N}$

$$\partial B_1(0) \subset \bigcup_{j \in \mathbb{N}} B_{\frac{1}{k}}(\alpha_j) .$$

Then on letting

$$E_1 := B_{\frac{1}{k}}(\alpha_1) , \quad E_j := B_{\frac{1}{k}}(\alpha_j) \setminus \bigcup_{1 \leq l < j} E_l \quad \text{for } j > 1$$

we obtain a cover of $\partial B_1(0)$ consisting of disjoint Borel sets. (The same construction can be used for the Borel sets in the special case.) Now define the sets $E_{i,j}$ as above and set

$$f_{k,l} := \sum_{i=1}^{k} \sum_{j=1}^{l} \frac{i-1}{k} \mathcal{X}_{E_{i,j}} g\alpha_j , \quad \text{and} \quad A_{i,l} := \bigcup_{j > l} E_{i,j} .$$

As $A_{i,l+1} \subset A_{i,l}$ and $\bigcap_{l \in \mathbb{N}} A_{i,l} = \emptyset$, it holds for $l_1 < l_2$, recall A3.17, that

$$|f_{k,l_2} - f_{k,l_1}| \leq \sum_{i=1}^{k} \mathcal{X}_{A_{i,l_1}} g \longrightarrow 0 \quad \text{in } L(\mu; \mathbb{R}) \text{ as } l_1 \to \infty.$$

(Since $l_1 < l_2$, the property $l_1 \to \infty$ implies $l_2 \to \infty$.) Moreover, $f_{k,l}$ converges as $l \to \infty$ pointwise to

$$f_k := \sum_{i=1}^{k} \sum_{j=1}^{\infty} \frac{i-1}{k} \mathcal{X}_{E_{i,j}} g\alpha_j \,.$$

As $f_{k,l} \in L(\mu; Y)$, it follows from A3.12(2) that $f_k \in L(\mu; Y)$. The remainder of the proof proceeds as before. $\qquad \square$

We now prove Fatou's lemma and Lebesgue's convergence theorem, which will play a fundamental role in later proofs of results using Lebesgue spaces.

A3.20 Fatou's lemma. Let $f_j \in L(\mu; \mathbb{R})$ with $f_j \geq 0$ almost everywhere and

$$\liminf_{j \to \infty} \int_S f_j \, d\mu < \infty \,.$$

Then $\liminf_{j \to \infty} f_j \in L(\mu; \mathbb{R})$ and

$$\int_S \liminf_{j \to \infty} f_j \, d\mu \leq \liminf_{j \to \infty} \int_S f_j \, d\mu \,.$$

Proof. For $k \in \mathbb{N}$ it holds that almost everywhere

$$0 \leq g_k := \inf_{i \geq k} f_i \leq f_j \quad \text{for every } j \geq k \,.$$

It follows from 3.12(4) that g_k is measurable and hence, by A3.19(2), that g_k is integrable. Then A3.12(1) yields that

$$\int_S g_k \, d\mu \leq \int_S f_j \, d\mu \quad \text{for } j \geq k \,.$$

It follows that

$$\int_S g_k \, d\mu \leq \liminf_{j \to \infty} \int_S f_j \, d\mu < \infty \,.$$

In addition, the g_k are monotonically nondecreasing in k (almost everywhere) and hence, by A3.12(1), the same holds true for their integrals. This yields that these integrals converge, and so, for $k < l$,

$$\int_S |g_k - g_l| \, d\mu = \int_S g_l \, d\mu - \int_S g_k \, d\mu \longrightarrow 0 \quad \text{as } k, l \to \infty,$$

i.e. $(g_k)_{k \in \mathbb{N}}$ is a Cauchy sequence in $L(\mu; \mathbb{R})$. On noting that pointwise

$$\liminf_{j \to \infty} f_j = \lim_{k \to \infty} g_k \,,$$

the desired result follows from A3.12(2). $\qquad \square$

A3.21 Dominated convergence theorem (Lebesgue's convergence theorem). Let $f_j, f : S \to Y$ be μ-measurable and let $g \in L(\mu; \mathbb{R})$. If

$$|f_j| \le g \quad \mu\text{-almost everywhere} \quad \text{for all } j \in \mathbb{N},$$
$$f_j \to f \quad \mu\text{-almost everywhere} \quad \text{as } j \to \infty,$$

then $f_j, f \in L(\mu; Y)$ and

$$f_j \to f \quad \text{in } L(\mu; Y) \text{ as } j \to \infty.$$

Proof. It also holds that $|f| \le g$ almost everywhere. Then the majorant criterion A3.19(2) yields that $f_j, f \in L(\mu; Y)$. Define

$$g_j := g - \frac{1}{2}|f_j - f|.$$

It holds that $g_j \ge 0$ almost everywhere and (recall A3.12(1))

$$\int_S g_j \, d\mu \le \int_S g \, d\mu < \infty.$$

On noting that $\lim_{j \to \infty} g_j = g$ almost everywhere, it follows from Fatou's lemma that

$$\int_S g \, d\mu \le \liminf_{j \to \infty} \int_S g_j \, d\mu = \int_S g \, d\mu - \frac{1}{2} \limsup_{j \to \infty} \int_S |f_j - f| \, d\mu,$$

and hence

$$\limsup_{j \to \infty} \int_S |f_j - f| \, d\mu = 0.$$

$\square$

Two further essential theorems for the Lebesgue measure L^n in $\mathbb{R}^n$ are **Fubini's theorem** and the **change-of-variables theorem** for C^1-diffeomorphisms. These theorems are an elementary part of analysis and will not be presented here. However, their knowledge will be assumed from now on in this book. A proof of Fubini's theorem for regular measures is given in Appendix A6.

4 Subsets of function spaces

In this chapter, we consider subsets of the function spaces introduced in Chapter 3. Two fundamental properties of these subsets are convexity and compactness, which in applications are important. We first consider convex subsets (see 4.1–4.4), and in particular we prove the projection theorem in Hilbert spaces. Then we investigate compact subsets of metric spaces (see 4.6–4.16) and give a complete characterization of compact sets in C^0- and L^p-spaces (see 4.12 and 4.16). These characterizations are frequently used in applications, for example, to derive existence results for partial differential equations.

Convex subsets

It should be noted that the following definition only uses the vector space structure. Here we assume that the vector space is a $\mathbb{K}$-space.

4.1 Convex sets. Let X be a vector space over $\mathbb{K}$. For $A \subset X$ the **convex hull** of A is defined by

$$\operatorname{conv}(A) := \left\{ \sum_{i=1}^{k} a_i x_i \; ; \; k \in \mathbb{N}, \ x_i \in A, \ a_i \in \mathbb{R}, \ a_i \geq 0, \ \sum_{i=1}^{k} a_i = 1 \right\}.$$

The set A is called **convex** if $A = \operatorname{conv}(A)$, which is equivalent to

$$x, y \in A, \ a \in \mathbb{R}, \ 0 < a < 1 \quad \Longrightarrow \quad (1 - a)x + ay \in A$$

(see the example on the left-hand side of Fig. 4.1). For every set $A \subset X$ one has that $\operatorname{conv}(A)$ is convex and is the smallest convex set that contains A. Moreover, any intersection of convex sets is convex.

For examples of convex sets see E4.2.

4.2 Convex functions. If $A \subset X$ is convex, then $f : A \to \mathbb{R} \cup \{+\infty\}$ is called a **convex function** if

$$f\big((1 - a)x + ay\big) \leq (1 - a)f(x) + af(y)$$
$$\text{for all } x, y \in A \text{ and } a \in \mathbb{R} \text{ with } 0 \leq a \leq 1.$$

(Here the right-hand side is set to $+\infty$ whenever $f(x) = +\infty$ or $f(y) = +\infty$.)
Then, on setting $f(x) := +\infty$ for $x \in X \setminus A$, it holds that $f : X \to \mathbb{R} \cup \{+\infty\}$
is convex. A function $f : A \to \mathbb{R} \cup \{-\infty\}$ is called **concave** if $-f$ is a convex
function. If $f : X \to \mathbb{R} \cup \{+\infty\}$ is convex, then

$$\{x \in X ; \ f(x) \in \mathbb{R}\} \quad \text{is convex in } X$$

and (see the right-hand side of Fig. 4.1)

$$B := \{(x, \xi) \in X \times \mathbb{R} ; \ \xi \geq f(x)\} \quad \text{is convex in } X \times \mathbb{R}.$$

Conversely, every convex set $B \subset X \times \mathbb{R}$ defines a convex function $f : X \to \mathbb{R} \cup \{+\infty\}$ via

$$f(x) := \inf \left\{ \xi \in \mathbb{R} \cup \{+\infty\} ; \ (x, \xi) \in B \right\},$$

provided that for every x, the set over which the infimum is taken is either
empty (the infimum is then defined to be $+\infty$) or bounded from below.

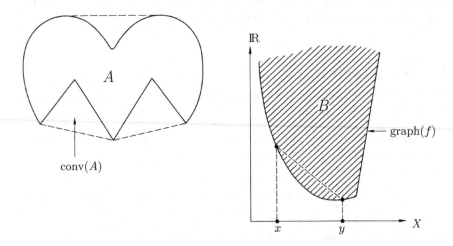

Fig. 4.1. *Convex sets*

One of the best known variational problems is the following: Given a
point $x \in X$ and a set $A \subset X$, find points $y \in A$ such that the distance
$y \mapsto \|x - y\|_X$ is minimal. We now prove that for closed convex sets A in a
Hilbert space X this variational problem admits a unique solution.

4.3 Projection theorem. Let X be a Hilbert space and let $A \subset X$ be
nonempty, closed and convex. Then there exists a unique map $P : X \to A$
such that

$$\|x - P(x)\|_X = \text{dist}(x, A) = \inf_{y \in A} \|x - y\|_X \quad \text{for all } x \in X. \tag{4-11}$$

For $x \in X$ an equivalent characterization of $P(x) \in A$ is given by

$$\operatorname{Re}\left(x - P(x), a - P(x)\right)_X \leq 0 \quad \text{for all } a \in A. \tag{4-12}$$

The map $P : X \to A$ is called the *(orthogonal) projection* from X to A.
Remark: We will use this theorem in 6.1 and in 9.17 and 9.18 for subspaces
$A \subset X$.

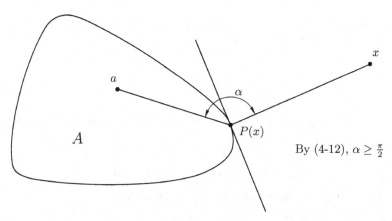

a

α

A

$P(x)$

By (4-12), $\alpha \geq \frac{\pi}{2}$

x

Fig. 4.2. *Orthogonal projection*

Proof. Throughout we write $\|\cdot\|$ in place of $\|\cdot\|_X$. For $x \in X$, on recalling
the definition of the distance, there exists a sequence $(a_k)_{k \in \mathbb{N}}$ in A such that

$$\|x - a_k\| \longrightarrow \operatorname{dist}(x, A) =: d \quad \text{as } k \to \infty.$$

Hence $(a_k)_{k \in \mathbb{N}}$ is also called a *minimal sequence*. Now it follows from the
parallelogram law in 2.2(4) that

$$\|(x - a_k) - (x - a_l)\|^2 + \|(x - a_k) + (x - a_l)\|^2$$
$$= 2\left(\|x - a_k\|^2 + \|x - a_l\|^2\right),$$

and so

$$\|a_l - a_k\|^2 = 2\left(\|x - a_k\|^2 + \|x - a_l\|^2 - 2\|x - \tfrac{1}{2}(a_k + a_l)\|^2\right).$$

As $\frac{1}{2}(a_k + a_l) \in A$ due to the convexity of A, this can be estimated by

$$\leq 2\left(\|x - a_k\|^2 + \|x - a_l\|^2 - 2d^2\right) \longrightarrow 0 \quad \text{as } k, l \to \infty. \tag{4-13}$$

Hence $(a_k)_{k \in \mathbb{N}}$ is a Cauchy sequence in X. Since X is complete and A is
closed, it follows that there exists

$$y := \lim_{k \to \infty} a_k \in A$$

and the continuity of the norm yields that

$$\|x - y\| = \lim_{k \to \infty} \|x - a_k\| = d.$$

If the same holds for $\widetilde{y} \in A$, then similarly to (4-13) it follows that

$$\|y - \widetilde{y}\|^2 \le 2(\|x - y\|^2 + \|x - \widetilde{y}\|^2 - 2d^2) = 0,$$

i.e. $y = \widetilde{y}$, and hence $P(x) := y$ is uniquely defined by (4-11).

Moreover, we have for $a \in A$ and $0 \le \varepsilon \le 1$, on noting that $(1 - \varepsilon)P(x) + \varepsilon a \in A$,

$$\|x - P(x)\|^2 = d^2 \le \|x - ((1 - \varepsilon)P(x) + \varepsilon a)\|^2$$
$$= \|(x - P(x)) - \varepsilon(a - P(x))\|^2$$
$$= \|x - P(x)\|^2 - 2\varepsilon \mathrm{Re}\,(x - P(x),\, a - P(x))_X + \mathcal{O}(\varepsilon^2),$$

which implies that

$$\mathrm{Re}\,(x - P(x),\, a - P(x))_X \le 0.$$

Conversely, if the above is satisfied, we conclude that

$$\|x - a\|^2 = \|x - P(x) + P(x) - a\|^2$$
$$= \|x - P(x)\|^2 + 2\mathrm{Re}\,(x - P(x),\, P(x) - a)_X + \|P(x) - a\|^2$$
$$\ge \|x - P(x)\|^2.$$

$\square$

4.4 Remark. In 4.3 one has the following:

(1) If $A \subset X$ is nonempty, closed and an **affine subspace**, i.e.

$$x, y \in A,\ \alpha \in \mathbb{K} \quad \Longrightarrow \quad (1 - \alpha)x + \alpha y \in A,$$

then P is **affine linear**, i.e.

$$P((1 - \alpha)x + \alpha y) = (1 - \alpha)P(x) + \alpha P(y) \quad \text{for all } x, y \in X,\ \alpha \in \mathbb{K}.$$

Moreover, for any given $a_0 \in A$ the point $P(x) \in A$ is characterized by

$$(x - P(x),\, a - a_0)_X = 0 \quad \text{for all } a \in A. \tag{4-14}$$

(2) If $A \subset X$ is nonempty, closed and a **subspace**, i.e.

$$x, y \in A,\ \alpha, \beta \in \mathbb{K} \quad \Longrightarrow \quad \alpha x + \beta y \in A,$$

then P is linear and the point $P(x) \in A$ is characterized by

$$x - P(x) \in A^\perp. \tag{4-15}$$

Proof (1). Let $x \in X$. For $a \in A$, $\alpha \in \mathbb{K}$ we have that $\tilde{a} := (1-\alpha)P(x)+\alpha a \in A$ with

$$(x - P(x), \tilde{a} - P(x))_X = \overline{\alpha}(x - P(x), a - P(x))_X,$$

and hence, by (4-12),

$$0 \geq \text{Re}(x - P(x), \tilde{a} - P(x))_X = \text{Re}\Big(\overline{\alpha}(x - P(x), a - P(x))_X\Big)$$

for all $\alpha \in \mathbb{K}$. This implies that

$$(x - P(x), a - P(x))_X = 0 \quad \text{for all } a \in A. \tag{4-16}$$

Subtracting the same equation with a_0 in place of a then yields (4-14). Conversely, if $P(x) \in A$ satisfies (4-14), then choosing $a_0 = P(x)$ yields (4-16) and hence (4-12).

Now let $x, y \in X$ and $\alpha \in \mathbb{K}$. It follows from the characterization of $P(x), P(y) \in A$ in (4-14), on setting $z := (1 - \alpha)x + \alpha y$, that

$$\big(z - ((1 - \alpha)P(x) + \alpha P(y)), a - a_0\big)_X = 0 \quad \text{for all } a \in A.$$

Now the characterization of $P(z)$ in (4-14) yields that $(1-\alpha)P(x)+\alpha P(y) = P(z)$, and hence P is affine linear. □

Proof (2). Setting $a_0 = 0$ in (4-14) yields $x - P(x) \in A^{\perp}$, and hence the characterization (4-15) of $P(x)$. Moreover, for $\alpha \in \mathbb{K}$ we have that $\alpha x - \alpha P(x) \in A^{\perp}$. Now the characterization of $P(\alpha x)$ in (4-15) gives $\alpha P(x) = P(\alpha x)$, which together with (1) implies the linearity of P. Conversely, (4-15) immediately yields (4-14). □

In Banach spaces, the norm in general does not attain the infimum over closed convex sets (see E4.3), but the infimum can be approximated to an arbitrary accuracy. In the case of a subspace this allows us to prove the following result.

4.5 Almost orthogonal element. Let X be a normed space and let $Y \subset X$ be a closed proper subspace. In addition, let $0 < \theta < 1$ (if X is a Hilbert space, then $\theta = 1$ is also allowed). Then there exists an $x_\theta \in X$ such that

$$\|x_\theta\|_X = 1 \quad \text{and} \quad \theta \leq \text{dist}(x_\theta, Y) \leq 1.$$

Proof. We write $\|\cdot\|$ in place of $\|\cdot\|_X$. Choose $x \in X \setminus Y$. Since Y is closed, $\text{dist}(x, Y) > 0$. Together with $\theta < 1$ this yields that there exists a $y_\theta \in Y$ such that

$$0 < \|x - y_\theta\| \leq \frac{1}{\theta}\text{dist}(x, Y).$$

Let

$$x_\theta := \frac{x - y_\theta}{\|x - y_\theta\|}.$$

Then for all $y \in Y$

$$\| x_\theta - y \| = \frac{1}{\| x - y_\theta \|} \| x - (y_\theta + \| x - y_\theta \| y) \| .$$

As $y_\theta + \| x - y_\theta \| y \in Y$, this can be estimated by

$$\geq \frac{\operatorname{dist}(x, Y)}{\| x - y_\theta \|} \geq \frac{\operatorname{dist}(x, Y)}{\frac{1}{\theta} \cdot \operatorname{dist}(x, Y)} = \theta .$$

This shows that $\operatorname{dist}(x_\theta, Y) \geq \theta$. In addition, it follows immediately from $0 \in Y$ that $\operatorname{dist}(x_\theta, Y) \leq \| x_\theta \| = 1$.

If X is a Hilbert space and $\theta = 1$, then set $y_1 = P(x)$, where P is the orthogonal projection onto Y from 4.3.   □

Compact subsets

A second class of subsets $A \subset X$ for which the above variational problem is solvable are compact subsets (see 4.11). Several possible notions of compactness are defined in the following theorem. The most general definition is the covering compactness, which can also be formulated in topological spaces. As we will show, in metric spaces this notion is equivalent to sequential compactness: This follows from the fact that then for each $x \in X$ there exists a countable neighbourhood basis, e.g. $\left(B_{1/k}(x) \right)_{k \in \mathbb{N}}$. Here we call a system $(U_i)_{i \in I}$ in a topological space $(X, \mathcal{T})$ a **neighbourhood basis** at the point $x \in X$ if

$$x \in U_i \in \mathcal{T} \text{ for all } i \in I ,$$
$$x \in U \in \mathcal{T} \quad \Longrightarrow \quad U_i \subset U \text{ for an } i \in I . \tag{4-17}$$

One of the most important results in metric spaces is:

4.6 Compactness. For every subset A of a metric space (X, d) the following are equivalent:

(1) A is **covering compact**, i.e.

> Every open cover of A contains a finite subcover.

(2) A is **sequentially compact**, i.e.

Every sequence in A contains a convergent subsequence with limit in A.

(3) (A, d) is complete and A is **precompact**, i.e.

For every $\varepsilon > 0$ there exists a finite cover of A consisting of ε-balls.

Definition: We call a subset $A \subset X$ of a metric space **compact** if A satisfies one of these three equivalent properties.

Proof (1)⟹(2). If the sequence $(x_k)_{k \in \mathbb{N}}$ in A does not have a cluster point in A, then for each $y \in A$ there exists an $r_y > 0$ such that

$$N_y := \{k \in \mathbb{N} \,; \, x_k \in B_{r_y}(y) \cap A\}$$

is finite. As the balls $\left(B_{r_y}(y)\right)_{y \in A}$ form an open cover of A, it follows from (1) that there exist finitely many points $y_1, \dots, y_n \in A$ such that

$$A \subset \bigcup_{i=1}^{n} B_{r_{y_i}}(y_i) \ .$$

Since $x_k \in A$ for all $k \in \mathbb{N}$, it would follow that $\mathbb{N} = \bigcup_{i=1}^{n} N_{y_i}$ is finite, a contradiction. □

Proof (2)⟹(3). First we prove the completeness. It follows from (2) that any Cauchy sequence in A has a cluster point in A. On the other hand, in general any Cauchy sequence can have at most one cluster point. This implies that the Cauchy sequence has a limit in A (see the remark in 2.21(3)). Hence (A, d) is complete.

Now we prove the precompactness. If for an $\varepsilon > 0$ there exists no finite ε-cover of A, then we can inductively find $x_k \in A$ such that

$$x_{k+1} \in A \setminus \bigcup_{i=1}^{k} B_\varepsilon(x_i) \ .$$

Then $(x_k)_{k \in \mathbb{N}}$ has no cluster point, which contradicts (2). □

Proof (3)⟹(1). Let $(U_i)_{i \in I}$ be an open cover of A, i.e. I is a set, $U_i \subset X$ are open for $i \in I$, and $A \subset \bigcup_{i \in I} U_i$. Let

$$\mathcal{B} := \left\{ B \subset A \,; \, J \subset I, \ B \subset \bigcup_{i \in J} U_i \implies J \text{ is infinite} \right\} .$$

We want to show that $A \notin \mathcal{B}$. It follows from the precompactness of A that:

$$B \in \mathcal{B} \text{ and } \varepsilon > 0 \implies \text{There exists a cover } A \subset \bigcup_{i=1}^{n_\varepsilon} B_\varepsilon(x_i)$$

$$\implies B_\varepsilon(x_i) \cap B \in \mathcal{B} \text{ for an } i \text{ (depending on } \varepsilon).$$

We now assume that $A \in \mathcal{B}$. Then it follows inductively for $k \in \mathbb{N}$ (set $\varepsilon = \frac{1}{k}$) that there exist points $x_k \in X$ and sets B_k with $B_1 := A$ and

$$B_k := B_{\frac{1}{k}}(x_k) \cap B_{k-1} \in \mathcal{B} \quad \text{for } k \geq 2.$$

Choose $y_k \in B_k$. Then for $k \leq l$ both y_k and y_l belong to $B_{\frac{1}{k}}(x_k)$, and so $d(y_k, y_l) \leq \frac{2}{k}$, which means that $(y_k)_{k \in \mathbb{N}}$ is a Cauchy sequence in A. As A is complete, there exists a $y \in A$ such that

$$\varepsilon_k := d(y_k, y) \to 0 \quad \text{as } k \to \infty.$$

On noting that $y \in U_{i_0}$ for some i_0, for k sufficiently large we have

$$B_k \subset B_{\frac{1}{k}}(x_k) \subset B_{\frac{2}{k}}(y_k) \subset B_{\frac{2}{k}+\varepsilon_k}(y) \subset U_{i_0},$$

i.e. $B_k \notin \mathcal{B}$, which is a contradiction. □

4.7 Remarks. Let (X, d) be a metric space. Then:

(1) Subsets of precompact sets are precompact.

(2) $A \subset X$ precompact $\implies$ A bounded.

(3) $A \subset X$ precompact $\implies$ $\overline{A}$ closed and precompact.

(4) $A \subset X$ compact $\implies$ A closed.

(5) If X is a complete metric space, then for $A \subset X$:

$$A \text{ precompact} \quad \Longleftrightarrow \quad \overline{A} \text{ compact.}$$

(6) If $X = \mathbb{K}^n$ as a normed space:

$$A \subset \mathbb{K}^n \text{ precompact} \quad \Longleftrightarrow \quad A \text{ bounded.}$$

(7) *Heine-Borel theorem.* If $X = \mathbb{K}^n$ as a normed space:

$$A \subset \mathbb{K}^n \text{ compact} \quad \Longleftrightarrow \quad A \text{ bounded and closed.}$$

(8) If $A, A_i \subset X$ and $\delta_i > 0$ for $i \in \mathbb{N}$ then:

$$\left.\begin{array}{l} A \subset B_{\delta_i}(A_i), \\ A_i \text{ precompact for } i \in \mathbb{N}, \\ \delta_i \to 0 \text{ as } i \to \infty \end{array}\right\} \quad \Longrightarrow \quad A \text{ precompact.}$$

Proof (1) *to* (4). Use the statements in 4.6(3). □

Proof (5)$\Leftarrow$. Follows from 4.6(3) and (1). □

Proof (5)$\Rightarrow$. It follows from (3) and E2.8(1) that 4.6(3) is satisfied for $\overline{A}$. □

Proof (6)$\Leftarrow$. We prove this with respect to the Euclidean norm on $\mathbb{K}^n$. Let $A \subset B_R(0)$. For $\mathbb{K} = \mathbb{R}$ it holds for all $m \in \mathbb{N}$

$$B_R(0) \subset \bigcup_{\substack{q \in \mathbb{Z}^n \\ |q|_\infty \le m}} B_{c_n \cdot \varepsilon}(\varepsilon q) \quad \text{where} \quad c_n = \sqrt{n}, \ \varepsilon := \frac{R}{m}.$$

(For $\mathbb{K} = \mathbb{C}$ the union has to be taken over $q = q_1 + iq_2$ with $q_k \in \mathbb{Z}^n$, $|q_k|_\infty \le m$ for $k = 1, 2$.) For the ∞-norm $x \mapsto |x|_\infty$ we can set $c_n = 1$. It follows from 4.8 that the claim is true for any norm on $\mathbb{K}^n$. □

Proof (7)$\Rightarrow$. As $\mathbb{K}^n$ is complete, this follows from (4), 4.6(3) and (2). $\square$

Proof (7)$\Leftarrow$. As $\mathbb{K}^n$ is complete, this follows from (6) and 4.6(3). $\square$

Proof (8). Let $\varepsilon > 0$. Choose $i \in \mathbb{N}$ with $\delta_i \leq \varepsilon$. As A_i is precompact, there exist finitely many points $x_1, \ldots, x_m \in X$ such that

$$A_i \subset \bigcup_{j=1}^m B_\varepsilon(x_j) , \quad \text{and hence} \quad A \subset B_{\delta_i}(A_i) \subset \bigcup_{j=1}^m B_{2\varepsilon}(x_j) .$$

$\square$

4.8 Lemma. If X is a finite-dimensional $\mathbb{K}$-vector space, then all the norms on X are pairwise equivalent.

Remark: Let $n \in \mathbb{N}$. Every n-dimensional $\mathbb{K}$-vector space X is linearly equivalent to $\mathbb{K}^n$, i.e. there exists a linear and bijective map from X to $\mathbb{K}^n$. If X is a normed space, this map is continuous in both directions.

In the infinite-dimensional case lemma 4.8 does not hold (see the theorem after E9.2).

Proof. Let $n := \dim X$ and let $\{e_1, \ldots, e_n\}$ be a basis of X, i.e. every $x \in X$ has a unique representation

$$x = \sum_{i=1}^n x_i e_i$$

(the coefficients x_i depend linearly on x, that is, $x \mapsto (x_1, \ldots, x_n)$ is a linear, and bijective, map from X to $\mathbb{K}^n$). Then

$$\|x\|_\infty := \max_{i=1,\ldots,n} |x_i| \tag{4-18}$$

defines a norm on X. Let $x \mapsto \|x\|$ be an arbitrary additional norm. The claim follows if we can show that these two norms can be bounded by each other as in 2.15(2). Now

$$\|x\| \leq \sum_{i=1}^n |x_i| \, \|e_i\| \leq \left(\sum_{i=1}^n \|e_i\| \right) \cdot \|x\|_\infty .$$

On the other hand, if we assume that the corresponding converse bound does not hold, then for each $\varepsilon > 0$ there exists an $x^\varepsilon \in X$ such that $\|x^\varepsilon\| < \varepsilon \|x^\varepsilon\|_\infty$, which means that $x^\varepsilon \neq 0$ and so we can assume that $\|x^\varepsilon\|_\infty = 1$ (otherwise consider $\frac{x^\varepsilon}{\|x^\varepsilon\|_\infty}$ in place of x^ε). Therefore

$$\|x^\varepsilon\| < \varepsilon \quad \text{and} \quad \|x^\varepsilon\|_\infty = 1 . \tag{4-19}$$

Hence there exist an i_0 and a countable subsequence $\varepsilon \to 0$ such that

$$\left|x_{i_0}^\varepsilon\right| = 1. \tag{4-20}$$

Moreover, for every i the set $\{x_i^\varepsilon \; ; \; \varepsilon > 0\}$ is bounded, and hence precompact in $\mathbb{K}$ (see 4.7(6)). That means that we can choose a further subsequence $\varepsilon \to 0$ such that, for $i = 1, \ldots, n$,

$$x_i^\varepsilon \to \xi_i \quad \text{as } \varepsilon \to 0 \tag{4-21}$$

with certain numbers $\xi_i \in \mathbb{K}$. Let

$$x := \sum_{i=1}^n \xi_i e_i \, .$$

Then it follows from (4-19) and (4-21) that

$$\|x\| \le \|x^\varepsilon\| + \|x - x^\varepsilon\| \le \varepsilon + \left(\sum_{i=1}^n \|e_i\| \right) \cdot \max_{i=1,\ldots,n} |\xi_i - x_i^\varepsilon| \longrightarrow 0$$

as $\varepsilon \to 0$. Hence, $x = 0$, i.e. $\xi_i = 0$ for $i = 1, \ldots, n$. On the other hand, it follows from (4-20) that $|\xi_{i_0}| = 1$, a contradiction. □

As an example we have seen in the Introduction that the C^0-norm and the L^2-norm are not equivalent on $X = C^0([-1, 1])$. In particular, X is not complete with respect to the latter norm (see also E7.3). For finite-dimensional spaces we obtain the following conclusion from 4.8:

4.9 Lemma. Every finite-dimensional subspace of a normed space is complete and hence a closed subspace.

Proof. Let X be a normed space with norm $\|\cdot\|_X$. Let $\{e_1, \ldots, e_n\}$ be a basis of a subspace $Y \subset X$ with $\dim Y = n$. It follows from 4.8 that $\|\cdot\|_X$ and $\|\cdot\|_\infty$, defined by

$$\|x\|_\infty := \max_i |x_i|, \quad \text{where} \quad x = \sum_{i=1}^n x_i e_i,$$

are equivalent norms on Y. Therefore, if $\left(x^k\right)_{k \in \mathbb{N}}$ is a Cauchy sequence in Y and

$$x^k = \sum_{i=1}^n x_i^k e_i,$$

then for every $i \in \{1, \ldots, n\}$ the sequence $\left(x_i^k\right)_{k \in \mathbb{N}}$ is a Cauchy sequence in $\mathbb{K}$ and hence has a limit $\xi_i \in \mathbb{K}$. It follows that with respect to the $\|\cdot\|_\infty$-norm

$$x^k = \sum_{i=1}^n x_i^k e_i \longrightarrow \sum_{i=1}^n \xi_i e_i \in Y \quad \text{as } i \to \infty$$

and then, by 4.8, also with respect to $\|\cdot\|_X$. Hence Y is complete. The closedness of Y then follows from E2.8(2). □

4.10 Theorem. For every normed space X it holds that:

$$\overline{B_1(0)} \text{ compact} \quad \Longleftrightarrow \quad \dim X < \infty .$$

Remark: In the above we can replace the closed unit ball with an arbitrary closed ball $\overline{B_R(x)} \subset X$. Consequently, the assertions 4.7(6) and 4.7(7) hold in every finite-dimensional normed space, and they are independent of the choice of norm.

Proof $\Rightarrow$. It follows from 4.6(3) that there exists a cover $\overline{B_1(0)} \subset \bigcup_{j=1}^{n} B_{\frac{1}{2}}(y_j)$. Let $Y := \text{span}\{y_j ;\ j = 1, \ldots, n\}$. By 4.9, Y is closed in X. If we assume that $Y \subset X$ is a proper subspace, then 4.5 yields that for every $0 < \theta < 1$ there exists an $x_\theta \in X$ with $\|x_\theta\|_X = 1$ and $\text{dist}(x_\theta, Y) \geq \theta$. In addition, there exists a j such that $x_\theta \in B_{\frac{1}{2}}(y_j)$, i.e.

$$\text{dist}(x_\theta, Y) \leq \|x_\theta - y_j\|_X < \tfrac{1}{2} ,$$

which is a contradiction, as we can choose $\theta \geq \frac{1}{2}$. $\qquad\square$

Proof $\Leftarrow$. We have that $\overline{B_1(0)} \subset X$ is closed and, on noting 4.9, that X is complete. Hence, by 4.7(5), we need to show that $A := B_1(0)$ is precompact. The set B is bounded with respect to the X-norm, and 4.8 yields that it is bounded with respect to the ∞-norm as defined in 4.8.

Thus we have to show that with respect to the ∞-norm the bounded set B is precompact. Now B can be covered by finitely many balls of the form (the balls are chosen with respect to the ∞-norm)

$$B_\varepsilon(\varepsilon z_q) \quad \text{with} \quad z_q = \textstyle\sum_{j=1}^{n} q_j e_j , \quad q = (q_1, \ldots, q_n) \in \mathbb{Z}^n ,$$

see the proof of 4.7(6) for $\mathbb{K} = \mathbb{R}$, and similar balls for $\mathbb{K} = \mathbb{C}$. $\qquad\square$

For compact sets the variational problem from the beginning of this chapter always has a solution:

4.11 Lemma. Let (X, d) be a metric space and let $A \subset X$ be compact. Then for $x \in X$ there exists an $a \in A$ such that

$$d(x, a) = \text{dist}(x, A) .$$

Remark: In general a is not unique, for instance for the $|\cdot|_\infty$-norm in $X = \mathbb{R}^2$ and $A = \{x \in \mathbb{R}^2 ;\ |x|_\infty \leq 1\}$.

Proof. Choose a **minimal sequence** $(a_k)_{k \in \mathbb{N}}$ in A, i.e. $a_k \in A$ such that $d(x, a_k) \to \text{dist}(x, A)$ as $k \to \infty$. As A is compact, there exist a subsequence $(a_{k_i})_{i \in \mathbb{N}}$ and an $a \in A$ such that $a_{k_i} \to a$ as $i \to \infty$, and so $d(x, a_{k_i}) \to d(x, a)$. Hence $d(x, a) = \text{dist}(x, A)$. $\qquad\square$

Compact sets of function spaces

We have seen that in finite-dimensional vector spaces the compact subsets are precisely the bounded closed sets. A characterization of compacts sets is also possible in function spaces. We consider this for the standard spaces C^0 and L^p in the case where the image space Y is finite-dimensional. As usual, $y \mapsto |y|$ denotes the norm of Y.

4.12 Arzelà-Ascoli theorem. Let $S \subset \mathbb{R}^n$ be compact and $A \subset C^0(S; Y)$, where Y is finite-dimensional. Then:

$$A \text{ is precompact} \quad \Longleftrightarrow \quad A \text{ is bounded and equicontinuous.}$$

The set A is called **equicontinuous** if

(1) $\sup\limits_{f \in A} \sup\limits_{x \in S} |f(x)| < \infty$,

(2) $\sup\limits_{f \in A} |f(x) - f(y)| \longrightarrow 0 \quad$ for $x, y \in S$ with $|x - y| \to 0$.

Example: Bounded sets $A \subset C^{0,\alpha}(S; \mathbb{K}^m)$ considered as subsets in $C^0(S; \mathbb{K}^m)$ are bounded and equicontinuous, and hence precompact sets $A \subset C^0(S; \mathbb{K}^m)$ (see also E4.15).

Remark: The following proof immediately carries over to compact metric spaces (S, d), upon replacing $|x - y|$ by $d(x, y)$.

Warning: The theorem does not hold for $C^0(S; Y)$, if Y is an infinite-dimensional Banach space. Then an additional condition is needed, which guarantees that for $f \in A$ the image $f(S) \subset Y$ is precompact (see also remark 5.4(2)).

Proof $\Rightarrow$. The precompactness of A yields that for $\varepsilon > 0$ there exists a cover $A \subset \bigcup_{i=1}^{n_\varepsilon} \mathrm{B}_\varepsilon(f_i^\varepsilon)$. For $f \in A$ it then holds that $f \in \mathrm{B}_\varepsilon(f_{i_0}^\varepsilon)$ for some i_0, and hence

$$\|f\|_{\sup} \le \varepsilon + \|f_{i_0}^\varepsilon\|_{\sup} \le \varepsilon + \max_{i=1,\dots,n_\varepsilon} \|f_i^\varepsilon\|_{\sup} < \infty,$$

and similarly

$$|f(x) - f(y)| \le 2\varepsilon + \max_{i=1,\dots,n_\varepsilon} |f_i^\varepsilon(x) - f_i^\varepsilon(y)|,$$

where, since functions in $C^0(S; Y)$ are uniformly continuous, the second term becomes arbitrarily small on choosing $|x - y|$ sufficiently small. This yields the desired result. $\qquad\square$

Proof $\Leftarrow$. We use the Heine-Borel theorem 4.7(7) and that Y is finite-dimensional (see 4.8). Let

$$R := \sup_{f \in A} \sup_{x \in S} |f(x)|.$$

For a given $\varepsilon > 0$, choose covers

$$\overline{B_R(0)} \subset \bigcup_{i=1}^{k} B_\varepsilon(\xi_i) \quad \text{and} \quad S \subset \bigcup_{j=1}^{l} B_\varepsilon(x_j)$$

with $\xi_i \in Y$, $i = 1,\ldots,k$ and $x_j \in \mathbb{R}^n$, $j = 1,\ldots,l$, where k and l depend on ε. For mappings $\sigma : \{1,\ldots,l\} \to \{1,\ldots,k\}$ we define

$$A_\sigma := \left\{ f \in A ; \left| f(x_j) - \xi_{\sigma(j)} \right| < \varepsilon \text{ for } j = 1,\ldots,l \right\} .$$

For each σ such that A_σ is nonempty, choose an $f_\sigma \in A_\sigma$.

Now for every function $f \in A$ there exists a σ such that $f \in A_\sigma$. If $x \in S$, then $x \in B_\varepsilon(x_j)$ for some j, and so

$$|f(x) - f_\sigma(x)| \le |f(x) - f(x_j)| + |f_\sigma(x) - f_\sigma(x_j)|$$
$$+ \left| f(x_j) - \xi_{\sigma(j)} \right| + \left| f_\sigma(x_j) - \xi_{\sigma(j)} \right|$$
$$< 2 \sup_{|y-z|\le\varepsilon} \sup_{g\in A} |g(y) - g(z)| + 2\varepsilon =: r_\varepsilon ,$$

and hence $\| f - f_\sigma \|_{\sup} \le r_\varepsilon$. This shows that

$$A \subset \bigcup_{\sigma \,:\, A_\sigma \neq \emptyset} B_{2r_\varepsilon}(f_\sigma) ,$$

where the number of balls over which the union is taken depends on ε. As A is equicontinuous, we have that $r_\varepsilon \to 0$ as $\varepsilon \to 0$, which yields the precompactness of A. $\qquad\square$

The characterization of compact sets in L^p-spaces (see 4.16) will be obtained with the help of the Arzelà-Ascoli theorem. To this end, we approximate L^p-functions with smooth functions by means of convolutions. In the proofs we will make use of Fubini's theorem, which we assume to be known (see the remark at the end of Appendix A3).

4.13 Convolution. Let $\varphi \in L^1(\mathbb{R}^n)$ and $1 \le p \le \infty$. Let Y be a Banach space.

(1) If $f : \mathbb{R}^n \times \mathbb{R}^n \to Y$ is Lebesgue measurable, then

$$F(x) := \int_{\mathbb{R}^n} \varphi(x - y) f(x, y) \, dy = \int_{\mathbb{R}^n} \varphi(y) f(x, x - y) \, dy \qquad (4\text{-}22)$$

defines a function $F \in L^p(\mathbb{R}^n; Y)$ with

$$\| F \|_{L^p} \le \| \varphi \|_{L^1} \cdot \sup_{h\in\operatorname{supp}(\varphi)} \| f(\cdot + h, \cdot) \|_{L^p} , \qquad (4\text{-}23)$$

provided that the supremum on the right-hand side of this estimate exists and is finite.

Notation: Here $f(\cdot + h, \cdot)$ denotes the function $x \mapsto f(x + h, x)$.

(2) In the following let $f(x, y)$ be independent of x, i.e. we consider a function $y \mapsto f(y) \in Y$. Then if $f \in L^p(\mathbb{R}^n; Y)$ the rule

$$(\varphi * f)(x) := \int_{\mathbb{R}^n} \varphi(x - y) f(y) \, dy = F(x)$$

defines a function $\varphi * f \in L^p(\mathbb{R}^n; Y)$, the **convolution** of φ and f. The above estimate then becomes the **convolution estimate**

$$\|\varphi * f\|_{L^p} \le \|\varphi\|_{L^1} \cdot \|f\|_{L^p}. \tag{4-24}$$

(3) It is $\mathrm{supp}(\varphi * f) \subset \mathrm{clos}\,(\{x + y; \ x \in \mathrm{supp}(\varphi), \ y \in \mathrm{supp}(f)\})$.

(4) If in addition $\varphi \in C_0^\infty(\mathbb{R}^n)$, it follows that $\varphi * f \in C^\infty(\mathbb{R}^n; Y)$, and the partial derivatives for multi-indices s are given by

$$\partial^s(\varphi * f) = (\partial^s \varphi) * f.$$

(5) It is $L^1(\mathbb{R}^n) = L^1(\mathbb{R}^n; \mathbb{K})$ is a commutative Banach algebra with the convolution as product.

Proof (1). We first assume that all of the following integrals exist. Then

$$|F(x)| \le \int_{\mathbb{R}^n} |\varphi(y)| \cdot |f(x, x - y)| \, dy.$$

For $p = \infty$ the claim follows immediately on noting that $\|f(\cdot, \cdot - h)\|_{L^p} = \|f(\cdot + h, \cdot)\|_{L^p}$. For $p < \infty$ we have that

$$\int_{\mathbb{R}^n} |F(x)|^p \, dx \le \int_{\mathbb{R}^n} \left(\int_{\mathbb{R}^n} |\varphi(y)| \cdot |f(x, x - y)| \, dy \right)^p dx.$$

Fubini's theorem yields for $p = 1$ that this is

$$= \int_{\mathbb{R}^n} |\varphi(y)| \left(\int_{\mathbb{R}^n} |f(x, x - y)| \, dx \right) dy,$$

which again yields the claim. Now let $1 < p < \infty$ and $\frac{1}{p} + \frac{1}{p'} = 1$. Then it follows from the Hölder inequality and Fubini's theorem that

$$\int_{\mathbb{R}^n} |F(x)|^p \, dx \le \int_{\mathbb{R}^n} \left(\int_{\mathbb{R}^n} |\varphi(y)|^{\frac{1}{p'}} |\varphi(y)|^{\frac{1}{p}} |f(x, x - y)| \, dy \right)^p dx$$

$$\le \int_{\mathbb{R}^n} \left(\left(\int_{\mathbb{R}^n} |\varphi(y)| \, dy \right)^{\frac{p}{p'}} \int_{\mathbb{R}^n} |\varphi(y)| |f(x, x - y)|^p \, dy \right) dx$$

$$= \|\varphi\|_{L^1}^{\frac{p}{p'}} \int_{\mathbb{R}^n} |\varphi(y)| \left(\int_{\mathbb{R}^n} |f(x, x - y)|^p \, dx \right) dy$$

$$\le \|\varphi\|_{L^1}^{\frac{p}{p'}+1} \cdot \sup_{y \in \mathrm{supp}(\varphi)} \int_{\mathbb{R}^n} |f(x, x - y)|^p \, dx,$$

which, since $\frac{p}{p'}+1 = p$, yields the desired result. The existence of the integrals can now be justified retrospectively, which yields in particular that $F(x)$ is well defined by (4-22) for almost all x.

In detail: The assumptions yield that $(x,y) \mapsto f(x, x-y)$ is Lebesgue measurable on $\mathbb{R}^n \times \mathbb{R}^n$, that the functions $x \mapsto f(x, x-y)$ are in $L^p(\mathbb{R}^n; Y)$ and that the supremum of the L^p-norms is finite. In the last inequality we apply the majorant criterion and Tonelli's theorem (a converse of Fubini's theorem), which yields that $(x,y) \mapsto |\varphi(y)| \cdot |f(x, x-y)|^p$ is Lebesgue integrable on $\mathbb{R}^n \times \mathbb{R}^n$. In the second inequality we apply for almost all x the Hölder inequality, and so $y \mapsto |\varphi(y)| \cdot |f(x, x-y)|$ is for such x in $L^1(\mathbb{R}^n)$. Integration over x yields, on using the majorant criterion, that $x \mapsto \int_{\mathbb{R}^n} |\varphi(y)| \cdot |f(x, x-y)|\, dy$ is in $L^p(\mathbb{R}^n)$. Analogously, the first inequality then shows that $y \mapsto \varphi(y) \cdot f(x, x-y)$ is for x as above in $L^1(\mathbb{R}^n; Y)$, and so $F(x)$ is well defined, and moreover $x \mapsto F(x)$ belongs to $L^p(\mathbb{R}^n; Y)$. □

Proof (4). If $\varphi \in C_0^0(\mathbb{R}^n)$, then $\varphi(x - \cdot) - \varphi(x_0 - \cdot) \to 0$ converges uniformly. Moreover, if $f \in L^p(\mathbb{R}^n; Y)$ it follows from Lebesgue's convergence theorem that

$$F(x) - F(x_0) = \int_{\mathbb{R}^n} \big(\varphi(x - y) - \varphi(x_0 - y)\big) f(y)\, dy \longrightarrow 0 \quad \text{as } x \to x_0.$$

If $\varphi \in C_0^1(\mathbb{R}^n)$, it holds for unit vectors $e \in \mathbb{R}^n$ and real numbers $h \neq 0$ that

$$\frac{1}{h}\big(F(x_0 + he) - F(x_0)\big) = \int_{\mathbb{R}^n} \frac{1}{h}\big(\varphi(x_0 + he - y) - \varphi(x_0 - y)\big) f(y)\, dy.$$

Since $\frac{1}{h}\big(\varphi(x_0 + he - \cdot) - \varphi(x_0 - \cdot)\big)$ converges uniformly to $\partial_e \varphi(x_0 - \cdot)$ as $h \to 0$, it follows once again from Lebesgue's convergence theorem that $\partial_e F(x_0)$ exists and that

$$\partial_e F(x_0) = \int_{\mathbb{R}^n} \partial_e \varphi(x_0 - y) f(y)\, dy.$$

Hence we have shown that $\partial_e(\varphi * f)(x) = \big((\partial_e \varphi) * f\big)(x)$. The result for higher derivatives then follows inductively. □

Proof (5). The commutativity follows from (4-22). The inequality (2-9) is the convolution estimate (4-24) for $p = 1$. Similarly to the proof above, the associativity $(f_1 * f_2) * f_3 = f_1 * (f_2 * f_3)$ for $f_1, f_2, f_3 \in L^1(\mathbb{R}^n)$ can be proved with Fubini's theorem. □

We will now show that $\varphi_k * f \to f$ in $L^p(\mathbb{R}^n; Y)$ as $k \to \infty$, if φ_k are nonnegative functions with integral 1 (i.e. they are **probability densities**) with the property that the support of φ_k shrinks to $\{0\}$ as $k \to \infty$ (see also 5.18(5)).

4.14 Dirac sequence.

(1) A sequence $(\varphi_k)_{k\in\mathbb{N}}$ in $L^1(\mathbb{R}^n)$ is called a **(general) Dirac sequence** if

$$\varphi_k \geq 0, \quad \int_{\mathbb{R}^n} \varphi_k \, dL^n = 1, \quad \int_{\mathbb{R}^n\setminus B_\varrho(0)} \varphi_k \, dL^n \to 0 \quad \text{as } k \to \infty$$

for every $\varrho > 0$. The last condition is, for instance, satisfied if $\mathrm{supp}(\varphi_k) \subset B_{\varrho_k}(0)$ with $\varrho_k \to 0$ as $k \to \infty$.

(2) Let $\varphi \in L^1(\mathbb{R}^n)$ with

$$\varphi \geq 0 \quad \text{and} \quad \int_{\mathbb{R}^n} \varphi \, dL^n = 1.$$

On defining for $\varepsilon > 0$

$$\varphi_\varepsilon(x) := \varepsilon^{-n}\varphi\left(\frac{x}{\varepsilon}\right),$$

it holds for every $\varrho > 0$ that

$$\int_{\mathbb{R}^n} \varphi_\varepsilon \, dL^n = 1 \quad \text{and} \quad \int_{\mathbb{R}^n\setminus B_\varrho(0)} \varphi_\varepsilon \, dL^n \to 0 \quad \text{as } \varepsilon \to 0.$$

This implies that for every null sequence $(\varepsilon_k)_{k\in\mathbb{N}}$, the sequence $(\varphi_{\varepsilon_k})_{k\in\mathbb{N}}$ defines a general Dirac sequence in the sense of (1). Accordingly, we call the family of functions $(\varphi_\varepsilon)_{\varepsilon>0}$ a **Dirac sequence** for φ.

(3) In applications one often chooses in (2) a function $\varphi \in C_0^\infty(B_1(0))$ (extended to $\mathbb{R}^n$ by 0), so that $\mathrm{supp}(\varphi_\varepsilon) \subset B_\varepsilon(0)$. We then also call $(\varphi_\varepsilon)_{\varepsilon>0}$ a **standard Dirac sequence**.

Notation: Here we observe that $(\varphi_\varepsilon)_{\varepsilon>0}$ is an abbreviation for $(\varphi_\varepsilon)_{\varepsilon\in]0,\infty[}$ (see the note in 2.18), as we are dealing with the map $\varepsilon \mapsto \varphi_\varepsilon$ from $]0,\infty[$ to $L^1(\mathbb{R}^n)$.

With the help of Dirac sequences, we can prove the following frequently used approximation results for functions in $L^p(\mathbb{R}^n; Y)$.

4.15 Theorem. Let $1 \leq p < \infty$.

(1) If $f \in L^p(\mathbb{R}^n; Y)$, then

$$\|f(\,\cdot\,+h) - f\|_{L^p(\mathbb{R}^n)} \to 0 \quad \text{for } h \in \mathbb{R}^n \text{ with } |h| \to 0.$$

Here $f(\,\cdot\,+h)$ denotes the function $x \mapsto f(x+h)$.

(2) If $f \in L^p(\mathbb{R}^n; Y)$ and $(\varphi_k)_{k\in\mathbb{N}}$ is a Dirac sequence, then

$$\varphi_k * f \to f \quad \text{in } L^p(\mathbb{R}^n; Y) \text{ as } k \to \infty.$$

(3) If $\Omega \subset \mathbb{R}^n$ is open, then $C_0^\infty(\Omega; Y)$ is dense in $L^p(\Omega; Y)$.

Proof (1). By 3.26(2), we can choose $f_j \in C_0^0(\mathbb{R}^n; Y)$ with $\|f - f_j\|_{L^p} \to 0$ as $j \to \infty$. Then

$$\|f(\cdot + h) - f\|_{L^p}$$
$$\leq \|f(\cdot + h) - f_j(\cdot + h)\|_{L^p} + \|f - f_j\|_{L^p} + \|f_j(\cdot + h) - f_j\|_{L^p}$$
$$\leq 2\|f - f_j\|_{L^p} + L^n(\text{supp}(f_j(\cdot + h) - f_j))^{\frac{1}{p}} \|f_j(\cdot + h) - f_j\|_{\sup}.$$

For every j the function f_j is uniformly continuous, and so the second term converges to zero as $h \to 0$. The first term converges to zero as $j \to \infty$.   □

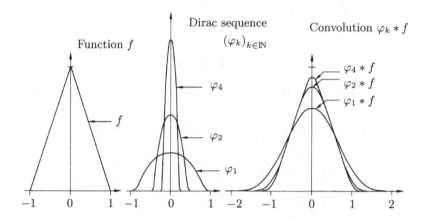

Fig. 4.3. *Convolution with a Dirac sequence*

Proof (2). Since the integrals of the φ_k are normalized,

$$(\varphi_k * f)(x) - f(x) = (\varphi_k * (f - f(x)))(x)$$
$$= \int_{\mathbb{R}^n} \varphi_k(x - y)(f(y) - f(x)) \, dy.$$

If we decompose φ_k for $\delta > 0$ into $\varphi_{k\delta} := \mathcal{X}_{B_\delta(0)} \cdot \varphi_k$ and $\psi_{k\delta} := \mathcal{X}_{\mathbb{R}^n \setminus B_\delta(0)} \cdot \varphi_k$, then it follows from 4.13(1) that

$$\|\varphi_k * f - f\|_{L^p(\mathbb{R}^n)} \leq \underbrace{\left(\int_{\mathbb{R}^n} \varphi_{k\delta} \, dL^n \right)}_{\leq 1} \cdot \underbrace{\sup_{|h| \leq \delta} \|f - f(\cdot + h)\|_{L^p(\mathbb{R}^n)}}_{\to 0 \text{ as } \delta \to 0, \text{ recall } (1)}$$

$$+ \underbrace{\left(\int_{\mathbb{R}^n} \psi_{k\delta} \, dL^n \right)}_{\substack{\to 0 \text{ as } k \to \infty \\ \text{for every } \delta}} \cdot \underbrace{\sup_{h \in \mathbb{R}^n} \|f - f(\cdot + h)\|_{L^p(\mathbb{R}^n)}}_{\leq 2\|f\|_{L^p(\mathbb{R}^n)}},$$

which proves (2).   □

Proof (3). We extend f to $\mathbb{R}^n \setminus \Omega$ by 0. If we choose $(\varphi_\varepsilon)_{\varepsilon>0}$ to be a standard Dirac sequence as in 4.14(3), then $\varphi_\varepsilon * f \in C^\infty(\mathbb{R}^n)$ by 4.13(4), but in general they do not have a compact support in Ω. That is why we cut off f on sets D_δ, which have a positive distance to $\partial\Omega$ and in addition are bounded: For $\delta > 0$ consider the sets

$$\Omega_\delta := \{x \in \mathbb{R}^n ; \, B_\delta(x) \subset \Omega\} \subset \Omega, \quad D_\delta := \Omega_\delta \cap B_{\frac{1}{\delta}}(0) , \qquad (4\text{-}25)$$

and define

$$(T_{\varepsilon,\delta}f)(x) := \int_{D_\delta} \varphi_\varepsilon(x - y)f(y)\, dy = \big(\varphi_\varepsilon * (\mathcal{X}_{D_\delta}f)\big)(x) \quad \text{for } x \in \mathbb{R}^n.$$

By 4.13(3) and 4.13(4),

$$T_{\varepsilon,\delta}f \in C_0^\infty(B_\varepsilon(D_\delta)) \subset C_0^\infty(\Omega)$$

for $\varepsilon \le \delta$, as $B_\delta(D_\delta) \subset \Omega$, and

$$(T_{\varepsilon,\delta}f - f)(x) = \int_{\mathbb{R}^n} \varphi_\varepsilon(x - y)\,(f(y) - f(x))\, dy$$
$$- \int_{\mathbb{R}^n} \varphi_\varepsilon(x - y)\mathcal{X}_{\Omega \setminus D_\delta}(y)f(y)\, dy .$$

It follows from 4.13(1) that

$$\|T_{\varepsilon,\delta}f - f\|_{L^p(\Omega)} \le \underbrace{\sup_{|h| \le \varepsilon} \|f(\bullet + h) - f\|_{L^p(\mathbb{R}^n)}}_{\to\, 0 \text{ as } \varepsilon \to 0, \text{ recall } (1)} + \underbrace{\|f\|_{L^p(\Omega \setminus D_\delta)}}_{\to\, 0 \text{ as } \delta \to 0} ,$$

which yields the desired result.   □

With the help of approximation by convolution, we will now prove a characterization of precompact subsets in $L^p(\mathbb{R}^n)$, which is very effective in applications, and which was originally proved by M. Riesz [MRiesz]. A further characterization, given by Fréchet and Kolmogorov, is the approximation of precompact sets by finite-dimensional ones (see [DunfordSchwartz: IV 8.18]).

4.16 Theorem (M. Riesz). Let $1 \le p < \infty$ and Y be finite-dimensional. Then $A \subset L^p(\mathbb{R}^n; Y)$ is precompact if and only if

(1) $\sup\limits_{f \in A} \|f\|_{L^p(\mathbb{R}^n)} < \infty$,

(2) $\sup\limits_{f \in A} \|f(\bullet + h) - f\|_{L^p(\mathbb{R}^n)} \longrightarrow 0 \quad$ for $h \in \mathbb{R}^n$ with $|h| \to 0$,

(3) $\sup\limits_{f \in A} \|f\|_{L^p(\mathbb{R}^n \setminus B_R(0))} \longrightarrow 0 \quad$ as $R \nearrow \infty$.

Remark: For the space $L^p(S)$ with a measurable set $S \subset \mathbb{R}^n$, see E4.21.

Proof ⇒. By the definition of precompactness in 4.6(3), for $\varepsilon > 0$ there exists a cover

$$A \subset \bigcup_{i=1}^{n_\varepsilon} B_\varepsilon(g_i^\varepsilon) \quad \text{with } n_\varepsilon \in \mathbb{N}, \; g_i^\varepsilon \in L^p(\mathbb{R}^n; Y).$$

For $f \in A$ we then have that $f \in B_\varepsilon\left(g_{i_f}^\varepsilon\right)$ for some i_f. It holds that

$$\|f\|_{L^p(\mathbb{R}^n)} \leq \varepsilon + \left\|g_{i_f}^\varepsilon\right\|_{L^p(\mathbb{R}^n)} \leq \varepsilon + \max_{i=1,\dots,n_\varepsilon} \|g_i^\varepsilon\|_{L^p(\mathbb{R}^n)} < \infty,$$

which implies (1). Similarly, it follows that

$$\|f(\cdot + h) - f\|_{L^p(\mathbb{R}^n)} \leq 2\varepsilon + \max_{i=1,\dots,n_\varepsilon} \|g_i^\varepsilon(\cdot + h) - g_i^\varepsilon\|_{L^p(\mathbb{R}^n)},$$

$$\|f\|_{L^p(\mathbb{R}^n \setminus B_R(0))} \leq \varepsilon + \max_{i=1,\dots,n_\varepsilon} \|g_i^\varepsilon\|_{L^p(\mathbb{R}^n \setminus B_R(0))},$$

with the second terms becoming small if h gets small and R gets large, respectively (see 4.15(1) and A3.17(2)). This proves (2) and (3).   □

Proof ⇐. Let $(\varphi_\varepsilon)_{\varepsilon>0}$ be a standard Dirac sequence and for small $\varepsilon > 0$ let $R_\varepsilon > 0$ be large. For $f \in A$ we define

$$(T_\varepsilon f)(x) := \int_{B_{R_\varepsilon}(0)} \varphi_\varepsilon(x - y) f(y) \, dy = \left(\varphi_\varepsilon * \left(\mathcal{X}_{B_{R_\varepsilon}(0)} f\right)\right)(x).$$

It follows from 4.13(2) that $T_\varepsilon f \in L^p(\mathbb{R}^n; Y)$. Moreover,

$$(T_\varepsilon f - f)(x) = \int_{\mathbb{R}^n} \varphi_\varepsilon(x - y) \mathcal{X}_{B_{R_\varepsilon}(0)}(y) \, (f(y) - f(x)) \, dy$$

$$- \int_{\mathbb{R}^n \setminus B_{R_\varepsilon}(x)} \varphi_\varepsilon(y) \, dy \cdot f(x).$$

As $\varphi_\varepsilon = 0$ outside of $B_\varepsilon(0)$, the second integral vanishes if $B_\varepsilon(0) \subset B_{R_\varepsilon}(x)$, i.e. if $|x| \leq R_\varepsilon - \varepsilon$. Then it follows from 4.13(1) that

$$\|T_\varepsilon f - f\|_{L^p(\mathbb{R}^n)}$$

$$\leq \sup_{|h|\leq\varepsilon} \|f - f(\cdot + h)\|_{L^p(\mathbb{R}^n)} + \|f\|_{L^p(\mathbb{R}^n \setminus B_{R_\varepsilon-\varepsilon}(0))}$$

$$\leq \sup_{|h|\leq\varepsilon} \sup_{g\in A} \|g - g(\cdot + h)\|_{L^p(\mathbb{R}^n)} + \sup_{g\in A} \|g\|_{L^p(\mathbb{R}^n \setminus B_{R_\varepsilon-\varepsilon}(0))}$$

$$=: \kappa_\varepsilon.$$

Combining (2) and (3) yields that $\kappa_\varepsilon \to 0$ as $\varepsilon \to 0$, if $R_\varepsilon \to \infty$ as $\varepsilon \to 0$. Moreover, it follows from 4.13(3) and 4.13(4) that $T_\varepsilon f \in C_0^\infty(B_{R_\varepsilon+\varepsilon}(0); Y)$. Hence, by using (1) and a Hölder inequality, where $\frac{1}{p} + \frac{1}{p'} = 1$, we obtain

$$\|T_\varepsilon f\|_{\sup} \leq \|\varphi_\varepsilon\|_{L^{p'}(\mathbb{R}^n)} \|f\|_{L^p(\mathbb{R}^n)} \leq C(\varepsilon),$$

$$\|\nabla(T_\varepsilon f)\|_{\sup} \leq \|\nabla\varphi_\varepsilon\|_{L^{p'}(\mathbb{R}^n)} \|f\|_{L^p(\mathbb{R}^n)} \leq C(\varepsilon),$$

with a constant $C(\varepsilon)$ that is independent of f. Hence the functions $T_\varepsilon f$ for $f \in A$ are bounded in $C^{0,1}(\overline{\mathrm{B}}_{R_\varepsilon+\varepsilon}(0); Y)$. Since Y is finite-dimensional the Arzelà-Ascoli theorem 4.12 implies that for $\delta > 0$ there exist functions g_i, $i = 1, \ldots, n(\varepsilon, \delta)$, in $C^0(\overline{\mathrm{B}}_{R_\varepsilon+\varepsilon}(0); Y)$, such that

$$A_\varepsilon := \{T_\varepsilon f \,;\; f \in A\} \subset \bigcup_{i=1}^{n(\varepsilon,\delta)} \mathrm{B}_\delta(g_i) \quad \text{with respect to the } C^0\text{-norm.}$$

Since the L^p-norm on $\mathrm{B}_{R_\varepsilon+\varepsilon}(0)$ can be estimated by the C^0-norm, it follows that

$$A_\varepsilon \subset \bigcup_{i=1}^{n(\varepsilon,\delta)} \mathrm{B}_\varrho(g_i) \quad \text{with respect to the } L^p\text{-norm,}$$

where $\varrho = \delta \cdot \|1\|_{L^p(\mathrm{B}_{R_\varepsilon+\varepsilon}(0))}$. Regarding g_i as elements in $L^p(\mathbb{R}^n; Y)$, by continuing g_i outside of $\overline{\mathrm{B}}_{R_\varepsilon+\varepsilon}(0)$ by 0, we obtain that

$$A \subset \bigcup_{i=1}^{n(\varepsilon,\delta)} \mathrm{B}_{\varrho+\kappa_\varepsilon}(g_i) \quad \text{with respect to the } L^p\text{-norm.}$$

If, for every ε, we now choose δ sufficiently small, such that e.g. $\varrho \leq \varepsilon$, then $\varrho + \kappa_\varepsilon \to 0$ as $\varepsilon \to 0$, which implies the precompactness of A.   □

Dense subsets

In the following, we consider some important examples of dense and separable subsets (see 2.13 for the definition) in function spaces. If A is a dense subset of X, then every element of in X can be approximated to an arbitrary accuracy by an element from A. For instance, we have seen in 4.15(3) that L^p-functions can be approximated with respect to the Lebesgue measure by C_0^∞-functions. In 4.24 we will show that $W^{m,p}$-functions can be approximated by C^∞-functions, where the crucial ingredient in the proof will be once again the convolution of functions.

Separable spaces are spaces that contain a countable dense subset, i.e. a dense subset that is countable. (These concepts were already defined in 2.13.) Separable Banach spaces play a special role in applications, because in numerical computations numbers and functions can only be represented by a finite number of bits. And this number is limited by the actual equipment (see also Chapter 9). In 4.18 we will list the most important separable function spaces. But first some general results.

4.17 Separable sets. Let X be a metric space.

(1) If $A_i \subset X$ for $i \in \mathbb{N}$ are separable, then so is $\overline{\bigcup_{i \in \mathbb{N}} A_i}$.

(2) If X is separable, then so is every subset of X.

(3) If X is a finite-dimensional $\mathbb{K}$-vector space, then X is separable.

(4) If X is a $\mathbb{K}$-vector space and $E \subset X$ is separable, then so is

$$\mathrm{span}(E) := \Big\{ \sum_{i=1}^{k} \alpha_i x_i \ ; \ k \in \mathbb{N}, \ x_i \in E, \ \alpha_i \in \mathbb{K} \text{ for } i = 1, \dots, k \Big\},$$

the *linear hull* of E.

Proof (2). Let $\{x_k ; \ k \in \mathbb{N}\}$ be dense in X and let $A \subset X$ be nonempty. For $k, l \in \mathbb{N}$ there exists, by the definition of the distance, an $a_{k,l} \in A$ such that

$$d_X(x_k, a_{k,l}) \le \mathrm{dist}(x_k, A) + \frac{1}{l}.$$

The denseness of the above sequence yields that for $a \in A$ and $\varepsilon > 0$ there exists an x_{k_ε} such that $d_X(a, x_{k_\varepsilon}) \le \varepsilon$. Then

$$d_X(a, a_{k_\varepsilon,l}) \le d_X(a, x_{k_\varepsilon}) + d_X(x_{k_\varepsilon}, a_{k_\varepsilon,l}) \le 2d_X(x_{k_\varepsilon}, a) + \frac{1}{l} \le 3\varepsilon$$

for sufficiently large l. This shows that $\{a_{k,l} ; \ (k, l) \in \mathbb{N}^2\}$ is a dense subset of A. The fact that the index set $\mathbb{N}^2$ is countable then yields the desired result. $\qquad\square$

Proof (4). Let A be a countable dense subset of E, i.e. $A \subset E$ is countable with $E \subset \overline{A}$. Then

$$\Big\{ \sum_{i=1}^{k} \alpha_i a_i \ ; \ k \in \mathbb{N}, \ \alpha_i \in Q, \ a_i \in A \text{ for } i = 1, \dots, k \Big\}$$

is a countable dense subset of $\mathrm{span}(E)$, where $Q = \mathbb{Q}$ in the case $\mathbb{K} = \mathbb{R}$, and where $Q = \{\alpha \in \mathbb{C} ; \ \mathrm{Re}\alpha \in \mathbb{Q}, \ \mathrm{Im}\alpha \in \mathbb{Q}\}$ in the case $\mathbb{K} = \mathbb{C}$. $\qquad\square$

4.18 Examples of separable spaces.

(1) The set $\mathbb{R}^n$ is separable, and also $\mathbb{C}^n$.

(2) For $1 \le p < \infty$ the set $\ell^p(\mathbb{K})$ is separable, but $\ell^\infty(\mathbb{K})$ is not.

(3) If $S \subset \mathbb{R}^n$ is closed and bounded, then $C^0(S)$ is separable.

(4) If $S \subset \mathbb{R}^n$ is Lebesgue measurable, then $L^p(S)$ is separable for $1 \le p < \infty$. If S is not a null set, then $L^\infty(S)$ is not separable.

(5) If $\Omega \subset \mathbb{R}^n$ is open, bounded and $m \ge 0$, then $C^m(\overline{\Omega})$ is separable.

(6) If $\Omega \subset \mathbb{R}^n$ is open, $m \ge 0$ and $1 \le p < \infty$, then $W^{m,p}(\Omega)$ is separable.

Remark: Assertions (2)–(6) remain valid if the image space $\mathbb{K}$ is replaced with a separable Banach space Y.

Proof (1). $\mathbb{Q}^n$ is countable and dense in $\mathbb{R}^n$.   □

Proof (2). For $p < \infty$, with Q as in the proof of 4.17(4) and with $\mathbf{e}_i$ as in 2.23,

$$\left\{ \sum_{i=1}^{k} \alpha_i \mathbf{e}_i \; ; \; k \in \mathbb{N}, \; \alpha_i \in Q \text{ for } 1 \le i \le k \right\}$$

is a countable dense subset of $\ell^p(\mathbb{K})$.

For $p = \infty$ let $\left(a^k\right)_{k \in \mathbb{N}}$ be a sequence in $\overline{B_1(0)} \subset \ell^\infty(\mathbb{K})$ with $a^k = \left(a_i^k\right)_{i \in \mathbb{N}}$ and $a_i^k \in \mathbb{R}$ for all $k, i \in \mathbb{N}$. Let

$$b_i := \begin{cases} a_i^i - 1 & \text{if } a_i^i \ge 0, \\ a_i^i + 1 & \text{if } a_i^i < 0. \end{cases}$$

Then $b := (b_i)_{i \in \mathbb{N}} \in \overline{B_1(0)}$ and $\left\| b - a^k \right\|_{\ell^\infty} \ge \left| b_k - a_k^k \right| = 1$. This shows that $\overline{B_1(0)}$ is not separable, and 4.17(2) then yields that neither is $\ell^\infty(\mathbb{K})$.   □

Proof (3). If $S \subset \mathbb{R}^n$ is compact (by 4.7(7), bounded and closed), then $C^0(S)$ is a Banach space with the supremum norm. For $\varepsilon > 0$ let

$$Q_\varepsilon(z) := \{x \in \mathbb{R}^n \; ; \; z_i \le x_i \le z_i + \varepsilon \text{ for } i = 1, \ldots, n\} \text{ for } z \in \varepsilon \mathbb{Z}^n \, ,$$

$$S \subset S_\varepsilon := \bigcup_{z \in M_\varepsilon} Q_\varepsilon(z) \quad \text{with } M_\varepsilon := \{z \in \varepsilon \mathbb{Z}^n \; ; \; Q_\varepsilon(z) \cap S \ne \emptyset\} \, .$$

Consider lattice points $y \in S_\varepsilon \cap (\varepsilon \mathbb{Z}^n)$ and choose $x_{\varepsilon,y} \in S$ such that $|y - x_{\varepsilon,y}|_\infty \le \varepsilon$. For $f \in C^0(S)$ define $f_\varepsilon(y) := f(x_{\varepsilon,y})$ and extend f_ε by *multilinear interpolation*, i.e.

$$f_\varepsilon(x) := \sum_{\gamma \, \in \, \{0,1\}^n} \left(\prod_{j \, : \, \gamma_j = 0} (1 - t_j) \prod_{j \, : \, \gamma_j = 1} t_j \right) f_\varepsilon(z + \varepsilon\gamma)$$

$$\text{for} \quad x = z + \varepsilon \sum_{i=1}^{n} t_i \mathbf{e}_i \in Q_\varepsilon(z), \; z \in M_\varepsilon \, .$$

This defines a function $f_\varepsilon \in C^0(S)$ with

$$\| f_\varepsilon - f \|_{C^0(S)} \le \sup\{|f(x_1) - f(x_2)| \; ; \; x_1, x_2 \in S, \; |x_1 - x_2|_\infty \le 2\varepsilon\} \, ,$$

which converges to 0 as $\varepsilon \to 0$, because f is uniformly continuous on S. On setting $\varepsilon = \frac{1}{k}$, $k \in \mathbb{N}$, and on approximating in each case the finitely many values $f_\varepsilon(y)$ by rational numbers, we obtain the desired result. (Alternatively, the result can be shown by using polynomials to approximate f, see 9.10.)   □

Proof (4). For $p < \infty$ first consider the case $S = \mathbb{R}^n$. For $f \in L^p(\mathbb{R}^n)$ we define its *piecewise constant interpolation* by

$$f_\varepsilon(x) := \sum_{z \in \varepsilon \mathbb{Z}^n} \chi_{Q_\varepsilon(z)}(x)\, \alpha_{\varepsilon,z}\,, \quad \alpha_{\varepsilon,z} := \frac{1}{\varepsilon^n} \int_{Q_\varepsilon(z)} f(y)\, \mathrm{d}y\,,$$

where the $Q_\varepsilon(z)$ are defined as in (3). The interpolant f_ε approximates f since

$$\int_{\mathbb{R}^n} |f_\varepsilon(x) - f(x)|^p\, \mathrm{d}x = \sum_z \int_{Q_\varepsilon(z)} |\alpha_{\varepsilon,z} - f(x)|^p\, \mathrm{d}x$$

$$= \sum_z \int_{Q_\varepsilon(z)} \left| \frac{1}{\varepsilon^n} \int_{Q_\varepsilon(z)} (f(y) - f(x))\, \mathrm{d}y \right|^p \mathrm{d}x$$

$$= \sum_z \frac{1}{\varepsilon^n} \int_{Q_\varepsilon(z)} \int_{Q_\varepsilon(z)} |f(y) - f(x)|^p\, \mathrm{d}y\, \mathrm{d}x$$

$$\leq \sup_{h:|h| \leq \varepsilon} \int_{\mathbb{R}^n} |f(y+h) - f(y)|^p\, \mathrm{d}y = \sup_{h:|h| \leq \varepsilon} \|f(\cdot + h) - f\|_{L^p}^p\,,$$

which by 4.15(1) converges to zero as $\varepsilon \to 0$. The desired result now follows by choosing countably many $\varepsilon > 0$ and rational approximations of $\alpha_{\varepsilon,z}$, similarly to the proof of (3). For arbitrary S continue functions to $\mathbb{R}^n$ by 0 and restrict the approximations to S. (An alternative proof can be based on approximating f with step functions using 3.26(1), and approximating the steps by A3.14(1) with cuboids. This reduces the problem to the approximation of finitely many real parameters. Another alternative proof approximates f by continuous functions using 3.26(2), and then applies (3).)

For $p = \infty$, on recalling that $L^n(S) > 0$, there exist measurable disjoint sets $S_j \subset S$, $j \in \mathbb{N}$, with $L^n(S_j) > 0$ such that $S = \bigcup_{j \in \mathbb{N}} S_j$. Similarly to the proof of (2), let $(f_k)_{k \in \mathbb{N}}$ be a sequence in $\overline{B_1(0)} \subset L^\infty(S)$. In addition, let $a_i^k := \operatorname*{ess\,sup}_{S_i} f_k$, let b_i as in the proof above, and define $g \in L^\infty(S)$ by $g(x) := b_j$ for $x \in S_j$ and $j \in \mathbb{N}$. It follows that $\|g - f_k\|_{L^\infty} \geq 1$. □

Proof (5). For $f \in C^m(\overline{\Omega})$, define

$$T(f)(x) := (\partial^s f(x))_{|s| \leq m}\,.$$

Then $T : C^m(\overline{\Omega}) \to C^0(\overline{\Omega}; Y)$, where

$$Y := \{(y_s)_{|s| \leq m}\,;\ y_s \in \mathbb{R} \text{ for } |s| \leq m\}$$

is a Euclidean space. It follows from (3) (see the above remark) that $C^0(\overline{\Omega}; Y)$ is separable. Hence, on noting 4.17(2), $T(C^m(\overline{\Omega}))$ is separable. Combining the fact that T is linear and that $\|T(f)\|_{C^0}$ can be estimated from above and from below by $\|f\|_{C^m}$ yields the separability of $C^m(\overline{\Omega})$. □

Proof (6). As in (5), now with $T : W^{m,p}(\Omega) \to L^p(\Omega; Y)$, on utilizing (4). □

In order to approximate differentiable functions by C^∞-functions we make use of convolutions, and we need a tool which guarantees that functions can be "partitioned" or "localized".

4.19 Cut-off function. Let $\Omega \subset \mathbb{R}^n$ be open and let $K \subset \mathbb{R}^n$ be compact with $B_\delta(K) \subset \Omega$, where $\delta > 0$. Then there exists a **cut-off function** $\eta \in C_0^\infty(\Omega)$ satisfying

$$0 \leq \eta \leq 1 \ , \ \eta = 1 \text{ on } K \ , \ \text{supp}(\eta) \subset B_\delta(K) \ ,$$
$$|\partial^s \eta| \leq C_{n,s} \cdot \delta^{-|s|} \text{ for all multi-indices } s.$$

Here $C_{n,s}$ are constants that depend only on n and s.

Proof. Let $(\varphi_\varepsilon)_{\varepsilon > 0}$ be a standard Dirac sequence. Then $\eta := \varphi_{\frac{\delta}{4}} * \mathcal{X}_{B_{\frac{\delta}{2}}(K)}$ has the desired properties. $\qquad\square$

4.20 Partition of unity. Let $S \subset \mathbb{R}^n$ be nonempty and $N \subset \mathbb{N}$.

(1) We call $(U_i)_{i \in N}$ an **open cover** of S if U_i are (nonempty) open sets with $S \subset \bigcup_{i \in N} U_i$. (It is also possible to require that, in addition, $U_i \cap S \neq \emptyset$ for $i \in N$).

(2) The cover is called **locally finite** if for any $x \in \bigcup_{i \in N} U_i$ there exists a ball $\overline{B_\varepsilon(x)}$ such that $\{i \in N \, ; \ U_i \cap \overline{B_\varepsilon(x)} \neq \emptyset\}$ is finite, i.e.

$$\forall \, x \in \bigcup_{i \in N} U_i \, : \, \exists \, \varepsilon > 0 : \, \Big(\ \{i \in N \, ; \ U_i \cap \overline{B_\varepsilon(x)} \neq \emptyset\} \text{ is finite} \ \Big).$$

(This condition is only relevant for nonfinite N.)

(3) We call $(\eta_j)_{j \in N}$ a **partition of unity** for S subject to a locally finite open cover $(U_j)_{j \in N}$ of S if

$$\eta_j \in C_0^\infty(U_j), \quad \eta_j \geq 0 \ , \quad \text{and} \quad \sum_{j \in N} \eta_j(x) = 1 \text{ for } x \in S.$$

Here the sum locally contains only finitely many nonzero terms, where η_j is defined to be 0 outside U_j. Hence instead of $\eta_j \in C_0^\infty(U_j)$ one can also say $\eta_j \in C^\infty(\mathbb{R}^n)$ with compact support $\text{supp}(\eta_j) \subset U_j$.

Proposition: Let $\Omega \subset \mathbb{R}^n$ be open and let

$$K_j \subset U_j \subset \overline{U_j} \subset \Omega \text{ for } j \in \mathbb{N}, \quad K_j \text{ and } \overline{U_j} \text{ compact,}$$

such that $(U_j)_{j \in \mathbb{N}}$ is a locally finite open cover of Ω, with $K_j \cap K_i = \emptyset$ for $j \neq i$, i.e. the K_j are pairwise disjoint. Then there exists a partition of unity $(\eta_j)_{j \in \mathbb{N}}$ for Ω subject to this cover with the additional property that $\eta_j(x) = 1$ for $x \in K_j$.

Remark: In the assumptions of the proposition some, or all, of the compact sets K_j may be empty.

Conclusion: Let $K \subset \mathbb{R}^n$ be compact and let $(U_j)_{j=1,\ldots,k}$ be a finite open cover of K. Then there exists a partition of unity $(\eta_j)_{j=1,\ldots,k}$ for K subject to this cover.

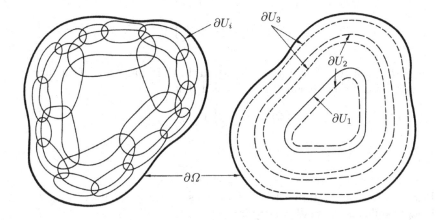

Fig. 4.4. *Cover of an open set*

Proof of the proposition. Firstly, we modify the sets U_j to

$$V_j := U_j \setminus \bigcup_{i\,:\,i \neq j} K_i = \bigcap_{i\,:\,i \neq j} (U_j \setminus K_i)\,.$$

Then it clearly holds that

$$K_j \subset V_j \subset U_j, \text{ with } V_j \cap K_i = \emptyset \text{ for } i \neq j.$$

Moreover, we claim that $(V_j)_{j \in \mathbb{N}}$ is also a locally finite open cover of Ω. In order to show this, we make use of the compactness of $\overline{U_j}$. As the original cover is locally finite, $\{i \in \mathbb{N}\,;\ U_i \cap \overline{U_j} \neq \emptyset\}$ is finite. Hence also $K_i \cap \overline{U_j} \neq \emptyset$ holds only for finitely many i, say, at most for $i = 1, \ldots, m_j$. Consequently,

$$V_j = U_j \setminus \bigcup_{i\,:\,i \leq m_j, i \neq j} K_i$$

is nonempty and open. For $x \in \Omega$ it holds that $x \in U_j$ for some j. Then we have that either $x \in V_j$, or, by the definition of V_j, that $x \in K_i \subset V_i$ for some $i \neq j$. This proves the covering property of the V_j.

Secondly, we define open sets W_j with

$$K_j \subset W_j \subset \overline{W_j} \subset V_j$$

such that $(W_j)_{j \in \mathbb{N}}$ is still an open cover of Ω. To this end, we construct W_m, by induction on m, satisfying

$$\bigcup_{j<m} W_j \cup \bigcup_{j \geq m} V_j = \Omega\,. \tag{4-26}$$

Letting $m \to \infty$ this then yields the desired covering property, because the cover $(V_j)_{j \in \mathbb{N}}$ is locally finite. Now let $m \geq 1$ and assume that W_j for $1 \leq j < m$ have already been constructed. Then

$$\partial V_m \subset \bigcup_{j<m} W_j \cup \bigcup_{j>m} V_j ,$$

where ∂V_m is compact and where the set in the right-hand side of the inclusion is open. Hence there exists a $\delta_m > 0$ such that

$$\overline{B_{\delta_m}(\partial V_m)} \subset \bigcup_{j<m} W_j \cup \bigcup_{j>m} V_j .$$

Let

$$W_m := V_m \setminus \overline{B_{\delta_m}(\partial V_m)} .$$

It follows from $V_m \neq \emptyset$ that $W_m \neq \emptyset$ for δ_m sufficiently small. Also $K_m \subset W_m$ if δ_m is sufficiently small. Therefore,

$$V_m \subset W_m \cup \overline{B_{\delta_m}(\partial V_m)} \subset \bigcup_{j<m+1} W_j \cup \bigcup_{j \geq m+1} V_j ,$$

which implies (4-26) with m replaced by $m+1$, and so concludes the induction.

Thirdly, we define the corresponding cut-off functions. Since $\overline{W_j}$ is a compact subset of the open set V_j, it follows from 4.19 that there exists a function $\widetilde{\eta}_j \in C_0^\infty(V_j)$ with $0 \leq \widetilde{\eta}_j \leq 1$ and $\widetilde{\eta}_j = 1$ on $\overline{W_j}$. The covering property of the W_j then yields that

$$\sum_{j \in \mathbb{N}} \widetilde{\eta}_j(x) > 0 \quad \text{for all } x \in \Omega,$$

where locally in Ω the sum contains only finitely many nonzero terms. Define

$$\eta_j(x) := \frac{\widetilde{\eta}_j(x)}{\sum_{j \in \mathbb{N}} \widetilde{\eta}_j(x)} .$$

By construction the η_j have the desired properties.   □

Proof of the conclusion. We reduce this to the result derived above for the case $\Omega = \mathbb{R}^n$, by extending the cover of K to a cover of $\mathbb{R}^n$. Choose radii $0 < R_0 < R_1 < \ldots$ with $R_i \to \infty$ as $i \to \infty$ such that $K \subset B_{R_0}(0)$. Define $U_{k+1} := B_{R_1}(0) \setminus K$ and $U_{k+i} := B_{R_i}(0) \setminus \overline{B_{R_{i-2}}(0)}$ for $i \geq 2$. This yields a locally finite open cover $(U_j)_{j \in \mathbb{N}}$ of $\mathbb{R}^n$. Now apply the previously shown result with $K_j = \emptyset$ for $j \in \mathbb{N}$ and obtain a corresponding partition of unity $(\eta_j)_{j \in \mathbb{N}}$. By construction, $\eta_j(x) = 0$ for $x \in K$ and $j > k$. This yields the desired result.   □

4.21 Examples of partition of unity. The results in 4.20 can be applied to the following situations:

(1) Let $x_j \in \mathbb{R}$ with $x_j < x_{j+1}$ for $j \in \mathbb{Z}$ and let $\delta_j > 0$ such that

$$\emptyset \neq K_j := [x_j + \delta_j, x_{j+1} - \delta_j] \subset U_j := \,]x_j, x_{j+1}[\,.$$

This yields a partition of unity as in Fig. 4.5.

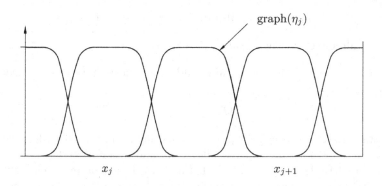

graph(η_j)

x_j x_{j+1}

Fig. 4.5. *A partition of unity in* $\mathbb{R}$

(2) For a given $\varepsilon > 0$ consider the cover $(B_{2\varepsilon}(x))_{x \in \varepsilon \mathbb{Z}^n}$ of $\mathbb{R}^n$, where the balls are formed with respect to the ∞-norm. This is a uniform cover of the whole space and it yields a partition of unity for $\mathbb{R}^n$.

(3) Let $\Omega \subset \mathbb{R}^n$ be open and bounded. Consider the cover $(U_i)_{i \in \mathbb{N}}$ of Ω, where

$$U_i := \left\{ x \in \Omega \; ; \; \tfrac{\delta}{2} \cdot 2^{-i} < \mathrm{dist}(x, \partial\Omega) < 2\delta \cdot 2^{-i} \right\}, \quad \delta := \mathrm{diam}(\Omega)\,.$$

This yields a partition of unity $(\eta_i)_{i \in \mathbb{N}}$ for Ω with $\eta_i \in C_0^\infty(\Omega)$. The cover is of the type shown on the right-hand side of Fig. 4.4.

(4) Let $\Omega \subset \mathbb{R}^n$ be open and bounded. For $k \in \mathbb{N}$ and $x \in 2^{-k}\mathbb{Z}^n$ let $U_{k,x} := B_{2^{1-k}}(x)$ with respect to the ∞-norm. Define inductively

$$M_k := \left\{ x \in 2^{-k}\mathbb{Z}^n \; ; \; \overline{U_{k,x}} \subset \Omega, \; x \notin U_{l,y} \text{ for all } y \in M_l \text{ with } l < k \right\}$$

and consider the cover $(U_{k,x})_{x \in M_k, k \in \mathbb{N}}$ of Ω. This again yields a partition of unity $(\eta_i)_{i \in \mathbb{N}}$ for Ω with $\eta_i \in C_0^\infty(\Omega)$. The cover is of the type shown on the left-hand side of Fig. 4.4.

In a further application of convolution with a Dirac sequence, we will now show in 4.24 that Sobolev functions can be approximated by smooth functions. We defined these functions in 3.27 through the existence of weak derivatives. The fact that these weak derivatives are uniquely defined follows from the following

4.22 Fundamental lemma of calculus of variations. Let $\Omega \subset \mathbb{R}^n$ be open and let Y be a Banach space. Then for $g \in L^1(\Omega; Y)$ the following are equivalent:

(1) $\int_\Omega \zeta g \, \mathrm{d}L^n = 0$ for all $\zeta \in C_0^\infty(\Omega)$.

(2) $\int_E g \, \mathrm{d}L^n = 0$ for all measurable bounded sets E with $\overline{E} \subset \Omega$.

(3) $g = 0$ almost everywhere in Ω.

Proof (1)$\Rightarrow$(2). Let E be measurable and bounded with $\overline{E} \subset \Omega$. Consider the functions

$$\zeta_\varepsilon(x) := \int_E \varphi_\varepsilon(x - y) \, \mathrm{d}y = (\varphi_\varepsilon * \mathcal{X}_E)(x)$$

with a standard Dirac sequence $(\varphi_\varepsilon)_{\varepsilon>0}$. Combining 4.13(3) and 4.13(4) yields that $\zeta_\varepsilon \in C_0^\infty(\Omega)$ for sufficiently small $\varepsilon > 0$, and in addition $0 \le \zeta_\varepsilon \le 1$. It follows from 4.15(2) that $\zeta_\varepsilon \to \mathcal{X}_E$ in $L^1(\mathbb{R}^n)$ and hence, on recalling 3.22(1), there exists a subsequence $\varepsilon \to 0$ such that $\zeta_\varepsilon \to \mathcal{X}_E$ almost everywhere in $\mathbb{R}^n$. Lebesgue's convergence theorem then yields that $\zeta_\varepsilon g \to \mathcal{X}_E g$ in $L^1(\Omega; Y)$ and hence

$$0 = \int_\Omega \zeta_\varepsilon g \, \mathrm{d}L^n \longrightarrow \int_\Omega \mathcal{X}_E g \, \mathrm{d}L^n = \int_E g \, \mathrm{d}L^n.$$

$\square$

Proof (2)$\Rightarrow$(3). Let $(\varphi_\varepsilon)_{\varepsilon>0}$ be a Dirac sequence for

$$\varphi(x) := L^n(B_1(0))^{-1} \mathcal{X}_{B_1(0)}(x).$$

For $x \in \Omega$ and small ε,

$$g_\varepsilon(x) := (\varphi_\varepsilon * \mathcal{X}_\Omega g)(x) = \frac{1}{L^n(B_\varepsilon(0))} \int_{B_\varepsilon(x)} g \, \mathrm{d}L^n = 0.$$

By 4.15(2), $g_\varepsilon \to g$ in $L^1(\Omega; Y)$ and then for a subsequence $\varepsilon \to 0$ almost everywhere in Ω. $\square$

4.23 Local approximation of Sobolev functions. Let $\Omega \subset \mathbb{R}^n$ be open and let $f \in W^{m,p}(\Omega)$ with $1 \le p < \infty$. Choose a standard Dirac sequence $(\varphi_\varepsilon)_{\varepsilon>0}$ and define

$$(T_\varepsilon f)(x) := \int_\Omega \varphi_\varepsilon(x - y) f(y) \, \mathrm{d}y = (\varphi_\varepsilon * \mathcal{X}_\Omega f)(x).$$

For open sets $D \subset \Omega$ with $\delta := \mathrm{dist}(D, \partial\Omega) > 0$ it holds that

$$T_\varepsilon f \in W^{m,p}(D) \cap C^\infty(D) \quad \text{for } \varepsilon < \delta$$

and that $T_\varepsilon f \to f$ in $W^{m,p}(D)$ as $\varepsilon \to 0$.

Proof. It follows from 4.13(4) that $T_\varepsilon f \in C^\infty(\mathbb{R}^n)$ and that for $|s| \le m$

$$\partial^s(T_\varepsilon f)(x) = \int_\Omega \partial^s \varphi_\varepsilon(x-y)f(y)\,\mathrm{d}y = (-1)^{|s|}\int_\Omega \frac{\partial^s}{\partial y^s}\big(\varphi_\varepsilon(x-y)\big)f(y)\,\mathrm{d}y\,.$$

Now for $x \in \Omega$ with $\mathrm{dist}(x,\partial\Omega) > \varepsilon$ the function $y \mapsto \varphi_\varepsilon(x-y)$ is in $C_0^\infty(\Omega)$, and hence, on recalling the definition of the Sobolev space, the right-hand side is

$$= \int_\Omega \varphi_\varepsilon(x-y)\partial^s f(y)\,\mathrm{d}y = T_\varepsilon\left(\partial^s f\right)(x)\,.$$

Hence it follows for every open set $D \subset \Omega$ with $\mathrm{dist}(D,\partial\Omega) > 0$, on choosing $\varepsilon < \mathrm{dist}(D,\partial\Omega)$, that $T_\varepsilon f \in W^{m,p}(D)$ with $\partial^s(T_\varepsilon f)(x) = T_\varepsilon\left(\partial^s f\right)(x)$ for $|s| \le m$. Consequently,

$$\left\|\partial^s(T_\varepsilon f) - \partial^s f\right\|_{L^p(D)} = \left\|T_\varepsilon(\partial^s f) - \partial^s f\right\|_{L^p(D)} \longrightarrow 0 \quad \text{as } \varepsilon \to 0$$

on recalling 4.15(2), i.e. $T_\varepsilon f \to f$ in $W^{m,p}(D)$. $\qquad\square$

We now prove the approximation property presented in 3.28.

4.24 Theorem. For $1 \le p < \infty$ the space $W^{m,p}(\Omega) \cap C^\infty(\Omega)$ is dense in $W^{m,p}(\Omega)$.

Proof. Let $(U_k)_{k\in\mathbb{N}}$ be a locally finite open cover of Ω (see 4.21(3), 4.21(4)), such that $\overline{U_k} \subset \Omega$ are compact. It follows from 4.20 that there exists a corresponding partition of unity $(\eta_k)_{k\in\mathbb{N}}$. Moreover, let $c_k > 0$ (to be defined below) and $\varepsilon > 0$. By 4.23, for $f \in W^{m,p}(\Omega)$ there exist $f_{k,\varepsilon} \in C^\infty(U_k)$ with

$$\|f - f_{k,\varepsilon}\|_{W^{m,p}(U_k)} \le \varepsilon c_k\,.$$

Let

$$f_\varepsilon := \sum_{k\in\mathbb{N}} \eta_k f_{k,\varepsilon}\,, \quad \text{so that} \quad f_\varepsilon - f = \sum_{k\in\mathbb{N}} \eta_k(f_{k,\varepsilon} - f)\,,$$

where locally in Ω the sums contain only finitely many nonzero terms. For each term in the sums we can compute the weak derivatives with the help of the product rule, since for $\zeta \in C_0^\infty(\Omega)$

$$\int_\Omega \eta_k \partial_i \zeta f \,\mathrm{d}L^n = \int_\Omega \big(\partial_i(\eta_k\zeta) - \zeta\partial_i\eta_k\big)f\,\mathrm{d}L^n = -\int_\Omega (\eta_k\zeta\partial_i f + \zeta\partial_i\eta_k f)\,\mathrm{d}L^n\,.$$

Hence, $\eta_k f \in W^{1,p}(\Omega)$ with

$$\partial_i(\eta_k f) = \eta_k\partial_i f + (\partial_i\eta_k)f\,.$$

On repeating this calculation of the partial derivative, it follows inductively that $\eta_k f \in W^{m,p}(\Omega)$ and that for $|s| \le m$ the Leibniz rule holds:

$$\partial^s(\eta_k f) = \sum_{0 \le r \le s} \binom{s}{r} (\partial^{s-r}\eta_k)\partial^r f \,.$$

Clarification: The Leibniz rule (4-27), below, for smooth functions is well known from analysis. Here we prove it for the product of a smooth function and a Sobolev function. Later in 4.25 we will derive it for the product of two Sobolev functions, where the proof will make use of the result we are currently in the process of proving.

We obtain that

$$\partial^s f_\varepsilon - \partial^s f = \sum_{0 \le r \le s} \binom{s}{r} \sum_k (\partial^{s-r}\eta_k)(\partial^r f_{k,\varepsilon} - \partial^r f) \,,$$

and hence that

$$\|\partial^s f_\varepsilon - \partial^s f\|_{L^p(\Omega)} \le C \sum_k \|\eta_k\|_{C^m(\overline{\Omega})} \|f_{k,\varepsilon} - f\|_{W^{m,p}(U_k)}$$

$$\le C\varepsilon \sum_k c_k \|\eta_k\|_{C^m(\overline{\Omega})} \le C\varepsilon \,,$$

where the constants C depend only on m and n, if we choose c_k at the beginning of the proof for instance such that $c_k \le 2^{-k}\big(\|\eta_k\|_{C^m(\overline{\Omega})} + 1\big)^{-1}$.
□

The approximability of Sobolev functions can be used to prove results for these functions, such as the following generalization of the product rule employed in the previous proof.

4.25 Product rule for Sobolev functions. Let $\Omega \subset \mathbb{R}^n$ be open. Let $1 \le p \le \infty$ with $\frac{1}{p} + \frac{1}{p'} = 1$. If $f \in W^{m,p}(\Omega)$ and $g \in W^{m,p'}(\Omega)$, then $f \cdot g \in W^{m,1}(\Omega)$ and the weak derivatives of $f \cdot g$ can be computed with the product rule

$$\partial^s(fg) = \sum_{0 \le r \le s} \binom{s}{r}(\partial^{s-r}f)\partial^r g \qquad (\textbf{\textit{Leibniz rule}}). \qquad (4\text{-}27)$$

Proof. A symmetry argument yields that we may assume $p < \infty$. Then it follows from 4.24 that there exist functions $f_k \in W^{m,p}(\Omega) \cap C^\infty(\Omega)$ with $f_k \to f$ in $W^{m,p}(\Omega)$. Similarly to the proof of 4.24, for $\zeta \in C_0^\infty(\Omega)$ we have

$$\int_\Omega \partial_i \zeta \, f_k g \, d\mathcal{L}^n = -\int_\Omega \zeta(g\partial_i f_k + f_k \partial_i g) \, d\mathcal{L}^n \,.$$

Using the Hölder inequality and letting $k \to \infty$ we obtain

$$\int_\Omega \partial_i \zeta f g \, d\mathcal{L}^n = -\int_\Omega \zeta(g\partial_i f + f\partial_i g) \, d\mathcal{L}^n \,,$$

i.e. the desired result for $m = 1$. For $m > 1$ replace ζ inductively by its derivatives.
□

4.26 Chain rule for Sobolev functions. Let $\Omega, \widetilde{\Omega} \subset \mathbb{R}^n$ be open and let $\tau : \widetilde{\Omega} \to \Omega$ be a C^1-*diffeomorphism*, i.e. τ is bijective with $\tau \in C^1(\widetilde{\Omega})$ and $\tau^{-1} \in C^1(\Omega)$, with bounded derivative matrices $D\tau$ and $D\tau^{-1}$. If $f \in W^{1,p}(\Omega)$, $1 \le p \le \infty$, then $f \circ \tau \in W^{1,p}(\widetilde{\Omega})$ and the weak derivatives of $f \circ \tau$ can be computed with the **chain rule**

$$\partial_i(f \circ \tau) = \sum_{j=1}^{n} (\partial_j f) \circ \tau \; \partial_i \tau_j . \tag{4-28}$$

Remark: A corresponding $W^{m,p}$-version also holds. However, the general formula for the n-dimensional chain rule of arbitrary order is difficult to write down concisely. In applications a recursive formula (see e.g. (10-11)) usually suffices.

Proof. For $p < \infty$ choose f_k as in the previous proof. Then $f_k \circ \tau \in C^1(\widetilde{\Omega})$, with

$$\partial_i(f_k \circ \tau) = \sum_{j=1}^{n} (\partial_j f_k) \circ \tau \; \partial_i \tau_j . \tag{4-29}$$

The transformation theorem (for C^1-functions) yields that

$$\int_\Omega |\partial_j f_k - \partial_j f_l|^p \, \mathrm{d}L^n = \int_{\widetilde{\Omega}} |(\partial_j f_k) \circ \tau - (\partial_j f_l) \circ \tau|^p |\det D\tau| \, \mathrm{d}L^n .$$

As $|\det D\tau|$ is strictly positive, we obtain that $((\partial_j f_k) \circ \tau)_{k \in \mathbb{N}}$ is a Cauchy sequence in $L^p(\widetilde{\Omega})$. Moreover, it follows from 3.22(1) that for a subsequence $\partial_j f_k \to \partial_j f$ almost everywhere in Ω, and hence also $(\partial_j f_k) \circ \tau \to (\partial_j f) \circ \tau$ almost everywhere in $\widetilde{\Omega}$ (on noting lemma 4.27, below), since $\tau^{-1} : \widetilde{\Omega} \to \mathbb{R}^n$ is locally Lipschitz continuous. Hence we have that $(\partial_j f_k) \circ \tau \to (\partial_j f) \circ \tau$ in $L^p(\widetilde{\Omega})$. Similarly, it follows that $f_k \circ \tau \to f \circ \tau$ in $L^p(\widetilde{\Omega})$. On letting $k \to \infty$, the desired result follows from the chain rule (4-29). $\qquad\square$

4.27 Lemma. Let $D \subset \mathbb{R}^n$ be open and bounded and let $\tau \in C^{0,1}(\overline{D}; \mathbb{R}^n)$. For $N \subset D$:

$$L^n(N) = 0 \quad \Longrightarrow \quad L^n(\tau(N)) = 0 .$$

Proof. We claim (for the proof see below) that for $\varepsilon > 0$ there exists a cover $(B_{r_j}(x_j))_{j \in \mathbb{N}}$ of N with $x_j \in \overline{N}$, $B_{r_j}(x_j) \subset D$, and

$$\sum_{j \in \mathbb{N}} L^n(B_{r_j}(x_j)) \le \varepsilon . \tag{4-30}$$

Then the $\tau(B_{r_j}(x_j) \cap N)$ form a cover of $\tau(N)$, and $\tau(B_{r_j}(x_j)) \subset B_{l \cdot r_j}(\tau(x_j))$, with l denoting the Lipschitz constant of τ. Hence

$$\mathrm{L}^n\big(\tau(N)\big) \le \sum_{j\in\mathbb{N}} \mathrm{L}^n\big(\mathrm{B}_{l\cdot r_j}(\tau(x_j))\big) = \sum_{j\in\mathbb{N}} l^n \mathrm{L}^n\big(\mathrm{B}_{r_j}(x_j)\big) \le l^n\varepsilon.$$

For the proof of (4-30), first consider a cover of N with cuboids as in A3.4. On appropriately partitioning these cuboids if necessary, we obtain a new cover as in A3.4, now consisting of cuboids Q_j, $Q_j \cap N \ne \emptyset$ for $j \in \mathbb{N}$, which are close to being cubes, i.e.

$$\overline{Q_j} = \bigtimes_{i=1}^{n} [a_{ji}, b_{ji}] \quad \text{with} \quad 2 \cdot \min_{i=1,\dots,n}(b_{ji} - a_{ji}) \ge s_j := \max_{i=1,\dots,n}(b_{ji} - a_{ji}).$$

Choose $x_j \in Q_j \cap N$. Then $Q_j \subset \overline{\mathrm{B}_{r_j}(x_j)}$, where $r_j := \sqrt{n} \cdot s_j$, and

$$\mathrm{L}^n\big(\overline{\mathrm{B}_{r_j}(x_j)}\big) = \kappa_n \cdot r_j^n \le \kappa_n(2\sqrt{n})^n \cdot \mathrm{L}^n(Q_j)$$

with $\kappa_n := \mathrm{L}^n(\mathrm{B}_1(0))$. □

This lemma implies that a Lipschitz map τ transforms L^n-measurable sets into L^n-measurable sets. As a consequence, if f is an L^n-measurable function, then also $f \circ \tau$ is L^n-measurable.

E4 Exercises

E4.1 Subsets of C^0 and L^1. Let $I :=]-1,1[$. Find the interior and the closed hull of

(1) $A := \{f \in C^0(\overline{I}) ; \ f > 0\}$.

(2) $A := \{f \in L^1(I) ; \ f > 0 \text{ almost everywhere}\}$.

Solution (1). For $f \in A$ it holds that $\inf_{\overline{I}} f > 0$. This yields that $\mathring{A} = A$. Moreover,

$$\overline{A} = \{f \in C^0(\overline{I}) ; \ f \ge 0\}.$$

□

Solution (2). Similarly to (1), it follows that $\overline{A} = \{f \in L^1(I) ; \ f \ge 0 \text{ almost everywhere}\}$. We now show that $\mathring{A} = \emptyset$. For every $f \in A$ there exists an $M > 0$ such that the set $\{f \le M\} := \{x \in I ; \ f(x) \le M\}$ has positive measure. For $m \in \mathbb{N}$ we now partition I (except for finitely many points) into the intervals $I_{mi} :=]\frac{i-1}{m}, \frac{i}{m}[$, $i = 1,\dots,m$. Then $\{f \le M\} \cap I_{mi}$ also has positive measure for at least one i. For this i let $I_m^* := \{f \le M\} \cap I_{mi}$. On I_m^* we have that $f - 2M\mathcal{X}_{I_m^*} \le -M < 0$, and so $f - 2M\mathcal{X}_{I_m^*} \notin A$. On noting that $\big\|\mathcal{X}_{I_m^*}\big\|_{L^1} \le \frac{1}{m}$, it holds that

$$f - 2M\mathcal{X}_{I_m^*} \longrightarrow f \quad \text{in } L^1(I) \text{ as } m \to \infty.$$

This shows that $f \notin \mathring{A}$. □

E4.2 Convex sets. Let X be a vector space over $\mathbb{K}$. Show that:

(1) Every affine subspace $U \subset X$ is convex.

(2) If $X \neq \{0\}$ is a normed space, then for $x \in X$ and $r > 0$ the balls $B_r(x)$ and $\overline{B}_r(x)$ are convex, but $\partial B_r(x)$ is not convex.

(3) If X is a Hilbert space and $x \in X$, then for $r \geq 0$ and $a \in \mathbb{K}$ the *strip* $\{y \in X ;\ |(x, y)_X - a| \leq r\}$, and for $a \in \mathbb{R}$ the *affine half-space* $\{y \in X ;\ \mathrm{Re}\,(x, y)_X \leq a\}$, are convex.

(4) If $X = C^0(S; \mathbb{R})$ and $g \in X$, then $\{f \in X ;\ f(x) \geq g(x) \text{ for } x \in S\}$ is convex.

(5) If $X = L^p(\mu; \mathbb{R})$ and $g \in X$, then $\{f \in X;\ f \geq g \ \mu\text{-almost everywhere}\}$ is convex.

E4.3 Distance in a Banach space. In a Banach space, in general the distance to a closed subspace is not attained.

Solution. The space $X := \{f \in C^0([0,1]) ;\ f(0) = 0\}$ with the C^0-norm is a Banach space and $Y := \{f \in X ;\ \int_0^1 f(x)\,\mathrm{d}x = 0\}$ is a closed proper subspace. We claim that

$$\mathrm{dist}(f, Y) = \left| \int_0^1 f(x)\,\mathrm{d}x \right| \quad \text{for all } f \in X. \tag{E4-1}$$

To see this, note that for all $g \in Y$

$$\left| \int_0^1 f(x)\,\mathrm{d}x \right| = \left| \int_0^1 (f(x) - g(x))\,\mathrm{d}x \right| \leq \|f - g\|_{C^0} \,.$$

Moreover, $h_n(x) := \left(1 + \frac{1}{n}\right)x^{\frac{1}{n}}$ satisfies $\int_0^1 h_n(x)\,\mathrm{d}x = 1$, and so

$$g_n := f - \left(\int_0^1 f(x)\,\mathrm{d}x\right)h_n \in Y$$

with

$$\|f - g_n\|_{C^0} = \left| \int_0^1 f(x)\,\mathrm{d}x \right| \cdot \left(1 + \tfrac{1}{n}\right) \to \left| \int_0^1 f(x)\,\mathrm{d}x \right| \quad \text{as } n \to \infty.$$

Hence we have shown (E4-1). In addition,

$$\left| \int_0^1 h(x)\,\mathrm{d}x \right| < \|h\|_{C^0} \quad \text{for all } h \in X \setminus \{0\}, \tag{E4-2}$$

since $h(0) = 0$. Now let $f \in X \setminus Y$. If the distance to Y were attained by a $g_0 \in Y$, then it would follow that

$$\left| \int_0^1 f(x)\,\mathrm{d}x \right| = \mathrm{dist}(f, Y) \quad \text{(recall (E4-1))}$$

$$= \|f - g_0\|_{C^0} > \left| \int_0^1 (f(x) - g_0(x))\,\mathrm{d}x \right| \quad \text{(recall (E4-2))}$$

$$= \left| \int_0^1 f(x)\,\mathrm{d}x \right|,$$

a contradiction. $\qquad\qquad\qquad\qquad\qquad\qquad\qquad\qquad\qquad\qquad\qquad\qquad$ □

E4.4 Strictly convex spaces. Let X be a normed $\mathbb{K}$-vector space with norm $x \mapsto \|x\|$. Then the following are equivalent:

(1) X is *strictly normed*, i.e. for all $x, y \in X$

$$\|x + y\| = \|x\| + \|y\| \quad\Longrightarrow\quad \{x, y\} \text{ is linearly dependent.}$$

(2) $B_1(0) \subset X$ is *strictly convex*, i.e. for all $x, y \in X$

$$\|x\| = \|y\| = 1, \ x \neq y \quad\Longrightarrow\quad \left\|\tfrac{1}{2}(x + y)\right\| < 1.$$

(3) Every closed convex set $K \subset X$ contains at most one element with minimal norm, i.e.

$$x_j \in K \text{ with } \|x_j\| = \inf_{x \in K} \|x\| \text{ for } j = 1, 2 \quad\Longrightarrow\quad x_1 = x_2.$$

Solution (1)⇒(2). Assume that $\left\|\tfrac{1}{2}(x + y)\right\| \geq 1$. Then

$$2 \leq \|x + y\| \leq \|x\| + \|y\| = 2,$$

and so equality must hold. Hence it follows from (1) that there exist $\alpha, \beta \in \mathbb{K}$ with $(\alpha, \beta) \neq 0$ and $\alpha x + \beta y = 0$. By symmetry, we may assume that $\beta \neq 0$, and hence upon scaling that $\beta = 1$. Then $y = -\alpha x$, and so

$$2 = \|x + y\| = |1 - \alpha| \cdot \|x\| = |\alpha - 1| \quad \text{and} \quad 1 = \|y\| = |\alpha| \cdot \|x\| = |\alpha|,$$

which implies $\alpha = -1$, i.e. $y = x$, a contradiction. $\qquad\qquad\qquad\qquad$ □

Solution (2)⇒(3). Let $\varrho := \inf_{x \in K} \|x\|$. If $\varrho = 0$, then $x_1 = 0, x_2 = 0$. Hence let $\varrho > 0$. We have that $\tfrac{1}{2}(x_1 + x_2) \in K$, and so $\left\|\tfrac{1}{2}(x_1 + x_2)\right\| \geq \varrho$. Assuming that $x_1 \neq x_2$, we could apply (2) for $\tfrac{1}{\varrho}x_1, \tfrac{1}{\varrho}x_2$ and would obtain that $\left\|\tfrac{1}{2}(x_1 + x_2)\right\| < \varrho$, a contradiction. $\qquad\qquad\qquad\qquad\qquad$ □

Solution (3)⇒(1). Let $x, y \in X$ be linearly independent. Consider $\widetilde{x} := \tfrac{1}{\|x\|}x$ and $\widetilde{y} := \tfrac{1}{\|y\|}y$, together with the closed convex set

$$K := \{(1 - s)\widetilde{x} + s\widetilde{y}; \ s \in [0, 1]\}.$$

Put $\varphi(s) := \|(1 - s)\widetilde{x} + s\widetilde{y}\|$. Then $\varphi : [0, 1] \to \mathbb{R}$ is convex (follows from the triangle inequality) and continuous. Since φ is continuous and nonnegative, there exists an absolute minimum s_0 of φ on $[0, 1]$. On noting that $\varphi(0) = \varphi(1) = 1$, it follows from (3) that $0 < s_0 < 1$ and $\varphi(s_0) < 1$. The convexity of φ yields that $\varphi(s) < 1$ for all $s \in \,]0, 1[$. Moreover,

$$\frac{1}{2}(x + y) = r\big((1 - s)\widetilde{x} + s\widetilde{y}\big) \quad \text{with} \quad \begin{cases} r := \dfrac{1}{2}(\|x\| + \|y\|), \\[2mm] s := \dfrac{\|y\|}{2r} \in \,]0, 1[. \end{cases}$$

This implies that

$$1 > \varphi(s) = \left\| \frac{1}{2r}(x+y) \right\| \quad \text{and so} \quad \|x+y\| < \|x\| + \|y\|.$$

This shows (1). □

E4.5 Separation theorem in $\mathbb{R}^n$. Let $A, B \subset \mathbb{R}^n$ be nonempty, closed convex sets with $A \cap B = \emptyset$. Then there exist $x_0, e \in \mathbb{R}^n$ with $|e| = 1$ and

$$(a - x_0) \bullet e \geq 0 \text{ for } a \in A \quad \text{and} \quad (b - x_0) \bullet e \leq 0 \text{ for } b \in B.$$

Remark: For the infinite-dimensional case, see 8.12.

Solution. Let $A_k := A \cap \overline{B_k(0)}$ and $B_k := B \cap \overline{B_k(0)}$. Since A_k and B_k are compact and disjoint, there exist $a_k \in A_k$, $b_k \in B_k$ with

$$|a_k - b_k| = \text{dist}(A_k, B_k) > 0.$$

As A_k and B_k are convex, it follows, similarly to the projection theorem 4.3, that

$$(b_k - a_k) \bullet (a - a_k) \leq 0 \quad \text{for } a \in A_k,$$
$$(a_k - b_k) \bullet (b - b_k) \leq 0 \quad \text{for } b \in B_k.$$

On setting

$$e_k := \frac{a_k - b_k}{|a_k - b_k|}, \quad \alpha_k := \frac{a_k + b_k}{2} \bullet e_k$$

we obtain that

$$a \bullet e_k \geq a_k \bullet e_k \geq \alpha_k \quad \text{for } a \in A_k,$$
$$b \bullet e_k \leq b_k \bullet e_k \leq \alpha_k \quad \text{for } b \in B_k.$$

As $|e_k| = 1$, there exists an $e \in \mathbb{R}^n$ with $|e| = 1$ such that $e_k \to e$ for a subsequence $k \to \infty$. Choosing arbitrary but fixed $a_0 \in A$, $b_0 \in B$, it follows that $a_0 \in A_k$, $b_0 \in B_k$ for k sufficiently large, and so

$$-|b_0| \leq b_0 \bullet e_k \leq \alpha_k \leq a_0 \bullet e_k \leq |a_0|.$$

Hence the α_k are bounded and we can choose a subsequence such that $\alpha_k \to \alpha \in \mathbb{R}$ as $k \to \infty$. On noting that $a \bullet e_k \geq \alpha_k$ for $a \in A_j$, if $k \geq j$, it follows that $a \bullet e \geq \alpha$ for $a \in A_j$, and hence also for all $a \in A$. Similarly, we obtain that $b \bullet e \leq \alpha$ for all $b \in B$. Now choose $x_0 := \alpha e$. □

E4.6 Convex functions. Let $\Omega \subset \mathbb{R}^n$ be open and convex. Then every convex function $f : \Omega \to \mathbb{R}$ is locally Lipschitz continuous, i.e. $f \in C^{0,1}(S)$ for all compact sets $S \subset \Omega$.

Solution. Let $\overline{B_r(x_0)} \subset \Omega$, where the balls are taken with respect to the sum norm $|\cdot|_1$. Then

$$\overline{B_r(x_0)} = \text{conv } M, \quad M := \{x_0 \pm re_i \, ; \, i = 1, \ldots, n\}$$

(this can be shown by induction on n). Hence every $z \in \overline{B_r(x_0)}$ can be represented as

$$z = \sum_{i=1}^{m} \lambda_i z_i \quad \text{with } m \in \mathbb{N}, \, z_i \in M, \, \lambda_i \geq 0, \, \sum_{i=1}^{m} \lambda_i = 1.$$

The convexity of f implies that

$$f(z) \leq \sum_{i=1}^{m} \lambda_i f(z_i) \leq C := \max_{\xi \in M} |f(\xi)|,$$

i.e. f is bounded from above on $\overline{B_r(x_0)}$. Now let $x \in \overline{B_{\frac{r}{2}}(x_0)} \setminus \{x_0\}$. Choose $\alpha \geq 1$ such that $z := x_0 - \alpha(x - x_0) \in \partial B_r(x_0)$. Then

$$x_0 = \frac{1}{1+\alpha} z + \frac{\alpha}{1+\alpha} x,$$

and so

$$f(x_0) \leq \frac{1}{1+\alpha} f(z) + \frac{\alpha}{1+\alpha} f(x),$$

which implies

$$f(x) \geq \frac{1+\alpha}{\alpha} f(x_0) - \frac{1}{\alpha} f(z) \geq -2|f(x_0)| - C,$$

i.e. f is also bounded from below on $\overline{B_{\frac{r}{2}}(x_0)}$. For $x, y \in \overline{B_{\frac{r}{2}}(x_0)}$, $x \neq y$, we now choose $\alpha \geq 1$ such that $z := x + \alpha(y - x) \in \partial B_r(x_0)$. Then

$$y = \frac{1}{\alpha} z + \left(1 - \frac{1}{\alpha}\right) x,$$

and so

$$f(y) \leq \frac{1}{\alpha} f(z) + \left(1 - \frac{1}{\alpha}\right) f(x),$$

which, on recalling the above estimate, implies that

$$f(y) - f(x) \leq \frac{1}{\alpha}(f(z) - f(x)) \leq \frac{1}{\alpha} 2 \left(C + |f(x_0)|\right).$$

In addition,

$$\frac{1}{\alpha} = \frac{|y - x|_1}{|z - x|_1} \leq \frac{2}{r} |y - x|_1.$$

Then a symmetry argument yields that

$$|f(y) - f(x)| \le \frac{4}{r}(C + |f(x_0)|)\,|y - x|_1\,.$$

Hence f is Lipschitz continuous on $\overline{B_{\frac{r}{2}}(x_0)}$. Now cover compact subsets $S \subset \Omega$ with appropriately chosen finitely many balls (for $x \in S$ choose a ball $B_{\frac{r_x}{4}}(x)$ with $B_{r_x}(x) \subset \Omega$). Conclude that $f \in C^{0,1}(S)$.   □

E4.7 Characterization of convex functions. Let $\Omega \subset \mathbb{R}^n$ be open and convex. Then

(1) $f \in C^1(\Omega; \mathbb{R})$ is convex if and only if the **Weierstraß E-function** satisfies

$$E(x, y) := f(y) - f(x) - \nabla f(x) \bullet (y - x) \ge 0 \quad \text{for all } x, y \in \Omega\,.$$

(2) $f \in C^1(\Omega; \mathbb{R})$ is convex if and only if ∇f is **monotone**, i.e

$$(\nabla f(x) - \nabla f(y)) \bullet (x - y) \ge 0 \quad \text{for all } x, y \in \Omega\,.$$

(3) $f \in C^2(\Omega; \mathbb{R})$ is convex if and only if $D^2 f$ is **positive semidefinite**, i.e.

$$\sum_{i,j=1}^{n} \partial_{ij} f(x)\xi_i\xi_j \ge 0 \quad \text{for all } \xi \in \mathbb{R}^n,\ x \in \Omega\,.$$

Solution (1) *and* (2). For $x_0, x_1 \in \Omega$ define

$$g(t) := f(x_t) \quad \text{with} \quad x_t := (1 - t)x_0 + tx_1 \in \Omega \quad \text{for } 0 \le t \le 1\,.$$

Then we have that

$$g'(t) = \nabla f(x_t) \bullet (x_1 - x_0)\,.$$

If f is convex, then so is g, since for $0 \le \alpha \le 1$

$$g\big((1 - \alpha)s + \alpha t\big) = f\big((1 - \alpha)x_s + \alpha x_t\big) \le (1 - \alpha)f(x_s) + \alpha f(x_t)\,.$$

In particular, for $0 < \varepsilon < 1$ we have that

$$g(\varepsilon) \le (1 - \varepsilon)g(0) + \varepsilon g(1) \quad \text{and so} \quad \frac{g(\varepsilon) - g(0)}{\varepsilon} \le g(1) - g(0)\,.$$

On letting $\varepsilon \to 0$ we obtain that $g'(0) \le g(1) - g(0)$, i.e. $E(x_0, x_1) \ge 0$.
    If the Weierstraß function is nonnegative, it follows that

$$0 \le E(x_0, x_1) + E(x_1, x_0) = \big(\nabla f(x_1) - \nabla f(x_0)\big) \bullet (x_1 - x_0)\,,$$

i.e. ∇f is monotone. If ∇f is monotone, then so is g', since for $0 \le s < t \le 1$,

$$g'(t) - g'(s) = \frac{\big(\nabla f(x_t) - \nabla f(x_s)\big) \bullet (x_t - x_s)}{t - s} \ge 0\,.$$

We obtain for $0 \le \alpha \le 1$ that

$$(1 - \alpha)\big(g(\alpha) - g(0)\big) = (1 - \alpha) \int_0^\alpha g'(t)\, dt \le (1 - \alpha)\alpha g'(\alpha)$$

$$\le \alpha \int_\alpha^1 g'(t)\, dt = \alpha\big(g(1) - g(\alpha)\big)\,,$$

i.e.

$$g(\alpha) \le (1 - \alpha)g(0) + \alpha g(1) \quad \text{or} \quad f(x_\alpha) \le (1 - \alpha)f(x_0) + \alpha f(x_1)\,.$$

As this holds for all x_0, x_1, α, we have shown the convexity of f. □

Solution (3). If f is convex, then it follows from (2) that for $x \in \Omega$, $\xi \in \mathbb{R}^n$ and $\varepsilon > 0$ sufficiently small

$$0 \le \frac{1}{\varepsilon}\xi \bullet \big(\nabla f(x + \varepsilon\xi) - \nabla f(x)\big) \longrightarrow \xi \bullet D^2 f(x)\xi \quad \text{as } \varepsilon \searrow 0\,.$$

Conversely, if $D^2 f$ is positive semidefinite, then for $x, y \in \Omega$

$$\big(\nabla f(y) - \nabla f(x)\big) \bullet (x - y) = \int_0^1 (x - y) \bullet D^2 f\big((1 - t)x + ty\big)(x - y)\, dt \ge 0\,.$$

Hence (2) yields that f is a convex function. □

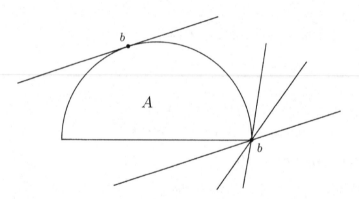

Fig. 4.6. *Supporting lines in* $\mathbb{R}^2$

E4.8 Supporting planes. Let $A \subset \mathbb{R}^n$ be convex. Then for $b \in \partial A$ there exists an $e \in \mathbb{R}^n$ with $|e| = 1$ such that

$$(a - b) \bullet e \ge 0 \quad \text{for all } a \in A\,.$$

Remark: We then call $\{x \in \mathbb{R}^n \,;\, (x - b) \bullet e = 0\}$ a **supporting plane** to A at the point b. This supporting plane need not be uniquely determined. It is unique in the case where A is the set above the graph of a convex C^1-function. The supporting plane inequality is then given by the nonnegativity of the E-function in E4.7(1).

Hint: See also the separation theorem 8.12.

Solution. There exist $b_k \in \mathbb{R}^n \setminus \overline{A}$ for $k \in \mathbb{N}$ with $b_k \to b$ as $k \to \infty$. Fix k and set $B := \{b_k\}$. It follows from E4.5 that there exist points $x_k, e_k \in \mathbb{R}^n$ with $|e_k| = 1$ such that

$$(a - x_k) \bullet e_k \geq 0 \text{ for } a \in A \quad \text{and} \quad (b_k - x_k) \bullet e_k \leq 0.$$

It follows that $(a - b_k) \bullet e_k \geq 0$ for $a \in A$. Now choose a subsequence $k \to \infty$ with $e_k \to e$. $\qquad\square$

E4.9 Jensen's inequality. Let $\Phi : \mathbb{R}^m \to \mathbb{R}$ be convex with $\Phi \geq 0$, and let $(S, \mathcal{B}, \mu)$ be a measure space with $\mu(S) = 1$. Then for any $f \in L^1(\mu; \mathbb{R}^m)$

$$\Phi\left(\int_S f \, d\mu\right) \leq \int_S \Phi \circ f \, d\mu.$$

Note: The right-hand side is to be set to $+\infty$ if $\Phi \circ f$ is not integrable.

Solution. By E4.6, Φ is Lipschitz continuous, and so $\Phi \circ f$ is measurable. Moreover, it holds for $y_0 \in \mathbb{R}^m$ that $(y_0, \Phi(y_0)) \in \partial A$, where $A := \{(y, \xi) \in \mathbb{R}^{m+1} ; \xi \geq \Phi(y)\}$. It follows from E4.8 that there exists an $e = (e', e_{m+1}) \in \mathbb{R}^{m+1} \setminus \{0\}$ with

$$(\xi - \Phi(y_0))e_{m+1} + (y - y_0) \bullet e' \geq 0 \quad \text{for all } (y, \xi) \in A.$$

We may assume that $e_{m+1} = 1$. (To see this, let $y = y_0$ and $\xi = \Phi(y_0) + 1$, which implies that $e_{m+1} \geq 0$. Assuming that $e_{m+1} = 0$ yields that $(y - y_0) \bullet e' = 0$ for all $y \in \mathbb{R}^m$, and so $e' = 0$, which contradicts $e \neq 0$.) It follows that

$$\Phi(y) \geq \Phi(y_0) - (y - y_0) \bullet e' \quad \text{for all } y \in \mathbb{R}^m.$$

In this inequality we let for $x \in S$

$$y := f(x) \quad \text{and} \quad y_0 := \int_S f \, d\mu.$$

Then integrating over $S_m := \{x \in S ; |f(x)| \leq m\}$ we obtain

$$\int_{S_m} \Phi \circ f \, d\mu \geq \mu(S_m)\Phi(y_0) - \left(\int_{S_m} f \, d\mu - \mu(S_m)y_0\right) \bullet e' \longrightarrow \Phi(y_0)$$

as $m \to \infty$, whence

$$\Phi\left(\int_S f \, d\mu\right) \leq \liminf_{m \to \infty} \int_{S_m} \Phi \circ f \, d\mu.$$

$\qquad\square$

E4.10 L^p-inequalities. Let μ be as in E4.9, i.e. $\mu(S) = 1$. For μ-measurable functions $f : S \to Y$ apply E4.9 appropriately in order to obtain an inequality for the integral of $|f|^p$ for $0 < p < 1$ and $1 < p < \infty$, respectively.

Solution. For $0 < p < 1$ apply E4.9 to $\Phi(z) := |z|^{\frac{1}{p}}$. For $f \in L^1(\mu; Y)$ it then holds that $|f|^p \in L^1(\mu)$, since $\mu(S) < \infty$ and $|f|^p \leq \max(1, |f|)$. We obtain

$$\int_S |f|^p \, d\mu \leq \left(\int_S |f| \, d\mu \right)^p \quad \text{and so} \quad \|f\|_{L^p} \leq \|f\|_{L^1} .$$

For $1 < p < \infty$ apply E4.9 to $\Phi(z) := |z|^p$. For $f \in L^p(\mu; Y)$ it then holds that $f \in L^1(\mu; Y)$, on noting $\mu(S) < \infty$ and the Hölder inequality. We obtain

$$\left(\int_S |f| \, d\mu \right)^p \leq \int_S |f|^p \, d\mu \quad \text{and so} \quad \|f\|_{L^1} \leq \|f\|_{L^p} .$$

$\square$

E4.11 The space L^p for $p < 1$. Let $0 < p < 1$. Then $L^p(]0, 1[)$ equipped with the Fréchet metric

$$\varrho(f) := \int_0^1 |f(x)|^p \, dx$$

is a metric space. Show that:

(1) The convex hull of any ball $B_r(0)$ is the whole space.

(2) There exists no norm on $L^p(]0, 1[)$ which induces the same topology as the metric ϱ.

Solution. Because $(s + t)^p \leq s^p + t^p$ for all $s, t \geq 0$ the metric ϱ is a Fréchet metric. For $f \in L^p(]0, 1[)$ we have that

$$x \mapsto \int_0^x |f(x)|^p \, dx$$

is continuous, and hence for $n \in \mathbb{N}$ there exist numbers $0 = x_0 < x_1 < \cdots < x_{n-1} < x_n = 1$ such that

$$\int_0^{x_i} |f(x)|^p \, dx = \frac{i}{n} \int_0^1 |f(x)|^p \, dx .$$

It then follows that

$$f = \sum_{i=1}^n \frac{1}{n} f_i \text{ in } L^p(]0, 1[) \quad \text{where} \quad f_i := n f \cdot \chi_{]x_{i-1}, x_i[} , \tag{E4-3}$$

and

$$\int_0^1 |f_i(x)|^p \, dx = n^p \int_{x_{i-1}}^{x_i} |f(x)|^p \, dx = n^{p-1} \int_0^1 |f(x)|^p \, dx < r$$

for n sufficiently large, hence (with respect to the metric) $f_i \in B_r^{\text{metric}}(0)$. Therefore $f \in \text{conv} \, B_r^{\text{metric}}(0)$, on noting (E4-3). This yields the desired result (1).

In order to show (2), we assume that there exists a norm $\|\cdot\|$ that is equivalent to ϱ. Then the ball $B_1^{\text{norm}}(0)$ must be open with respect to ϱ, and so

$$B_r^{\mathrm{metric}}(0) \subset B_1^{\mathrm{norm}}(0)$$

for an $r > 0$. Then it follows from (1) that

$$L^p(]0,1[) = \mathrm{conv}\, B_r^{\mathrm{metric}}(0) \subset \mathrm{conv}\, B_1^{\mathrm{norm}}(0) = B_1^{\mathrm{norm}}(0) \ ,$$

a contradiction. $\qquad\qquad\square$

E4.12 Cross product of normed vector spaces. Let $(X_1, \|\cdot\|_1)$ and $(X_2, \|\cdot\|_2)$ be normed $\mathbb{K}$-vector spaces. In addition, let $|\cdot|$ be a **monotone norm** on $\mathbb{R}^2$, i.e. for $(a_1, a_2), (b_1, b_2) \in \mathbb{R}^2$

$$|a_1| \le |a_2|,\ |b_1| \le |b_2| \quad\Longrightarrow\quad |(a_1, a_2)| \le |(b_1, b_2)| \, .$$

Show that:

(1) $\|(x_1, x_2)\| := |(\|x_1\|_1, \|x_2\|_2)|$ for $x_1 \in X_1$, $x_2 \in X_2$ defines a norm on $X_1 \times X_2$.
(2) All the norms on $X_1 \times X_2$ which are defined by a monotone norm on $\mathbb{R}^2$ as in (1) are equivalent.

Solution. For (2) use 4.8 on $\mathbb{R}^2$. $\qquad\qquad\square$

E4.13 Compact sets in ℓ^2. Determine whether the following subsets of $\ell^2(\mathbb{R})$ are bounded and/or compact.

$$E_1 := \left\{ x \in \ell^2(\mathbb{R}) \ ; \ |x_i| \le \tfrac{1}{\sqrt{i}} \text{ for all } i \right\} ,$$
$$E_2 := \left\{ x \in \ell^2(\mathbb{R}) \ ; \ \textstyle\sum_{i=1}^{\infty} x_i^2 \le 1 \right\} ,$$
$$E_3 := \left\{ x \in \ell^2(\mathbb{R}) \ ; \ |x_i| \le \tfrac{1}{i} \text{ for all } i \right\} \quad (\textbf{Hilbert cube}).$$

Solution. The set E_1 is not bounded, and hence also not compact, because for $n \in \mathbb{N}$ we have that

$$x^n := \textstyle\sum_{i=1}^{n} \tfrac{1}{\sqrt{i}} e_i \in E_1 \quad \text{with} \quad \|x^n\|_{\ell^2}^2 = \sum_{i=1}^{n} \tfrac{1}{i} \longrightarrow \infty \quad \text{as } n \to \infty.$$

By definition, $E_2 = \overline{B_1(0)}$ is bounded, but it is not compact. To see this, observe that $(e_n)_{n \in \mathbb{N}}$ is a sequence in E_2, which, since $\|e_n - e_m\|_{\ell^2} = \sqrt{2}$ for $n \ne m$, does not contain a convergent subsequence.

The set E_3 is compact: If $(x^n)_{n \in \mathbb{N}}$ is a sequence in E_3, then for all $i \in \mathbb{N}$ the sequence $(x_i^n)_{n \in \mathbb{N}}$ is bounded, since $|x_i^n| \le \tfrac{1}{i}$. On applying a diagonalization procedure, we obtain a subsequence $n \to \infty$, such that $x_i^n \to x_i$ as $n \to \infty$ for all $i \in \mathbb{N}$, with certain numbers x_i. Then $|x_i| \le \tfrac{1}{i}$, and so $x := (x_i)_{i \in \mathbb{N}} \in E_3$ with

$$\|x^n - x\|_{\ell^2}^2 \le \underbrace{\sum_{i=1}^{j} |x_i^n - x_i|^2}_{\substack{\to\, 0 \text{ as } n \to \infty \\ \text{for every } j}} + \underbrace{\sum_{i=j+1}^{\infty} \frac{4}{i^2}}_{\to\, 0 \text{ as } j \to \infty} ,$$

i.e. $x^n \to x$ in $\ell^2(\mathbb{R})$ as $n \to \infty$. ☐

E4.14 Bounded and compact sets in $L^1(]0,1[)$. Determine whether the following sets are bounded and precompact in $L^1(]0,1[)$, respectively:

$$E_1 := \{ f :]0,1[\to \mathbb{R} \; ; \; f(x) = x^{-\alpha}, \; 0 \le \alpha < 1 \},$$
$$E_2 := \{ f :]0,1[\to \mathbb{R} \; ; \; f(x) = x^{-\alpha}, \; -\infty < \alpha \le 1 - \delta \} \text{ where } \delta > 0,$$
$$E_3 := \{ f :]0,1[\to \mathbb{R} \; ; \; f(x) = \sin(\omega x), \; \omega \in \mathbb{R} \}.$$

Solution. The set E_1 is not bounded, since

$$\int_0^1 x^{-\alpha} \, \mathrm{d}x = \left[\frac{1}{1-\alpha} x^{1-\alpha} \right]_{x=0}^{x=1} = \frac{1}{1-\alpha} \longrightarrow \infty \qquad \text{as } \alpha \nearrow 1. \qquad \text{(E4-4)}$$

The same calculation yields the boundedness of E_2.

The precompactness of E_2 can, for instance, be shown as follows: Let $f_\alpha(x) := x^{-\alpha}$ for $-\infty < \alpha \le 1 - \delta$. For every x we have that $f_\alpha(x)$ depends continuously on α, and $0 \le f_\alpha \le f_{1-\delta}$. By Lebesgue's convergence theorem $\alpha \mapsto f_\alpha$ is a continuous map into $L^1(]0,1[)$. On defining $f_{-\infty} := 0$, it follows from (E4-4) that it is even continuous on $[-\infty, 1-\delta]$, which we consider as a compact interval on the extended real line (see 2.8(2)). Hence the image $E_2 \cup \{0\}$ of this continuous map is also compact.

The set E_3 is obviously bounded, but it is not precompact. To see this, observe that, for $h > 0$ with $\frac{1}{2h} \in \mathbb{N}$ and $\omega = \frac{\pi}{2h}$,

$$\int_0^1 |\sin(\omega(x+h)) - \sin(\omega x)| \, \mathrm{d}x = \int_0^1 |\cos(\omega x) - \sin(\omega x)| \, \mathrm{d}x$$
$$= \frac{1}{\omega} \int_0^\omega |\cos y - \sin y| \, \mathrm{d}y = \frac{1}{\pi} \int_0^\pi |\cos y - \sin y| \, \mathrm{d}y > 0.$$

Hence condition 4.16(2) is violated. ☐

E4.15 Comparison of Hölder spaces. Let $S \subset \mathbb{R}^n$ be a compact set and let $0 < \alpha < \beta \le 1$. Then bounded sets in $C^{0,\beta}(S)$ are precompact in $C^{0,\alpha}(S)$ (a more general result is 10.9).

Solution. Let $(f_k)_{k \in \mathbb{N}}$ be a bounded sequence in $C^{0,\beta}(S)$, with $\|f_k\|_{C^{0,\beta}} \le R$ for all k, say. Then $\{f_k \; ; \; k \in \mathbb{N}\}$ is bounded in $C^0(S)$ and equicontinuous. It follows from the Arzelà-Ascoli theorem that there exists an $f \in C^0(S)$ such that $f_k \to f$ in $C^0(S)$ for a subsequence $k \to \infty$. For $x, y \in S$ it then holds that

$$\frac{|(f_k - f)(x) - (f_k - f)(y)|}{|x - y|^\alpha}$$

for $|x - y| \ge \delta$ can be estimated by

$$\le 2\delta^{-\alpha} \|f_k - f\|_{C^0}$$

and for $0 < |x - y| < \delta$ satisfies

$$= \lim_{l \to \infty} \frac{|(f_k - f_l)(x) - (f_k - f_l)(y)|}{|x - y|^\alpha} \le \delta^{\beta - \alpha} \|f_k - f_l\|_{C^{0,\beta}} \le 2R\delta^{\beta - \alpha}.$$

Therefore,

$$\|f_k - f\|_{C^{0,\alpha}} \le (1 + 2\delta^{-\alpha}) \underbrace{\|f_k - f\|_{C^0}}_{\to 0 \text{ for } k \to \infty} + 2R \underbrace{\delta^{\beta - \alpha}}_{\to 0 \text{ for } \delta \to 0},$$

and hence $f_k \to f$ in $C^{0,\alpha}(S)$ as $k \to \infty$. $\qquad\qquad\square$

E4.16 Compactness with respect to the Hausdorff metric. Let $\mathcal{A}$ be as in E2.9 with $X = \mathbb{R}^n$ and let $R > 0$. Then

$$\{ A \in \mathcal{A} ;\ A \subset \overline{B_R(0)} \} \quad \text{is compact in } \mathcal{A}.$$

Solution. Let $(A_m)_{m \in \mathbb{N}}$ be a sequence in $\mathcal{A}$ with $A_m \subset B := \overline{B_R(0)}$. Then the functions

$$f_m(x) := \text{dist}(x, A_m)$$

are bounded in $C^{0,1}(B)$ (see E2.2(1)). By the Arzelà-Ascoli theorem, there exists an $f \in C^0(B)$ such that $f_m \to f$ in $C^0(B)$ for a subsequence $m \to \infty$. Let

$$A := \{x \in B ;\ f(x) = 0\}.$$

Then A is closed and nonempty, since assuming $A = \emptyset$ would yield that f is strictly positive on B, and hence also f_m for m sufficiently large, a contradiction.

In addition, we have that $f(x) = \text{dist}(x, A)$. To see this, observe that for $a \in A$ it holds, as $m \to \infty$,

$$f(x) = f(x) - f(a) \longleftarrow \text{dist}(x, A_m) - \text{dist}(a, A_m) \le |x - a|$$

by E2.2(1), and hence $f(x) \le \text{dist}(x, A)$. If $x \in B \setminus A$ and $r < \text{dist}(x, A)$, then f is strictly positive on $\overline{B_r(x)} \cap B$. Hence so is f_m for m sufficiently large, i.e. $\overline{B_r(x)} \cap B \cap A_m = \emptyset$, which, since $A_m \subset B$, implies that $\overline{B_r(x)} \cap A_m = \emptyset$, and so

$$r \le \text{dist}(x, A_m) \longrightarrow f(x).$$

This shows that $f(x) \ge \text{dist}(x, A)$. It follows from E2.9 (with $M = B$) that

$$d_H(A_m, A) = \sup_{x \in B} |\text{dist}(x, A_m) - \text{dist}(x, A)|$$

$$= \|f_m - f\|_{C^0} \longrightarrow 0 \quad \text{as } m \to \infty.$$

$\qquad\qquad\square$

E4.17 Uniform continuity. If X is a compact metric space and Y is a metric space, then every continuous function $f : X \to Y$ is **uniformly continuous**, i.e.

$$d_Y\big(f(y), f(x)\big) \longrightarrow 0 \quad \text{for } x, y \in X \text{ with } d_X(y, x) \to 0,$$

which, using quantifiers, can be written as

$$\forall\, \varepsilon > 0 : \exists\, \delta > 0 : \forall\, x, y \in X : d_X(x, y) \le \delta \Longrightarrow d_Y\big(f(x), f(y)\big) \le \varepsilon.$$

Solution. For $\varepsilon > 0$ and $x \in X$ there exists a $\delta_x > 0$ with

$$d_Y\big(f(y), f(x)\big) \le \varepsilon \quad \text{for } y \in B_{2\delta_x}(x).$$

As $(B_{\delta_x}(x))_{x \in X}$ is an open cover of X, there exists a finite collection of balls $B_{\delta_{x_i}}(x_i)$, $i = 1, \dots, m_\varepsilon$, that cover the compact set X. Let

$$\delta := \min\{\delta_{x_i} \; ; \; i = 1, \dots, m_\varepsilon\}.$$

For any two points $y, z \in X$ with $d_X(y, z) \le \delta$ it then holds that $y \in B_{\delta_{x_i}}(x_i)$ for an i, and so $y, z \in B_{2\delta_{x_i}}(x_i)$, which implies that

$$d_Y\big(f(y), f(z)\big) \le d_Y\big(f(y), f(x_i)\big) + d_Y\big(f(z), f(x_i)\big) \le 2\varepsilon.$$

$\square$

E4.18 Continuous extension. Let X be a metric space, let $A \subset X$ be dense and let Y be a complete metric space. Then every uniformly continuous function $f : A \to Y$ admits a unique uniformly continuous extension $\widetilde{f} : X \to Y$.

Remark: For the linear case, see also E5.3.

Solution. Since f is uniformly continuous on A, it holds for $x \in X$ that

$$d_Y\big(f(y_1), f(y_2)\big) \longrightarrow 0 \quad \text{for } y_1, y_2 \in A \text{ with } y_1, y_2 \to x.$$

Since Y is complete, it follows that for $x \in \overline{A} = X$ there exists

$$\widetilde{f}(x) := \lim_{y \in A, \; y \to x} f(y) \quad \text{in } Y,$$

and $\widetilde{f}(x) = f(x)$ for $x \in A$. Then $\widetilde{f}$ is also uniformly continuous. To see this, observe that for points $x_1, x_2 \in X$ with $0 < d_X(x_1, x_2) \le \delta$ there exist points $y_1, y_2 \in A$ such that $d_Y\big(\widetilde{f}(x_i), f(y_i)\big) \le \delta$ and $d_X(x_i, y_i) \le \frac{\delta}{2}$, and hence $d_X(y_1, y_2) \le 2\delta$. This implies that

$$d_Y\big(\widetilde{f}(x_1), \widetilde{f}(x_2)\big) \le 2\delta + d_Y\big(f(y_1), f(y_2)\big)$$
$$\le 2\delta + \sup_{\substack{z_1, z_2 \in A \\ d_X(z_1, z_2) \le 2\delta}} d_Y\big(f(z_1), f(z_2)\big) \longrightarrow 0$$

as $\delta \to 0$.

$\square$

E4.19 Dini's theorem. Let X be a compact metric space. In addition, let $f_k \in C^0(X; \mathbb{R})$ and let $f_k(x) \searrow 0$ (monotonically!) as $k \to \infty$ for all $x \in X$. Then $\|f_k\|_{C^0} \longrightarrow 0$ as $k \to \infty$.

Solution. Otherwise there exist a subsequence $k \to \infty$ and points $x_k \in X$ such that $f_k(x_k) \geq \varepsilon > 0$ for some $\varepsilon > 0$. As X is compact, it follows that $x_k \to x \in X$ for a subsequence $k \to \infty$.

Choose m with $f_m(x) \leq \frac{\varepsilon}{4}$. As f_m is continuous, there exists a $\delta > 0$ such that $f_m(y) \leq \frac{\varepsilon}{2}$ for $y \in X$ with $d(y, x) \leq \delta$.

We have for sufficiently large k that $d(x_k, x) \leq \delta$ and, if $k \geq m$, it follows that $f_k(x_k) \leq f_m(x_k) \leq \frac{\varepsilon}{2}$, a contradiction. $\qquad\square$

E4.20 Nonapproximability in the space $C^{0,\alpha}$. Let $0 < \alpha \leq 1$ and $I \subset \mathbb{R}$ be a nontrivial compact interval. Then $C^1(I)$ is not (!) dense in $C^{0,\alpha}(I)$.

Solution. Without loss of generality, let $I = [-1, 1]$. The fundamental theorem of calculus yields that $C^1(I) \subset C^{0,\alpha}(I)$. Let $f(x) := |x|^\alpha$. Then $f \in C^{0,\alpha}(I)$ and for $g \in C^1(I)$ and $x \in I \setminus \{0\}$ we have that

$$\|f - g\|_{C^{0,\alpha}} \geq \frac{|(f-g)(x) - (f-g)(0)|}{|x|^\alpha} = \left|1 - \frac{g(x) - g(0)}{|x|^\alpha}\right|$$

$$= \left|1 - \frac{x}{|x|^\alpha} \int_0^1 g'(sx)\, ds\right|.$$

For $\alpha < 1$ this is

$$\geq 1 - \|g'\|_{C^0} \cdot |x|^{1-\alpha} \to 1 \qquad \text{as } x \to 0,$$

and so $\|f - g\|_{C^{0,\alpha}} \geq 1$. For $\alpha = 1$ this converges

$$\longrightarrow \begin{cases} |1 - g'(0)| & \text{as } x \searrow 0, \\ |1 + g'(0)| & \text{as } x \nearrow 0. \end{cases}$$

This implies

$$\|f - g\|_{C^{0,1}} \geq \max\left(|1 - g'(0)|, |1 + g'(0)|\right) \geq 1.$$

$\qquad\square$

E4.21 Compact subsets of L^p. Let $1 \leq p < \infty$ and let $S \subset \mathbb{R}^n$ be bounded and measurable. Then $A \subset L^p(S)$ is precompact, if there exist measurable sets $S_k \subset \mathbb{R}^n$ and numbers $\varepsilon_k > 0$ with

$$S_k \subset S_{k+1} \subset S, \qquad \bigcup_{k \in \mathbb{N}} S_k = S, \qquad \varepsilon_k \searrow 0 \quad \text{as } k \to \infty,$$

such that the following conditions are satisfied:

(1) It holds that

$$\sup_{f \in A} \int_S |f(x)|^p \, dx < \infty.$$

(2) For every $k \in \mathbb{N}$ it holds for $h \in \mathbb{R}^n$ that

$$\sup_{f \in A} \int_{\{x \,;\, x, x+h \in S_k\}} |f(x+h) - f(x)|^p \, dx \longrightarrow 0 \quad \text{as } |h| \to 0.$$

(3) It holds that

$$\sup_{f \in A} \int_{S \cap B_{\varepsilon_k}(\mathbb{R}^n \setminus S_k)} |f(x)|^p \, dx \longrightarrow 0 \quad \text{as } k \to \infty.$$

Note: The assertion remains valid, if S is unbounded but the S_k are bounded.

Solution. We extend functions $f \in L^p(S)$ outside of S by 0 and obtain functions $\widetilde{f} \in L^p(\mathbb{R}^n)$. In this way, the set A becomes a subset $\widetilde{A} \subset L^p(\mathbb{R}^n)$.

If we assume that (2) holds with the integral over the whole of $\mathbb{R}^n$ for the extended functions, then it follows immediately that $\widetilde{A}$ satisfies the properties of the Riesz theorem 4.16, where we note that S is bounded. Hence $\widetilde{A}$ is then a precompact subset of $L^p(\mathbb{R}^n)$. Since for $f \in A$ and $g \in L^p(\mathbb{R}^n)$

$$\|f - \mathcal{X}_S \, g\|_{L^p(S)} \le \left\| \widetilde{f} - g \right\|_{L^p(\mathbb{R}^n)},$$

we conclude that then also A is a precompact subset of $L^p(S)$.

Now consider (2) as given in the problem. Then

$$\int_{\mathbb{R}^n} \left| \widetilde{f}(x+h) - \widetilde{f}(x) \right|^p dx \le \int_{\{x \,;\, x, x+h \in S_k\}} |f(x+h) - f(x)|^p \, dx$$

$$+ \int_{\{x \,;\, x \notin S_k\}} \left| \widetilde{f}(x+h) - \widetilde{f}(x) \right|^p dx + \int_{\{\widetilde{x} \,;\, \widetilde{x} \notin S_k\}} \left| \widetilde{f}(\widetilde{x}) - \widetilde{f}(\widetilde{x} - h) \right|^p d\widetilde{x}.$$

The first integral is controlled by condition (2). The second integral for $|h| < \varepsilon_k$ can be estimated by

$$\le 2^{p-1} \left(\int_{B_{\varepsilon_k}(\mathbb{R}^n \setminus S_k)} \left| \widetilde{f}(x) \right|^p dx + \int_{\mathbb{R}^n \setminus S_k} \left| \widetilde{f}(x) \right|^p dx \right)$$

$$\le 2 \cdot 2^{p-1} \int_{B_{\varepsilon_k}(\mathbb{R}^n \setminus S_k)} \left| \widetilde{f}(x) \right|^p dx,$$

and hence it can be controlled by condition (3). The third integral can be estimated in exactly the same way. Hence 4.16(2) holds for the integrals over $\mathbb{R}^n$, and so together with (1) the set A is a precompact subset of $L^p(S)$.

Remark: Since S is bounded, condition 4.16(3) is not relevant. $\square$

5 Linear operators

In this chapter, X, Y, Z, etc. usually denote normed $\mathbb{K}$-vector spaces. We consider linear maps T from X to Y, where, following the notation for matrices, we usually write Tx instead of $T(x)$, and similarly ST instead of $S \circ T$ for linear maps $T : X \to Y$ and $S : Y \to Z$. In functional analysis, only the continuous linear maps are of importance (see E9.2), which are those linear maps for which $T(x)$ can be estimated by x:

5.1 Lemma. If $T : X \to Y$ is linear and $x_0 \in X$, then the following are equivalent:

(1) T is continuous.

(2) T is continuous at x_0.

(3) $\sup_{\|x\|_X \leq 1} \|Tx\|_Y < \infty$.

(4) There exists a constant C with $\|Tx\|_Y \leq C\|x\|_X$ for all $x \in X$.

Property 5.1(4) written with quantifiers reads

$$\exists\, C \geq 0 : \big(\, \forall\, x \in X : \|Tx\|_Y \leq C\|x\|_X \,\big)$$

Proof (2)$\Rightarrow$(3). There exists an $\varepsilon > 0$ such that $T\big(\overline{\mathrm{B}_\varepsilon(x_0)}\big) \subset \mathrm{B}_1(T(x_0))$. Let $x \in \overline{\mathrm{B}_1(0)}$. Then $x_0 + \varepsilon x \in \overline{\mathrm{B}_\varepsilon(x_0)}$, and hence

$$T(x_0) + \varepsilon T(x) = T(x_0 + \varepsilon x) \in \mathrm{B}_1(T(x_0))\,,$$

which implies that $T(x) \in \mathrm{B}_{\frac{1}{\varepsilon}}(0)$. $\qquad\square$

Proof (3)$\Rightarrow$(4). Let C be the supremum in (3). Then for $x \neq 0$

$$\|T(x)\|_Y = \|x\|_X \cdot \left\| T\Big(\frac{x}{\|x\|_X}\Big) \right\|_Y \leq \|x\|_X \cdot C\,.$$

$\qquad\square$

Proof (4)$\Rightarrow$(1). For $x, x_1 \in X$ we have that

$$\|T(x) - T(x_1)\|_Y = \|T(x - x_1)\|_Y \leq C\|x - x_1\|_X \longrightarrow 0 \qquad \text{as } x \to x_1\,,$$

i.e. T is continuous at x_1. This is true for all x_1. $\qquad\square$

5.2 Linear operators. We define

$$\mathcal{L}(X;Y) := \big\{ T : X \to Y \ ; \ T \text{ is linear and continuous} \big\}.$$

We call maps in $\mathcal{L}(X;Y)$ **linear operators**. This is true in general for topological vector spaces X and Y (see 5.23). In the literature, if they are normed spaces, elements of $\mathcal{L}(X;Y)$ are often also called **bounded operators**. If X and Y are normed spaces, on recalling 5.1(3), we define for every linear operator $T \in \mathcal{L}(X;Y)$ the **operator norm** of T by

$$\|T\|_{\mathcal{L}(X;Y)} := \sup_{\|x\|_X \le 1} \|Tx\|_Y < \infty. \tag{5-5}$$

In the following, we often use the abbreviation $\|T\|$ for $\|T\|_{\mathcal{L}(X;Y)}$. It follows from the proof of 5.1 that $\|T\|_{\mathcal{L}(X;Y)}$ is the smallest number satisfying

$$\|Tx\|_Y \le \|T\|_{\mathcal{L}(X;Y)} \|x\|_X \quad \text{for all } x \in X. \tag{5-6}$$

We set $\mathcal{L}(X) := \mathcal{L}(X;X)$ and denote the **identity** on X by Id (or by I). Clearly, Id $\in \mathcal{L}(X)$.

5.3 Theorem. Let X, Y, and Z be normed spaces.

(1) $\mathcal{L}(X;Y)$ equipped with $\|\cdot\|_{\mathcal{L}(X;Y)}$ in (5-5) is a normed space.

(2) $\mathcal{L}(X;Y)$ is a Banach space if Y is a Banach space.

(3) If $T \in \mathcal{L}(X;Y)$ and $S \in \mathcal{L}(Y;Z)$, then $ST \in \mathcal{L}(X;Z)$ and

$$\|ST\|_{\mathcal{L}(X;Z)} \le \|S\|_{\mathcal{L}(Y;Z)} \cdot \|T\|_{\mathcal{L}(X;Y)}.$$

(4) $\mathcal{L}(X)$ is a Banach algebra if X is a Banach space. Here the product in $\mathcal{L}(X)$ is given by the composition of operators.

Proof (1). For $T_1, T_2 \in \mathcal{L}(X;Y)$ and $x \in X$

$$\|(T_1 + T_2)x\|_Y \le \|T_1 x\|_Y + \|T_2 x\|_Y \le \big(\|T_1\| + \|T_2\|\big)\|x\|_X.$$

Hence $T_1 + T_2 \in \mathcal{L}(X;Y)$ with $\|T_1 + T_2\| \le \|T_1\| + \|T_2\|$, i.e. the operator norm satisfies the triangle inequality. $\qquad\square$

Proof (2). If $(T_k)_{k \in \mathbb{N}}$ is a Cauchy sequence in $\mathcal{L}(X;Y)$, then for $x \in X$, since $\|T_k x - T_l x\|_Y \le \|T_k - T_l\| \cdot \|x\|_X$, the sequence $(T_k x)_{k \in \mathbb{N}}$ is a Cauchy sequence in Y. As Y is complete, we have that

$$Tx := \lim_{k \to \infty} T_k x \quad \text{in } Y$$

exists pointwise, and it follows easily that $T : X \to Y$ is linear. It then follows that

$$\|(T - T_j)x\|_Y = \lim_{k \to \infty} \|(T_k - T_j)x\|_Y \le \liminf_{k \to \infty} \|T_k - T_j\| \cdot \|x\|_X,$$

and so $T - T_j \in \mathscr{L}(X;Y)$, by 5.1(4), and

$$\|T - T_j\|_{\mathscr{L}(X;Y)} \leq \liminf_{k \to \infty} \|T_k - T_j\|_{\mathscr{L}(X;Y)} \longrightarrow \quad \text{as } j \to \infty$$

(cf. the proof of completeness of $C^0(S;Y)$ in 3.2). □

Proof (3). On noting that

$$\|S(Tx)\|_Z \leq \|S\| \cdot \|Tx\|_Y \leq \|S\| \cdot \|T\| \cdot \|x\|_X \,,$$

we have that $ST \in \mathscr{L}(X;Z)$ with $\|ST\| \leq \|S\| \cdot \|T\|$. □

Proof (4). Follows from (3) and (2). □

5.4 Remarks.

(1) If X is finite-dimensional, then every linear map $T : X \to Y$ is continuous, i.e. in $\mathscr{L}(X;Y)$. For noncontinuous linear maps, see E9.2.

(2) Every $T \in \mathscr{L}(X;Y)$ is Lipschitz continuous, since

$$\|T(x) - T(y)\|_Y \leq \|T\| \cdot \|x - y\|_X \,.$$

If follows that for $R > 0$ and $M > 0$

$$A := \left\{ T|_{\overline{B_R(0)}} \;;\; T \in \mathscr{L}(X;Y), \; \|T\|_{\mathscr{L}(X;Y)} \leq M \right\}$$

is a bounded and equicontinuous subset of $C^0\big(\overline{B_R(0)};Y\big)$. However, the Arzelà-Ascoli theorem is not valid in this context. Observe that A as a subset of $C^0\big(\overline{B_R(0)};Y\big)$ is not (!) precompact, unless X and Y are finite-dimensional. Only then are the domain and the image set of these continuous functions precompact, which played an essential role in the proof of 4.12.

(3) Linear operators occur as Fréchet derivatives of nonlinear maps $F : X \to Y$. We define $T \in \mathscr{L}(X;Y)$ to be the **Fréchet derivative** of F at $x \in X$, if

$$\frac{F(y) - F(x) - T(y - x)}{\|y - x\|_X} \longrightarrow 0 \quad \text{in } Y \text{ as } y \to x \text{ in } X \text{ with } y \neq x.$$

This is the linear approximation property of the mapping $y \mapsto F(y)$ near x, given by the mapping $y \mapsto F(x) + T(y - x)$. Using quantifiers this definition reads

$$\forall \, \varepsilon > 0 : \exists \, \delta > 0 : \forall \, y \in X :$$
$$\|y - x\|_X \leq \delta \implies \|F(y) - F(x) - T(y - x)\|_Y \leq \varepsilon \cdot \|y - x\|_X \,.$$

Proof (1). If n is the dimension of X and $\{e_1, \ldots, e_n\}$ is a basis of X, then for $x = \sum_{i=1}^{n} x_i e_i \in X$

$$\|Tx\|_Y \leq \sum_{i=1}^{n} |x_i| \|Te_i\|_Y \leq \left(\sum_{i=1}^{n} \|Te_i\|_Y \right) \cdot \max_{i=1,\ldots,n} |x_i| \,.$$

If we take, for instance,

$$\|x\| := \max_{i=1,\ldots,n} |x_i|$$

as the norm in X (recall lemma 4.8), then, by 5.1, the inequality proves the continuity of T with

$$\|T\|_{\mathscr{L}(X;Y)} \le \sum_{i=1}^{n} \|Te_i\|_Y .$$

□

We now give a list of special linear operators and some notation. The detailed study of the properties of each class of linear operators will be the subject of the following chapters.

5.5 Definitions.

(1) The space $X' := \mathscr{L}(X; \mathbb{K})$ is the **dual space** to X. The elements of X' are also called **linear functionals**. This is true for general topological vector spaces. If X a normed space, then the norm from (5-5) for $T \in X'$ is

$$\|T\|_{X'} := \sup_{\|x\|_X \le 1} |Tx| . \tag{5-7}$$

(2) The set of **compact (linear) operators** from X to Y is defined by

$$\mathscr{K}(X;Y) := \left\{ T \in \mathscr{L}(X;Y) \; ; \; \overline{T(B_1(0))} \text{ is compact} \right\} .$$

If Y is complete, then we can replace "$\overline{T(B_1(0))}$ is compact" in the definition by "$T(B_1(0))$ is precompact" (see 4.7(5)).

(3) A linear map $P : X \to X$ is called a **(linear) projection** if $P^2 = P$. We denote the set of **continuous (linear) projections** by

$$\mathscr{P}(X) := \left\{ P \in \mathscr{L}(X) \; ; \; P^2 = P \right\} .$$

(4) For $T \in \mathscr{L}(X;Y)$ we denote by

$$\mathscr{N}(T) \quad (\text{or } \ker(T)) \quad := \{x \in X \; ; \; Tx = 0\}$$

the **null space** (or **kernel**) of T. The continuity of T immediately yields that $\mathscr{N}(T)$ is a closed subspace. The **range** (or **image**) of T is defined by

$$\mathscr{R}(T) \quad (\text{or } \text{im}(T)) \quad := \{Tx \in Y \; ; \; x \in X\}.$$

The subspace $\mathscr{R}(T)$ in general is not closed (see the example 5.6(3)). We will often denote the image of a linear map also as $T(X) = \mathscr{R}(T)$.

(5) $T \in \mathscr{L}(X;Y)$ is called a **(linear continuous) embedding** of X into Y if T is injective, i.e. if $\mathscr{N}(T) = \{0\}$.

Observe: In general, the term embedding is used only for very special maps T, see for example the embedding theorems in Chapter 10.

(6) Let X and Y be complete spaces. If $T \in \mathscr{L}(X;Y)$ is bijective, then $T^{-1} \in \mathscr{L}(Y;X)$ (see the inverse mapping theorem 7.8, which plays an essential role in functional analysis). Then T is called an *invertible (linear) operator*, or a *(linear continuous) isomorphism*.

(7) $T \in \mathscr{L}(X;Y)$ is called an *isometry* (see the definition in 2.24) if

$$\|Tx\|_Y = \|x\|_X \quad \text{for all } x \in X.$$

(8) If $T \in \mathscr{L}(X;Y)$, then

$$(T'y')(x) := y'(Tx) \quad \text{for } y' \in Y', \ x \in X$$

defines a linear map $T' : Y' \to X'$, the *adjoint map* of T. We also call T' the *adjoint operator* of T, because $T' \in \mathscr{L}(Y',X')$.

Proof (8). For $x \in X$ and $y' \in Y'$,

$$|(T'y')(x)| = |y'(Tx)| \leq \|y'\|_{Y'} \|Tx\|_Y \leq \|y'\|_{Y'} \cdot \|T\| \cdot \|x\|_X \,,$$

so that, by (5-7),

$$\|T'y'\|_{X'} \leq \|y'\|_{Y'} \cdot \|T\|\,,$$

hence, by (5-5), $T' \in \mathscr{L}(Y',X')$ with $\|T'\| \leq \|T\|$ (see also 12.1, where we will show that $\|T'\| = \|T\|$). □

Dual spaces will be investigated in Chapter 6. In particular, we will characterize the dual spaces of $C^0(S)$ and $L^p(\mu)$, i.e. we will introduce measure and function spaces, respectively, that are isomorphic to these dual spaces. Continuous linear projections will be considered in Chapter 9. In Chapter 10, we will present the most important types of compact operators, and Chapter 11 will be devoted to the spectral theorem for compact operators. Results on adjoint maps can be found in Chapter 12.

We now give some examples of linear operators.

5.6 Examples.

(1) Let $S \subset \mathbb{R}^n$ be compact and let $(S,\mathcal{B},\mu)$ be a measure space with $\mu(S) < \infty$, and such that $\mathcal{B}$ contains the Borel sets of S. Then $C^0(S) \subset L^1(\mu)$ and

$$T_\mu f := \int_S f \, \mathrm{d}\mu \quad \text{for } f \in C^0(S)$$

defines a functional $T_\mu \in C^0(S)'$ (see 6.22 and theorem 6.23). For example, if $\mu = \delta_x$ is the Dirac measure for $x \in S$, then $T_{\delta_x} f = f(x)$.

(2) Examples of operators in $\mathscr{L}\big(C^0(S)\big)$, $S \subset \mathbb{R}^n$ compact, are the multiplication operators

$$(T_g f)(x) := f(x)g(x) \quad \text{for } f \in C^0(S)\,,$$

for a fixed $g \in C^0(S)$.

(3) An example of an operator $T \in \mathscr{L}(C^0(S); C^1(S))$ with $S = [0, 1]$ is

$$(Tf)(x) := \int_0^x f(\xi) \, d\xi \quad \text{for } f \in C^0(S).$$

One may also consider T as an operator in $\mathscr{L}(C^0(S))$. Then $\mathscr{R}(T)$ is not closed in $C^0(S)$, since $\mathscr{R}(T) = \{g \in C^1(S) ; \; g(0) = 0\}$ is a proper subset of the closure $\overline{\mathscr{R}(T)} = \{g \in C^0(S) ; \; g(0) = 0\}$. Similarly, T can be defined as an operator in $\mathscr{L}(L^1(S))$. Then $\mathscr{R}(T) = \{g \in W^{1,1}(]0,1[) ; \; g(0) = 0\}$ (see E3.6), which is a proper dense subset of $\overline{\mathscr{R}(T)} = L^1(S)$.

(4) Let $1 \leq p \leq \infty$ and let $\frac{1}{p} + \frac{1}{p'} = 1$. Then for $g \in L^{p'}(\mu)$ the Hölder inequality yields that

$$T_g f := \int_S f\bar{g} \, d\mu \quad \text{for } f \in L^p(\mu)$$

defines a functional $T_g \in L^p(\mu)'$ (see theorem 6.12).

(5) If p, p' are as in (4) and $g^s \in L^{p'}(\Omega)$ for $|s| \leq m$ with $g = (g^s)_{|s| \leq m}$, then

$$T_g f := \sum_{|s| \leq m} \int_\Omega \partial^s f \cdot \overline{g^s} \, dL^n \quad \text{for } f \in W^{m,p}(\Omega)$$

defines a functional $T_g \in W^{m,p}(\Omega)'$.

(6) Let p be as in (4) and let $(\varphi_k)_{k \in \mathbb{N}}$ be a Dirac sequence. Then 4.13(2) yields that

$$T_k f(x) := \int_{\mathbb{R}^n} \varphi_k(x - y) f(y) \, dy = (\varphi_k * f)(x)$$

defines an operator $T_k \in \mathscr{L}(L^p(\mathbb{R}^n))$ with $\|T_k\| \leq 1$. It follows from 4.15(2) that, if $p < \infty$,

$$(T_k - \mathrm{Id})f \to 0 \quad \text{in } L^p(\mathbb{R}^n) \text{ as } k \to \infty$$

for every $f \in L^p(\mathbb{R}^n)$. However, T_k does not converge in the operator norm (see E5.6).

We now prove some fundamental properties of linear operators.

5.7 Neumann series. Let X be a Banach space and let $T \in \mathscr{L}(X)$ with

$$\limsup_{m \to \infty} \|T^m\|^{\frac{1}{m}} < 1$$

(in particular, this is satisfied if $\|T\| < 1$). Then $\mathrm{Id} - T$ is bijective and $(\mathrm{Id} - T)^{-1} \in \mathscr{L}(X)$ with

$$(\mathrm{Id} - T)^{-1} = \sum_{n=0}^\infty T^n \quad \text{in } \mathscr{L}(X).$$

Proof. For $k \in \mathbb{N}$ let $S_k := \sum_{n=0}^{k} T^n$. Choose $m \in \mathbb{N}$ and $\theta < 1$ with $\|T^n\| \le \theta^n$ for $n \ge m$. Then for $m \le k < l$

$$\|S_l - S_k\| = \left\| \sum_{n=k+1}^{l} T^n \right\| \le \sum_{n=k+1}^{l} \|T^n\| \le \sum_{n=k+1}^{\infty} \theta^n \longrightarrow 0 \quad \text{as } k \to \infty.$$

Since $\mathscr{L}(X)$ is complete, there exists the limit

$$S := \lim_{k \to \infty} S_k \quad \text{in } \mathscr{L}(X).$$

It follows that as $k \to \infty$

$$(\mathrm{Id} - T)S \longleftarrow (\mathrm{Id} - T)S_k$$
$$= \sum_{n=0}^{k} (T^n - T^{n+1}) = \mathrm{Id} - T^{k+1} \longrightarrow \mathrm{Id} \quad \text{in } \mathscr{L}(X),$$

because for $k \ge m$ we have that $\|T^{k+1}\| \le \theta^{k+1} \to 0$ as $k \to \infty$. Similarly, one can show that $S(\mathrm{Id} - T) = \mathrm{Id}$. Hence S is the inverse of $\mathrm{Id} - T$. $\quad\square$

As a consequence, we obtain that in the space of linear operators, perturbations of invertible operators are again invertible.

5.8 Theorem on invertible operators. Let X, Y be Banach spaces. Then the set of invertible operators in $\mathscr{L}(X; Y)$ is an open subset. More precisely: If $X \ne \{0\}$ and $Y \ne \{0\}$, then for $T, S \in \mathscr{L}(X; Y)$ we have that

$$\left. \begin{array}{l} T \text{ invertible,} \\ \|S - T\| < \|T^{-1}\|^{-1} \end{array} \right\} \implies S \text{ invertible.}$$

Proof. Let $R := T - S$. Then $S = T(\mathrm{Id} - T^{-1}R) = (\mathrm{Id} - RT^{-1})T$, where $\|T^{-1}R\| \le \|T^{-1}\| \cdot \|R\| < 1$, and similarly $\|RT^{-1}\| < 1$. Applying 5.7 yields the desired result. $\quad\square$

5.9 Analytic functions of operators. Let

$$f(z) := \sum_{n=0}^{\infty} a_n z^n$$

be a power series in $\mathbb{K}$ with radius of convergence $\varrho > 0$. Let X be a Banach space over $\mathbb{K}$. If $T \in \mathscr{L}(X)$, then

$$\limsup_{m \to \infty} \|T^m\|^{\frac{1}{m}} < \varrho \implies f(T) := \sum_{n=0}^{\infty} a_n T^n \quad \text{exists in } \mathscr{L}(X).$$

Proof. There exists an r with $0 < r < \varrho$ and an $n \in \mathbb{N}$ with $\|T^m\| \leq r^m$ for $m \geq n$. For $n \leq m \leq k$ it then holds that

$$\left\| \sum_{i=m}^{k} a_i T^i \right\| \leq \sum_{i=m}^{k} |a_i| \|T^i\| \leq \sum_{i=m}^{\infty} |a_i| r^i \longrightarrow 0 \quad \text{as } m \to \infty$$

thanks to the assumption on the power series. $\qquad\square$

5.10 Examples. Let X be a Banach space.

(1) *Exponential function.* For all $T \in \mathscr{L}(X)$ we define

$$\exp(T) \ \left(\text{or } \ \mathrm{e}^T \right) \ := \sum_{n=0}^{\infty} \frac{1}{n!} T^n \ \in \mathscr{L}(X).$$

For $T, S \in \mathscr{L}(X)$

$$S\,T = T\,S \quad \Longrightarrow \quad \mathrm{e}^{T+S} = \mathrm{e}^T \, \mathrm{e}^S.$$

(2) *Evolution equation.* For $T \in \mathscr{L}(X)$ the function $A(s) := \mathrm{e}^{sT}$ for $s \in \mathbb{R}$ defines an $A \in C^\infty(\mathbb{R}; \mathscr{L}(X))$ with

$$\frac{\mathrm{d}}{\mathrm{d}s} A(s) = T\,A(s) = A(s)\,T.$$

(3) *Logarithm.* For $T \in \mathscr{L}(X)$ with $\|\mathrm{Id} - T\| < 1$ we define

$$\log(T) := -\sum_{n=1}^{\infty} \frac{1}{n}(\mathrm{Id} - T)^n \ \in \mathscr{L}(X).$$

(4) For $T \in \mathscr{L}(X)$ with $\|T\| < 1$ the function $A(s) := \log(\mathrm{Id} - sT)$ for $|s| < 1$ defines an $A \in C^\infty(\,]-1, 1[\,; \mathscr{L}(X))$ with

$$\frac{\mathrm{d}}{\mathrm{d}s} A(s) = -T\,(\mathrm{Id} - sT)^{-1} = -(\mathrm{Id} - sT)^{-1}\,T$$

and $\exp(A(s)) = \mathrm{Id} - sT$.

The following theorem shows that linear operators commute with the integral (and hence it is a linear version of Jensen's inequality in E4.9).

5.11 Theorem. Let $(S, \mathcal{B}, \mu)$ be a measure space and let Y and Z be Banach spaces. If $f \in L^1(\mu; Y)$ and $T \in \mathscr{L}(Y; Z)$, then $T \circ f \in L^1(\mu; Z)$ and

$$T\left(\int_S f \, \mathrm{d}\mu \right) = \int_S T \circ f \, \mathrm{d}\mu.$$

Explanation: Setting $I_Y f := \int_S f \, \mathrm{d}\mu$ defines $I_Y \in \mathscr{L}(L^1(\mu; Y); Y)$, and similarly I_Z. In addition, let $\widetilde{T}$ be the operator corresponding to T lifted to functions, i.e. $(\widetilde{T}f)(x) := T(f(x))$ defines $\widetilde{T} \in \mathscr{L}(L^1(\mu; Y); L^1(\mu; Z))$. The theorem then says that

$$T\,I_Y = I_Z\,\tilde{T}\,,$$

i.e. in this sense, the integral commutes with linear operators.

Proof. Approximate f in $L^1(\mu;Y)$ with step functions

$$f_k = \sum_{i=1}^{n_k} \mathcal{X}_{E_{ki}}\alpha_{ki} \quad \text{with } \alpha_{ki} \in Y \text{ and } \mu(E_{ki}) < \infty,$$

with E_{ki}, $i = 1, \ldots, n_k$, being pairwise disjoint. Then as $k \to \infty$

$$T\left(\int_S f\,d\mu\right) \longleftarrow T\left(\int_S f_k\,d\mu\right) = T\left(\sum_i \mu(E_{ki})\alpha_{ki}\right)$$

$$= \sum_i \mu(E_{ki})T\alpha_{ki} = \int_S T\circ f_k\,d\mu\,.$$

Since

$$\int_S \|T\circ f_k - T\circ f_l\|_Z\,d\mu \le \|T\|\int_S \|f_k - f_l\|_Y\,d\mu \longrightarrow 0 \quad \text{as } k,l \to \infty,$$

we have that $(T\circ f_k)_{k\in\mathbb{N}}$ is a Cauchy sequence in $L^1(\mu;Z)$. It follows that there exists a $g \in L^1(\mu;Z)$ such that

$$T\circ f_k \longrightarrow g \text{ in } L^1(\mu;Z)$$

as $k \to \infty$, and hence also

$$\int_S T\circ f_k\,d\mu \longrightarrow \int_S g\,d\mu\,.$$

For a subsequence $k \to \infty$ it holds that $T\circ f_k \to g$ almost everywhere in S, and for a further subsequence $k \to \infty$ we have that $f_k \to f$ and hence also $T\circ f_k \to T\circ f$ almost everywhere in S. Consequently, $g = T\circ f$ almost everywhere. $\qquad\square$

The linear operators between function spaces that are most important in applications are differential and integral operators.

5.12 Hilbert-Schmidt integral operators. Let $\Omega_1 \subset \mathbb{R}^{n_1}$, $\Omega_2 \subset \mathbb{R}^{n_2}$ be Lebesgue measurable, $1 < p < \infty$ and $1 < q < \infty$, and let $K : \Omega_1 \times \Omega_2 \to \mathbb{K}$ be Lebesgue measurable with

$$\|K\| := \left(\int_{\Omega_1}\left(\int_{\Omega_2} |K(x,y)|^{p'}\,dy\right)^{\frac{q}{p'}}dx\right)^{\frac{1}{q}} < \infty, \tag{5-8}$$

where $\frac{1}{p} + \frac{1}{p'} = 1$. Then

$$(Tf)(x) := \int_{\Omega_2} K(x,y)f(y)\,dy$$

defines an operator $T \in \mathscr{L}\big(L^p(\Omega_2; \mathbb{K}); L^q(\Omega_1; \mathbb{K})\big)$ with $\|T\| \leq \|K\|$. We call K the **integral kernel** of the operator T.

Remark: In 10.15 we will show that T is a compact operator.

Proof. We first assume that all of the following integrals exist. Then using the Hölder inequality we have that

$$\int_{\Omega_1} |Tf(x)|^q\,dx = \int_{\Omega_1} \left| \int_{\Omega_2} K(x,y)f(y)\,dy \right|^q dx$$

$$\leq \int_{\Omega_1} \left| \int_{\Omega_2} |K(x,y)|^{p'}\,dy \right|^{\frac{q}{p'}} \cdot \left(\int_{\Omega_2} |f(y)|^p\,dy \right)^{\frac{q}{p}} dx = \|K\|^q \cdot \|f\|^q_{L^p(\Omega_2)},$$

which yields the desired result. The existence of the integrals can now be justified retrospectively, similarly to the proof of 4.13(1), and it follows in particular that $Tf \in L^q(\Omega_1)$. Here we note that the assumption (5-8) states that $K(x,\cdot) \in L^{p'}(\Omega_2)$ for almost all $x \in \Omega_1$, and that the function $x \mapsto \|K(x,\cdot)\|_{L^{p'}(\Omega_2)}$ lies in $L^q(\Omega_1)$. □

Now we introduce the set of locally integrable functions.

5.13 Definition. Let $\Omega \subset \mathbb{R}^n$ be open.

(1) We let $D \subset\subset \Omega$ be a shorthand notation for a set $D \subset \mathbb{R}^n$ which is precompact with $\overline{D} \subset \Omega$.

Remark: One also says that D is a **relatively precompact** subset of Ω, which means that the closure of D is compact in the relative topology of Ω.

(2) For $1 \leq p \leq \infty$, let

$$L^p_{\text{loc}}(\Omega) := \big\{ f : \Omega \to \mathbb{K} \ ; \ f\big|_D \in L^p(D) \text{ for all } D \subset\subset \Omega \big\},$$

the vector space of **locally in Ω p-integrable** functions.

(3) Equipped with the Fréchet metric

$$\varrho(f) := \sum_{i \in \mathbb{N}} 2^{-i} \frac{\|f\|_{L^p(K_i)}}{1 + \|f\|_{L^p(K_i)}} \qquad \text{for } f \in L^p_{\text{loc}}(\Omega)$$

this is a complete metric space. Here $(K_i)_{i \in \mathbb{N}}$ is a sequence of compact sets, which is an exhaustion of Ω (see (3-2)).

(4) Analogously we define $W^{m,p}_{\text{loc}}(\Omega)$, i.e.

$$W^{m,p}_{\text{loc}}(\Omega) := \{f : \Omega \to \mathbb{K}; \ f\big|_D \in W^{m,p}(D) \text{ for all open sets } D \subset\subset \Omega\}.$$

With this we state the following.

5.14 Linear differential operators. Let $\Omega \subset \mathbb{R}^n$ be open and assume $a_s : \Omega \to \mathbb{K}$ for multi-indices s with $|s| \le m$. Then

$$(Tf)(x) := \sum_{|s| \le m} a_s(x)\partial^s f(x)$$

defines an operator

(1) $T \in \mathscr{L}\big(C^m(\Omega); C^0(\Omega)\big)$, if $a_s \in C^0(\Omega)$ for $|s| \le m$.
Remark: $T \in \mathscr{L}\big(C^m(\overline{\Omega}); C^0(\overline{\Omega})\big)$, if all $a_s \in C^0(\overline{\Omega})$ and Ω is bounded.
(2) $T \in \mathscr{L}\big(C^{m,\alpha}(\Omega); C^{0,\alpha}(\Omega)\big)$ with $0 < \alpha \le 1$ provided $a_s \in C^{0,\alpha}(\Omega)$ for $|s| \le m$.
Remark: $T \in \mathscr{L}\big(C^{m,\alpha}(\overline{\Omega}); C^{0,\alpha}(\overline{\Omega})\big)$, if $a_s \in C^{0,\alpha}(\overline{\Omega})$ and Ω is bounded.
(3) $T \in \mathscr{L}\big(W^{m,p}_{\mathrm{loc}}(\Omega); L^p_{\mathrm{loc}}(\Omega)\big)$ with $1 \le p \le \infty$, provided $a_s \in L^\infty_{\mathrm{loc}}(\Omega)$ for $|s| \le m$.
Remark: $T \in \mathscr{L}\big(W^{m,p}(\Omega); L^p(\Omega)\big)$, if $a_s \in L^\infty(\Omega)$.

In each case we call T a **linear differential operator** of **order** m, and we call a_s for $|s| \le m$ the **coefficients** of the differential operator.

Distributions

We now want to consider the functionals in 5.6 in a more general setting. To this end, we restrict the functionals to the common vector space $C_0^\infty(\Omega)$ (here set $S := \overline{\Omega}$ in 5.6). Hence we consider functions and measures only in Ω, i.e. as in 5.14 without boundary conditions. This leads to the following

5.15 Notation. Let $\Omega \subset \mathbb{R}^n$ be open.

(1) Let $(\Omega, \mathcal{B}, \mu)$ be a measure space such that $\mathcal{B}$ contains the Borel sets of Ω and such that μ is finite on compact subsets. Then

$$[\mu](\zeta) \ (\text{or } \mathrm{T}_\Omega(\mu)(\zeta)) \ := \int_\Omega \zeta \, d\mu \quad \text{for } \zeta \in C_0^\infty(\Omega)$$

defines a linear map $[\mu] \ \big(\text{or } \mathrm{T}_\Omega(\mu)\big) : C_0^\infty(\Omega) \to \mathbb{K}$.
Remark: With the notation in 5.6(1) we have that $[\mu] = \mathrm{T}_\Omega(\mu) = T_\mu\big|_{C_0^\infty(\Omega)}$.

Note: The integral in this definition is the Riemann integral (see 6.22). Hence for the measures considered here one has $C_0^0(\Omega) \subset L^1(\mu)$.
(2) Let $f \in L^1_{\mathrm{loc}}(\Omega)$. Then

$$[f](\zeta) \ (\text{or } \mathrm{T}_\Omega(f)(\zeta)) \ := \int_\Omega \zeta \cdot f \, d\mathrm{L}^n \quad \text{for } \zeta \in C_0^\infty(\Omega)$$

defines a linear map $[f]$ (or $\mathrm{T}_\Omega(f)) : C_0^\infty(\Omega) \to \mathbb{K}$.

Observe: This is a special case of (1), on setting $\mu(E) := \int_E f \, \mathrm{dL}^n$ for Lebesgue measurable sets $E \subset\subset \Omega$ (see the definition 5.13(1)).

Remark: With the notation in 5.6(4) one has $[f] = \mathrm{T}_\Omega(f) = T_{\overline{f}}|_{C_0^\infty(\Omega)}$.

5.16 Lemma. Let $\Omega \subset \mathbb{R}^n$ be open and consider the map in 5.15(2)

$$f \mapsto [f] = \mathrm{T}_\Omega(f) \text{ from } L_{\mathrm{loc}}^1(\Omega) \text{ to } \{T : C_0^\infty(\Omega) \to \mathbb{K}; \ T \text{ linear}\}.$$

(1) This map is linear and injective.

(2) The function f can be reconstructed from $[f] = \mathrm{T}_\Omega(f)$.

(3) The definition of the weak derivatives $\partial^s f$ of a function $f \in W_{\mathrm{loc}}^{m,1}(\Omega)$ in (3-17) can now be written as

$$(-1)^{|s|}[f](\partial^s \zeta) = [\partial^s f](\zeta) \quad \text{for } \zeta \in C_0^\infty(\Omega), \ |s| \leq m. \tag{5-9}$$

Proof (1). This follows from 4.22 (applied to sets $D \subset\subset \Omega$, or note that the fundamental lemma holds in $L_{\mathrm{loc}}^1(\Omega)$). $\qquad \square$

Proof (2). To see this, choose $\zeta_\varepsilon = \varphi_\varepsilon * \mathcal{X}_E$ with $E \subset\subset \Omega$ as in the proof of 4.22. Then $[f](\zeta_\varepsilon) \to \int_E f \, \mathrm{dL}^n$ as $\varepsilon \to 0$. Now choose $E = \mathrm{B}_\varepsilon(x)$ with $x \in \Omega$ and obtain for (a subsequence) $\varepsilon \to 0$ that

$$(\varphi_\varepsilon * f)(x) = \mathrm{L}^n(\mathrm{B}_\varepsilon(x))^{-1} \int_{\mathrm{B}_\varepsilon(x)} f \, \mathrm{dL}^n \to f(x)$$

for L^n-almost all x. Here we have used 4.15(2). $\qquad \square$

This means that knowledge of all the values $[f](\zeta)$ with $\zeta \in C_0^\infty(\Omega)$ provides full information on the function f almost everywhere in Ω. Hence we also call $C_0^\infty(\Omega)$ the space of **test functions**. We transfer this to linear maps $T : C_0^\infty(\Omega) \to \mathbb{K}$, where the main property is motivated by the structure of the identity (5-9).

5.17 Distributions. Let $\Omega \subset \mathbb{R}^n$ be open and let $T : C_0^\infty(\Omega) \to \mathbb{K}$ be linear.

(1) For all multi-indices s, the **distributional derivative** $\partial^s T$ is the linear map $\partial^s T : C_0^\infty(\Omega) \to \mathbb{K}$ defined by

$$(\partial^s T)(\zeta) := (-1)^{|s|} T(\partial^s \zeta) \quad \text{for } \zeta \in C_0^\infty(\Omega). \tag{5-10}$$

(2) We call the linear map T a **distribution** on Ω, and use the notation

$$T \in \mathscr{D}'(\Omega),$$

if for all open sets $D \subset\subset \Omega$ there exist a constant C_D and a $k_D \in \mathbb{N} \cup \{0\}$ such that

$$|T(\zeta)| \leq C_D \|\zeta\|_{C^{k_D}(\overline{D})} \quad \text{for all } \zeta \in C_0^\infty(\Omega) \text{ with supp}(\zeta) \subset D. \qquad (5\text{-}11)$$

If $k = k_D$ can be chosen independently of D, then k (if chosen minimally) is called the **order** of T.

(3) If T is a distribution, then so is $\partial^s T$ for all multi-indices s. If T is a distribution of order k, then $\partial^s T$ is a distribution of order $k + |s|$.

Proof (3). We have $|(\partial^s T)(\zeta)| \leq C_D \|\partial^s \zeta\|_{C^{k_D}(\overline{D})} \leq C_D \|\zeta\|_{C^{k_D+|s|}(\overline{D})}.$ $\square$

5.18 Examples.

(1) For $f \in W^{m,p}(\Omega)$ and $|s| \leq m$

$$\partial^s[f] = [\partial^s f] \quad \text{in } \mathscr{D}'(\Omega). \qquad (5\text{-}12)$$

Hence the definition of $W^{m,p}(\Omega)$ can also be formulated as follows: A function $f \in L^p(\Omega)$ is in $W^{m,p}(\Omega)$ if all its distributional derivatives up to order m can be identified with functions in $L^p(\Omega)$.

(2) For $f \in L^1_{\text{loc}}(\Omega)$ and $\zeta \in C_0^\infty(D)$ with $D \subset\subset \Omega$

$$[f](\zeta) = \int_\Omega \zeta \cdot f \, d\mathcal{L}^n \quad \text{with} \quad |[f](\zeta)| \leq \|f\|_{L^1(D)} \cdot \|\zeta\|_{C^0(\overline{D})}.$$

It follows that $[f] \in \mathscr{D}'(\Omega)$ and is of order 0.

(3) For μ is as in 5.15(1) and for $\zeta \in C_0^\infty(D)$ with $D \subset\subset \Omega$

$$[\mu](\zeta) = \int_\Omega \zeta \, d\mu \quad \text{with} \quad |[\mu](\zeta)| \leq \mu(D) \|\zeta\|_{C^0(\overline{D})}.$$

It follows that $[\mu] \in \mathscr{D}'(\Omega)$ and is of order 0.

(4) As an example, let $\Omega = \mathbb{R}$ and, given $c_-, c_+ \in \mathbb{R}$, let

$$f(x) := \begin{cases} c_+ & \text{for } x > 0, \\ c_- & \text{for } x < 0. \end{cases}$$

By (2), $[f]$ is a distribution of order 0. With the definitions in 5.17(1) and 5.15 it follows that

$$[f]'(\zeta) = -[f](\zeta') = (c_+ - c_-)\zeta(0) = (c_+ - c_-)[\delta_0](\zeta),$$

where δ_0 is the Dirac measure at the point 0. Hence $[f]'$ is also a distribution of order 0. In addition,

$$[f]''(\zeta) = -[f]'(\zeta') = -(c_+ - c_-)\zeta'(0).$$

Hence $[f]''$ is a distribution of order 1, if $c_- \neq c_+$.

(5) Let $(\varphi_k)_{k\in\mathbb{N}}$ be a general Dirac sequence and let δ_0 be the Dirac measure at $0 \in \mathbb{R}^n$. Then it holds as $k \to \infty$ that

$$[\varphi_k](\zeta) \longrightarrow [\delta_0](\zeta) \quad \text{for all } \zeta \in C_0^\infty(\mathbb{R}^n),$$

i.e. $[\varphi_k]$ converges to $[\delta_0]$ as $k \to \infty$ pointwise as a linear map. The name Dirac sequence originates from this property.

(6) As a further example, let $f(x) := \log|x|$ for $x \in \mathbb{R}^n \setminus \{0\}$. Then $f \in L_{loc}^1(\mathbb{R}^n)$, and so, by (2), $[f]$ is a distribution of order 0 on $\mathbb{R}^n$. For $1 \leq i \leq n$

$$(\partial_i[f])(\zeta) = \begin{cases} \displaystyle\int_{\mathbb{R}^n} \zeta(x)\frac{x_i}{|x|^2}\,\mathrm{d}x & \text{for } n \geq 2, \\[2ex] \displaystyle\lim_{\varepsilon \searrow 0} \int_{\mathbb{R}\setminus[-\varepsilon,\varepsilon]} \zeta(x)\frac{1}{x}\,\mathrm{d}x & \text{for } n = 1. \end{cases}$$

In order to prove this, verify with the help of Gauß's theorem that as $\varepsilon \searrow 0$

$$(\partial_i[f])(\zeta) = -[f](\partial_i\zeta) \quad\longleftarrow\quad -\int_{\mathbb{R}^n\setminus B_\varepsilon(0)} \partial_i\zeta \cdot f\,\mathrm{d}L^n$$

$$= \int_{\mathbb{R}^n\setminus B_\varepsilon(0)} \zeta\partial_i f\,\mathrm{d}L^n + \int_{\partial B_\varepsilon(0)} \nu_i\zeta f\,\mathrm{d}H^{n-1},$$

where $\nu_i(x) = \frac{x_i}{|x|}$ is the i-th component of the outer normal to the set $B_\varepsilon(0)$ (see A8.5(3) for the general situation). It can be seen that the second integral converges to zero as $\varepsilon \searrow 0$. In the case $n \geq 2$ the function $x \mapsto x_i|x|^{-2}$ is in $L_{loc}^1(\mathbb{R}^n)$, but not for $n = 1$. Hence for $n \geq 2$ it holds that $\partial_i[f]$ is a distribution of order 0, while for $n = 1$ it can be shown that it is a distribution of order 1.

The essential estimate (5-11) is used in order to approximate distributions with C^∞-functions by means of convolutions.

5.19 Approximation of distributions. Let $\Omega \subset \mathbb{R}^n$ and let $T \in \mathscr{D}'(\Omega)$. For $\varphi \in C_0^\infty(B_r(0))$ and $x \in \Omega$ with $B_r(x) \subset \Omega$,

$$(\varphi * T)(x) := T(\varphi(x - \cdot)) \tag{5-13}$$

is well defined, since $\varphi(x - \cdot) \in C_0^\infty(\Omega)$. Moreover, it holds that:

(1) For $T = [f]$ with $f \in L_{loc}^1(\Omega)$ it follows that

$$(\varphi * [f])(x) = (\varphi * f)(x) \quad \text{if } B_r(x) \subset \Omega.$$

(2) If $D \subset\subset \Omega$ with $B_r(D) \subset \Omega$, then $\varphi * T \in C^\infty(D)$, with derivatives $\partial^s(\varphi * T) = (\partial^s\varphi) * T$.

(3) Let $D \subset\subset \Omega$ and let $(\varphi_\varepsilon)_{\varepsilon>0}$ be a standard Dirac sequence. For small ε we have that $\varphi_\varepsilon * T \in C^\infty(D)$ and for all $\zeta \in C_0^\infty(D)$

$$[\varphi_\varepsilon * T](\zeta) \longrightarrow T(\zeta) \quad \text{as } \varepsilon \to 0.$$

Proof (1). It holds that

$$(\varphi * [f])(x) = [f](\varphi(x - \cdot)) = \int_\Omega \varphi(x - y)f(y)\, dy = (\varphi * f)(x),$$

since $\operatorname{supp}(\varphi(x - \cdot)) \subset \Omega$ (formally set $f = 0$ in the exterior of Ω). $\qquad\square$

Proof (2). Let k_D be chosen for T and D as in (5-11). On introducing the difference quotients $\partial_i^h \psi(x) := \frac{1}{h}(\psi(x + h\mathbf{e}_i) - \psi(x))$, the linearity of T yields that

$$\partial_i^h(\varphi * T)(x) = T\left(\partial_i^h \varphi(x - \cdot)\right).$$

We have that $\partial_i^h \varphi(x - \cdot) \to \partial_i \varphi(x - \cdot)$ in $C^{k_D}(\overline{D})$ as $h \to 0$, and hence it follows from (5-11) that

$$T(\partial_i^h \varphi(x - \cdot)) \longrightarrow T(\partial_i \varphi(x - \cdot)) = \left((\partial_i \varphi) * T\right)(x).$$

This shows that the partial derivative $\partial_i(\varphi * T)(x) = \left((\partial_i \varphi) * T\right)(x)$ exists. The desired result for higher derivatives now follows by induction on the order of the derivative. $\qquad\square$

Proof (3). We have that

$$[\varphi_\varepsilon * T](\zeta) = \int_\Omega \zeta(x) \underbrace{(\varphi_\varepsilon * T)(x)}_{= T(\varphi_\varepsilon(x - \cdot))}\, dx.$$

Now it holds that (the proof is given below)

$$\int_\Omega \zeta(x) T\left(\varphi_\varepsilon(x - \cdot)\right) dx = T\left(\int_\Omega \zeta(x)\varphi_\varepsilon(x - \cdot)\, dx\right). \qquad (5\text{-}14)$$

The argument of T on the right-hand side is $\zeta_\varepsilon(\cdot)$, if $\zeta_\varepsilon := \varphi_\varepsilon^- * \zeta$ with $\varphi_\varepsilon^-(y) := \varphi_\varepsilon(-y)$. Since $\zeta_\varepsilon \to \zeta$ in $C^{k_D}(\overline{D})$ as $\varepsilon \to 0$, it follows that $T(\zeta_\varepsilon) \to T(\zeta)$, if k_D for T and D is chosen as in (5-11), and so we have shown that

$$[\varphi_\varepsilon * T](\zeta) = T(\zeta_\varepsilon) \longrightarrow T(\zeta) \quad \text{as } \varepsilon \to 0.$$

The identity (5-14) is closely related to theorem 5.11 and the proof is analogous: Approximate ζ uniformly by step functions ζ_j with a common compact support in D. Then (5-14) holds for ζ_j because of the linearity of T. The left-hand side converges as $j \to \infty$, since $T(\varphi_\varepsilon(x - \cdot))$ is continuous, recall (2). The right-hand side converges using the same argument as above, since $\varphi_\varepsilon^- * \zeta_j \to \varphi_\varepsilon^- * \zeta$ in $C^{k_D}(\overline{D})$. $\qquad\square$

For functional analysis purposes, the following result is of importance: The vector space $C_0^\infty(\Omega)$ can be equipped with a topology $\mathcal{T}$ in such a way that T is a distribution if and only if T lies in the corresponding dual space, i.e. if $T : C_0^\infty(\Omega) \to \mathbb{K}$ is linear and continuous with respect to the topology $\mathcal{T}$. We denote $C_0^\infty(\Omega)$, equipped with the topology $\mathcal{T}$, by $\mathscr{D}(\Omega)$ (see 5.21). The dual space $\mathscr{D}(\Omega)'$ is then the same as $\mathscr{D}'(\Omega)$ (see 5.23).

5.20 Topology on $C_0^\infty(\Omega)$. Let $\Omega \subset \mathbb{R}^n$ be open. Define

$$p(\zeta) := \sum_{k=0}^{\infty} 2^{-k} \frac{\|\zeta\|_{C^k(\overline{D})}}{1 + \|\zeta\|_{C^k(\overline{D})}} \quad \text{for } \zeta \in C_0^\infty(\Omega) \text{ with supp}(\zeta) \subset D \subset\subset \Omega,$$

where the right-hand side is independent of the choice of D. Choose an open cover $(D_j)_{j \in \mathbb{N}}$ of Ω with sets $D_j \subset\subset D_{j+1} \subset \Omega$ for all $j \in \mathbb{N}$. For every sequence $\varepsilon = (\varepsilon_j)_{j \in \mathbb{N}}$ with $\varepsilon_j > 0$ for $j \in \mathbb{N}$ define

$$U_\varepsilon := \text{conv}\left(\bigcup_{j \in \mathbb{N}} \{ \zeta \in C_0^\infty(\Omega) \; ; \; \text{supp}(\zeta) \subset D_j \text{ and } p(\zeta) < \varepsilon_j \} \right).$$

Finally, define

$$\mathcal{T} := \{ U \subset C_0^\infty(\Omega) \; ; \; \text{for } \zeta \in U \text{ there exists an } \varepsilon \text{ with } \zeta + U_\varepsilon \subset U \}.$$

Then:

(1) p is a Fréchet metric with $p(r\zeta) \le r p(\zeta)$ for $r \ge 1$.

(2) For all ε it holds that $U_\varepsilon \in \mathcal{T}$.

(3) $\mathcal{T}$ is a topology. Hence the sets U_ε form a neighbourhood basis (see the definition (4-17)) of 0 with respect to $\mathcal{T}$.

(4) $\mathcal{T}$ is independent of the choice of cover $(D_j)_{j \in \mathbb{N}}$.

We remark that $\mathcal{T}$ is stronger than the topology induced by p. This follows from the fact that the p-ball $B_\varrho(0) \subset C_0^\infty(\Omega)$ is a neighbourhood in the $\mathcal{T}$-topology, namely, $B_\varrho(0) = U_\varepsilon$ with $\varepsilon = (\varepsilon_j)_{j \in \mathbb{N}}$ and $\varepsilon_j = \varrho$.

Proof (2). Let $\zeta \in U_\varepsilon$. Consider a finite convex combination

$$\zeta = \sum_{k=1}^{k_0} \alpha_k \zeta_k \in U_\varepsilon \quad \text{with } k_0 \in \mathbb{N}, \; \alpha_k > 0, \; \sum_{k=1}^{k_0} \alpha_k = 1, \qquad (5\text{-}15)$$

where $\zeta_k \in C_0^\infty(D_{j_k})$ with $p(\zeta_k) < \varepsilon_{j_k}$. Choose $0 < \theta < 1$ such that $p(\zeta_k) < \theta \varepsilon_{j_k}$ for all $k = 1, \ldots, k_0$, and set $\delta = (\delta_j)_{j \in \mathbb{N}}$ with $\delta_j := (1 - \theta)\varepsilon_j$. We claim that $\zeta + U_\delta \subset U_\varepsilon$. To see this, let

$$\eta = \sum_{l=1}^{l_0} \beta_l \eta_l \in U_\delta \quad \text{with } l_0 \in \mathbb{N}, \; \beta_l > 0, \; \sum_{l=1}^{l_0} \beta_l = 1,$$

where $\eta_l \in C_0^\infty(D_{m_l})$ with $p(\eta_l) < \delta_{m_l}$. Then, on noting (1),

$$p(\tfrac{1}{\theta}\zeta_k) \le \tfrac{1}{\theta} p(\zeta_k) < \varepsilon_{j_k} \quad \text{and} \quad p(\tfrac{1}{1-\theta}\eta_l) \le \tfrac{1}{1-\theta} p(\eta_l) < \varepsilon_{m_l},$$

i.e. $\frac{1}{\theta}\zeta_k$ and $\frac{1}{1-\theta}\eta_l$ are elements of U_ε. Hence the convexity of U_ε yields that

$$\zeta + \eta = \theta \sum_{k=1}^{k_0} \alpha_k \cdot \frac{1}{\theta}\zeta_k + (1-\theta) \sum_{l=1}^{l_0} \beta_l \cdot \frac{1}{1-\theta}\eta_l \in U_\varepsilon.$$

This shows that $U_\varepsilon \in \mathcal{T}$.   □

Proof (3). We need to show that $U^1 \cap U^2 \in \mathcal{T}$, if $U^1, U^2 \in \mathcal{T}$. But this follows from $U_\varepsilon \subset U_{\varepsilon^1} \cap U_{\varepsilon^2}$, where $\varepsilon_j := \min(\varepsilon_j^1, \varepsilon_j^2)$ for $j \in \mathbb{N}$.   □

Proof (4). Let $\left(\widetilde{D}_j\right)_{j \in \mathbb{N}}$ be another cover and let $\widetilde{U}_{\widetilde{\varepsilon}}$ with $\widetilde{\varepsilon} = (\widetilde{\varepsilon}_j)_{j \in \mathbb{N}}$ be a set defined as above, now with respect to this cover. Since $\overline{D}_j$ is compact with $\overline{D}_j \subset \Omega$, for each $j \in \mathbb{N}$ there exists an $m_j \in \mathbb{N}$ with $D_j \subset \widetilde{D}_{m_j}$. Setting $\varepsilon_j := \widetilde{\varepsilon}_{m_j}$ for $j \in \mathbb{N}$ and $\varepsilon = (\varepsilon_j)_{j \in \mathbb{N}}$ then yields that $U_\varepsilon \subset \widetilde{U}_{\widetilde{\varepsilon}}$.   □

5.21 The space $\mathscr{D}(\Omega)$. We denote the vector space $C_0^\infty(\Omega)$, equipped with the topology $\mathcal{T}$ from 5.20, by $\mathscr{D}(\Omega)$. Then $\mathscr{D}(\Omega)$ is a *locally convex topological vector space*, i.e. it holds that:

(1) $\mathscr{D}(\Omega)$ with $\mathcal{T}$ is a Hausdorff space.

(2) $\mathscr{D}(\Omega)$ is a vector space and addition and scalar multiplication are continuous (as maps from $\mathscr{D}(\Omega) \times \mathscr{D}(\Omega)$ to $\mathscr{D}(\Omega)$ and from $\mathbb{K} \times \mathscr{D}(\Omega)$ to $\mathscr{D}(\Omega)$, respectively).

(3) For $\zeta \in U$ with $U \in \mathcal{T}$ there exists a convex set $V \in \mathcal{T}$ with $\zeta \in V \subset U$.

Proof (3). By their definition, the sets U_ε in 5.20 are convex.   □

Proof (2). We claim for every U_ε that $U_\delta + U_\delta \subset U_\varepsilon$, where $\delta = (\delta_j)_{j \in \mathbb{N}}$ with $\delta_j := \frac{1}{2}\varepsilon_j$, which implies the continuity of the addition. For the proof let

$$\zeta_l \in C_0^\infty(D_{j_l}) \quad \text{with } p(\zeta_l) < \delta_{j_l} \text{ for } l = 1, 2.$$

We have that $\zeta_1 + \zeta_2 = \frac{1}{2}(2\zeta_1 + 2\zeta_2)$ with $p(2\zeta_l) \leq 2p(\zeta_l) \leq 2\delta_{j_l} = \varepsilon_{j_l}$, and so $\zeta_1 + \zeta_2 \in U_\varepsilon$, as U_ε is convex. Then the same also holds for arbitrary elements $\zeta_1, \zeta_2 \in U_\delta$.

In order to show the continuity of the scalar multiplication at the point $(\alpha_0, \zeta_0) \in \mathbb{K} \times \mathscr{D}(\Omega)$, let U_ε be given. Let $\zeta_0 \in C_0^\infty(D_{j_0})$ and write

$$\alpha\zeta - \alpha_0\zeta_0 = \frac{1}{2}\big(2(\alpha - \alpha_0)\zeta_0 + 2\alpha(\zeta - \zeta_0)\big).$$

Let $|\alpha - \alpha_0| < \gamma \leq \frac{1}{2}$ and let $\zeta - \zeta_0 \in C_0^\infty(D_j)$ with $p(\zeta - \zeta_0) < \delta_j$, where γ, δ_j need to chosen. Now it holds that $\|2\gamma\zeta_0\|_{C^k(\overline{D}_{j_0})} \to 0$ as $\gamma \to 0$ for all $k \in \mathbb{N}$, and so it follows (as in 2.23(2)) that

$$p\big(2(\alpha - \alpha_0)\zeta_0\big) \leq p(2\gamma\zeta_0) \to 0 \quad \text{as } \gamma \to 0.$$

If we now choose $\gamma \leq \frac{1}{2}$ with $p(2\gamma\zeta_0) < \varepsilon_{j_0}$, then $2(\alpha - \alpha_0)\zeta_0 \in U_\varepsilon$. In addition, since $|2\alpha| \leq 2(|\alpha_0| + \gamma) \leq 2|\alpha_0| + 1$,

$$p\big(2\alpha(\zeta - \zeta_0)\big) \le (1 + 2|\alpha_0|)p(\zeta - \zeta_0) < \varepsilon_j\,,$$

if we set $\delta_j := (1 + 2|\alpha_0|)^{-1}\varepsilon_j$. This implies that also $2\alpha(\zeta - \zeta_0) \in U_\varepsilon$, and hence $\alpha\zeta \in \alpha_0\zeta_0 + U_\varepsilon$. Then the same also follows for all $\zeta \in \zeta_0 + U_\delta$, where $\delta := (\delta_j)_{j\in\mathbb{N}}$. $\qquad\square$

Proof (1). Let $\zeta^1, \zeta^2 \in \mathscr{D}(\Omega)$ with $\zeta^1 \ne \zeta^2$ and $\zeta := \zeta^1 - \zeta^2$. We claim that

$$(\zeta^1 + U_\varepsilon) \cap (\zeta^2 + U_\varepsilon) = \emptyset\,,$$

if $\varepsilon = (\varrho)_{j\in\mathbb{N}}$ and $\varrho > 0$ is sufficiently small. Indeed, if $\eta^1, \eta^2 \in U_\varepsilon$ with $\zeta^1 + \eta^1 = \zeta^2 + \eta^2$, then also $-\eta^1 \in U_\varepsilon$, and so

$$\zeta = \zeta^1 - \zeta^2 = (-\eta^1) + \eta^2 \in U_\varepsilon + U_\varepsilon \subset U_{2\varepsilon}\,,$$

on recalling the proof of (2). Now write ζ as a convex combination as in (5-15), so that

$$\frac{\|\zeta_k\|_{C^0}}{1 + \|\zeta_k\|_{C^0}} \le p(\zeta_k) < 2\varrho\,.$$

This implies, if $\varrho < \frac{1}{2}$, that

$$0 \ne \|\zeta\|_{C^0} \le \sum_{k=1}^{k_0} \alpha_k\|\zeta_k\|_{C^0} \le \max_{k=1,\dots,k_0} \|\zeta_k\|_{C^0} < \frac{2\varrho}{1-2\varrho}\,,$$

which is not possible, if ϱ depending on ζ was chosen sufficiently small. $\qquad\square$

5.22 Lemma. For every sequence $(\zeta_m)_{m\in\mathbb{N}}$ in $\mathscr{D}(\Omega)$ it holds that:

$$\zeta_m \to 0 \quad \text{as } m \to \infty \text{ in } \mathscr{D}(\Omega)$$

if and only if

(1) There exists an open $D \subset\subset \Omega$ such that $\zeta_m \in C_0^\infty(D)$ for all m.
(2) For all $D \subset\subset \Omega$ and all $k \in \mathbb{N}$ it holds that $\|\zeta_m\|_{C^k(\overline{D})} \to 0$ as $m \to \infty$.

Proof $\Leftarrow$. On noting that $\overline{D}$ is compact and $\overline{D} \subset \Omega$, the cover in 5.20 contains a D_j such that $D \subset D_j$. Then for a given ε it follows from (2) (as in 2.23(2)) that $p(\zeta_m) < \varepsilon_j$ for large m, and so $\zeta_m \in U_\varepsilon$. $\qquad\square$

Proof $\Rightarrow$. If we assume that (1) is not satisfied, then there exist an open cover $(D_j)_{j\in\mathbb{N}}$ of Ω with $D_j \subset\subset \Omega$ and $D_{j-1} \subset D_j$, as well as $x_j \in D_j \setminus \overline{D_{j-1}}$ and a subsequence $m_j \to \infty$, such that $\zeta_{m_j}(x_j) \ne 0$. Then

$$U := \left\{ \zeta \in \mathscr{D}(\Omega) \,;\, \sum_{j\in\mathbb{N}} \frac{2}{|\zeta_{m_j}(x_j)|}\|\zeta\|_{C^0(\overline{D_j}\setminus D_{j-1})} \le 1 \right\}$$

is a convex subset of $\mathscr{D}(\Omega)$. On noting that for all j

$$\left\{ \zeta \in C_0^\infty(D_j) \,;\, p(\zeta) < \varepsilon_j \right\} \subset U\,, \quad \text{where} \quad \varepsilon_j := \left(1 + \sum_{i\le j} \frac{2}{|\zeta_{m_i}(x_i)|}\right)^{-1}\,,$$

we have that $U_\varepsilon \subset U$, if $\varepsilon = (\varepsilon_j)_{j \in \mathbb{N}}$ and U_ε is defined with respect to the cover $(D_j)_{j \in \mathbb{N}}$. The definition of the topology and the fact that $\zeta_m \to 0$ in $\mathscr{D}(\Omega)$ as $m \to \infty$ yield that $\zeta_m \in U_\varepsilon$ for large m. But it follows from the construction of U that the ζ_{m_j} do not lie in U, a contradiction. This shows (1).

Now for $k \in \mathbb{N}$ and $\delta > 0$ choose $\varepsilon = (\varepsilon_j)_{j \in \mathbb{N}}$ with $2^k \varepsilon_j = \left(1 + \frac{1}{\delta}\right)^{-1} > 0$ for all j, which yields that

$$U_\varepsilon \subset \left\{ \zeta \in C_0^\infty(\Omega) \,;\, \|\zeta\|_{C^k} \leq \delta \right\}.$$

For large m we have that $\zeta_m \in U_\varepsilon$, and so $\|\zeta_m\|_{C^k} \leq \delta$. This shows (2). $\square$

5.23 The dual space of $\mathscr{D}(\Omega)$. Consider (see 5.5(1)) the dual space

$$\mathscr{D}(\Omega)' = \{T : \mathscr{D}(\Omega) \to \mathbb{K} \,;\, T \text{ is linear and continuous}\}$$

of $\mathscr{D}(\Omega)$. Then (with the notation in 5.17(2))

$$\mathscr{D}(\Omega)' = \mathscr{D}'(\Omega).$$

Proof $\subset$. Let $T \in \mathscr{D}(\Omega)'$. If $T \notin \mathscr{D}'(\Omega)$, then there exist a $D \subset\subset \Omega$ and $\zeta_m \in C_0^\infty(D)$ with

$$1 = |T\zeta_m| > m\|\zeta_m\|_{C^m(\overline{D})} \quad \text{for } m \in \mathbb{N}.$$

For all $k \in \mathbb{N}$ it then follows that $\|\zeta_m\|_{C^k(\overline{D})} \to 0$ as $m \to \infty$, and so 5.22 yields $\zeta_m \to 0$ as $m \to \infty$ in $\mathscr{D}(\Omega)$. Now the continuity of T implies that $T\zeta_m \to 0$ as $m \to \infty$, which is a contradiction. $\square$

Proof $\supset$. Let $T \in \mathscr{D}'(\Omega)$, let $(D_j)_{j \in \mathbb{N}}$ be the exhaustion from 5.20 and let

$$|T\zeta| \leq C_j \|\zeta\|_{C^{k_j}(\overline{D_j})} \quad \text{for } \zeta \in C_0^\infty(D_j).$$

For $\delta > 0$ let $\varepsilon = (\varepsilon_j)_{j \in \mathbb{N}}$ be defined by $\varepsilon_j := 2^{-k_j} \frac{\delta}{C_j + \delta}$. Then

$$\zeta \in C_0^\infty(D_j) \text{ with } p(\zeta) < \varepsilon_j \implies |T\zeta| \leq C_j \|\zeta\|_{C^{k_j}(\overline{D_j})} \leq \delta.$$

As T is linear, it follows that $|T\zeta| \leq \delta$ for all $\zeta \in U_\varepsilon$ (with U_ε as in 5.20). This proves the continuity of T. $\square$

E5 Exercises

E5.1 Commutator. Let X be a nontrivial normed vector space and let $P, Q : X \to X$ be linear maps with $PQ - QP = \mathrm{Id}$. Then P and Q cannot both be continuous. (This relation, which appears in quantum mechanics, is called the Heisenberg relation.)

Solution. It follows inductively for $n \in \mathbb{N}$ that

$$PQ^n - Q^n P = nQ^{n-1}, \qquad (\text{E5-1})$$

on noting that for such n we have that

$$PQ^{n+1} - Q^{n+1}P = \underbrace{(PQ^n - Q^n P)}_{=nQ^{n-1}} Q + Q^n \underbrace{(PQ - QP)}_{=1Q^0=\mathrm{Id}}$$

$$= nQ^{n-1}Q + Q^n = (n+1)Q^n .$$

Assuming that $P, Q \in \mathscr{L}(X)$, it follows from (E5-1) that

$$n\|Q^{n-1}\| \leq 2\|P\| \cdot \|Q^n\| \leq 2\|P\| \cdot \|Q\| \cdot \|Q^{n-1}\|,$$

and hence $Q^{n-1} = 0$ for large n, that is, for $n > 2\|P\| \cdot \|Q\|$. It follows inductively from (E5-1) that $Q^{n-m} = 0$ for $m = 1, \ldots, n$, i.e. $\mathrm{Id} = Q^0 = 0$, a contradiction if $X \neq \{0\}$. □

E5.2 Nonexistence of the inverse. For noncomplete normed spaces, the inverse in 5.7 in general does not exist.

Solution. We give a counterexample. Let $Y := \ell^2(\mathbb{R})$ and let

$$X := \left\{ x = (x_i)_{i \in \mathbb{N}} \in \mathbb{R}^\mathbb{N} ; \text{ only finitely many } x_i \neq 0 \right\} \subset \ell^2(\mathbb{R}) = Y,$$

i.e. X is equipped with the Y-norm. Let $\varepsilon > 0$. For the **shift operator**

$$(Tx)_i := \begin{cases} 0 & \text{for } i = 1, \\ \varepsilon x_{i-1} & \text{for } i > 1, \end{cases}$$

it holds that $T \in \mathscr{L}(Y)$ and $\|T\| = \varepsilon$. Hence for $\varepsilon < 1$ we can apply 5.7 for Y and T, and obtain, for instance, that

$$(\mathrm{Id} - T)^{-1}e_1 = \sum_{n=0}^\infty T^n e_1 = (\varepsilon^{i-1})_{i \in \mathbb{N}} \notin X.$$

On the other hand, $Tx \in X$ for $x \in X$. Hence 5.7 is not valid for X and $T|_X$ (X is not complete and $\overline{X} = Y$). □

E5.3 Unique extension of linear maps. Let $Z \subset X$ be a dense subspace and let $T \in \mathscr{L}(Z; Y)$. Then there exists a unique continuous extension $\widetilde{T}$ of T to X. Moreover, $\widetilde{T} \in \mathscr{L}(X; Y)$.

Solution. T is uniformly continuous on Z (in fact Lipschitz continuous with Lipschitz constant $\|T\|$). Hence, on recalling E4.18,

$$\widetilde{T}x := \lim_{z \in Z \,:\, z \to x} Tz \quad \text{for } x \in X$$

defines a unique continuous extension of T to X. In addition, the linearity of T carries over to $\widetilde{T}$.   □

E5.4 Limit of linear maps. Let $(T_k)_{k \in \mathbb{N}}$ be a bounded sequence in $\mathscr{L}(X;Y)$ and let $D \subset X$ be dense. If there exists

$$\lim_{k \to \infty} T_k x \quad \text{for } x \in D, \tag{E5-2}$$

then there exists

$$Tx := \lim_{k \to \infty} T_k x \quad \text{for all } x \in X$$

and $T \in \mathscr{L}(X;Y)$.

Solution. Let $\|T_k\| \leq C < \infty$ for all k and let $Z := \operatorname{span}(D)$. Then it follows from (E5-2) that

$$Tz := \lim_{k \to \infty} T_k z$$

exists for all $z \in Z$, and that T is linear on Z. Since

$$\|Tz\| = \lim_{k \to \infty} \|T_k z\| \leq C\|z\|,$$

it holds that $T \in \mathscr{L}(Z;Y)$. Let $\widetilde{T} \in \mathscr{L}(X;Y)$ be the unique extension of T to X from E5.3. Then it holds for all $x \in X$ and $z \in Z$ that

$$\left\|\widetilde{T}x - T_k x\right\| \leq \left\|\widetilde{T}z - T_k z\right\| + \left(\left\|\widetilde{T}\right\| + C\right)\|x - z\|$$

$$\longrightarrow \left(\left\|\widetilde{T}\right\| + C\right)\|x - z\| \quad \text{as } k \to \infty.$$

As $\overline{Z} = X$, we can choose $\|x - z\|$ arbitrarily small. This shows that

$$\widetilde{T}x = \lim_{k \to \infty} T_k x \quad \text{for all } x \in X.$$

□

E5.5 Pointwise convergence of operators. Let $T, T_k \in \mathscr{L}(X;Y)$, $k \in \mathbb{N}$, with $\|T_k\| \leq C < \infty$ and let $D \subset X$ be dense. If for all $x \in D$

$$T_k x \longrightarrow Tx \quad \text{as } k \to \infty,$$

then this also holds for all $x \in X$.

Solution. See the second part of the solution of E5.4.   □

E5.6 Convergence of operators. Let T_k be defined as in 5.6(6) with $1 \le p < \infty$. Does it hold that $T_k \longrightarrow \mathrm{Id}$ as $k \to \infty$ in the space $\mathscr{L}\big(L^p(\mathrm{I\!R}^n)\big)$?

Solution. No! As an example, let $n = 1$ and $\varphi_k = \psi_{\varepsilon_k}$ with $\varepsilon_k \to 0$ as $k \to \infty$, where $\psi_\varepsilon(x) := \frac{1}{2\varepsilon}$ for $|x| < \varepsilon$ and $\psi_\varepsilon(x) := 0$ for $|x| > \varepsilon$. Then consider $T_k \varphi_k = \psi_{\varepsilon_k} * \psi_{\varepsilon_k}$. Direct calculations yield that

$$\psi_\varepsilon * \psi_\varepsilon(x) = \max\big(0, \frac{1}{2}\varepsilon(1 - \frac{|x|}{2\varepsilon})\big),$$

$$\|\psi_\varepsilon\|_{L^p} = (2\varepsilon)^{\frac{1}{p}-1},$$

$$\|\psi_\varepsilon * \psi_\varepsilon - \psi_\varepsilon\|_{L^p} = (1+p)^{-\frac{1}{p}} \cdot (4\varepsilon)^{\frac{1}{p}-1}.$$

Consequently,

$$\|T_k - \mathrm{Id}\| \ge \frac{\|T_k \psi_{\varepsilon_k} - \psi_{\varepsilon_k}\|_{L^p}}{\|\psi_{\varepsilon_k}\|_{L^p}} = \frac{1}{2}\Big(\frac{1+p}{2}\Big)^{-\frac{1}{p}} > 0.$$

$\square$

6 Linear functionals

In this chapter we deal with the representations of dual spaces, i.e. we will state canonical isomorphisms between the most important dual spaces and already known spaces. We will use this to solve boundary value problems for partial differential equations.

The most important case is that of a Hilbert space, for which the dual space is isomorphic to the space itself (theorem 6.1). As a consequence, we obtain the Lax-Milgram theorem (see 6.2), with the help of which elliptic boundary value problems can be solved (see 6.4 – 6.9).

In the second part, we state representations of the dual spaces of $L^p(\mu)$ for $p < \infty$ (see 6.12) and of $C^0(S)$ (see 6.23). The proof of 6.23 will employ the Hahn-Banach theorem (see 6.14 – 6.15). This theorem states that continuous linear maps $f : Y \to \mathbb{K}$ can be extended from a subspace $Y \subset X$ to the full space X such that the norm of the map is maintained, which is one of the general principles of functional analysis. A constructive proof of the Hahn-Banach theorem for separable spaces X will be given in 9.2.

Lax-Milgram's theorem

We start with an existence theory, which is based on the following result.

6.1 Riesz representation theorem. If X is a Hilbert space, then

$$J(x)(y) := (y\,,\,x)_X \quad \text{for } x, y \in X$$

defines an isometric conjugate linear isomorphism $J : X \to X'$.

Notation: In the remainder of this book we will also denote this isomorphism by $R_X : X \to X'$.

Definition: Here a map J is called **conjugate linear** if for all $x, y \in X$ and $\alpha \in \mathbb{K}$ it holds that $J(\alpha x + y) = \overline{\alpha}J(x) + J(y)$. In the case $\mathbb{K} = \mathbb{R}$ this reduces to J being linear.

Proof. By the Cauchy-Schwarz inequality,

$$|J(x)(y)| \le \|x\|_X \cdot \|y\|_Y\,,$$

i.e. $J(x) \in X'$ with $\|J(x)\|_{X'} \le \|x\|_X$. On noting that $|J(x)(x)| = \|x\|_X^2$, we see that $\|J(x)\|_{X'} \ge \|x\|_X$. Hence J is isometric, and in particular injective.

Now the crucial step is to show that J is surjective. Let $0 \neq x_0' \in X'$ and let P be the orthogonal projection from 4.3 onto the closed null space $\mathcal{N}(x_0')$. Choose $e \in X$ with $x_0'(e) = 1$ and define

$$x_0 := e - Pe, \quad \text{hence also } x_0'(x_0) = 1,$$

and in particular $x_0 \neq 0$. Now it follows from 4.3 (see 4.4(2)) that

$$(y, x_0)_X = 0 \quad \text{for all } y \in \mathcal{N}(x_0'). \tag{6-3}$$

For all $x \in X$,

$$x = \underbrace{x - x_0'(x)x_0}_{\in \mathcal{N}(x_0')} + x_0'(x)x_0 \, ,$$

which together with (6-3) yields that

$$(x, x_0)_X = (x_0'(x)x_0, x_0)_X = x_0'(x)\|x_0\|^2 \, ,$$

and hence

$$x_0'(x) = \left(x, \frac{x_0}{\|x_0\|^2} \right)_X = J\left(\frac{x_0}{\|x_0\|^2} \right)(x) \, .$$

$\square$

An application of the Riesz representation theorem is the

6.2 Lax-Milgram theorem. Let X be a Hilbert space over $\mathbb{K}$ and let $a : X \times X \to \mathbb{K}$ be sesquilinear. Assume that there exist constants c_0 and C_0 with $0 < c_0 \leq C_0 < \infty$ such that for all $x, y \in X$

(1) $|a(x,y)| \leq C_0\|x\|_X\|y\|_X$ (*Continuity*),

(2) $\mathrm{Re}\, a(x, x) \geq c_0\|x\|_X^2$ (*Coercivity*).

Then there exists a unique map $A : X \to X$ with

$$a(y, x) = (y, Ax)_X \quad \text{for all } x, y \in X.$$

In addition, $A \in \mathscr{L}(X)$ is an invertible operator with

$$\|A\| \leq C_0 \quad \text{and} \quad \|A^{-1}\| \leq \frac{1}{c_0} \, .$$

Proof. For every $x \in X$ it follows from (1) that the function $a(\cdot, x)$ lies in X' and satisfies

$$\|a(\cdot, x)\|_{X'} \leq C_0\|x\|_X \, .$$

Hence, by the Riesz representation theorem 6.1, there exists a unique element $A(x) \in X$ such that

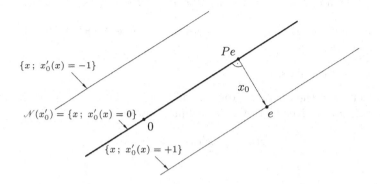

Fig. 6.1. *Proof of the Riesz representation theorem*

$$a(y, x) = (y, A(x))_X \quad \text{for all } y \in X$$

and moreover

$$\|A(x)\|_X = \|a(\cdot, x)\|_{X'} \le C_0 \|x\|_X .$$

Since a and the scalar product are conjugate linear in the second argument, it follows that A is linear. Hence $A \in \mathscr{L}(X)$ with $\|A\| \le C_0$. Moreover,

$$c_0 \|x\|_X^2 \le \operatorname{Re} a(x, x) = \operatorname{Re}(x, A(x))_X \le \|x\|_X \cdot \|Ax\|_X ,$$

and so

$$c_0 \|x\|_X \le \|A(x)\|_X \quad \text{for all } x \in X, \tag{6-4}$$

which implies that $\mathscr{N}(A) = \{0\}$. In addition, it follows that the image space $\mathscr{R}(A)$ is closed, on noting that for $x_k, x \in X$

$$A(x_k) \to y \quad \text{as } k \to \infty$$
$$\implies \quad \|x_k - x_l\|_X \le \tfrac{1}{c_0} \|A(x_k - x_l)\|_X \quad \text{(recall (6-4))}$$
$$= \tfrac{1}{c_0} \|A(x_k) - A(x_l)\|_X \to 0 \quad \text{as } k, l \to \infty$$
$$\implies \quad x_k \to x \quad \text{as } k \to \infty \text{ for an } x \in X$$
$$\implies \quad A(x_k) \to A(x) \quad \text{(as } A \text{ is continuous)}$$
$$\implies \quad y = Ax \in \mathscr{R}(A).$$

It remains to show that $\mathscr{R}(A) = X$. If $\mathscr{R}(A) \ne X$, then, on recalling that $\mathscr{R}(A)$ is a closed subspace, the projection theorem 4.3 yields that there exists an $x_0 \in X \setminus \mathscr{R}(A)$ such that (recall 4.4(2))

$$(y, x_0)_X = 0 \quad \text{for all } y \in \mathscr{R}(A)$$

(choose an $\tilde{x}_0 \in X \setminus \mathscr{R}(A)$ and set $x_0 := \tilde{x}_0 - P\tilde{x}_0$, where P is the orthogonal projection onto $\mathscr{R}(A)$). This yields, on setting $y = A(x_0)$, that

$$0 = \mathrm{Re}\,(A(x_0)\,,\,x_0)_X = \mathrm{Re}\,(x_0\,,\,A(x_0))_X = \mathrm{Re}\,a(x_0,x_0) \geq c_0\|x_0\|_X^2 > 0\,,$$

a contradiction. Hence we have shown that A is bijective. It follows from (6-4) that $\|A^{-1}\| \leq \frac{1}{c_0}$. $\qquad\qquad\square$

6.3 Consequences.

(1) Let A be the operator from 6.2 and let R_X be the isometry from theorem 6.1. For a given $x' \in X'$, the unique solution of

$$a(y,x) = x'(y) \quad \text{for all } y \in X \tag{6-5}$$

is then $x := A^{-1}R_X^{-1}x'$.

(2) The solution in (1) has the **stability property**

$$\|x\|_X \leq \frac{1}{c_0}\|x'\|_{X'}\,. \tag{6-6}$$

Interpretation: If we consider two "right-hand sides" x_1' and x_2' and the corresponding solutions x_1 and x_2 in (1), then it follows from (6-6), due to the linearity of the problem ($x_1 - x_2$ is the solution to $x_1' - x_2'$), that

$$\|x_1 - x_2\|_X \leq \tfrac{1}{c_0}\|x_1' - x_2'\|_{X'}\,.$$

Hence the error in the solutions can be estimated by the error in the data. This justifies the term stability.

(3) Formulated for the operator A, the Lax-Milgram theorem reads as follows: Let X be a Hilbert space and let $A \in \mathcal{L}(X)$ be **coercive**, i.e. there exists a constant $c_0 > 0$ such that

$$\mathrm{Re}\,(x\,,\,Ax)_X \geq c_0\|x\|_X^2 \quad \text{for all } x \in X.$$

Then A is invertible, with $\|A^{-1}\| \leq \frac{1}{c_0}$.

(4) If a in 6.2 is a scalar product, then the solution x in statement (1) is, in addition, the uniquely determined absolute minimum of the functional

$$E(y) := \tfrac{1}{2}a(y,y) - \mathrm{Re}\,x'(y)\,.$$

Proof (1) *and* (2). By the definition of A and R_X, for all $x,y \in X$

$$a(y,x) = (y\,,\,Ax)_X = (R_X Ax)(y)\,,$$

and $R_X A : X \to X'$ is bijective. If $x = (R_X A)^{-1}x'$, then it follows from (6-4) that

$$c_0\|x\|_X \leq \|Ax\|_X = \|R_X^{-1}x'\|_X = \|x'\|_{X'}\,.$$

$\qquad\qquad\square$

Proof (3). The product $a(y,x) := (y\,,\,Ax)_X$ satisfies the properties in 6.2 with $C_0 = \|A\|$. Moreover, A is the operator corresponding to a from 6.2. $\quad\square$

Proof (4). Let $y \in X$. Then

$$
\begin{aligned}
E(y) - E(x) &= \tfrac{1}{2}\big(a(y,y) - a(x,x)\big) - \operatorname{Re} x'(y - x) \\
&= \tfrac{1}{2}\big(a(y,y) - a(x,x)\big) - \operatorname{Re} a(y - x, x) \\
&= \tfrac{1}{2}\big(a(y,y) - a(y,x) - a(x,y) + a(x,x)\big) \\
&= \tfrac{1}{2}a(y - x, y - x) \geq \tfrac{c_0}{2}\|y - x\|_X^2 \,.
\end{aligned}
$$

$\square$

The Lax-Milgram theorem has applications for integral operators (see E6.3) and for differential operators, which will be discussed in the following. First we consider the classical case in spaces of continuous functions.

6.4 Elliptic boundary value problems. Let $\Omega \subset \mathbb{R}^n$ be open and bounded and let $\mathbb{K} = \mathbb{R}$. We want to find functions $u \in C^2(\Omega)$ satisfying the differential equation

$$
-\sum_{i=1}^{n} \partial_i \Big(\sum_{j=1}^{n} a_{ij}\partial_j u + h_i\Big) + bu + f = 0 \quad \text{in } \Omega. \tag{6-7}
$$

Here $a_{ij}, h_i \in C^1(\Omega)$ for $i, j = 1, \dots, n$ and $f, b \in C^0(\Omega)$ are given real-valued functions, and we assume that there exists a $c_0 > 0$ such that for all $x \in \Omega$,

$$
\sum_{i,j=1}^{n} a_{ij}(x)\xi_i\xi_j \geq c_0|\xi|^2 \quad \text{for all } \xi \in \mathbb{R}^n. \tag{6-8}
$$

We then say that the matrix $(a_{ij}(x))_{i,j}$ is **uniformly elliptic** in x. (For every $c > 0$, the set of points $\xi \in \mathbb{R}^n$, for which $\sum_{i,j} a_{ij}(x)\xi_i\xi_j = c$, is an ellipsoid.) Let us emphasize here that the matrix $(a_{ij}(x))_{i,j}$ need not be symmetric.

It turns out that, under certain assumptions, there exists a unique function u solving (6-7), which in addition satisfies suitable boundary conditions on $\partial\Omega$. The two most frequently occurring boundary conditions in mathematical physics are:

(1) **Dirichlet boundary condition.** Let $g \in C^0(\partial\Omega)$ be given. Find a function $u \in C^2(\Omega) \cap C^0(\overline{\Omega})$ which solves the following **Dirichlet boundary value problem:**

$$
u \text{ satisfies (6-7) in } \Omega, \quad u = g \text{ on } \partial\Omega.
$$

(2) **Neumann boundary condition.** We assume that Ω has a C^1-boundary, i.e. that the boundary $\partial\Omega$ can be locally represented as the graph of a C^1-function in an appropriately chosen coordinate system (as in A8.2). In addition, we assume that $a_{ij}, h_i \in C^0(\overline{\Omega})$. Let $g \in C^0(\partial\Omega)$ be given. Find a function $u \in C^2(\Omega) \cap C^1(\overline{\Omega})$ which solves the following **Neumann boundary value problem:**

$$u \text{ satisfies (6-7) in } \Omega \, , \quad -\sum_{i=1}^{n} \nu_i \Big(\sum_{j=1}^{n} a_{ij} \partial_j u + h_i \Big) = g \text{ on } \partial\Omega.$$

Here $\nu = (\nu_i)_{i=1,\ldots,n}$ is the **outer normal** to $\partial\Omega$.

Remark: For the boundary value problem (1) to be at all solvable, there must exist some function $u_0 \in C^2(\Omega) \cap C^0(\overline{\Omega})$ with $u_0 = g$ on $\partial\Omega$. Then the boundary value problem can be transformed to one for $\tilde{u} := u - u_0$, by replacing g with $\tilde{g} := 0$, h_i with $\tilde{h}_i := h_i + \sum_j a_{ij}\partial_j u_0$, and f with $\tilde{f} := f + bu_0$. Analogously, for (2) there must exist a function $u_0 \in C^2(\Omega) \cap C^1(\overline{\Omega})$ with $-\sum_i \nu_i(\sum_j a_{ij}\partial_j u_0 + h_i) = g$ on $\partial\Omega$. Then the boundary value problem can be transformed to one for $\tilde{u} := u - u_0$, by replacing g with $\tilde{g} := 0$, h_i with $\tilde{h}_i := 0$, and f with $\tilde{f} := f - \sum_i \partial_i(\sum_j a_{ij}\partial_j u_0 + h_i) + bu_0$. We then call the boundary conditions **homogeneous**.

We now give an equivalent definition of the boundary value problem with the help of test functions (this gives a connection to distributions, which were treated at the end of section 5).

In the Dirichlet case, if we multiply the differential equation (6-7) by functions $\zeta \in C_0^\infty(\Omega)$, then we obtain after integration by parts that

$$\int_\Omega \Big(\sum_i \partial_i \zeta \Big(\sum_j a_{ij}\partial_j u + h_i \Big) + \zeta(bu + f) \Big) \, \mathrm{dL}^n = 0. \tag{6-9}$$

Conversely, if this integral identity is satisfied for all $\zeta \in C_0^\infty(\Omega)$, then we obtain, on reversing the integration by parts, that

$$\int_\Omega \zeta w \, \mathrm{dL}^n = 0 \quad \text{with} \quad w := -\sum_i \partial_i \Big(\sum_j a_{ij}\partial_j u + h_i \Big) + bu + f.$$

If we assume that $w(x_0) \neq 0$ for some $x_0 \in \Omega$, then we can choose an $\varepsilon_0 > 0$ with $w > 0$ or $w < 0$ in $\mathrm{B}_{\varepsilon_0}(x_0) \subset \Omega$, and then a nontrivial $\zeta \in C_0^\infty(\mathrm{B}_{\varepsilon_0}(x_0))$ with $\zeta \geq 0$, in order to obtain a contradiction. Hence it follows that $w = 0$ in Ω (this also follows directly from 4.22), i.e. the differential equation (6-7) holds in Ω.

Similarly, in the Neumann case, if we multiply the differential equation (6-7) by functions $\zeta \in C^\infty(\overline{\Omega})$, on assuming that $a_{ij}, h_i \in C^1(\overline{\Omega})$, we obtain after integration by parts that

$$\int_\Omega \Big(\sum_i \partial_i \zeta \Big(\sum_j a_{ij}\partial_j u + h_i \Big) + \zeta(bu + f) \Big) \, \mathrm{dL}^n + \int_{\partial\Omega} \zeta g \, \mathrm{dH}^{n-1} = 0. \tag{6-10}$$

Conversely, if this holds for all $\zeta \in C^\infty(\overline{\Omega})$, then as before we obtain the differential equation in Ω (here it is sufficient to consider $\zeta \in C_0^\infty(\Omega)$), and then it holds for $\zeta \in C^\infty(\overline{\Omega})$ that

$$\int_{\partial\Omega} \zeta w \, \mathrm{dH}^{n-1} = 0 \quad \text{with} \quad w := \sum_i \nu_i \left(\sum_j a_{ij} \partial_j u + h_i \right) + g.$$

Similarly to the argumentation above, it now follows that the Neumann boundary condition is satisfied.

The basic idea for the solution of these boundary value problems with the help of Hilbert space methods is to interpret the integral terms in (6-9) and (6-10) as an L^2-bilinear form, and enlarge the spaces for test functions and solutions accordingly. As the test function appears with ζ and $\partial_i \zeta$, the appropriate test space for (6-9) is the closure of $C_0^\infty(\Omega)$ in the space $W^{1,2}(\Omega)$, i.e. the space $W_0^{1,2}(\Omega)$ (see 3.29). Since functions in $W_0^{1,2}(\Omega)$, when Ω has a C^1-boundary, have in a weak sense boundary values 0 (see A8.10), $W_0^{1,2}(\Omega)$ is also the appropriate enlarged solution space. For (6-10) the appropriate test space is the closure of $C^\infty(\overline{\Omega})$ in the space $W^{1,2}(\Omega)$, i.e. for sets Ω with a C^1-boundary the space $W^{1,2}(\Omega)$ itself (see A8.7), which is also the appropriately enlarged solution space.

For the resulting weak formulations of the problem it is no longer necessary to assume that the data a_{ij}, h_i, b, f of the problem are continuous functions in Ω. However, it is necessary to make assumptions on their integrability, for instance as formulated in the following:

6.5 Weak boundary value problems. With $\mathbb{K} = \mathbb{R}$ it is assumed in the following that $\Omega \subset \mathbb{R}^n$ is open and bounded, that $a_{ij} \in L^\infty(\Omega)$ satisfy the ellipticity condition (6-8) for almost all $x \in \Omega$, and that $b \in L^\infty(\Omega)$ and $h_i, f \in L^2(\Omega)$. The weak formulation of the boundary value problem in 6.4 is defined as follows (where we consider only the case $g = 0$):

(1) We call $u : \Omega \to \mathbb{R}$ a **weak solution** of the **Dirichlet problem** if

$u \in W_0^{1,2}(\Omega)$ and

$$\int_\Omega \left(\sum_i \partial_i \zeta \left(\sum_j a_{ij} \partial_j u + h_i \right) + \zeta(bu + f) \right) \mathrm{dL}^n = 0$$

for all $\zeta \in W_0^{1,2}(\Omega)$.

Here, as remarked above, if Ω has a C^1-boundary, then the condition $u \in W_0^{1,2}(\Omega)$ in a weak sense contains the homogeneous boundary conditions, and it is irrelevant whether ζ varies in the space $W_0^{1,2}(\Omega)$, or only in the space $C_0^\infty(\Omega)$.

(2) We call $u : \Omega \to \mathbb{R}$ a **weak solution** of the **Neumann problem** if

$u \in W^{1,2}(\Omega)$ and

$$\int_\Omega \left(\sum_i \partial_i \zeta \left(\sum_j a_{ij} \partial_j u + h_i \right) + \zeta(bu + f) \right) \mathrm{dL}^n = 0$$

for all $\zeta \in W^{1,2}(\Omega)$.

Here, as explained above, if Ω has a C^1-boundary, then the integral term in a weak sense contains the homogeneous boundary conditions (for $g = 0$ in 6.4(2) the boundary integral in (6-10) vanishes), and it is irrelevant whether ζ varies in the space $W^{1,2}(\Omega)$, or only in the space $C^\infty(\overline{\Omega})$.

We will now prove the existence of solutions to these weak boundary value problems.

6.6 Existence theorem for the Neumann problem. Let the assumptions in 6.5 hold and let $b_0 > 0$ with $b(x) \geq b_0$ for almost all $x \in \Omega$. Then there exists a unique solution $u \in W^{1,2}(\Omega)$ for the Neumann problem in 6.5(2). Moreover,

$$\|u\|_{W^{1,2}} \leq C(\|h\|_{L^2} + \|f\|_{L^2}),$$

with a constant C that is independent of h and f.

Proof. For $u, v \in W^{1,2}(\Omega)$ we define

$$a(u,v) := \sum_{i,j} \int_\Omega \partial_i u \cdot a_{ij} \partial_j v \, \mathrm{dL}^n + \int_\Omega u \cdot bv \, \mathrm{dL}^n . \qquad (6\text{-}11)$$

(We mention that in general a does not need to be a scalar product, for $(a_{ij})_{ij}$ can be asymmetric.) Then a is bilinear, with

$$|a(u,v)| \leq \sum_{i,j} \|a_{ij}\|_{L^\infty} \|\partial_i u\|_{L^2} \|\partial_j v\|_{L^2} + \|b\|_{L^\infty} \|u\|_{L^2} \|v\|_{L^2}$$

$$\leq C\|u\|_{W^{1,2}} \|v\|_{W^{1,2}} \quad \text{with } C := \sum_{i,j} \|a_{ij}\|_{L^\infty} + \|b\|_{L^\infty} .$$

In addition, it follows from the assumptions on a_{ij} and b that

$$a(u,u) \geq c_0 \int_\Omega |\nabla u|^2 \, \mathrm{dL}^n + b_0 \int_\Omega |u|^2 \, \mathrm{dL}^n \geq c \cdot \|u\|_{W^{1,2}}^2$$

with $c := \min(c_0, b_0)$. Hence a satisfies the assumptions of the Lax-Milgram theorem 6.2 on the Hilbert space $W^{1,2}(\Omega)$. We want to find a $u \in W^{1,2}(\Omega)$ such that

$$a(v,u) = F(v) \quad \text{for all } v \in W^{1,2}(\Omega),$$

where

$$F(v) := -\int_\Omega \left(\sum_i \partial_i v \cdot h_i + vf \right) \mathrm{dL}^n . \qquad (6\text{-}12)$$

It follows from 6.3(1) that there exists a unique such u if F belongs to the dual space of $W^{1,2}(\Omega)$. But this is the case, since F is linear, with

$$|F(v)| \leq \|h\|_{L^2} \|\nabla v\|_{L^2} + \|f\|_{L^2} \|v\|_{L^2} \leq (\|h\|_{L^2} + \|f\|_{L^2}) \|v\|_{W^{1,2}} .$$

In addition, the solution u can be estimated by the data, since, by 6.3(2),

$$\|u\|_{W^{1,2}} \le \frac{1}{c}\|F\| \le \frac{1}{c}\left(\|h\|_{L^2} + \|f\|_{L^2}\right).$$

□

The Dirichlet problem can also be solved in the case $b = 0$. Here we need the following

6.7 Poincaré inequality. If $\Omega \subset \mathbb{R}^n$ is open and bounded, then there exists a constant C_0 (which depends on Ω), such that

$$\int_{\Omega} |u|^2\, dL^n \le C_0 \int_{\Omega} |\nabla u|^2\, dL^n \quad \text{for all } u \in W_0^{1,2}(\Omega).$$

Note: See also 8.16 and E10.10.

Proof. On noting that both sides of the inequality depend continuously on u in the $W^{1,2}$-norm, and on recalling the definition of $W_0^{1,2}(\Omega)$, it is sufficient to prove the estimate for functions $u \in C_0^\infty(\Omega)$. In the case $n = 1$, let $\Omega \subset [a, b] \subset \mathbb{R}$. Then the Hölder inequality yields for $a \le x \le b$, on setting $u = 0$ in $\mathbb{R} \setminus \Omega$, that

$$|u(x)|^2 = |u(x) - u(a)|^2 = \left| \int_a^x \partial_x u(y)\, dy \right|^2$$

$$\le (x - a) \int_a^x |\partial_x u(y)|^2\, dy \le (b - a) \int_a^b |\partial_x u(y)|^2\, dy.$$

Integration over x gives

$$\int_a^b |u|^2\, dL^1 \le (b - a)^2 \int_a^b |\partial_x u|^2\, dL^1. \tag{6-13}$$

In the case $n > 1$, let $\Omega \subset [a, b] \times \mathbb{R}^{n-1}$. Then we obtain (6-13) by integrating over x_1. Integration over the remaining $n - 1$ coordinates then yields the desired result. (Hence the Poincaré inequality also holds for infinite slab domains.)   □

6.8 Existence theorem for the Dirichlet problem. Let the assumptions in 6.5 hold and let $b \ge 0$. Then there exists a unique weak solution $u \in W_0^{1,2}(\Omega)$ for the Dirichlet problem in 6.5(1). Moreover,

$$\|u\|_{W^{1,2}} \le C\left(\|h\|_{L^2} + \|f\|_{L^2}\right)$$

with a constant C that is independent of h and f.

Proof. Consider the bilinear form a in (6-11), now on the Hilbert space $W_0^{1,2}(\Omega)$. As in the proof of 6.6,

$$|a(u,v)| \le C\|u\|_{W^{1,2}}\|v\|_{W^{1,2}}$$

and the assumptions on the coefficients yield that

$$a(u,u) \ge c_0 \int_\Omega |\nabla u|^2 \, d\mathbf{L}^n = c_0 \, \|\nabla u\|_{L^2}^2 \quad \text{for } u \in W_0^{1,2}(\Omega).$$

Then it follows, with the constant C_0 from 6.7, that

$$\|u\|_{W^{1,2}}^2 = \|u\|_{L^2}^2 + \|\nabla u\|_{L^2}^2 \le (C_0 + 1)\|\nabla u\|_{L^2}^2 \le \frac{C_0 + 1}{c_0} a(u,u),$$

and so $a(u,u) \ge c \, \|u\|_{W^{1,2}}^2$ with $c = c_0 \cdot (C_0 + 1)^{-1}$. Hence a satisfies the assumptions of the Lax-Milgram theorem 6.2 on the Hilbert space $W_0^{1,2}(\Omega)$. The functional F in (6-12), restricted to the space $W_0^{1,2}(\Omega)$, then lies in its dual space. Hence, by 6.3(1), there exists a unique $u \in W_0^{1,2}(\Omega)$ with

$$a(v,u) = F(v) \quad \text{for all } v \in W_0^{1,2}(\Omega).$$

The estimate follows again from 6.3(2) (see the proof 6.6). $\square$

6.9 Remark (Regularity of the solution). Based on the existence proofs in 6.6 and 6.8 for weak solutions of the boundary value problem, it is possible to show a posteriori that a weak solution is indeed a classical solution of the boundary value problem in the sense of 6.4, provided the data a_{ij}, h_i, b, f and $\partial\Omega$ satisfy certain regularity conditions (by the regularity theory for partial differential equations, see e.g. [GilbargTrudinger]). If we assume, for instance, that $a_{ij} \in C^{m,1}(\Omega)$, $h_i \in W^{m+1,2}(\Omega)$ and $f \in W^{m,2}(\Omega)$ with $m \ge 0$, then it follows that $u \in W_{loc}^{m+2,2}(\Omega)$ (see Friedrichs' theorem A12.2). If in addition $\partial\Omega$ is locally given by graphs of $C^{m+1,1}$-functions, then one can correspondingly show that $u \in W^{m+2,2}(\Omega)$ (see A12.3). These two theorems constitute the L^2-regularity theory. This compares with the L^p-theory, which is based on the Calderón-Zygmund inequality in 10.20, and the Schauder theory, which on the basis of the Hölder-Korn-Lichtenstein inequality in 10.19 gives regularity results in Hölder spaces.

Radon-Nikodým's theorem

After we have shown in 6.1 that the dual space of a Hilbert space is canonically isomorphic to the Hilbert space itself, we now want to consider specific Banach spaces, $L^p(\mu)$ and $C^0(S)$, and characterize their dual spaces. (a list of dual spaces can be found in [DunfordSchwartz: IV 15, S. 374-379]). First we state a characterization of $L^p(\mu)'$, for which we will need the Radon-Nikodým theorem 6.11.

6.10 Definition (Variational measure). Let $\mathcal{B}$ be a ring over a set S (see A3.1) and let $\lambda : \mathcal{B} \to \mathbb{K}^m$ be additive. For $E \in \mathcal{B}$ define

$$|\lambda|(E) := \sup \Big\{ \sum_{i=1}^{k} |\lambda(E_i)| \; ; \; k \in \mathbb{N}, \; E_i \in \mathcal{B} \text{ pairwise disjoint}, \; E_i \subset E \Big\}.$$

It holds that $|\lambda| : \mathcal{B} \to [0, \infty]$ is additive. We also call $|\lambda|$ the *variational measure* for λ. In addition, in the case where $\mathcal{B}$ contains the set S, we call

$$\|\lambda\|_{\mathrm{var}} := |\lambda|(S)$$

the *total variation* of λ. The measure λ is called a *bounded measure* if $\|\lambda\|_{\mathrm{var}} < \infty$.

Proof. We prove the additivity of $|\lambda|$. If $B_1, B_2 \in \mathcal{B}$ are disjoint, then it is easy to see that

$$|\lambda|(B_1) + |\lambda|(B_2) \le |\lambda|(B_1 \cup B_2).$$

Moreover, for $\varepsilon > 0$ choose disjoint $E_i \in \mathcal{B}$, $i = 1, \ldots, k$, with $E_i \subset B_1 \cup B_2$, such that

$$|\lambda|(B_1 \cup B_2) - \varepsilon \le \sum_{i=1}^{k} |\lambda(E_i)| = \sum_{i=1}^{k} |\lambda(E_i \cap B_1) + \lambda(E_i \cap B_2)|$$
$$\le |\lambda|(B_1) + |\lambda|(B_2).$$

$\square$

6.11 Radon-Nikodým theorem. Let $(S, \mathcal{B}, \mu)$ be a σ-finite measure space and let

$$\nu : \mathcal{B} \to \mathbb{K} \quad \text{be } \sigma\text{-additive with } \|\nu\|_{\mathrm{var}} < \infty.$$

In addition, let ν be *absolutely continuous* with respect to μ, i.e. for all $E \in \mathcal{B}$

$$\mu(E) = 0 \quad \Longrightarrow \quad \nu(E) = 0.$$

Then there exists a unique function $f \in L^1(\mu)$ such that

$$\nu(E) = \int_E f \, \mathrm{d}\mu \quad \text{for all } E \in \mathcal{B}.$$

Remark: The function f is called the *Radon-Nikodým derivative* of ν with respect to μ, and is also denoted by $\frac{\mathrm{d}\nu}{\mathrm{d}\mu}$.

Proof. Let $f_1, f_2 \in L^1(\mu)$ be two such representing functions and let $f := f_1 - f_2$. Let $E := \{x \in S; \; f(x) \bullet e \ge \delta\}\}$, where $e \in \mathbb{K} \setminus \{0\}$ and $\delta > 0$. Then (recall 5.11)

$$0 = \left(\int_E f_1 \, d\mu - \int_E f_2 \, d\mu \right) \bullet e = \int_E f \bullet e \, d\mu \geq \delta \mu(E) \,,$$

and so $\mu(E) = 0$ for all e, δ, which implies that $f_1 = f_2$ μ-almost everywhere. This proves the uniqueness.

In order to prove the existence, we may assume that ν is real-valued (otherwise consider the real and imaginary part separately). It follows from the Hahn decomposition (see A6.2) that we may further assume that ν is nonnegative. Then $(S, \mathcal{B}, \mu + \nu)$ is also a measure space, since for $N \in \mathcal{B}$ and $E \subset S$

$$(\mu + \nu)(N) = 0, \ E \subset N$$
$$\implies \ \mu(N) = 0, \ E \subset N \ \implies \ E \in \mathcal{B}, \ \mu(E) = 0 \,.$$

Now ν induces a measure space $(S, \widehat{\mathcal{B}}, \nu)$ with $\mathcal{B} \subset \widehat{\mathcal{B}}$, where the sets from $\widehat{\mathcal{B}}$ are unions of sets from $\mathcal{B}$ with ν-null sets. Since $\nu \leq \mu + \nu$, it holds that $L^1(\mu + \nu)$ is contained in $L^1(\nu)$. On recalling that $\nu(S) < \infty$, it follows from the Hölder inequality that $L^2(\nu) \subset L^1(\nu)$. Hence if $g \in L^2(\mu + \nu)$, then

$$\left| \int_S g \, d\nu \right| \leq \sqrt{\nu(S)} \|g\|_{L^2(\nu)} \leq \sqrt{\nu(S)} \|g\|_{L^2(\mu+\nu)} \,.$$

As $L^2(\mu + \nu)$ is a Hilbert space, the Riesz representation theorem 6.1 then implies that there exists an $h \in L^2(\mu + \nu)$ such that, for all $g \in L^2(\mu + \nu)$,

$$\int_S g \, d\nu = (g, h)_{L^2(\mu+\nu)} = \int_S gh \, d(\mu + \nu) \,,$$

i.e.

$$\int_S g(1 - h) \, d\nu = \int_S gh \, d\mu \quad \text{for all } g \in L^2(\mu + \nu). \tag{6-14}$$

We now show that

$$0 \leq h < 1 \quad (\mu + \nu)\text{-almost everywhere.}$$

On setting $g = \mathcal{X}_{\{h<0\} \cap S_m}$, where $\{h < 0\} := \{x \in S; \ h(x) < 0\}$ and S_m is as in 3.9(4), it follows from (6-14) that

$$0 \leq \int_{\{h<0\} \cap S_m} (1 - h) \, d\nu = \int_{\{h<0\} \cap S_m} h \, d\mu \leq -\varepsilon \mu \left(\{h < -\varepsilon\} \cap S_m \right) \,.$$

This implies that $\mu \left(\{h < -\varepsilon\} \cap S_m \right) = 0$ for all $\varepsilon > 0$ and all m, and hence also $\mu \left(\{h < 0\} \right) = 0$. Since ν is absolutely continuous with respect to μ, it follows that also $\nu \left(\{h < 0\} \right) = 0$. Similarly, it follows from (6-14) that, when $g = \mathcal{X}_{\{h \geq 1\} \cap S_m}$,

$$0 \geq \int_{\{h \geq 1\} \cap S_m} (1 - h) \, d\nu = \int_{\{h \geq 1\} \cap S_m} h \, d\mu \geq \mu \left(\{h \geq 1\} \cap S_m \right) \,,$$

and so $\mu\left(\{h \geq 1\}\right) = 0$, which by assumption yields that $\nu\left(\{h \geq 1\}\right) = 0$. This shows that $0 \leq h < 1$ almost everywhere with respect to $\mu + \nu$. In particular, it follows that for $E \in \mathcal{B}$ with $\mu(E) < \infty$ we can in (6-14) choose

$$g = \frac{1 - h^k}{1 - h} \mathcal{X}_E = \left(\sum_{i=0}^{k-1} h^i\right) \mathcal{X}_E \in L^\infty(\mu + \nu),$$

which yields that

$$\int_E (1 - h^k)\, d\nu = \int_E \frac{h}{1 - h}(1 - h^k)\, d\mu.$$

On noting that $\mu+\nu$-almost everywhere $0 \leq (1-h^k)\mathcal{X}_E \nearrow \mathcal{X}_E \in L^1(\mu+\nu)$ as $k \nearrow \infty$, we conclude from the monotone convergence theorem that $\frac{h}{1-h}\mathcal{X}_E \in L^1(\mu)$ and

$$\nu(E) = \int_E \frac{h}{1 - h}\, d\mu,$$

i.e. $\frac{h}{1-h}$ is the desired function. The fact that $\frac{h}{1-h} \in L^1(\mu)$ follows again from the monotone convergence theorem, upon setting $E = \bigcup_{j \leq m} S_j$, taking the limit $m \to \infty$, and recalling that $\nu(S) < \infty$. (A purely measure theoretical proof of the Radon-Nikodým theorem can be found in e.g. [Halmos].)   □

6.12 Theorem (Dual space of L^p for $p < \infty$). Let $(S, \mathcal{B}, \mu)$ be a measure space and let $1 \leq p < \infty$ (the **dual exponent** p' is given by $\frac{1}{p} + \frac{1}{p'} = 1$, if $p = 1$ then $p' = \infty$). In the case $p = 1$, we assume in addition that μ is σ-finite. For $f \in L^{p'}(\mu)$ let

$$J(f)(g) := \int_S g\overline{f}\, d\mu \quad \text{for all } g \in L^p(\mu).$$

Then $J : L^{p'}(\mu) \to L^p(\mu)'$ is a conjugate linear isometric isomorphism. *Special case:* In the Hilbert space case $p = 2 = p'$, the isometry J coincides with the isometry in 6.1.

Proof. It follows from the Hölder inequality that J is well defined and that $\|J(f)\|_{(L^p)'} \leq \|f\|_{L^{p'}}$. Clearly, J is conjugate linear. Moreover, J is injective, since $J(f) = 0$ implies in the case $p > 1$ with $g := |f|^{p'-2}f \in L^p(\mu)$ that

$$0 = J(f)(g) = \int_S |f|^{p'}\, d\mu,$$

and so that $f = 0$ in $L^{p'}(\mu)$. In the case $p = 1$ set $g = \mathcal{X}_{S_m} f \in L^1(\mu)$ with S_m as in 3.9(4) and obtain that $f = 0$ almost everywhere in S_m. Letting $m \to \infty$ we conclude that $f = 0$ in $L^\infty(\mu)$.

Now let $F \in L^p(\mu)'$. We need to show that there exists an $f \in L^{p'}(\mu)$ with

$$F = J(f) \quad \text{and} \quad \|f\|_{L^{p'}} \leq \|F\|_{(L^p)'}.$$

First we consider the special case $\mu(S) < \infty$. Then

$$\nu(E) := F(\mathcal{X}_E) \quad \text{for } E \in \mathcal{B}$$

satisfies the assumptions of the Radon-Nikodým theorem. To see this, note that for disjoint sets $E_1, \ldots, E_m$ in $\mathcal{B}$ with $\nu(E_i) \neq 0$ it holds that

$$\sum_{i=1}^{m} |\nu(E_i)| = \sum_{i=1}^{m} \sigma_i \nu(E_i) \quad \text{with } \sigma_i := \frac{\overline{\nu(E_i)}}{|\nu(E_i)|}$$

$$= F\left(\sum_{i=1}^{m} \sigma_i \mathcal{X}_{E_i}\right) \leq \|F\|_{(L^p)'} \cdot \left\|\sum_{i=1}^{m} \sigma_i \mathcal{X}_{E_i}\right\|_{L^p} \tag{6-15}$$

$$= \|F\|_{(L^p)'} \cdot \left(\sum_{i=1}^{m} \mu(E_i)\right)^{\frac{1}{p}} \leq \|F\|_{(L^p)'} \cdot \mu(S)^{\frac{1}{p}},$$

i.e. $\|\nu\|_{\mathrm{var}} < \infty$. In addition, for $E = \bigcup_{i \in \mathbb{N}} E_i$ with $E_i \in \mathcal{B}$, $E_i \subset E_{i+1}$

$$|\nu(E) - \nu(E_i)| = \left|F(\mathcal{X}_{E \setminus E_i})\right| \leq \|F\|_{(L^p)'} \mu(E \setminus E_i)^{\frac{1}{p}} \to 0 \quad \text{as } i \to \infty,$$

i.e. ν is σ-additive. By the way, ν is absolutely continuous w.r.t. μ, since for μ-null sets E we have $\mathcal{X}_E = 0$ in $L^p(\mu)$, and therefore $\nu(E) = F(\mathcal{X}_E) = 0$.

Hence, by the Radon-Nikodým theorem 6.11, there exists a function $f \in L^1(\mu)$ with

$$F(\mathcal{X}_E) = \int_S \mathcal{X}_E \overline{f} \, d\mu \quad \text{for all } E \in \mathcal{B}.$$

It follows that

$$F(g) = \int_S g \overline{f} \, d\mu \tag{6-16}$$

for all functions $g \in L^\infty(\mu)$, because such functions can be uniformly approximated by finite linear combinations of characteristic functions $\mathcal{X}_E$ with measurable $E \subset S$ (see the note in 3.26(1)). Now for $m \in \mathbb{N}$ and $1 \leq q < \infty$ we choose in particular

$$g = \mathcal{X}_{A_m} |f|^{q-2} f, \quad \text{where } A_m := \{x \in S; \ 0 < |f(x)| \leq m\},$$

and obtain from (6-16) that

$$\int_{A_m} |f|^q \, d\mu = F(g) \leq \|F\|_{(L^p)'} \|g\|_{L^p} = \|F\|_{(L^p)'} \left(\int_{A_m} |f|^{p(q-1)} \, d\mu\right)^{\frac{1}{p}}.$$

In the case $p > 1$, setting $q = p'$ (so that $p(q - 1) = p'$), yields after cancellation that

$$\left(\int_{A_m} |f|^{p'} \, d\mu\right)^{\frac{1}{p'}} \leq \|F\|_{(L^p)'}.$$

On letting $m \to \infty$, it follows from the monotone convergence theorem that $f \in L^{p'}(\mu)$ and $\|f\|_{L^{p'}} \le \|F\|_{(L^p)'}$. In the case $p = 1$, choose $q \in \mathbb{N}$ and obtain inductively that

$$\int_{A_m} |f|^q \, d\mu \le \|F\|_{(L^p)'} \int_{A_m} |f|^{q-1} \, d\mu \le \|F\|_{(L^p)'}^q \cdot \mu(A_m),$$

i.e.

$$\left(\int_{A_m} |f|^q \, d\mu \right)^{\frac{1}{q}} \le \|F\|_{(L^p)'} \cdot \mu(A_m)^{\frac{1}{q}}.$$

Then, on letting $q \to \infty$, it follows from E3.4 (for the function $\mathcal{X}_{A_m} f$) that $|f| \le \|F\|_{(L^p)'}$ almost everywhere in A_m, which implies that $\|f\|_{L^\infty} \le \|F\|_{(L^p)'}$.

On noting that the functions g, for which (6-16) originally held, are dense in $L^p(\mu)$, it now follows from the Hölder inequality that (6-16) holds for all $g \in L^p(\mu)$, and so $F = J(f)$, which is what we wanted to show.

We now consider the case of a general measure space, and define $\widetilde{\mathcal{B}} := \{A \in \mathcal{B}\,;\, \mu(A) < \infty\}$. For $A \in \widetilde{\mathcal{B}}$ let

$$\mu_A(E) := \mu(A \cap E), \quad F_A(g) := F(\mathcal{X}_A g).$$

Then $\mu_A(S) < \infty$ with $\mu_A(S \setminus A) = 0$, and $F_A \in L^p(\mu_A)'$, with $\|F_A\|_{(L^p)'} \le \|F\|_{(L^p)'}$. Hence it follows from what we have shown so far that there exists a unique $f_A \in L^{p'}(\mu_A)$ with

$$F_A(g) = \int_S g \overline{f_A} \, d\mu_A \quad \text{for all } g \in L^p(\mu_A) \tag{6-17}$$

and $\|f_A\|_{L^{p'}} = \|F_A\|_{(L^p)'}$. On defining $f_A(x) := 0$ for $x \in S \setminus A$, we have that $f_A \in L^{p'}(\mu)$. As in the proof of the injectivity of J, it follows that $f_{A_1} = f_{A_2}$ μ-almost everywhere in $A_1 \cap A_2$ for $A_1, A_2 \in \widetilde{\mathcal{B}}$. Hence, $|f_{A_1}| \le |f_{A_2}|$ μ-almost everywhere if $A_1 \subset A_2$, and then

$$\|f_{A_1}\|_{L^{p'}} \le \|f_{A_2}\|_{L^{p'}} = \|F_{A_2}\|_{(L^p)'} \le \|F\|_{(L^p)'} < \infty.$$

It follows that there exist $B_m \in \widetilde{\mathcal{B}}$ with $B_m \subset B_{m+1}$ for $m \in \mathbb{N}$, such that

$$\|f_{B_m}\|_{L^{p'}} \longrightarrow s := \sup_{A \in \widetilde{\mathcal{B}}} \|f_A\|_{L^{p'}} \quad \text{as } m \to \infty.$$

If $p = 1$, then the B_m can be chosen such that $S_m \subset B_m$, where the S_m are as in 3.9(4). Then

$$B := \bigcup_{m \in \mathbb{N}} B_m, \quad f(x) := \begin{cases} f_{B_m}(x) & \text{for } x \in B_m,\ m \in \mathbb{N}, \\ 0 & \text{for } x \in S \setminus B, \end{cases}$$

(for $p > 1$ by the monotone convergence theorem) defines an $f \in L^{p'}(\mu)$ with

$$\|f\|_{L^{p'}} = s = \sup_{A \in \tilde{\mathcal{B}}} \|f_A\|_{L^{p'}} = \sup_{A \in \tilde{\mathcal{B}}} \|F_A\|_{(L^p)'} \leq \|F\|_{(L^p)'} \,.$$

Now

$$f_A = f \text{ almost everywhere in } A, \text{ if } A \in \mathcal{B} \text{ with } A \subset B,$$

since in $A \cap B_m$ it holds almost everywhere that $f = f_{B_m} = f_{A \cap B_m} = f_A$. We claim that

$$f_A = 0 \text{ almost everywhere in } S, \text{ if } A \in \mathcal{B} \text{ with } A \cap B = \emptyset.$$

In the case $p = 1$, this trivially follows from $B = S$. In the case $p > 1$, on noting that $A \cap B_m = \emptyset$, it follows that

$$|f_{A \cup B_m}|^{p'} = |f_A|^{p'} + |f_{B_m}|^{p'}, \quad \text{and so} \quad s^{p'} \geq \|f_A\|_{L^{p'}}^{p'} + \|f_{B_m}\|_{L^{p'}}^{p'} \,.$$

Letting $m \to \infty$ yields that $s^{p'} \geq \|f_A\|_{L^{p'}}^{p'} + s^{p'}$, and hence our claim.

Now let $g \in L^p(\mu)$ with $g = 0$ almost everywhere in $S \setminus A$ for an $A \in \mathcal{B}$. Then, by (6-17),

$$F(g) = F_A(g) = \int_S g\overline{f_A}\,\mathrm{d}\mu_A = \int_A g\overline{f_A}\,\mathrm{d}\mu\,.$$

Since, as shown above, $f_A = f_{A \setminus B} = 0$ in $A \setminus B$ and $f_A = f_{A \cap B} = f$ in $A \cap B$, this is in turn equal to

$$\int_{A \cap B} g\overline{f}\,\mathrm{d}\mu = \int_A g\overline{f}\,\mathrm{d}\mu = \int_S g\overline{f}\,\mathrm{d}\mu = J(f)(g)\,.$$

On noting that such functions g are dense in $L^p(\mu)$ (approximating g, for example, by $\mathcal{X}_{A_n}g$, $n \in \mathbb{N}$, with $A_n := \{x \in S;\ |g(x)| \geq \frac{1}{n}\}$), it follows that $F(g) = J(f)(g)$ for all $g \in L^p(\mu)$. □

With the help of the result in theorem 6.12, we can establish a distributional characterization of L^p-functions:

6.13 Corollary. Let $\Omega \subset \mathbb{R}^n$ be open and let $1 \leq p \leq \infty$. Then it holds for functions $f : \Omega \to \mathbb{K}$ that

$$f \in L^p(\Omega) \quad \Longleftrightarrow \quad \begin{cases} f \in L^1_{\mathrm{loc}}(\Omega) \text{ and there exists a } C \text{ with} \\ \left| \int_\Omega \zeta f\,\mathrm{d}L^n \right| \leq C\|\zeta\|_{L^{p'}(\Omega)} \text{ for all } \zeta \in C_0^\infty(\Omega). \end{cases}$$

The constant C on the right-hand side satisfies $\|f\|_{L^p(\Omega)} \leq C$.

Notation: Here $L^1_{\mathrm{loc}}(\Omega)$ is the space of locally integrable functions in Ω, defined in 5.13(2). Moreover, $1 \leq p' \leq \infty$ is the **dual exponent**, i.e. $\frac{1}{p} + \frac{1}{p'} = 1$.

Note: For a generalization of the result to Sobolev functions, see E6.7.

Proof ⇒. The Hölder inequality yields that

$$\left| \int_\Omega \zeta f \, dL^n \right| \le \|\zeta\|_{L^{p'}(\Omega)} \cdot \|f\|_{L^p(\Omega)} \,.$$

□

Proof ⇐. The estimate yields that on $C_0^\infty(\Omega)$ equipped with the $L^{p'}$-norm,

$$F(\zeta) := \int_\Omega \zeta f \, dL^n$$

is linear and continuous. In the case $p > 1$, we have that $C_0^\infty(\Omega)$ is dense in $L^{p'}(\Omega)$ (this follows from 4.15(3) as $p' < \infty$), and so F can be uniquely extended to $L^{p'}(\Omega)$, as a functional $F \in L^{p'}(\Omega)'$ (see E5.3). Hence it follows from 6.12 that there exists an $\widetilde{f} \in L^p(\Omega)$ with

$$F(g) = \int_\Omega g\widetilde{f} \, dL^n \quad \text{for all } g \in L^{p'}(\Omega).$$

Since

$$\int_\Omega \zeta f \, dL^n = \int_\Omega \zeta\widetilde{f} \, dL^n \quad \text{for all } \zeta \in C_0^\infty(\Omega),$$

$f = \widetilde{f}$ almost everywhere in Ω (see 4.22). In the case $p = 1$, set

$$g(x) := \begin{cases} \dfrac{\overline{f(x)}}{|f(x)|}, & \text{if } f(x) \ne 0, \\ 0, & \text{otherwise.} \end{cases}$$

Let $D \subset\subset \Omega$ and let $(\varphi_\varepsilon)_{\varepsilon>0}$ be a standard Dirac sequence. Then $\zeta_\varepsilon := \varphi_\varepsilon * (\mathcal{X}_D g) \in C_0^\infty(\Omega)$ for sufficiently small $\varepsilon > 0$, and

$$\left| \int_\Omega \zeta_\varepsilon f \, dL^n \right| \le C\|\zeta_\varepsilon\|_{L^\infty} \le C.$$

Letting $\varepsilon \to 0$, we obtain from Lebesgue's convergence theorem (as $\zeta_\varepsilon \to \mathcal{X}_D g$ almost everywhere for a subsequence $\varepsilon \to 0$) that

$$\int_D |f| \, dL^n = \left| \int_D gf \, dL^n \right| \le C,$$

where the constant C is independent of D. Hence $f \in L^1(\Omega)$.   □

Hahn-Banach's theorem

For the characterization of $C^0(S)'$ we will use the fact that functionals on $C^0(S)$ can be extended norm-preservingly to $B(S)$ (see the proof 6.23). The existence of such extensions in more general situations is guaranteed by the following two theorems.

6.14 Hahn-Banach theorem. Let X be an $\mathbb{R}$-vector space and let the following hold:

(1) $p : X \to \mathbb{R}$ is **sublinear**, i.e. for all $x, y \in X$ and $\alpha \in \mathbb{R}$,

$$p(x + y) \le p(x) + p(y) \quad \text{and} \quad p(\alpha x) = \alpha p(x) \text{ for } \alpha \ge 0.$$

(2) $f : Y \to \mathbb{R}$ is linear with a subspace $Y \subset X$.
(3) $f(x) \le p(x)$ for $x \in Y$.

Then there exists a linear map $F : X \to \mathbb{R}$ such that

$$F(x) = f(x) \text{ for } x \in Y \quad \text{and} \quad F(x) \le p(x) \text{ for } x \in X.$$

Proof. We consider the class of all extensions of f, that is,

$$\mathcal{M} := \big\{ (Z, g) \; ; \; Z \text{ subspace}, \, Y \subset Z \subset X,$$
$$g : Z \to \mathbb{R} \text{ linear}, \, g = f \text{ on } Y, \, g \le p \text{ on } Z \big\} \, .$$

Consider an arbitrary $(Z, g) \in \mathcal{M}$ with $Z \ne X$ and a $z_0 \in X \setminus Z$. We want to extend g at least to

$$Z_0 := \operatorname{span}(Z \cup \{z_0\}) = Z \oplus \operatorname{span}\{z_0\} \, .$$

We attempt the ansatz

$$g_0(z + \alpha z_0) := g(z) + c\alpha \quad \text{for } z \in Z \text{ and } \alpha \in \mathbb{R}.$$

Here c still needs to be suitably chosen, so that $(Z_0, g_0) \in \mathcal{M}$. Clearly, g_0 is linear on Z_0. Moreover, $g_0 = g = f$ on Y. It remains to show that

$$g(z) + c\alpha \le p(z + \alpha z_0) \quad \text{for } z \in Z \text{ and } \alpha \in \mathbb{R}.$$

Since $g \le p$ on Z, this is satisfied for $\alpha = 0$. For $\alpha > 0$ the inequality is equivalent to

$$c \le \tfrac{1}{\alpha}\big(p(z + \alpha z_0) - g(z)\big) = p\big(\tfrac{z}{\alpha} + z_0\big) - g\big(\tfrac{z}{\alpha}\big)$$

and for $\alpha < 0$ to

$$c \ge \tfrac{1}{\alpha}\big(p(z + \alpha z_0) - g(z)\big) = g\big(-\tfrac{z}{\alpha}\big) - p\big(-\tfrac{z}{\alpha} - z_0\big) \, .$$

Hence we need to find a number c such that

$$\sup_{z \in Z} \big(g(z) - p(z - z_0)\big) \le c \le \inf_{z \in Z} \big(p(z + z_0) - g(z)\big) \, .$$

This is possible, because for $z, z' \in Z$ we have

$$g(z') + g(z) = g(z' + z) \le p(z' + z)$$
$$= p(z' - z_0 + z + z_0) \le p(z' - z_0) + p(z + z_0) \, ,$$

and hence

$$g(z') - p(z' - z_0) \le p(z + z_0) - g(z).$$

We now hope that this extension procedure yields an $(X, F) \in \mathcal{M}$. To this end, we make use of

Zorn's lemma: Let $(\mathcal{M}, \le)$ be a nonempty *partially ordered set* (i.e. if $m_1 \le m_2$ and $m_2 \le m_3$, then $m_1 \le m_3$, and $m \le m$ for all $m \in \mathcal{M}$) such that every totally ordered subset $\mathcal{N}$ (i.e. for all $n_1, n_2 \in \mathcal{N}$ it holds that $n_1 \le n_2$ or $n_2 \le n_1$) has an upper bound (i.e. there exists an $m \in \mathcal{M}$ with $n \le m$ for all $n \in \mathcal{N}$). Then $\mathcal{M}$ contains a *maximal element* (i.e. there exists an $m_0 \in \mathcal{M}$ such that for all $m \in \mathcal{M}$ it holds that $m_0 \le m \Longrightarrow m \le m_0$).

In our case, an order is defined by

$$(Z_1, g_1) \le (Z_2, g_2) \quad :\Longleftrightarrow \quad Z_1 \subset Z_2 \text{ and } g_2 = g_1 \text{ on } Z_1.$$

We need to verify the assumptions of Zorn's lemma. Let $\mathcal{N} \subset \mathcal{M}$ be totally ordered and define

$$Z_* := \bigcup_{(Z, g) \in \mathcal{N}} Z,$$

$$g_*(x) := g(x), \quad \text{if } x \in Z \text{ and } (Z, g) \in \mathcal{N}.$$

We need to show that $(Z_*, g_*) \in \mathcal{M}$. Now $Y \subset Z_* \subset X$, and g_* is a well defined function, because

$$x \in Z_1 \cap Z_2 \,, \ (Z_1, g_1) \in \mathcal{N} \,, \ (Z_2, g_2) \in \mathcal{N}$$
$$\Longrightarrow (Z_1, g_1) \le (Z_2, g_2) \text{ or } (Z_2, g_2) \le (Z_1, g_1) \quad \text{(total order of } \mathcal{N})$$
$$\Longrightarrow Z_1 \subset Z_2 \text{ and } g_2 = g_1 \text{ on } Z_1 \quad \text{(in the first case)}$$
$$\Longrightarrow g_2(x) = g_1(x) \quad \text{(as } x \in Z_1).$$

The properties $g_* = f$ on Y and $g_* \le p$ on Z_* carry over. The linearity of Z_* and g_* can be seen as follows:

$$x, y \in Z_* \,, \ \alpha \in \mathbb{R}$$
$$\Longrightarrow \text{ There exist } (Z_x, g_x) \in \mathcal{N}, (Z_y, g_y) \in \mathcal{N} \text{ with } x \in Z_x \text{ and } y \in Z_y$$
$$\Longrightarrow (Z_x, g_x) \le (Z_y, g_y) \text{ or } (Z_y, g_y) \le (Z_x, g_x)$$
$$\Longrightarrow x, y \in Z_\xi \text{ with } \xi = y \text{ in the first and } \xi = x \text{ in the second case,}$$
$$\text{hence also } x + \alpha y \in Z_\xi \subset Z_* \text{ and}$$
$$g_*(x + \alpha y) = g_\xi(x + \alpha y) = g_\xi(x) + \alpha g_\xi(y) = g_*(x) + \alpha g_*(y).$$

Hence it follows from Zorn's lemma that $\mathcal{M}$ has a maximal element (Z, g). If we assume that $Z \ne X$, then the extension procedure from the beginning of the proof yields a $(Z_0, g_0) \in \mathcal{M}$ with

$$(Z, g) \le (Z_0, g_0) \quad \text{and} \quad Z_0 \ne Z,$$

which contradicts the maximality of (Z, g). $\square$

The Hahn-Banach theorem has the following version for linear functionals.

6.15 Hahn-Banach theorem (for linear functionals). Let X be a normed $\mathbb{K}$-vector space and Y be a subspace (with the norm of X !). Then for $y' \in Y'$ there exists an $x' \in X'$ with

$$x' = y' \text{ on } Y \quad \text{and} \quad \|x'\|_{X'} = \|y'\|_{Y'} .$$

Proof for $\mathbb{K} = \mathbb{R}$. Choose

$$p(x) := \|y'\|_{Y'} \|x\|_X \quad \text{for } x \in X$$

in 6.14, so that for $y \in Y$

$$y'(y) \le \|y'\|_{Y'} \|y\|_Y = \|y'\|_{Y'} \|y\|_X = p(y) .$$

Then, by 6.14, there exists a linear map $x' : X \to \mathbb{R}$ with

$$x' = y' \text{ on } Y \quad \text{and} \quad x' \le p \text{ on } X.$$

The second property implies that

$$\pm x'(x) = x'(\pm x) \le p(\pm x) = \|y'\|_{Y'} \|x\|_X ,$$

i.e. $x' \in X'$ with $\|x'\|_{X'} \le \|y'\|_{Y'}$, and the first property implies that

$$\|y'\|_{Y'} = \sup_{\substack{y \in Y \\ \|y\|_X \le 1}} |y'(y)| = \sup_{\substack{y \in Y \\ \|y\|_X \le 1}} |x'(y)| \le \|x'\|_{X'} .$$

$\square$

Proof for $\mathbb{K} = \mathbb{C}$. Consider X and Y as normed $\mathbb{R}$-vector spaces $X_{\mathbb{R}}$ and $Y_{\mathbb{R}}$ (i.e. scalar multiplication is defined only for real numbers, but the norms remain the same). Let $X'_{\mathbb{R}}$ and $Y'_{\mathbb{R}}$ be the corresponding dual spaces. For $y' \in Y'$ it then holds that

$$y'_{\text{re}} := \text{Re} y' \in Y'_{\mathbb{R}} \quad \text{with} \quad \|y'_{\text{re}}\|_{Y'_{\mathbb{R}}} \le \|y'\|_{Y'}$$

and

$$y'(x) = \text{Re} y'(x) + i \text{Im} y'(x) = y'_{\text{re}}(x) - i y'_{\text{re}}(ix) .$$

It follows from the real case treated above that there exists an extension x'_{re} of y'_{re} to $X_{\mathbb{R}}$ with $\|x'_{\text{re}}\|_{X'_{\mathbb{R}}} = \|y'_{\text{re}}\|_{Y'_{\mathbb{R}}}$. Define

$$x'(x) := x'_{\text{re}}(x) - i x'_{\text{re}}(ix) .$$

Then $x' = y'$ on Y, and $x' : X \to \mathbb{C}$ is $\mathbb{C}$-linear, because x' is $\mathbb{R}$-linear and for $x \in X$ we have that

$$x'(\mathrm{i}x) = x'_{\mathrm{re}}(\mathrm{i}x) - \mathrm{i}x'_{\mathrm{re}}(-x) = x'_{\mathrm{re}}(\mathrm{i}x) + \mathrm{i}x'_{\mathrm{re}}(x)$$
$$= \mathrm{i}\left(-\mathrm{i}x'_{\mathrm{re}}(\mathrm{i}x) + x'_{\mathrm{re}}(x)\right) = \mathrm{i}x'(x).$$

Now let $x \in X$. Then $x'(x) \in \mathbb{C}$ can be written as $x'(x) = re^{\mathrm{i}\theta}$ with $\theta \in \mathbb{R}$ and $r \geq 0$. Therefore,

$$|x'(x)| = r = \mathrm{Re}\left(e^{-\mathrm{i}\theta}x'(x)\right) = \mathrm{Re}x'(e^{-\mathrm{i}\theta}x)$$
$$= x'_{\mathrm{re}}(e^{-\mathrm{i}\theta}x) \leq \|x'_{\mathrm{re}}\|_{X'_{\mathbb{R}}}\|x\|_{X'},$$

and we recall that $\|x'_{\mathrm{re}}\|_{X'_{\mathbb{R}}} = \|y'_{\mathrm{re}}\|_{Y'_{\mathbb{R}}} \leq \|y'\|_{Y'}$. This shows that $x' \in X'$ with $\|x'\|_{X'} \leq \|y'\|_{Y'}$. As x' is an extension of y', it must also hold that $\|x'\|_{X'} \geq \|y'\|_{Y'}$. $\qquad\square$

As an application, we show that points in a normed space can be separated from subspaces with the help of linear functionals (see the generalization of suspaces to closed convex sets in 8.12). This separation property is often used in order to show that a given subspace is dense in the ambient space X.

6.16 Theorem. Let Y be a closed subspace of the normed space X and let $x_0 \notin Y$. Then there exists an $x' \in X'$ with

$$x' = 0 \text{ on } Y, \quad \|x'\|_{X'} = 1, \quad x'(x_0) = \mathrm{dist}(x_0, Y).$$

Remark: Then there also exists an $x' \in X'$ with

$$x' = 0 \text{ on } Y, \quad \|x'\|_{X'} = \frac{1}{\mathrm{dist}(x_0, Y)}, \quad x'(x_0) = 1.$$

Proof. On

$$Y_0 := \mathrm{span}\,(Y \cup \{x_0\}) = Y \oplus \mathrm{span}\{x_0\}$$

define

$$y'_0(y + \alpha x_0) := \alpha \cdot \mathrm{dist}(x_0, Y) \quad \text{for } y \in Y \text{ and } \alpha \in \mathbb{K}.$$

Then $y'_0 : Y_0 \to \mathbb{K}$ is linear and $y'_0 = 0$ on Y. We want to show that $y'_0 \in Y'_0$ with $\|y'_0\|_{Y'_0} = 1$, as 6.15 then yields the desired result.

Let $y \in Y$ and $\alpha \neq 0$. Then

$$\mathrm{dist}(x_0, Y) \leq \left\| x_0 - \frac{-y}{\alpha} \right\|_X,$$

and so

$$|y'_0(y + \alpha x_0)| \leq |\alpha| \left\| x_0 - \frac{-y}{\alpha} \right\|_X = \|\alpha x_0 + y\|_X,$$

and hence $y_0 \in Y'_0$ with $\|y'_0\|_{Y'_0} \leq 1$. The closedness of Y yields that $\mathrm{dist}(x_0, Y) > 0$, and so for $\varepsilon > 0$ we can choose a $y_\varepsilon \in Y$ such that

$$\|x_0 - y_\varepsilon\|_X \le (1 + \varepsilon)\mathrm{dist}(x_0, Y).$$

Then

$$y_0'(x_0 - y_\varepsilon) = \mathrm{dist}(x_0, Y) \ge \tfrac{1}{1+\varepsilon}\|x_0 - y_\varepsilon\|_X,$$

which, since $x_0 - y_\varepsilon \ne 0$, implies that $\|y_0'\|_{Y_0'} \ge \tfrac{1}{1+\varepsilon} \to 1$ as $\varepsilon \searrow 0$. $\square$

6.17 Corollaries. Let X be a normed space and let $x_0 \in X$. Then:

(1) If $x_0 \ne 0$, then there exists an $x_0' \in X'$ with

$$\|x_0'\|_{X'} = 1 \quad \text{and} \quad x_0'(x_0) = \|x_0\|_X.$$

(2) If $x'(x_0) = 0$ for all $x' \in X'$, then $x_0 = 0$.

(3) Setting $Tx' := x'(x_0)$ for $x' \in X'$ defines an element T of $\mathscr{L}(X'; \mathbb{K}) = (X')'$, the bidual space (see 8.2), with $\|T\| = \|x_0\|_X$.

Proof. (1) is the result in 6.16 with $Y = \{0\}$, and (2) follows from (1). In (3) we have that $|Tx'| \le \|x'\|_{X'}\|x_0\|_X$, and if $x_0 \ne 0$ it holds that $|Tx_0'| = \|x_0\|_X$ with x_0' as in (1). Hence $\|T\| = \|x_0\|_X$. $\square$

6.18 Remark. The result 6.16 may also be interpreted as a generalization of the projection theorem for Hilbert spaces in the linear case. To see this, assume that X is a Hilbert space and define

$$x'(x) := \left(x, \frac{x_0 - Px_0}{\|x_0 - Px_0\|} \right)_X,$$

where P is the orthogonal projection onto Y from 4.3. It follows from 4.4(2) that $x' = 0$ on Y and hence

$$x'(x_0) = x'(x_0 - Px_0) = \|x_0 - Px_0\|_X,$$

and moreover $|x'(x)| \le \|x\|_X$. Hence x' has all the properties in 6.16.

Riesz-Radon's theorem

As we have seen in 6.12 the dual space of the function space $L^p(\mu)$, if $1 \le p < \infty$, is isomorphic to a space that is again a function space. We will now show that the dual space of $C^0(S)$ is isomorphic to a space of measures. To this end, we need the following definitions (the notations are the same as in [DunfordSchwartz : IV 2]).

6.19 Definition (Borel sets). Let X be a topological space. The set of *Borel sets* is defined as the smallest σ-algebra that contains the closed subsets of X (or, equivalently, the open subsets of X).

6.20 Spaces of additive measures. Let $S \subset \mathbb{R}^n$ be equipped with the relative topology of $\mathbb{R}^n$ (see 2.11). Let $\mathcal{B}_0$ be the smallest Boolean algebra that contains the closed (or, equivalently, open) subsets of S, and let $\mathcal{B}_1$ be the set of Borel sets of S, i.e. the smallest σ-algebra containing $\mathcal{B}_0$. Then

$$ba(S; \mathbb{K}^m) := \{\lambda : \mathcal{B}_0 \to \mathbb{K}^m \, ; \; \lambda \text{ is additive and } \|\lambda\|_{\mathrm{var}} < \infty\} \, ,$$

$$ca(S; \mathbb{K}^m) := \{\lambda : \mathcal{B}_1 \to \mathbb{K}^m \, ; \; \lambda \text{ is } \sigma\text{-additive and } \|\lambda\|_{\mathrm{var}} < \infty\}$$

are $\mathbb{K}$-vector spaces and, equipped with the total variation as the norm, also Banach spaces. In the definition, ba stands for "bounded additive" and ca stands for "countably additive". As usual, we set $ba(S) := ba(S; \mathbb{K})$ and $ca(S) := ca(S; \mathbb{K})$.

Proof. We prove the completeness. Let $(\lambda_k)_{k \in \mathbb{N}}$ be a Cauchy sequence in $ba(S; \mathbb{K}^m)$. Then it holds for $E \in \mathcal{B}_0$ that

$$|\lambda_l(E) - \lambda_k(E)| \le \|\lambda_l - \lambda_k\|_{\mathrm{var}} \to 0 \quad \text{as } k, l \to \infty,$$

and so there exists

$$\lambda(E) := \lim_{l \to \infty} \lambda_l(E) \quad \text{for } E \in \mathcal{B}_0$$

and the additivity carries over to λ. In addition,

$$\|\lambda - \lambda_k\|_{\mathrm{var}} \le \liminf_{l \to \infty} \|\lambda_l - \lambda_k\|_{\mathrm{var}} \longrightarrow 0 \quad \text{as } k \to \infty.$$

Analogously, for Cauchy sequences in $ca(S; \mathbb{K}^m)$ there exists a limit λ on $\mathcal{B}_1$. If $E_i \in \mathcal{B}_1$ with $E_i \supset E_{i+1}$ and $\bigcap_{i \in \mathbb{N}} E_i = \emptyset$, then for $l \ge k$ and as $l \to \infty$

$$|\lambda(E_i)| \longleftarrow |\lambda_l(E_i)| \le \underbrace{|\lambda_k(E_i)|}_{\substack{\to 0 \text{ as } i \to \infty \\ \text{for every } k}} + \underbrace{\|\lambda_l - \lambda_k\|_{\mathrm{var}}}_{\to 0 \text{ as } l \ge k \to \infty} ,$$

i.e. λ is σ-additive. □

6.21 Spaces of regular measures. Let $S \subset \mathbb{R}^n$, $\mathcal{B}_0$, and $\mathcal{B}_1$ be as in 6.20. A measure λ in $ba(S; \mathbb{K}^m)$ or $ca(S; \mathbb{K}^m)$ is called **regular** if for all $E \in \mathcal{B}_0$ or $E \in \mathcal{B}_1$, respectively,

$$\inf \big\{ |\lambda|(U \setminus K) \, ; \; K \subset E \subset U, \; K \text{ is closed in } S$$
$$\text{and } U \text{ is open in } S \big\} = 0 \, .$$

Here $|\lambda|$ is the variational measure from 6.10 and in S we consider the relative topology from 2.11, i.e. a set $U \subset S$ is called open in S if it is of the form $U = S \cap V$ for an open set $V \subset \mathbb{R}^n$, and a set $K \subset S$ is called closed in S if $S \setminus K$ is open in S. We define

$$rba(S; \mathbb{K}^m) := \{\lambda \in ba(S; \mathbb{K}^m)\,;\ \lambda \text{ is regular}\}\,,$$
$$rca(S; \mathbb{K}^m) := \{\lambda \in ca(S; \mathbb{K}^m)\,;\ \lambda \text{ is regular}\}\,.$$

These sets are $\mathbb{K}$-vector spaces and, equipped with the total variation as the norm, also Banach spaces. In the definition, rba stands for "regular bounded additive" and rca stands for "regular countably additive". As usual, we set $rba(S) := rba(S; \mathbb{K})$ and $rca(S) := rca(S; \mathbb{K})$.

Proof. For the completeness we need to show that for regular measures λ_k it follows from $\lambda_k \to \lambda$ in $ba(S; \mathbb{K}^m)$ as $k \to \infty$ that λ is also regular. To prove this we note that for $K \subset E \subset U$, as in the definition of regularity,

$$|\lambda|(U \setminus K) \le |\lambda_k|(U \setminus K) + \|\lambda - \lambda_k\|_{\mathrm{var}}\,.$$

The first term on the right-hand side can be made arbitrarily small for every k, by choosing U and K appropriately. □

In the following we need the fact that for regular measures $\mu : \mathcal{B}_1 \to [0, \infty]$, continuous functions are integrable, i.e. that they lie in $L^1(\mu)$. The proof of this result is the construction of the Riemann integral, which for our purposes we give here for vector-valued measures $\lambda : \mathcal{B}_0 \to \mathbb{K}^m$.

6.22 Integral of continuous functions (Riemann integral). Let $\mathcal{B}_0$ be as in 6.20. In addition, assume that $\lambda : \mathcal{B}_0 \to \mathbb{K}^m$ is additive with $\|\lambda\|_{\mathrm{var}} < \infty$. For step functions

$$f = \sum_{i=1}^{k} \mathcal{X}_{E_i} \alpha_i\,, \quad k \in \mathbb{N},\ \alpha_i \in \mathbb{K},\ E_i \in \mathcal{B}_0\,,$$

it holds that

$$\int_S f \,\mathrm{d}\lambda := \sum_{i=1}^{k} \alpha_i \lambda(E_i)$$

is independent of the representation of f. Moreover, we have that (choose E_i in the representation of f disjoint)

$$\left| \int_S f \,\mathrm{d}\lambda \right| \le \|f\|_{\mathrm{sup}} \cdot \|\lambda\|_{\mathrm{var}}\,.$$

Every continuous and bounded function $f : S \to \mathbb{K}$ can be approximated by such step functions in the supremum norm. To see this, cover the bounded set $\overline{f(S)}$ with open sets U_i, $i = 1, \dots, l$, with diameter $\le \frac{1}{k}$. Then one can construct another cover by (cf. the proof of A3.19(2))

$$V_i := U_i \setminus \bigcup_{j<i} U_j \quad \text{for } i = 1, \dots, l,$$

where now the sets V_i are pairwise disjoint. In addition,

$$E_i := f^{-1}(V_i) = f^{-1}(U_i) \setminus \bigcup_{j<i} f^{-1}(U_j) \in \mathcal{B}_0 \, .$$

On choosing $\alpha_i \in V_i$, if V_i is nonempty, it follows that

$$\left\| \sum_{i=1}^{l} \alpha_i \mathcal{X}_{E_i} - f \right\|_{\sup} \leq \frac{1}{k} \, ,$$

which proves the desired approximation property.

Now, if $(f_k)_{k \in \mathbb{N}}$ is a sequence of step functions that converges uniformly to f, then it follows that

$$\left| \int_S f_k \, d\lambda - \int_S f_l \, d\lambda \right| \leq \| f_k - f_l \|_{\sup} \cdot \| \lambda \|_{\mathrm{var}} \longrightarrow 0 \quad \text{as } k, l \to \infty.$$

Hence there exists

$$\int_S f \, d\lambda := \lim_{k \to \infty} \int_S f_k \, d\lambda \, ,$$

and the limit is independent of the choice of approximating sequence $(f_k)_{k \in \mathbb{N}}$.

6.23 Riesz-Radon theorem (Dual space of C^0). Let $S \subset \mathbb{R}^n$ be compact. Then

$$J(\nu)(f) := \int_S f \, d\nu$$

defines a linear isometric isomorphism

$$J : rca(S) \to C^0(S)' .$$

Here $rca(S)$ is the space defined in 6.21 and the integral for continuous functions is defined as in 6.22.

Proof. For $\nu \in rca(S)$ and $f \in C^0(S)$ it follows from the definition of the Riemann integral that

$$|J(\nu)(f)| = \left| \int_S f \, d\nu \right| \leq \| f \|_{\sup} \cdot \| \nu \|_{\mathrm{var}} \, ,$$

and hence J is continuous. Moreover, J is isometric. To see this, note that for $\nu \in rca(S)$ and $\varepsilon > 0$ there exists a partitioning of S into Borel sets E_i, $i = 1, \ldots, m$, with

$$\| \nu \|_{\mathrm{var}} \leq \varepsilon + \sum_{i=1}^{m} |\nu(E_i)| \, .$$

As ν is regular, there exist compact sets $K_i \subset E_i$ with $|\nu|(E_i \setminus K_i) \leq \frac{\varepsilon}{m}$. Then $\mathrm{B}_\delta(K_i)$ are disjoint sets for sufficiently small $\delta > 0$, and

$$|\nu|\big(S \cap \mathrm{B}_\delta(K_i) \setminus K_i\big) \to 0 \quad \text{as } \delta \searrow 0,$$

which follows once again from the regularity of ν. On defining

$$f_i(x) := \max\left(1 - \tfrac{1}{\delta}\mathrm{dist}(x, K_i), 0\right)$$

and

$$\sigma_i := \begin{cases} \dfrac{\overline{\nu(K_i)}}{|\nu(K_i)|}, & \text{if } \nu(K_i) \neq 0, \\ 0, & \text{otherwise,} \end{cases}$$

it holds, if δ is sufficiently small, that

$$\left\|\sum_{i=1}^{m} \sigma_i f_i\right\|_{\sup} \leq 1$$

and

$$\left| J(\nu)\left(\sum_{i=1}^{m} \sigma_i f_i\right) \right| = \left| \sum_{i=1}^{m} \sigma_i \int_S f_i \, d\nu \right|$$

$$= \left| \sum_{i=1}^{m} \left(|\nu(K_i)| + \sigma_i \int_{S \cap B_\delta(K_i) \setminus K_i} f_i \, d\nu \right) \right|$$

$$\geq \sum_{i=1}^{m} |\nu(K_i)| - \sum_{i=1}^{m} |\nu|\left(S \cap B_\delta(K_i) \setminus K_i\right)$$

$$\geq \|\nu\|_{\mathrm{var}} - 2\varepsilon - \sum_{i=1}^{m} |\nu|\left(S \cap B_\delta(K_i) \setminus K_i\right)$$

$$\longrightarrow \|\nu\|_{\mathrm{var}} \quad \text{on letting } \delta \searrow 0 \text{ and then } \varepsilon \searrow 0.$$

Now the crucial step is to show that for $F \in C^0(S)'$ there exists a $\nu \in rca(S)$ with $J(\nu) = F$. It follows from the Hahn-Banach theorem that F can be extended norm-preservingly to $F \in B(S)'$ ($B(S)$ is the space defined in 3.1). Define

$$\lambda(E) := F(\mathcal{X}_E) \quad \text{for } E \subset S.$$

Then λ is additive and $\|\lambda\|_{\mathrm{var}} \leq \|F\|_{B(S)'}$, which follows as in (6-15). Therefore, by the definition of the Riemann integral,

$$F(f) = \int_S f \, d\lambda$$

for all $f \in C^0(S)$. Hence we want to find a $\nu \in rca(S)$ such that

$$\int_S f \, d\nu = \int_S f \, d\lambda \quad \text{for all } f \in C^0(S).$$

The proof that such a ν exists is given in Appendix A6 (see A6.6).   □

With the help of the result in theorem 6.23, we can provide a distributional characterization of regular measures.

6.24 Corollary. Let $\Omega \subset \mathbb{R}^n$ be open and bounded, let $C \geq 0$ and let

$$T : C_0^0(\Omega) \to \mathbb{K} \quad \text{be linear with} \quad |T(\zeta)| \leq C \cdot \|\zeta\|_{\sup} \text{ for all } \zeta \in C_0^0(\Omega).$$

Then there exists a unique $\lambda \in rca(\Omega)$ with

$$\|\lambda\|_{var} = \sup \left\{ |T(\zeta)| \; ; \; \zeta \in C_0^0(\Omega), \|\zeta\|_{\sup} = 1 \right\} \leq C,$$

$$T(\zeta) = \int_\Omega \zeta \, d\lambda \quad \text{for all } \zeta \in C_0^0(\Omega).$$

Remark: It is sufficient to assume that

$$T \in \mathscr{D}'(\Omega) \quad \text{with} \quad |T(\zeta)| \leq C \cdot \|\zeta\|_{\sup} \text{ for all } \zeta \in C_0^\infty(\Omega).$$

That is because T can then be uniquely extended to a linear map on $C_0^0(\Omega)$, which satisfies the above estimate (approximate functions in $C_0^0(\Omega)$ by means of convolutions).

Proof. Consider the open sets

$$\Omega_m := \left\{ x \in \Omega \; ; \; \operatorname{dist}(x, \partial\Omega) > \tfrac{1}{m} \right\}.$$

For $m \geq m_0$, with m_0 sufficiently large, Ω_m is nonempty and $S_m := \overline{\Omega_m} \subset \Omega_{m+1}$ is compact. For $m > m_0$ choose $\eta_m \in C_0^\infty(\Omega_m)$ with $0 \leq \eta_m \leq 1$ and $\eta_m = 1$ on S_{m-1}. Then

$$T_m(g) := T(\eta_m g) \quad \text{for } g \in C^0(S_m)$$

defines a $T_m \in C^0(S_m)'$ with

$$\|T_m\| \leq C_T := \sup\{|T(\zeta)| \; ; \; \zeta \in C_0^0(\Omega), \|\zeta\|_{\sup} = 1\} \leq C.$$

Hence it follows from 6.23 that there exist uniquely determined $\nu_m \in rca(S_m)$ with $\|\nu_m\|_{var} \leq C_T$ and

$$T_m(g) = \int_{S_m} g \, d\nu_m \quad \text{for } g \in C^0(S_m).$$

For $\zeta \in C_0^0(\Omega_m)$ and $l > m$ it holds that $\eta_l \zeta = \zeta$ (here we set $\zeta = 0$ outside of Ω_m), and so

$$\int_{S_m} \zeta \, d\nu_l = \int_{S_l} \zeta \, d\nu_l = T(\eta_l \zeta) = T(\zeta)$$

independently of l. We claim that

$$\nu_l(E) \text{ is independent of } l > m \text{ for Borel sets } E \subset S_{m-1}. \tag{6-18}$$

Indeed, let $K \subset S_{m-1}$ be compact. Then $\zeta_\delta(x) := \max(1 - \frac{1}{\delta}\text{dist}(x, K), 0)$ for small $\delta > 0$ defines a $\zeta_\delta \in C_0^0(\Omega_m)$. Since ν_l is a regular measure, $|\nu_l|(B_\delta(K) \setminus K) \searrow 0$ as $\delta \searrow 0$, and hence

$$\int_{S_m} \zeta_\delta \, d\nu_l \longrightarrow \nu_l(K) \quad \text{as } \delta \searrow 0,$$

i.e. (6-18) holds for compact sets in S_{m-1}. The regularity of ν_l then implies that (6-18) holds for all Borel sets. For Borel sets E with $\overline{E} \subset \Omega$ we have that $E \subset S_m$ for some $m \in \mathbb{N}$, and it follows from (6-18) that

$$\lambda(E) := \nu_l(E) \quad \text{for } l, m \in \mathbb{N} \text{ with } E \subset S_m, \ l \geq m+2$$

is well defined. For $\zeta \in C_0^0(\Omega)$ it holds that $\text{supp}(\zeta) \subset \Omega_m$ for some $m \in \mathbb{N}$ and

$$T(\zeta) = \int_{S_m} \zeta \, d\lambda$$

independently of m.

We need to show that λ can be extended to a $\lambda \in rca(\Omega)$. If E_i, $i = 1, \ldots, k$, are pairwise disjoint with $\overline{E_i} \subset \Omega$, then, as above, there exists an m with $E_i \subset S_m$ for $i = 1, \ldots, k$ and

$$\sum_{i=1}^{k} |\lambda(E_i)| = \sum_{i=1}^{k} |\nu_{m+2}(E_i)| \leq \|\nu_{m+2}\|_{\text{var}} \leq C_T.$$

In addition, for every Borel set $E \subset \Omega$ the limit

$$\lambda(E) := \lim_{m \to \infty} \lambda(E \cap S_m) \tag{6-19}$$

exists. To see this, let $E_m := E \cap S_m \setminus S_{m-1}$ for $m > m_0$ and $E_{m_0} := E \cap S_{m_0}$. Then

$$E \cap S_m = \bigcup_{i=m_0}^{m} E_i, \quad \lambda(E \cap S_m) = \sum_{i=m_0}^{m} \lambda(E_i)$$

and, as shown above,

$$\sum_{i=m_0}^{m} |\lambda(E_i)| \leq C_T.$$

Hence (6-19) defines an extension of λ to the Borel sets of Ω. Then it easily follows that $\lambda \in rca(\Omega)$ with $\|\lambda\|_{\text{var}} \leq C_T$. From the representation of T it then easily follows that $C_T \leq \|\lambda\|_{\text{var}}$. $\qquad \square$

As an application of theorem 6.23 (and in particular of 6.24), we consider the space $BV(\Omega)$. This space plays an important role in the functional analysis treatment of certain geometric differential equations, because it replaces the space $W^{1,p}(\Omega)$ for $p = 1$, which is not reflexive (see 8.11(4)). The functions in $BV(\Omega)$ have the advantage that their weak derivatives (see 6.25, below) can be interpreted as elements of a dual space. For existence proofs in reflexive spaces one employs theorem 8.10, however in the space $BV(\Omega)$ one can apply theorem 8.5.

6.25 Functions of bounded variation. Let $\Omega \subset \mathbb{R}^n$ be open and bounded. Consider pairs (f, λ) with $f \in L^1(\Omega)$ and $\lambda \in rca(\Omega; \mathbb{K}^n)$ such that the following rule of integration by parts holds:

$$\int_\Omega \partial_i \zeta \cdot f \, d\mathrm{L}^n + \int_\Omega \zeta \, d\lambda_i = 0 \quad \text{for all } \zeta \in C_0^\infty(\Omega) \tag{6-20}$$

for $i = 1, \ldots, n$. This is equivalent to

$$\partial_i[f] = [\lambda_i] \quad \text{in } \mathscr{D}'(\Omega)$$

for $i = 1, \ldots, n$.

Notation: The λ_i-integral is defined in 6.22, while the distributions $[f]$ and $[\lambda_i]$ are defined in 5.15.

In the spirit of the analogous definition in Sobolev spaces, we call $\partial_i f := \lambda_i$ the **weak derivative** of f. We have that:

(1) The set

$$BV(\Omega) := \big\{ f \in L^1(\Omega) \,;\ \text{there exists a } \lambda \in rca(\Omega; \mathbb{K}^n),$$
$$\text{such that (6-20) holds} \big\}$$

of functions of **bounded variation** is a $\mathbb{K}$-vector space, and it becomes a Banach space with the norm

$$\|f\|_{BV(\Omega)} := \|f\|_{L^1(\Omega)} + \|\lambda\|_{\mathrm{var}}.$$

(2) $W^{1,1}(\Omega) \subset BV(\Omega)$ with a continuous inclusion.

(3) $W^{1,1}(\Omega)$ is a proper subset of $BV(\Omega)$.

Proof (2). For $f \in W^{1,1}(\Omega)$ the corresponding measure $\lambda \in rca(\Omega; \mathbb{K}^n)$ is given by

$$\lambda(E) := \int_E \nabla f \, d\mathrm{L}^n.$$

Moreover, $\|\lambda\|_{\mathrm{var}} \le \|\nabla f\|_{L^1(\Omega)}$. $\square$

Proof (3). The fact that the space $BV(\Omega)$ is larger than $W^{1,1}(\Omega)$ follows from the existence of measures that have no representation as a function. For instance, for $\Omega = \,] -1, 1[\subset \mathbb{R}$ the **Heaviside function**

$$f(x) := \begin{cases} 1 & \text{for } x > 0, \\ 0 & \text{for } x < 0, \end{cases}$$

lies in $BV(]-1,1[)$ with

$$\int_{-1}^{1} \zeta' f \, dL^1 = -\zeta(0) = -\int_{-1}^{1} \zeta \, d\delta_0 \, ,$$

i.e. the weak derivative is the Dirac measure δ_0 at the point 0, and so

$$[f]' = [\delta_0] \quad \text{in } \mathscr{D}'(\Omega).$$

This example can be generalized to an arbitrary Ω.   □

The following theorem yields an equivalent definition of the space $BV(\Omega)$, which is formulated with the help of the distribution $[f] \in \mathscr{D}'(\Omega)$ for $f \in L^1(\Omega)$ (see 5.15). An additional possible definition in the case $n = 1$ is presented in E6.9.

6.26 Theorem. Let $\Omega \subset \mathbb{R}^n$ be open and bounded, and for $f \in L^1(\Omega)$ let

$$\|f\|_{\mathrm{grad}} := \sup \Big\{ \Big| \int_{\Omega} f \, \mathrm{div}\, g \, dL^n \Big| \; ; \; g \in C_0^\infty(\Omega; \mathbb{K}^n) \text{ with }$$

$$|g(x)| \le 1 \text{ for } x \in \Omega \Big\} \in [0, \infty] \, .$$

Here the **divergence** of a vector field is defined by

$$\mathrm{div}\, v := \sum_{i=1}^{n} \partial_i v_i \quad \text{for } v \in C^1(\Omega; \mathbb{K}^n).$$

Then

$$BV(\Omega) = \{ f \in L^1(\Omega) \; ; \; \|f\|_{\mathrm{grad}} < \infty \}$$

and for $f \in BV(\Omega)$ with $\nabla f := (\partial_i f)_{i=1,\ldots,n} \in rca(\Omega; \mathbb{K}^n)$,

$$\|f\|_{\mathrm{grad}} = \|\nabla f\|_{\mathrm{var}}.$$

Proof. For $g \in C_0^0(\Omega; \mathbb{K}^n)$ let

$$\int_{\Omega} g \bullet d\lambda := \sum_{i=1}^{n} \int_{\Omega} g_i \, d\overline{\lambda_i} \, , \quad \text{so that} \quad \Big| \int_{\Omega} g \bullet d\lambda \Big| \le \|g\|_{\mathrm{sup}} \cdot \|\lambda\|_{\mathrm{var}} \, ,$$

which follows by approximating g with step functions as in 6.22.

For $f \in BV(\Omega)$ with $\lambda_i := \partial_i f$ as in 6.25 and g as in the above definition of $\|f\|_{\mathrm{grad}}$ it then holds that

$$\Big| \int_{\Omega} f \, \mathrm{div}\, g \, dL^n \Big| = \Big| \sum_{i=1}^{n} \int_{\Omega} g_i \, d\lambda_i \Big| = \Big| \int_{\Omega} \overline{g} \bullet d\lambda \Big| \le \|\lambda\|_{\mathrm{var}} \, ,$$

and so $\|f\|_{\mathrm{grad}} \le \|\lambda\|_{\mathrm{var}}.$

Now let $f \in L^1(\Omega)$ with $\|f\|_{\text{grad}} < \infty$ and put

$$T_i(\zeta) := - \int_\Omega f \partial_i \zeta \, d\mathrm{L}^n = - \int_\Omega f \operatorname{div}(\zeta e_i) \, d\mathrm{L}^n \quad \text{for } \zeta \in C_0^\infty(\Omega).$$

By the definition of $\|f\|_{\text{grad}}$ it holds that $|T_i(\zeta)| \le \|\zeta\|_{\sup} \cdot \|f\|_{\text{grad}}$. This estimate shows that T_i can be uniquely extended onto $C_0^0(\Omega)$. Hence, by 6.24, there exists a $\lambda_i \in rca(\Omega)$ with

$$T_i(\zeta) = \int_\Omega \zeta \, d\lambda_i \quad \text{for } \zeta \in C_0^0(\Omega).$$

This shows that $f \in BV(\Omega)$ with $\partial_i f = \lambda_i$. On setting $\lambda := (\lambda_i)_{i=1,\ldots,n}$ it then holds for $g \in C_0^\infty(\Omega; \mathbb{K}^n)$ that

$$\overline{\int_\Omega g \bullet d\lambda} = \sum_{i=1}^n \int_\Omega \overline{g_i} \, d\lambda_i = \sum_{i=1}^n T_i(\overline{g_i}) = - \int_\Omega f \operatorname{div}(\overline{g}) \, d\mathrm{L}^n,$$

and so

$$\left| \int_\Omega g \bullet d\lambda \right| \le \|g\|_{\sup} \cdot \|f\|_{\text{grad}}.$$

Similarly to the proof of the isometry property in 6.23, this implies the inequality $\|\lambda\|_{\text{var}} \le \|f\|_{\text{grad}}$. $\square$

E6 Exercises

E6.1 Dual norm on $\mathbb{R}^n$. Let $\|\cdot\|$ be a norm on $\mathbb{R}^n$, i.e. we consider the normed space $(\mathbb{R}^n, \|\cdot\|)$.

(1) Show that

$$J(x)(y) := \sum_{i=1}^n y_i x_i \quad \text{for } x, y \in \mathbb{R}^n$$

defines a linear map $J : (\mathbb{R}^n, \|\cdot\|) \to (\mathbb{R}^n, \|\cdot\|)'$.

(2) Show that

$$\|x\|' := \|J(x)\| \quad \text{for } x \in \mathbb{R}^n$$

defines a norm on $\mathbb{R}^n$ (we call it the **dual norm** to $\|\cdot\|$).

(3) $J : (\mathbb{R}^n, \|\cdot\|') \to (\mathbb{R}^n, \|\cdot\|)'$ is an isometric isomorphism.

(4) For $1 \le p \le \infty$, find the dual norm to the p-norm in 2.5.

E6.2 Dual space of the cross product. Let X_1 and X_2 be normed spaces and

$$J : X_1' \times X_2' \to (X_1 \times X_2)',$$
$$J\big((x_1', x_2')\big)\big((x_1, x_2)\big) := x_1' x_1 + x_2' x_2.$$

Show that J is an isometric isomorphism if the norms in $X_1 \times X_2$ and $X_1' \times X_2'$ are defined as in E4.12(1) with respect to $|\cdot|$ and $|\cdot|'$, respectively.
Remark: Here $|\cdot|'$ is the dual norm to $|\cdot|$ from E6.1(2).
Show that this dual norm is also a monotone norm on $\mathbb{R}^2$.

E6.3 Integral equation. Let $K \in L^2(\Omega \times \Omega)$ and let $f \in L^2(\Omega)$, where $\Omega \subset \mathbb{R}^n$ is Lebesgue measurable. For $\lambda \in \mathbb{R}$ consider the *integral equation*

$$\int_\Omega K(x, y) u(y) \, dy = \lambda u(x) + f(x) \quad \text{for almost all } x \in \Omega.$$

Show that for $\lambda > \|K\|_{L^2(\Omega \times \Omega)}$ there exists a unique solution $u \in L^2(\Omega)$.

Solution. It follows from 5.12 that

$$(Tu)(x) := \int_\Omega K(x, y) u(y) \, dy$$

defines an operator $T \in \mathscr{L}\big(L^2(\Omega)\big)$ with $\|T\|_{\mathscr{L}(L^2(\Omega))} \leq \|K\|_{L^2(\Omega \times \Omega)}$. Then also $A := \lambda \mathrm{Id} - T \in \mathscr{L}\big(L^2(\Omega)\big)$ and for $u \in L^2(\Omega)$

$$\begin{aligned}
\mathrm{Re}\,(u,\, Au)_{L^2} &= \lambda \|u\|_{L^2}^2 - \mathrm{Re}\,(u,\, Tu)_{L^2} \\
&\geq \lambda \|u\|_{L^2}^2 - \|T\|_{\mathscr{L}(L^2(\Omega))} \cdot \|u\|_{L^2}^2 \\
&\geq \underbrace{\Big(\lambda - \|K\|_{L^2(\Omega \times \Omega)}\Big)}_{=:c_0 > 0} \|u\|_{L^2}^2 \,.
\end{aligned}$$

It follows from the Lax-Milgram theorem (see the equivalent result 6.3(3)) that A is invertible, and so $u := A^{-1}(-f)$ is the solution of the integral equation. $\qquad \qquad \square$

E6.4 Examples of elements from $C^0([0,1])'$. Show that the following maps T are linear and continuous on $C^0([0,1])$ and calculate their norm.

(1) $T : C^0([0,1]) \to C^0([0,1])$, for a given $g \in C^0([0,1])$ defined by

$$(Tf)(x) := g(x) \cdot f(x).$$

(2) $T : C^0([0,1]) \to \mathbb{K}$, with $\alpha_i \in \mathbb{R}$ and pairwise distinct $x_i \in [0,1]$, $i = 1, \ldots, m$, defined by

$$Tf := \sum_{i=1}^m \alpha_i f(x_i).$$

(3) $T : C^0([0,1]) \to \mathbb{K}$, with points x_i and coefficients α_i as in (2), defined by

$$Tf := \int_0^1 f(x)\,\mathrm{d}x - \sum_{i=1}^m \alpha_i f(x_i)\,.$$

Solution (1). On noting that $|(Tf)(x)| \leq \|g\|_{\sup}\|f\|_{\sup}$, we have that T is continuous, with $\|T\| \leq \|g\|_{\sup}$. As $\|Tg\|_{\sup} = \|g^2\|_{\sup} = \|g\|_{\sup}^2$, it holds that $\|T\| \geq \|g\|_{\sup}$. $\qquad\square$

Solution (2). Since

$$|Tf| \leq \sum_{i=1}^m |\alpha_i| \cdot \|f\|_{\sup}\,,$$

T is continuous, with $\|T\| \leq \sum_{i=1}^m |\alpha_i|$. As the x_i are pairwise distinct, there exists a continuous function f with $|f| \leq 1$ and $f(x_i) = \text{sign}(\alpha_i)$ for $i = 1, \dots, m$. Then

$$|Tf| = \sum_{i=1}^m |\alpha_i|\,, \quad \text{and so} \quad \|T\| \geq \sum_{i=1}^m |\alpha_i|\,.$$

$\qquad\square$

Solution (3). Since

$$|Tf| \leq \left(1 + \sum_{i=1}^m |\alpha_i|\right)\|f\|_{\sup}\,,$$

T is continuous, with $\|T\| \leq 1 + \sum_{i=1}^m |\alpha_i|$. Now for small $\delta > 0$, chosen so that $\delta < \frac{1}{2}|x_i - x_j|$ for all $i \neq j$, consider the continuous function

$$f(x) := \begin{cases} (1 - \frac{|x-x_i|}{\delta})\text{sign}(-\alpha_i) + \frac{|x-x_i|}{\delta} & \text{if } x \in I_{i\delta} \text{ for an } i, \\ 1 & \text{otherwise,} \end{cases}$$

where $I_{i\delta} := [x_i - \delta, x_i + \delta]$ are disjoint intervals. Then $\|f\|_{\sup} = 1$ and

$$|Tf| = \left| \int_0^1 (f(x) - 1)\,\mathrm{d}x + 1 + \sum_{i=1}^m |\alpha_i| \right|$$

$$= \left| \sum_{i=1}^m \left(\int_{[0,1] \cap I_{i\delta}} (f(x) - 1)\,\mathrm{d}x \right) + 1 + \sum_{i=1}^m |\alpha_i| \right|$$

$$\geq -4m\delta + 1 + \sum_{i=1}^m |\alpha_i|\,,$$

which shows that $\|T\| \geq 1 + \sum_{i=1}^m |\alpha_i|$.

Result: This means that no such **quadrature formula** can approximate the integral over $[0,1]$ for all (!) continuous functions. $\qquad\square$

E6.5 Dual space of $C^m(I)$. Let $I \subset \mathbb{R}$ be a closed interval and let $x_0 \in I$. Then, for $m \geq 1$,

$$J(\xi, \nu)(f) := \sum_{i=1}^m \xi_i f^{(i-1)}(x_0) + \int_I f^{(m)}\,\mathrm{d}\nu$$

defines an isomorphism $J : \mathbb{K}^m \times rca(I) \to C^m(I)'$.

Solution. It holds that

$$|J(\xi,\nu)(f)| \le \left(\max_{i=1,\dots,m} |\xi_i| + \|\nu\|_{\mathrm{var}} \right) \|f\|_{C^m(I)},$$

and hence J is continuous with $\|J\| \le 1$ if on $\mathbb{K}^m \times rca(I)$ we introduce the norm

$$\|(\xi,\nu)\| := \max_{i=1,\dots,m} |\xi_i| + \|\nu\|_{\mathrm{var}}$$

and if the C^m-norm is defined as in 3.6. Now for every function $f \in C^m(I)$ we have

$$f(x) = \sum_{i=0}^{m-1} \frac{1}{i!} f^{(i)}(x_0)(x - x_0)^i + \frac{1}{(m-1)!} \int_{x_0}^{x} f^{(m)}(y)(x-y)^{m-1} \, dy.$$

This can be shown by induction on m. First, note that for $m = 1$ this is the fundamental theorem of calculus. The following identity then proves the formula inductively:

$$\int_{x_0}^{x} f^{(m)}(y)(x-y)^{m-1} \, dy = -\frac{1}{m} \int_{x_0}^{x} f^{(m)}(y) \frac{d}{dy}(x-y)^m \, dy$$

$$= \frac{1}{m} f^{(m)}(x_0)(x-x_0)^m + \frac{1}{m} \int_{x_0}^{x} f^{(m+1)}(y)(x-y)^m \, dy.$$

Hence, for every $F \in C^m(I)'$ we have

$$Ff = \sum_{i=0}^{m-1} f^{(i)}(x_0) F p_i + FT f^{(m)},$$

where

$$p_i(x) := \frac{(x-x_0)^i}{i!} \quad \text{and} \quad Tg(x) := \int_{x_0}^{x} g(y) \frac{(x-y)^{m-1}}{(m-1)!} \, dy.$$

For $i = 0, \dots, m-1$ it follows inductively that

$$(Tg)^{(i)}(x) = \int_{x_0}^{x} g(y) \frac{(x-y)^{m-1-i}}{(m-1-i)!} \, dy,$$

since the integrand vanishes at the upper limit x. In particular,

$$(Tg)^{(m-1)}(x) = \int_{x_0}^{x} g(y) \, dy, \quad \text{and so} \quad (Tg)^{(m)}(x) = g(x).$$

Hence we have the estimate $\|Tg\|_{C^m(I)} \le C \cdot \|g\|_{C^0(I)}$ and it follows that $T \in \mathscr{L}(C^0(I); C^m(I))$, which implies that $FT \in C^0(I)'$. By theorem 6.23, there exists a $\nu \in rca(I)$ with $\|\nu\|_{\mathrm{var}} = \|FT\|$ and

$$FTg = \int_I g \, d\nu \quad \text{for } g \in C^0(I).$$

Setting $\xi_i := Fp_{i-1}$ for $i = 1, \ldots, m$, we have that

$$F = J(\xi, \nu)$$

and

$$\|(\xi, \nu)\| \leq \left(\max_{i=0,\ldots,m-1} \|p_i\|_{C^m(I)} + \|T\| \right) \|F\| .$$

This shows that J is surjective. If in addition we can show that J is injective, then this estimate yields that the inverse J^{-1} is also continuous. If $J(\xi, \nu) = 0$, then it holds for $i = 1, \ldots, m$ that

$$0 = J(\xi, \nu)p_{i-1} = \xi_i$$

and for all $g \in C^0(I)$ that

$$0 = J(\xi, \nu)Tg = \int_I g \, d\nu ,$$

which yields $\nu = 0$, thanks to theorem 6.23. Hence J is injective. $\qquad\square$

Remark: If

$$J_1(\xi)(z) := z \bullet \xi$$

is the isometry $J_1 : \mathbb{K}^m \to (\mathbb{K}^m)'$ from 6.1 and

$$J_2(\nu)(g) := \int_I g \, d\nu$$

is the isometry $J_2 : rca(I) \to C^0(I)'$ from 6.23, then it follows from E6.2 that

$$J_0(\xi, \nu)(z, g) := J_1(\xi)(z) + J_2(\nu)(g)$$

defines an isomorphism $J_0 : \mathbb{K}^m \times rca(I) \to (\mathbb{K}^m \times C^0(I))'$. Moreover,

$$S(f) := \left(\left(f^{(i)}(x_0) \right)_{i=0,\ldots,m-1} , f^{(m)} \right)$$

defines a continuous linear map from $C^m(I)$ to $\mathbb{K}^m \times C^0(I)$. With these definitions

$$J = S'J_0 ,$$

where S' is the adjoint map of S (see 5.5(8)). Hence J being an isomorphism is equivalent to the isomorphy of S' and, by theorem 12.5, equivalent to the isomorphy of S.

E6.6 Dual space of c_0 and c. Let

$$c_0 := \left\{ x \in \ell^\infty(\mathbb{R}) \; ; \; \lim_{i \to \infty} x_i = 0 \right\},$$

$$c := \left\{ x \in \ell^\infty(\mathbb{R}) \; ; \; \text{it exists } \lim_{i \to \infty} x_i \right\}.$$

The sets c_0 and c, equipped with the $\ell^\infty(\mathbb{R})$-norm, are Banach spaces. Characterize the dual spaces c_0' and c'.

Solution. For every $y \in \ell^1(\mathbb{R})$, setting

$$J(y)(x) := \sum_{i=1}^\infty y_i x_i \quad \text{for } x \in c_0$$

defines a $J(y) \in c_0'$ with $\|J(y)\| \leq \|y\|_{\ell^1}$, because

$$|J(y)(x)| \leq \sup_i |x_i| \cdot \sum_{i=1}^\infty |y_i| = \|x\|_{\ell^\infty} \|y\|_{\ell^1}.$$

If we define for $n \in \mathbb{N}$

$$x_i := \begin{cases} \text{sign}(y_i) & \text{for } i \leq n, \\ 0 & \text{for } i > n, \end{cases}$$

then $\left\| (x_i)_{i \in \mathbb{N}} \right\|_{\ell^\infty} = 1$ and

$$J(y)(x) = \sum_{i \leq n} |y_i| \to \|y\|_{\ell^1} \quad \text{as } n \to \infty.$$

Hence $J : \ell^1(\mathbb{R}) \to c_0'$ is isometric. Now let $F \in c_0'$. Since for all $x \in c_0$ we have that

$$x = \sum_{i=1}^\infty x_i e_i \quad \text{in the } \ell^\infty\text{-norm},$$

it follows that

$$F(x) = \sum_{i=1}^\infty x_i F e_i,$$

and so $F = J(y)$, where $y_i := F e_i$, provided that $y \in \ell^1(\mathbb{R})$. But this is indeed the case, since

$$\sum_{i \leq n} |y_i| = F\left(\sum_{i \leq n} \text{sign}(y_i)\, e_i \right) \leq \|F\| \cdot \left\| \sum_{i \leq n} \text{sign}(y_i)\, e_i \right\|_{\ell^\infty} = \|F\|.$$

This shows that J is an isomorphism. Then the dual space c' can be characterized as follows:

$$Sx := \left(\lim_{i \to \infty} x_i, x_1 - \lim_{i \to \infty} x_i, x_2 - \lim_{i \to \infty} x_i, \dots \right)$$

defines an $S \in \mathscr{L}(c; c_0)$, and S is in fact an isomorphism, with

$$S^{-1}x = (x_2 + x_1, x_3 + x_1, x_4 + x_1, \dots).$$

Therefore

$$\tilde{J}(y) := J(y)S$$

defines an isomorphism $\tilde{J} : \ell^1(\mathbb{R}) \to c'$. $\qquad\qquad\qquad\qquad\qquad$ □

E6.7 Characterization of Sobolev functions. Let $\Omega \subset \mathbb{R}^n$ be open. For $m \in \mathbb{N} \cup \{0\}$ and $1 < p \le \infty$ (if $m = 0$ then also for $p = 1$) it holds for functions $f : \Omega \to \mathbb{R}$ that

$$f \in W^{m,p}(\Omega) \quad \Longleftrightarrow \quad \begin{cases} f \in L^1_{\mathrm{loc}}(\Omega) \text{ and there exists a constant } C \text{ with} \\[2mm] \left| \int_\Omega f \partial^s \zeta \, \mathrm{dL}^n \right| \le C \|\zeta\|_{L^{p'}(\Omega)} \\[2mm] \text{for all } |s| \le m \text{ and all } \zeta \in C_0^\infty(\Omega). \end{cases}$$

Here p' is the dual exponent to p.

Note: For this characterization in the case $m = 0$, see 6.13. In case $m > 0$ we have to assume $p > 1$, see the space $BV(\Omega)$ and 6.26.

Solution ⇒.

$$\left| \int_\Omega f \partial^s \zeta \, \mathrm{dL}^n \right| = \left| \int_\Omega \partial^s f \cdot \zeta \, \mathrm{dL}^n \right| \le \|\partial^s f\|_{L^p(\Omega)} \|\zeta\|_{L^{p'}(\Omega)}.$$

$\qquad\qquad\qquad\qquad\qquad\qquad\qquad\qquad\qquad\qquad\qquad\qquad\qquad$ □

Solution ⇐. It follows from 6.13 that $f \in L^p(\Omega)$. For $0 < |s| \le m$ let

$$F_s(\zeta) := \int_\Omega f \partial^s \zeta \, \mathrm{dL}^n \quad \text{for } \zeta \in C_0^\infty(\Omega).$$

The estimate $|F_s(\zeta)| \le C\|\zeta\|_{L^{p'}(\Omega)}$ says, since $p' < \infty$, that F_s can be extended to a functional on $L^{p'}(\Omega)$. Then it follows from 6.12, again since $p' < \infty$, that there exists a function $f_s \in L^p(\Omega)$ with

$$F_s(g) = \int_\Omega g \cdot f_s \, \mathrm{dL}^n \quad \text{for } g \in L^{p'}(\Omega).$$

Therefore,

$$\int_\Omega f \partial^s \zeta \, \mathrm{dL}^n = \int_\Omega f_s \zeta \, \mathrm{dL}^n \quad \text{for } \zeta \in C_0^\infty(\Omega),$$

which yields that $f \in W^{m,p}(\Omega)$ (with $\partial^s f = (-1)^{|s|} f_s$). $\qquad$ □

E6.8 Positive functionals on C_0^0. Let $\Omega \subset \mathbb{R}^n$ be open and let $F : C_0^0(\Omega; \mathbb{R}) \to \mathbb{R}$ be a linear map with

$$f \ge 0 \text{ in } \Omega \quad \Longrightarrow \quad F(f) \ge 0.$$

Then there exists a nonnegative locally bounded regular σ-additive measure μ on the Borel sets of Ω (μ is then also called a **Radon measure**) such that

$$F(f) = \int_\Omega f \, \mathrm{d}\mu \quad \text{for all } f \in C_0^0(\Omega; \mathbb{R}).$$

Solution. Here $\mathbb{K} = \mathbb{R}$. Let $D \subset \Omega$ be open and bounded with

$$d := \tfrac{1}{2}\mathrm{dist}(D, \partial\Omega) > 0 .$$

In addition, let $S := \overline{B_d(D)}$. Choose a cut-off function $\eta \in C_0^0(\Omega)$ (see 4.19) with

$$0 \le \eta \le 1, \ \eta = 1 \text{ on } D, \ \eta = 0 \text{ outside of } B_d(D) ,$$

e.g.

$$\eta(x) = \max(0, 1 - \tfrac{1}{d}\mathrm{dist}(x, D)) .$$

For nonnegative functions $f \in C^0(S)$ we then have that $\eta f \in C_0^0(\Omega)$, with

$$0 \le \eta f \le \eta \sup_S f ,$$

and so

$$0 \le F(\eta f) \le F(\eta) \cdot \sup_S f .$$

Then it follows for all $f \in C^0(S)$, on setting $f^+ := \max(f, 0)$ and $f^- := \max(-f, 0)$, that

$$|F(\eta f)| = |F(\eta f^+) - F(\eta f^-)|$$
$$\le (\sup_S f^+ + \sup_S f^-)F(\eta) \le \|f\|_{C^0(S)} \cdot F(\eta) .$$

Hence $f \mapsto F(\eta f)$ is a continuous functional on $C^0(S)$, and 6.23 yields the existence of a $\mu \in rca(S)$ with

$$F(\eta f) = \int_S f \, d\mu \quad \text{for all } f \in C^0(S).$$

For $f \in C_0^0(D)$ it holds that $\eta f = f$, and hence

$$F(f) = \int_S f \, d\mu \quad \text{for all } f \in C_0^0(D).$$

We need to show that $\mu \ge 0$. As μ is regular, it is sufficient to show that $\mu(K) \ge 0$ for compact sets $K \subset D$. Now, define

$$\eta_\varepsilon(x) := \max(0, 1 - \tfrac{1}{\varepsilon}\mathrm{dist}(x, K)) ,$$

so we have $\eta_\varepsilon \in C_0^0(D)$ for sufficiently small ε. Since $1 \ge \eta_\varepsilon \searrow \mathcal{X}_K$ pointwise as $\varepsilon \searrow 0$, we obtain that

$$0 \le F(\eta_\varepsilon) = \int_S \eta_\varepsilon \, d\mu \longrightarrow \mu(K) .$$

A similar argument shows that $\widetilde{\mu} = \mu$ in D, if $\widetilde{\mu}$ is the measure in $rca(\widetilde{S})$ for a $\widetilde{D}$ as above with $D \subset \widetilde{D}$. Exhausting Ω with countably many (not necessarily connected) domains D then yields the desired result. $\square$

As an alternative to the space $BV(\Omega)$ in 6.25 we define the following:

E6.9 Functions of bounded variation. In the one-dimensional case we define for $S := [a, b] \subset \mathbb{R}$

$$\widetilde{BV}(S) := \left\{ f : [a, b] \to \mathbb{K} \; ; \; \|f\|_{\widetilde{BV}} := |f(a)| + \mathrm{var}(f, S) < \infty \right\} ,$$

where the **variation** of f on $[a, b]$ is defined by

$$\mathrm{var}(f, [a, b]) := \sup \left\{ \sum_{i=1}^{m} |f(a_i) - f(a_{i-1})| \; ; \right.$$
$$\left. m \in \mathbb{N}, \; a = a_0 < a_1 < \ldots < a_m = b \right\} .$$

Show that for $f \in \widetilde{BV}(S)$ it holds that:

(1) For $a \le x_1 < x_2 < x_3 \le b$,

$$\mathrm{var}(f, [x_1, x_3]) = \mathrm{var}(f, [x_1, x_2]) + \mathrm{var}(f, [x_2, x_3]) .$$

(2) The following limits exist

$$f_+(x) := \lim_{\varepsilon \searrow 0} f(x + \varepsilon) \quad \text{for } a \le x < b,$$
$$f_-(x) := \lim_{\varepsilon \searrow 0} f(x - \varepsilon) \quad \text{for } a < x \le b.$$

(3) Every function in $\widetilde{BV}(S)$ has at most countably many discontinuity points.

Solution (1). The "$\le$" part in the identity follows from adding x_2 to the interval partitionings of $[x_1, x_3]$. □

Solution (2). Noting that

$$|f(x)| \le |f(a)| + |f(x) - f(a)| \le \|f\|_{\widetilde{BV}}$$

yields that f is bounded. Hence for $x < b$ there exists a sequence $(\kappa_i)_{i \in \mathbb{N}}$ with $\kappa_i \searrow x$ for $i \to \infty$, such that

$$\xi := \lim_{i \to \infty} f(\kappa_i)$$

exists. Now it follows from (1) that for all m

$$\sum_{i=1}^{m} \mathrm{var}(f, [\kappa_{i+1}, \kappa_i]) = \mathrm{var}(f, [\kappa_{m+1}, \kappa_1]) \le \|f\|_{\widetilde{BV}} < \infty,$$

and hence

$$\sum_{i=1}^{\infty} \mathrm{var}(f, [\kappa_{i+1}, \kappa_i]) \le \|f\|_{\widetilde{BV}} < \infty,$$

which implies that $\text{var}(f, [\kappa_{i+1}, \kappa_i]) \to 0$ as $i \to \infty$. Hence also

$$\sup_{\kappa_{i+1} \leq y \leq \kappa_i} |f(y) - \xi|$$

$$\leq |f(\kappa_i) - \xi| + \sup_{\kappa_{i+1} \leq y \leq \kappa_i} |f(y) - f(\kappa_i)|$$

$$\leq |f(\kappa_i) - \xi| + \text{var}(f, [\kappa_{i+1}, \kappa_i]) \to 0 \quad \text{as } i \to \infty,$$

which shows that $\xi = f_+(x)$. □

Solution (3). If $a < x_1 < \ldots < x_m < b$ are discontinuity points of f, for which $|f_+(x_i) - f_-(x_i)| \geq \delta$, then it holds for small $\varepsilon \to 0$ that

$$\text{var}(f, S) \geq \sum_{i=1}^m |f(x_i + \varepsilon) - f(x_i - \varepsilon)|$$

$$\to \sum_{i=1}^m |f_+(x_i) - f_-(x_i)| \geq m\delta,$$

and so $m \leq \delta^{-1} \|f\|_{\widetilde{BV}}$. On choosing a null sequence for δ, it follows that the discontinuity points of f are countable. □

Riemann-Stieltjes integral: Let $S = [a, b] \subset \mathbb{R}$ and $f \in \widetilde{BV}(S)$. Consider for $g \in C^0(S)$ and for partitionings $a = s_0 < s_1 < \ldots < s_n = b$ the sum

$$\sum_{i=1}^n g(s_i)\big(f(s_i) - f(s_{i-1})\big).$$

If $(t_j)_{j=1,\ldots,m}$ is a finer partitioning of S, say $t_{k_i} = s_i$ with $k_{i-1} < k_i$, then, on setting $\delta_s := \max_i |s_i - s_{i-1}|$,

$$\left| \sum_{i=1}^n g(s_i)\big(f(s_i) - f(s_{i-1})\big) - \sum_{j=1}^m g(t_j)\big(f(t_j) - f(t_{j-1})\big) \right|$$

$$= \left| \sum_{i=1}^n \sum_{j=k_{i-1}+1}^{k_i} \big(g(s_i) - g(t_j)\big)\big(f(t_j) - f(t_{j-1})\big) \right|$$

$$\leq \sup_{|x_1 - x_2| \leq \delta_s} |g(x_1) - g(x_2)| \cdot \|f\|_{\widetilde{BV}} \longrightarrow 0 \quad \text{as } \delta_s \to 0.$$

Hence the ***Riemann-Stieltjes integral***

$$\int_S g \, df := \lim_{\delta_s \to 0} \sum_{i=1}^n g(s_i)\big(f(s_i) - f(s_{i-1})\big)$$

exists for $f \in \widetilde{BV}(S)$ and $g \in C^0(S)$.

E6.10 Representation of the Riemann-Stieltjes integral. Suppose that $f \in \widetilde{BV}(S)$. Then the following holds for the above defined integral.

(1) There exists a $\lambda \in rca(S)$ with

$$\int_S g \, df = \int_S g \, d\lambda \quad \text{for all } g \in C^0(S).$$

(2) The measure λ in (1) satisfies for $a \leq x < b$

$$\lambda([a, x]) = \lim_{\varepsilon \searrow 0} \left(f(x + \varepsilon) - f(a) \right).$$

Solution (1). The map

$$T_f(g) := \int_S g \, df$$

satisfies

$$\left| \int_S g \, df \right| \leq \|g\|_{C^0} \cdot \|f\|_{\widetilde{BV}}.$$

It follows that $T_f \in C^0(S)'$ and hence theorem 6.23 yields the existence of a $\lambda \in rca(S)$ such that

$$\int_S g \, d\lambda = T_f(g) = \int_S g \, df \quad \text{for all } g \in C^0(S)$$

and $\|\lambda\|_{\text{var}} = \|T_f\|_{C^0(S)'}$. □

Solution (2). For $a < x_0 < b$ and sufficiently small $\varepsilon > 0$, consider the continuous function

$$g_\varepsilon(x) := \begin{cases} 1 & \text{for } x \leq x_0 + \varepsilon, \\ 1 - \frac{x - x_0 - \varepsilon}{\varepsilon} & \text{for } x_0 + \varepsilon \leq x \leq x_0 + 2\varepsilon, \\ 0 & \text{for } x_0 + 2\varepsilon \leq x. \end{cases}$$

Then by the σ-additivity of $|\lambda|$

$$\int_{[a, x_0 + \varepsilon]} g_\varepsilon \, d\lambda = \lambda([a, x_0 + \varepsilon]) \longrightarrow \lambda([a, x_0]) \quad \text{as } \varepsilon \to 0$$

and the definition of the Riemann integral gives

$$\left| \int_S g_\varepsilon \, d\lambda - \lambda([a, x_0 + \varepsilon]) \right| \leq |\lambda|([x_0 + \varepsilon, x_0 + 2\varepsilon]) \longrightarrow 0$$

for a sequence $\varepsilon \to 0$, since $\|\lambda\|_{\text{var}} < \infty$. Moreover, by the definition of the Riemann-Stieltjes integral,

$$\int_{[a, x_0 + \varepsilon]} g_\varepsilon \, df = f(x_0 + \varepsilon) - f(a)$$

which converges to $\lim_{\varepsilon \searrow 0} \left(f(x_0 + \varepsilon) - f(a) \right)$, and

$$\left| \int_S g_\varepsilon \, df - \left(f(x_0 + \varepsilon) - f(a) \right) \right| \leq \text{var}(f, [x_0 + \varepsilon, x_0 + 2\varepsilon]) \longrightarrow 0$$

as $\varepsilon \to 0$. □

Consider the functions

$$f_\varepsilon(x) := \begin{cases} 1 \text{ for } |x| \le \varepsilon, \\ 0 \text{ otherwise}, \end{cases} \qquad f(x) := \begin{cases} 1 \text{ for } x = 0, \\ 0 \text{ otherwise}. \end{cases}$$

Then $f_\varepsilon \to f$ pointwise as $\varepsilon \to 0$ and $f \ne 0$ in $\widetilde{BV}([-1,1])$. Also,

$$\text{var}(f, [-1,1]) = 2, \text{ but } \int_{[-1,1]} g \, df = 0 \text{ for all } g \in C^0([-1,1]).$$

In fact, with respect to the L^1-measure we have $f_\varepsilon \to 0$ almost everywhere as $\varepsilon \to 0$. As a consequence one considers function spaces

$$BV_{rc}([a,b]) := \{ f \in \widetilde{BV}([a,b]) \; ; \; f(x) = f_+(x) \text{ for } a \le x < b,$$
$$f(b) = f_-(b) \} ,$$
$$BV_{lc}([a,b]) := \{ f \in \widetilde{BV}([a,b]) \; ; \; f(a) = f_+(a),$$
$$f(x) = f_-(x) \text{ for } a < x \le b \} ,$$

which consist of right-continuous and left-continuous functions, respectively. Both spaces are bijective (isomorphic) to $BV(]a,b[)$ in 6.25.

E6.11 Normalized BV functions. With $S := [a,b] \subset \mathbb{R}$ and the notations as in E6.9, let

$$NBV(S) := \{ f \in \widetilde{BV}(S) \; ; \; f(x) = f_+(x) \text{ for } a \le x < b,$$
$$f(a) = 0 \text{ and } f(b) = f_-(b) \}$$

be the space of *normalized functions of bounded variation*, equipped with the norm of $\widetilde{BV}(S)$. Show that

$$(J\lambda)(x) := \lambda([a,x]) \quad \text{for } a \le x \le b$$

defines an isometric isomorphism

$$J : \{\lambda \in rca([a,b]) \; ; \; \lambda(\{a\}) = 0, \; \lambda(\{b\}) = 0\} \to NBV([a,b]).$$

Solution. The σ-additivity of λ yields that $f := J\lambda$ is right-continuous. Since $\lambda(\{a\}) = 0$ it follows that $f(a) = 0$, and since $\lambda(\{b\}) = 0$ the σ-additivity gives that $f(x) \to f(b)$ as $x \nearrow b$.

Moreover, for every partitioning $a = a_0 < a_1 < \ldots < a_m = b$,

$$\sum_{i=1}^m |f(a_i) - f(a_{i-1})| = \sum_{i=1}^m |\lambda(]a_{i-1}, a_i])| \le \|\lambda\|_{\text{var}},$$

i.e. $\|f\|_{\widetilde{BV}} \le \|\lambda\|_{\text{var}}$.

In addition, J is injective. In order to prove surjectivity, we use the previous exercise, which for a given $f \in NBV([a,b])$ yields a $\lambda \in rca([a,b])$, for which $\|\lambda\|_{\mathrm{var}} \leq \mathrm{var}(f, [a,b]) = \|f\|_{\widetilde{BV}}$. It follows from E6.10(2) that $J\lambda = f$, since for $a \leq x < b$

$$(J\lambda)(x) = \lambda([a,x]) = \lim_{\varepsilon \searrow 0} (f(x+\varepsilon) - f(a)) = f(x)$$

and also $(J\lambda)(b) = \lambda([a,b]) = \lim_{\varepsilon \searrow 0} \lambda([a, b-\varepsilon]) = f(b)$. $\qquad\square$

A6 Results from measure theory

The purpose of this appendix is to complete the proof of the representation theorem 6.23 (see A6.6). The necessary construction of regular measures can be found in A6.3.

Subsequently, we also present versions of Luzin's theorem (see A6.7) and Fubini's theorem (see A6.10).

In the following two results, S is an arbitrary set.

A6.1 Jordan decomposition. Let $\mathcal{B}$ be a ring of subsets of the set S and let $\lambda : \mathcal{B} \to \mathbb{R}$ be additive and bounded. Then

$$\lambda^+ := \tfrac{1}{2}(|\lambda| + \lambda), \quad \lambda^- := \tfrac{1}{2}(|\lambda| - \lambda)$$

are additive, bounded and nonnegative on $\mathcal{B}$. It holds that

$$\lambda = \lambda^+ - \lambda^-, \quad |\lambda| = \lambda^+ + \lambda^-,$$

and, in addition,

$$\lambda^+(E) = \sup_{A \in \mathcal{B} : A \subset E} \lambda(A) \quad \text{and} \quad \lambda^-(E) = - \inf_{A \in \mathcal{B} : A \subset E} \lambda(A).$$

Proof. On recalling 6.10, we only need to show that the last identity holds for λ^+.

If $A \subset E$, then $|\lambda|(A) \geq |\lambda(A)|$, and so

$$\lambda^+(E) \geq \lambda^+(A) \geq \tfrac{1}{2}(|\lambda(A)| + \lambda(A)) \geq \lambda(A).$$

Now for a given $\varepsilon > 0$ choose disjoint sets $E_1, \ldots, E_m$ with $E_i \subset E$ and

$$|\lambda|(E) \leq \varepsilon + \sum_{i=1}^{m} |\lambda(E_i)|.$$

On setting $E_{m+1} := E \setminus \bigcup_{i=1}^{m} E_i$, we have

$$\lambda(E) = \sum_{i=1}^{m+1} \lambda(E_i),$$

and so

$$\lambda^+(E) = \frac{1}{2}\big(|\lambda|(E) + \lambda(E)\big) \le \frac{\varepsilon}{2} + \frac{1}{2}\sum_{i=1}^{m+1}\big(|\lambda(E_i)| + \lambda(E_i)\big)$$

$$= \frac{\varepsilon}{2} + \sum_{i\,:\,\lambda(E_i)>0}\lambda(E_i) = \frac{\varepsilon}{2} + \lambda\Big(\bigcup_{i\,:\,\lambda(E_i)>0} E_i\Big) \le \frac{\varepsilon}{2} + \sup_{A\in\mathcal{B}\,:\,A\subset E}\lambda(A)\,.$$

$\square$

A6.2 Hahn decomposition. Let $\mathcal{B}$ be a σ-ring on the set S and let $\nu :$ $\mathcal{B} \to \mathbb{R}$ be σ-additive and bounded. Then there exists an $E^+ \in \mathcal{B}$ such that

$$\nu(E\cap E^+) \ge 0 \quad \text{and} \quad \nu(E\setminus E^+) \le 0 \quad \text{for all } E\in\mathcal{B}.$$

Proof. We assume that there exists an $E \in \mathcal{B}$ with $\nu(E) > 0$ (otherwise choose $E^+ := \emptyset$). We now want to find an $E^+ \in \mathcal{B}$ such that

$$\nu(E^+) = s_0 := \sup_{E\in\mathcal{B}}\nu(E)\,. \tag{A6-1}$$

Such an E^+ satisfies the desired result. To see this, assume that $\nu(E\setminus E^+) > 0$ for some $E \in \mathcal{B}$. Then

$$\nu(E^+\cup E) = \nu(E^+) + \nu(E\setminus E^+) > \nu(E^+) = s_0\,,$$

which contradicts the definition of s_0. Similarly, if $\nu(E\cap E^+) < 0$ for some $E \in \mathcal{B}$, then

$$\nu(E^+\setminus E) = \nu(E^+) - \nu(E\cap E^+) > \nu(E^+) = s_0\,,$$

which again contradicts the definition of s_0.

For the construction of E^+, define for $k \in \mathbb{N}$

$$\mathcal{M}_k := \big\{ E \in \mathcal{B}\,;\, \nu(E) \ge \big(1 - \tfrac{1}{k}\big)s_0 \big\}$$

with the partial order

$$E_1 \le E_2 \quad :\Longleftrightarrow \quad \big(E_1 \supset E_2 \text{ and } \nu(E_1) < \nu(E_2) \big) \text{ or } E_1 = E_2\,.$$

Let $\mathcal{N} \subset \mathcal{M}_k$ be totally ordered and let

$$s := \sup_{E\in\mathcal{N}} \nu(E)\,.$$

Then there exist $E_i \in \mathcal{N}$, $i \in \mathbb{N}$, with

$$\nu(E_i) \le \nu(E_{i+1}) \to s \quad \text{as } i \to \infty\,. \tag{A6-2}$$

As $\mathcal{N}$ is totally ordered, it follows that $E_i \le E_{i+1}$ or $E_{i+1} \le E_i$. If $E_i \le E_{i+1}$ then (A6-2) implies $E_i \supset E_{i+1}$, and if $E_{i+1} \le E_i$ it implies $E_i = E_{i+1}$. Therefore the sets E_i are decreasing and

$$E_0 := \bigcap_{i \in \mathbb{N}} E_i \in \mathcal{M}_k \,, \quad \nu(E_0) = \lim_{i \to \infty} \nu(E_i) = s \,.$$

The found set $E_0 \in \mathcal{M}_k$ is an upper bound of $\mathcal{N}$. This follows from the fact that if $E \in \mathcal{N}$ with $E_0 \leq E$, then $E \subset E_0$ and $\nu(E) > \nu(E_0)$, or $E = E_0$, where the former case contradicts the definition of s, since $\nu(E) > \nu(E_0) = s$, therefore $E = E_0$.

Hence, by Zorn's lemma (see the proof of 6.14), there exists a maximal element $M_k^+ \in \mathcal{M}_k$. It satisfies

$$\nu(M_k^+) \geq \left(1 - \tfrac{1}{k}\right)s_0 \,,$$

and in addition it holds for all $A \in \mathcal{B}$ that

$$A \subset M_k^+ \quad \Longrightarrow \quad \nu(A) \geq 0 \,. \tag{A6-3}$$

To see this, assume that $\nu(A) < 0$. Then $\nu(M_k^+ \setminus A) > \nu(M_k^+)$, and so $M_k^+ \setminus A \in \mathcal{M}_k$ with $M_k^+ \setminus A \geq M_k^+$. Then the maximality of M_k^+ yields that $M_k^+ \setminus A \leq M_k^+$, a contradiction.

Then the property (A6-3) also holds with M_k^+ replaced by the sets

$$E_k^+ := \bigcup_{j \leq k} M_j^+ \,,$$

because if $A \in \mathcal{B}$, $A \subset E_k^+$, then $A_j := A \cap M_j^+ \setminus \bigcup_{i<j} M_i^+ \subset M_j^+$ form a partition of A, and hence

$$\nu(A) = \sum_{j=1}^{k} \nu(A_j) \geq 0 \,.$$

In particular,

$$\nu(E_k^+) \geq \nu(M_k^+) \geq \left(1 - \tfrac{1}{k}\right)s_0 \,.$$

Hence

$$E^+ := \bigcup_{k \in \mathbb{N}} E_k^+ \in \mathcal{B} \quad \text{with} \quad \nu(E^+) = \lim_{k \to \infty} \nu(E_k^+) = s_0 \,.$$

Therefore E^+ satisfies (A6-1). □

In the following, let $S \subset \mathbb{R}^n$ be a closed set and let $\mathcal{B}_0$, $\mathcal{B}_1$ for S be defined as in 6.20. Furthermore, let $ba(S)$ etc. be the spaces defined in 6.20 and 6.21.

A6.3 Lemma. Let $\lambda \in ba(S)$ be nonnegative and let

$$\mu(E) := \sup_{\substack{A \,:\, A \subset E \\ A \text{ closed}}} \inf_{\substack{U \,:\, A \subset U \\ U \text{ open}}} \lambda(U) \quad \text{for } E \in \mathcal{B}_0.$$

Then $\mu \in rba(S)$ and

$$\int_S f \, d\mu = \int_S f \, d\lambda \quad \text{for all } f \in C^0(S).$$

Proof. (All occurring sets are in $\mathcal{B}_0$.) μ is nonnegative and monotone, i.e. $E_1 \subset E_2$ implies that $\mu(E_1) \leq \mu(E_2)$. For closed sets A

$$\mu(A) = \inf_{\substack{U \,:\, A \subset U \\ U \text{ open}}} \lambda(U), \quad \text{and so} \quad \mu(E) = \sup_{\substack{A \,:\, A \subset E \\ A \text{ closed}}} \mu(A) \tag{A6-4}$$

for all E. Define

$$\mathcal{M} := \{B \in \mathcal{B}_0 \,;\, \mu(E) = \mu(E \cap B) + \mu(E \setminus B) \text{ for all } E \in \mathcal{B}_0\}.$$

We want to show that $\mathcal{M} = \mathcal{B}_0$. Obviously, $\emptyset, S \in \mathcal{M}$ and from $B \in \mathcal{M}$ it follows that $S \setminus B \in \mathcal{M}$. If $A, B \in \mathcal{M}$, then it follows that for all $E \in \mathcal{B}_0$

$$\mu(E \cap (A \cap B)) + \mu(E \setminus (A \cap B))$$
$$= \mu\big(\underbrace{E \cap (A \cap B)}_{=(E \cap B) \cap A}\big) + \mu\big(\underbrace{(E \setminus (A \cap B)) \cap B}_{=(E \cap B) \setminus A}\big) + \mu\big(\underbrace{(E \setminus (A \cap B)) \setminus B}_{=E \setminus B}\big)$$
$$= \mu(E \cap B) + \mu(E \setminus B) = \mu(E),$$

and so $A \cap B \in \mathcal{M}$. Hence $\mathcal{M}$ is a Boolean algebra. It remains to show that $\mathcal{M}$ contains the closed sets. If A_1, A_2 are closed and disjoint, then there exist open disjoint sets U_i with $A_i \subset U_i$. Then it holds for every open set $U \supset A_1 \cup A_2$ that

$$\lambda(U) \geq \lambda\big(U \cap (U_1 \cup U_2)\big) = \lambda(U \cap U_1) + \lambda(U \cap U_2) \geq \mu(A_1) + \mu(A_2),$$

and combining with (A6-4) yields that

$$\mu(A_1 \cup A_2) \geq \mu(A_1) + \mu(A_2).$$

Now let B be closed and let E be arbitrary. Then if $A_1 \subset E \cap B$, $A_2 \subset E \setminus B$ are closed sets,

$$\mu(A_1) + \mu(A_2) \leq \mu(A_1 \cup A_2) \leq \mu(E),$$

and so (A6-4) implies that

$$\mu(E \cap B) + \mu(E \setminus B) \leq \mu(E).$$

On the other hand, if $A \subset E$ is closed and U_1, U_2 are open with $A \cap B \subset U_1$ and $A \setminus U_1 \subset U_2$, then $A \subset U_1 \cup U_2$, and hence

$$\lambda(U_1) + \lambda(U_2) \geq \lambda(U_1 \cup U_2) \geq \mu(A).$$

Taking the infimum over all U_2, and noting that $A \setminus U_1$ is closed, we obtain

$$\lambda(U_1) + \mu(A \setminus U_1) \geq \mu(A).$$

Since $A \setminus U_1$ is a closed subset of $E \setminus B$, it follows that

$$\lambda(U_1) + \mu(E \setminus B) \geq \mu(A).$$

Now noting that $A \cap B$ is closed, and taking the infimum over all U_1, we obtain

$$\mu(A \cap B) + \mu(E \setminus B) \geq \mu(A),$$

and so, since $A \cap B$ is a closed subset of $E \cap B$,

$$\mu(E \cap B) + \mu(E \setminus B) \geq \mu(A).$$

On taking the supremum over all A, it finally follows that

$$\mu(E \cap B) + \mu(E \setminus B) \geq \mu(E).$$

This shows that $B \in \mathcal{M}$, and hence $\mathcal{M} = \mathcal{B}_0$.

It follows that μ is additive on $\mathcal{M}$, for if $E_1, E_2 \in \mathcal{M}$ are disjoint, then it holds for all E that

$$\mu(E) = \mu(E \cap E_1) + \mu(E \setminus E_1),$$

and for $E = E_1 \cup E_2$ we obtain that

$$\mu(E_1 \cup E_2) = \mu(E_1) + \mu(E_2).$$

Moreover, μ is regular, because for E and $\varepsilon > 0$ there exist closed sets $A_1 \subset E$ and $A_2 \subset S \setminus E$ with

$$\mu(E) \leq \mu(A_1) + \varepsilon \quad \text{and} \quad \mu(S \setminus E) \leq \mu(A_2) + \varepsilon.$$

Then $A_1 \subset E \subset S \setminus A_2$ and, on recalling that $|\mu| = \mu$, it follows that

$$|\mu|\big((S \setminus A_2) \setminus A_1\big) \leq 2\varepsilon.$$

It remains to show that the integral identity holds. Without loss of generality let $0 \leq f \leq 1$. For $n \in \mathbb{N}$ define

$$E_i := \left\{ \tfrac{i}{n} \leq f < \tfrac{i+1}{n} \right\} \in \mathcal{B}_0 \quad \text{for } i = 0, \dots, n.$$

For a given $\varepsilon > 0$ choose $A_i \subset E_i$ closed with $\mu(E_i \setminus A_i) \leq \varepsilon$. Since the A_i are disjoint and f is continuous, there exist disjoint open sets U_i with

$$A_i \subset U_i \quad \text{and} \quad \inf_{U_i} f \geq \frac{i}{n} - \varepsilon.$$

As $\mu(A_i) \leq \lambda(U_i)$, it follows that

$$\int_S f \, d\mu \leq \sum_i \frac{i+1}{n} \mu(E_i) \leq \frac{1}{n} \mu(S) + \sum_i \frac{i}{n} \mu(E_i)$$

$$\leq \frac{\mu(S)}{n} + n\varepsilon + \sum_i \frac{i}{n} \lambda(U_i)$$

$$\leq \underbrace{\frac{\mu(S)}{n}}_{\to\, 0 \text{ as } n \to \infty} + \underbrace{n\varepsilon + \varepsilon\lambda(S)}_{\substack{\to\, 0 \text{ as } \varepsilon \to 0 \\ \text{for any } n}} + \int_S f \, d\lambda.$$

Replacing f by $1 - f$ yields, on noting that $\mu(S) = \lambda(S)$, that

$$\mu(S) - \int_S f \, d\mu = \int_S (1 - f) \, d\mu \leq \int_S (1 - f) \, d\lambda = \lambda(S) - \int_S f \, d\lambda,$$

and hence the desired result. $\qquad\square$

A6.4 Corollary. For $\lambda \in ba(S)$ there exists a $\nu \in rba(S)$ such that

$$\int_S f \, d\lambda = \int_S f \, d\nu \quad \text{for all } f \in C^0(S).$$

Proof. Since we can split λ into a real and an imaginary part, we may assume without loss of generality that λ is real-valued. Let $\lambda = \lambda^+ - \lambda^-$ be the Jordan decomposition of λ and let $\mu^\pm$ be the measures from A6.3 corresponding to $\lambda^\pm$. Set $\nu := \mu^+ - \mu^-$. It obviously holds that $|\nu| \leq \mu^+ + \mu^-$, and so the regularity of $\mu^\pm$ implies that ν is regular. $\qquad\square$

A6.5 Lemma (Alexandrov). If $S \subset \mathbb{R}^n$ is compact, then

$$\nu \in rba(S) \quad \Longrightarrow \quad \nu \text{ is } \sigma\text{-additive (on } \mathcal{B}_0 \text{ !).}$$

Proof (Compare A3.3). Let $E_i \in \mathcal{B}_0$, $i \in \mathbb{N}$, be disjoint and let $E := \bigcup_i E_i \in \mathcal{B}_0$. As ν is regular, we can choose for $\varepsilon > 0$ a closed set A with $A \subset E$ and $|\nu|(E \setminus A) \leq \varepsilon$ and open sets U_i with $E_i \subset U_i$ and $|\nu|(U_i \setminus E_i) \leq \varepsilon 2^{-i}$. On noting that $(U_i)_{i \in \mathbb{N}}$ is a cover of A with A being compact, we see that

$$A \subset \bigcup_{i=1}^m U_i \quad \text{for an } m,$$

and hence, since $|\nu|$ is nonnegative and additive (see 6.10), that

$$|\nu|(E) \leq \varepsilon + |\nu|(A) \leq \varepsilon + \sum_{i=1}^m |\nu|(U_i) \leq \varepsilon + \varepsilon \sum_{i=1}^\infty 2^{-i} + \sum_{i=1}^\infty |\nu|(E_i).$$

In addition, for all m

$$|\nu|(E) \geq |\nu|\Big(\bigcup_{i=1}^m E_i\Big) = \sum_{i=1}^m |\nu|(E_i),$$

which proves that

$$|\nu|(E) = \sum_{i=1}^\infty |\nu|(E_i).$$

Similarly, for all m

$$|\nu|\Big(\bigcup_{i>m} E_i\Big) = \sum_{i>m} |\nu|(E_i) \longrightarrow 0 \quad \text{as } m \to \infty.$$

We conclude that

$$\left| \nu(E) - \sum_{i=1}^{m} \nu(E_i) \right| = \left| \nu\left(E \setminus \bigcup_{i \leq m} E_i \right) \right|$$

$$= \left| \nu\left(\bigcup_{i>m} E_i \right) \right| \leq |\nu|\left(\bigcup_{i>m} E_i \right) \longrightarrow 0 \quad \text{as } m \to \infty.$$

$\square$

A6.6 Lemma. Let $S \subset \mathbb{R}^n$ be compact. For $\lambda \in ba(S)$ there exists a $\nu \in rca(S)$ with

$$\int_S f \, d\nu = \int_S f \, d\lambda \quad \text{for all } f \in C^0(S).$$

Proof. We may assume without loss of generality that λ is real-valued and nonnegative (see the proof of A6.4). Let $\mu \in rba(S)$ be the measure corresponding to λ as in A6.3. It follows from lemma A6.5 that μ is σ-additive on $\mathcal{B}_0$. Then by A3.15 there exists an extension of $(\mathcal{B}_0, \mu)$ to $(\mathcal{B}, \bar{\mu})$ with a σ-algebra $\mathcal{B}$ and a σ-additive measure $\bar{\mu}$ on $\mathcal{B}$. As $\mathcal{B}_1$ is the smallest σ-algebra that contains $\mathcal{B}_0$, it follows that $\mathcal{B}_1 \subset \mathcal{B}$. Hence $\bar{\mu}$ is σ-additive on $\mathcal{B}_1$.

We now show that $\bar{\mu}$ is also regular. To this end, let

$$\mathcal{M} := \{ E \in \mathcal{B}_1 \,; \text{ For } \varepsilon > 0 \text{ there exist sets } A \text{ and } U \text{ with}$$
$$A \subset E \subset U, \ A \text{ closed}, \ U \text{ open}, \ \bar{\mu}(U \setminus A) \leq \varepsilon \}.$$

Clearly $\mathcal{M}$ is an algebra, and since $\bar{\mu}$ is an extension of μ, it holds that $\mathcal{B}_0 \subset \mathcal{M}$. Then it follows that $\mathcal{M} = \mathcal{B}_1$, if we can show that

$$E_i \in \mathcal{M} \text{ for } i \in \mathbb{N} \text{ with } E_i \subset E_{i+1} \implies E := \bigcup_{i \in \mathbb{N}} E_i \in \mathcal{M}.$$

To this end, choose a closed set A_i with $A_i \subset E_i$ and an open set U_i with $E_i \subset U_i$ such that $\bar{\mu}(U_i \setminus A_i) \leq \varepsilon 2^{-i}$. Then

$$\bigcup_{i \leq m} A_i \subset E \subset \bigcup_{i \in \mathbb{N}} U_i =: U$$

and

$$\bar{\mu}\left(U \setminus \bigcup_{i \leq m} A_i \right) \leq \bar{\mu}\left(U \setminus \bigcup_{i \leq m} U_i \right) + \bar{\mu}\left(\bigcup_{i \leq m} U_i \setminus \bigcup_{i \leq m} A_i \right).$$

The first term is smaller than ε, if we choose m sufficiently large, and the second term is

$$\leq \bar{\mu}\left(\bigcup_{i \leq m} (U_i \setminus A_i) \right) \leq \sum_{i \leq m} \bar{\mu}(U_i \setminus A_i) \leq \varepsilon.$$

The integral identity follows as in the proof of A6.3. $\square$

We present the following result on measurable functions. Here S can be replaced with any compact topological space.

A6.7 Luzin's theorem. Let $S \subset \mathbb{R}^n$ be compact, $\mu \in rca(S)$ be nonnegative, and Y be a Banach space. Then every μ-measurable function $f : S \to Y$ is **μ-almost continuous**, i.e. for every μ-measurable set E and every $\varepsilon > 0$ there exists a compact set $K \subset E$ with $\mu(E \setminus K) \le \varepsilon$ such that $f\big|_K$ is a continuous function on K.

Proof. First we recall that for every μ-measurable set E there exist an $\widetilde{E} \in \mathcal{B}_1$ and a μ-null set N with $E \setminus N = \widetilde{E} \setminus N$ (see A3.14(2)). Moreover, for every μ-null set N and every $\varepsilon > 0$ there exists an $N_\varepsilon \in \mathcal{B}_1$ with $N \subset N_\varepsilon$ and $\mu(N_\varepsilon) \le \varepsilon$ (see A3.4). As μ is regular, there exist a compact set $\widetilde{K} \subset \widetilde{E}$ and an open set $\widetilde{U} \supset \widetilde{E}$ with $\mu(\widetilde{U} \setminus \widetilde{K}) \le \varepsilon$, as well as an open set $V \supset N_\varepsilon$ with $\mu(V) \le 2\varepsilon$. Then $K := \widetilde{K} \setminus V \subset E$ is compact and $U := \widetilde{U} \cup V \supset E$ is open with $\mu(U \setminus K) \le 3\varepsilon$.

There exists a μ-null set N such that $f(S \setminus N)$ is separable (see 3.11(2)). Choose a countable dense subset $\{y_j \; ; \; j \in \mathbb{N}\}$ of $f(S \setminus N)$. For every i it holds that the sets $\mathrm{B}_{\frac{1}{i}}(y_j)$, $j \in \mathbb{N}$, form a cover of $f(S \setminus N)$, and hence also

$$B_{ij} := \mathrm{B}_{\frac{1}{i}}(y_j) \setminus \bigcup_{k < j} \mathrm{B}_{\frac{1}{i}}(y_k) \; .$$

This implies that

$$E_{ij} := E \cap f^{-1}(B_{ij}) \setminus N \quad \text{for } j \in \mathbb{N}$$

form a disjoint partitioning of $E \setminus N$ into μ-measurable sets. It follows from the remark at the beginning of the proof that there exist compact sets $K_{ij} \subset E_{ij}$ with $\mu(E_{ij} \setminus K_{ij}) \le \varepsilon 2^{-i-j-1}$. Consequently, $\mu\big(E \setminus \bigcup_j K_{ij}\big) \le \varepsilon 2^{-i-1}$, and hence there exists a j_i with

$$\mu(E \setminus K_i) \le \varepsilon 2^{-i}, \quad \text{where} \quad K_i := \bigcup_{j \le j_i} K_{ij} \, .$$

K_i is a compact subset of $E \setminus N$, and by construction it is the disjoint union of the compact sets K_{ij} for $j \le j_i$. Hence

$$g_i(x) := y_j \quad \text{for } x \in K_{ij} \text{ (if } K_{ij} \ne \emptyset)$$

defines a $g_i \in C^0(K_i; Y)$ with

$$\sup_{x \in K_i} \| g_i(x) - f(x) \|_Y \le \frac{1}{i} \, .$$

Set $K := \bigcap_i K_i$. Then the functions $g_i\big|_K \in C^0(K; Y)$, and on K they converge uniformly to f as $i \to \infty$, which yields that $f\big|_K \in C^0(K; Y)$. In addition, K is a compact subset of E and

$$\mu(E \setminus K) \le \sum_{i \in \mathbb{N}} \mu(E \setminus K_i) \le \varepsilon \, .$$

$\square$

We add now a functional analysis formulation of Fubini's theorem, where we restrict ourselves to the case of bounded regular measures.

A6.8 Product measure. Let $S^l \subset \mathbb{R}^{n_l}$ be compact, $l = 1, 2$, and let $(S^l, \mathcal{B}^l, \mu^l)$ be measure spaces. Let $\mathcal{B}^l$ contain the Borel sets of S^l and let $\mu^l \in rca(S^l)$. Define

$$\mathcal{B}^1 \times \mathcal{B}^2 := \{E^1 \times E^2 \,;\, E^1 \in \mathcal{B}^1 \text{ and } E^2 \in \mathcal{B}^2\},$$

$$(\mu^1 \times \mu^2)(E^1 \times E^2) := \mu^1(E^1) \cdot \mu^2(E^2) \quad \text{for } E^1 \times E^2 \in \mathcal{B}^1 \times \mathcal{B}^2.$$

Denote by $\mathcal{B}^0$ the Boolean algebra induced by $\mathcal{B}^1 \times \mathcal{B}^2$. Then $\mathcal{B}^0$ consists of finite disjoint unions of sets in $\mathcal{B}^1 \times \mathcal{B}^2$, and $\mu^1 \times \mu^2$ can be canonically extended to an additive measure on $\mathcal{B}^0$.

Proposition: $\mu^1 \times \mu^2$ is σ-subadditive on $\mathcal{B}^0$, so that all the properties in A3.1 are satisfied.

Proof of proposition. Let $E, E_i \in \mathcal{B}^0$, $i \in \mathbb{N}$, with $E \subset \bigcup_{i \in \mathbb{N}} E_i$. We have to show that for $\mu := \mu^1 \times \mu^2$ it holds that

$$\mu(E) \leq \sum_{i \in \mathbb{N}} \mu(E_i).$$

By the definitions of $\mathcal{B}^0$ and μ, we may assume that

$$E_i = E_i^1 \times E_i^2 \in \mathcal{B}^1 \times \mathcal{B}^2.$$

As the μ^l are regular, it follows that for $\varepsilon > 0$ there exist open sets $U_i^l \in \mathcal{B}^l$ with (see the beginning of the proof in A6.7)

$$E_i^l \subset U_i^l \quad \text{and} \quad \mu^l(U_i^l \setminus E_i^l) \leq \varepsilon 2^{-i}.$$

Then

$$\mu(U_i^1 \times U_i^2) \leq \mu(E_i^1 \times E_i^2) + \mu((U_i^1 \setminus E_i^1) \times U_i^2) + \mu(E_i^1 \times (U_i^2 \setminus E_i^2))$$
$$\leq \mu(E_i^1 \times E_i^2) + \mu^1(U_i^1 \setminus E_i^1)\mu^2(S^2) + \mu^1(S^1)\mu^2(U_i^2 \setminus E_i^2)$$
$$\leq \mu(E_i^1 \times E_i^2) + C2^{-i}\varepsilon \quad \text{with } C := \mu^1(S^1) + \mu^2(S^2).$$

Similarly, there exists a compact set $K \in \mathcal{B}^0$ with

$$K \subset E \quad \text{and} \quad \mu(E) \leq \mu(K) + \varepsilon.$$

(E is the disjoint union of elements in $\mathcal{B}^1 \times \mathcal{B}^2$, and each of these subsets can be approximated in measure by compact subsets to an arbitrary accuracy. K is then the disjoint union of Cartesian products of compact sets.) Since the sets $U_1^l \times U_2^l$ form a cover of the set K, there exists an m with

$$K \subset \bigcup_{i=1}^{m} U_i^1 \times U_i^2,$$

and hence

$$\mu(E) \leq \mu(K) + \varepsilon \leq \sum_{i=1}^{m} \mu(U_i^1 \times U_i^2) + \varepsilon \leq \sum_{i=1}^{m} \mu(E_i^1 \times E_i^2) + (C+1)\varepsilon .$$

□

Therefore the Lebesgue integral for $(S^1 \times S^2, \mathcal{B}^0, \mu^1 \times \mu^2)$ can be constructed as in Appendix A3. In particular, there exists a measure extension to a measure space $(S^1 \times S^2, \mathcal{B}, \mu^1 \times \mu^2)$. We now characterize the Lebesgue space $L^p(\mu^1 \times \mu^2; Y)$ with the help of iterated integration. But first we consider the following special case:

A6.9 Lemma. If N is a $\mu^1 \times \mu^2$-null set, then for μ^1-almost all $x_1 \in S^1$

$$\{x_2 \in S^2 ;\ (x_1, x_2) \in N\}$$

is a μ^2-null set.

Proof. It follows from the definition of null sets in A3.4 that for $\varepsilon > 0$ there exist sets $E_i^l \in \mathcal{B}^l$, $i \in \mathbb{N}$, $l = 1, 2$ with

$$N \subset \bigcup_{i \in \mathbb{N}} E_i^1 \times E_i^2 \quad \text{and} \quad \sum_{i \in \mathbb{N}} \mu(E_i^1 \times E_i^2) \leq \varepsilon ,$$

where $\mu := \mu^1 \times \mu^2$. Consider the functions

$$g_{\varepsilon n}(x_1)(x_2) := \sum_{i \leq n} \mathcal{X}_{E_i^1}(x_1) \mathcal{X}_{E_i^2}(x_2) .$$

For all x_1 we have that $g_{\varepsilon n}(x_1) \in L^1(\mu^2)$ satisfying the following equation

$$G_{\varepsilon n}(x_1) := \int_{S^2} g_{\varepsilon n}(x_1) \, d\mu^2 = \sum_{i \leq n} \mathcal{X}_{E_i^1}(x_1) \mu^2(E_i^2) .$$

The function $G_{\varepsilon n} \in L^1(\mu^1)$ with

$$\int_{S^1} G_{\varepsilon n} \, d\mu^1 = \sum_{i \leq n} \mu^1(E_i^1) \mu^2(E_i^2) \leq \varepsilon .$$

On noting that

$$G_{\varepsilon n}(x_1) \nearrow G_{\varepsilon}(x_1) := \sum_{i \in \mathbb{N}} \mathcal{X}_{E_i^1}(x_1) \mu^2(E_i^2)$$

as $n \nearrow \infty$, it follows from the monotone convergence theorem (see A3.12(3)) that $G_{\varepsilon} \in L^1(\mu^1)$ with

$$\int_{S^1} G_\varepsilon \, d\mu^1 = \lim_{n \to \infty} \int_{S^1} G_{\varepsilon n} \, d\mu^1 \le \varepsilon \,.$$

But this means that $G_\varepsilon \to 0$ in $L^1(\mu^1)$ as $\varepsilon \to 0$. Hence there exists a subsequence $\varepsilon \to 0$ such that $G_\varepsilon(x_1) \to 0$ for μ^1-almost all $x_1 \in S^1$. In the following we consider such x_1. On noting that for small ε and as $n \nearrow \infty$ we have that

$$\int_{S^2} g_{\varepsilon n}(x_1) \, d\mu^2 = G_{\varepsilon n}(x_1) \nearrow G_\varepsilon(x_1) < \infty$$

and

$$g_{\varepsilon n}(x_1)(x_2) \nearrow g_\varepsilon(x_1)(x_2) := \sum_{i \in \mathbb{N}} \mathcal{X}_{E_i^1}(x_1) \mathcal{X}_{E_i^2}(x_2) \,,$$

it follows once again from the monotone convergence theorem that the function $g_\varepsilon(x_1) \in L^1(\mu^2)$ satisfies

$$\int_{S^2} g_\varepsilon(x_1) \, d\mu^2 = G_\varepsilon(x_1) \,.$$

Therefore, $g_\varepsilon(x_1) \to 0$ in $L^1(\mu^2)$, and so there exists a subsequence $\varepsilon \to 0$ (depending on x_1!) with $g_\varepsilon(x_1)(x_2) \to 0$ for μ^2-almost all $x_2 \in S^2$. But noting that $g_\varepsilon(x_1)(x_2) \ge \mathcal{X}_N(x_1, x_2)$ implies that $\mathcal{X}_N(x_1, x_2) = 0$ for μ^2-almost all $x_2 \in S^2$. □

A6.10 Fubini's theorem. Let Y be a Banach space and let $1 \le p < \infty$. Consider the product measure in A6.8. Then

$$(Jf)(x_1)(x_2) := f(x_1, x_2)$$

defines a linear isometric isomorphism

$$J : L^p(\mu^1 \times \mu^2; Y) \longrightarrow L^p(\mu^1; L^p(\mu^2; Y)) \,.$$

In particular, for $f \in L^p(\mu^1 \times \mu^2; Y)$ there exists

$$F(x_1) := \int_{S^2} f(x_1, x_2) \, d\mu^2(x_2) \quad \text{for } \mu^1\text{-almost all } x_1 \in S^1$$

and $F \in L^p(\mu^1; Y)$ with

$$\int_{S^1} F(x_1) \, d\mu^1(x_1) = \int_{S^1 \times S^2} f(x_1, x_2) \, d(\mu^1 \times \mu^2)(x_1, x_2) \,.$$

A symmetry argument then yields that

$$\int_{S_1} \left(\int_{S_2} f(x_1, x_2) \, d\mu^2(x_2) \right) d\mu^1(x_1) = \int_{S_2} \left(\int_{S_1} f(x_1, x_2) \, d\mu^1(x_1) \right) d\mu^2(x_2) \,.$$

Proof. Let $f \in L^p(\mu^1 \times \mu^2; Y)$ (we suppress in the following proof the argument Y). Since $p < \infty$, it follows from the construction of the Lebesgue integral (see the proof of 3.26(1)) that f can be approximated by step functions

$$f_k = \sum_{i=1}^{n} \mathcal{X}_{E_i} \alpha_i \quad \text{with } E_i \in \mathcal{B}^0 \text{ and } \alpha_i \in Y,$$

where n, E_i and α_i depend on k. The definition of $\mathcal{B}^0$ then yields that f_k can also be represented as

$$f_k = \sum_{i,j=1}^{n} \mathcal{X}_{E_i^1 \times E_j^2} \alpha_{ij} \quad \text{with } E_i^1 \in \mathcal{B}^1, \ E_j^2 \in \mathcal{B}^2, \ \alpha_{ij} \in Y$$

with a new n, where both the E_i^1 and the E_j^2 are disjoint. Then for all x_1

$$(Jf_k)(x_1) = \sum_{i,j=1}^{n} \mathcal{X}_{E_i^1}(x_1) \mathcal{X}_{E_j^2} \alpha_{ij} \in L^p(\mu^2),$$

and $Jf_k \in L^p(\mu^1; L^p(\mu^2))$, with

$$\int_{S^1} \|Jf_k\|_{L^p(\mu^2)}^p \, d\mu^1 = \sum_{i=1}^{n} \mu^1(E_i^1) \left\| \sum_{j=1}^{n} \mathcal{X}_{E_j^2} \alpha_{ij} \right\|_{L^p(\mu^2)}^p$$

$$= \sum_{i,j=1}^{n} \mu^1(E_i^1) \mu^2(E_j^2) \|\alpha_{ij}\|_Y^p = \int_{S^1 \times S^2} \|f_k\|_Y^p \, d\mu,$$

where $\mu := \mu^1 \times \mu^2$. Similarly, we observe that

$$\int_{S^2} (Jf_k)(x_1) \, d\mu^2$$

as a function of x_1 lies in $L^1(\mu^1)$ and satisfies

$$\int_{S^1} \left(\int_{S^2} (Jf_k)(x_1) \, d\mu^2 \right) d\mu^1(x_1) = \int_{S^1 \times S^2} f_k \, d\mu.$$

These properties, which we have derived for f_k, are of course also valid for the step functions $f_k - f_l$. Therefore,

$$\|Jf_k - Jf_l\|_{L^p(\mu^1; L^p(\mu^2))} = \|f_k - f_l\|_{L^p(\mu)} \to 0 \quad \text{as } k, l \to \infty.$$

By completeness of $L^p(\mu^1; L^p(\mu^2))$, there exists an F such that

$$Jf_k \to F \quad \text{in } L^p(\mu^1; L^p(\mu^2)) \text{ as } k \to \infty.$$

Hence there exists a subsequence such that for μ^1-almost all x_1

$$Jf_k(x_1) \to F(x_1) \quad \text{in } L^p(\mu^2).$$

On the other hand, since $f_k \to f$ in $L^p(\mu)$, $\mu = \mu^1 \times \mu^2$, there exists a subsequence such that

$$f_k(x_1, x_2) \to f(x_1, x_2) \quad \text{for } \mu\text{-almost all } (x_1, x_2).$$

It follows from A6.9 that then for μ^1-almost all x_1

$$f_k(x_1, x_2) \to f(x_1, x_2) \quad \text{for } \mu^2\text{-almost all } x_2.$$

On recalling that $f_k(x_1, x_2) = Jf_k(x_1)(x_2)$, we then obtain that

$$F(x_1) = f(x_1, \cdot) \quad \text{in } L^p(\mu^2)$$

for μ^1-almost all x_1, i.e. $F = Jf$. In addition, it follows from the above that

$$\|Jf\|_{L^p(\mu^1; L^p(\mu^2))} = \|f\|_{L^p(\mu)}.$$

This shows that J is well defined and isometric. Consequently the image of J is closed. Hence, in order to show the surjectivity, it is sufficient to show that the image is dense. Every element in $L^p(\mu^1; L^p(\mu^2))$ can be approximated by linear combinations of functions $\mathcal{X}_{E^1} g$ with $E^1 \in \mathcal{B}^1$ and $g \in L^p(\mu^2)$, and similarly g can be approximated by linear combinations of $\mathcal{X}_{E^2} \alpha$ with $E^2 \in \mathcal{B}^2$ and $\alpha \in Y$. But functions $F(x_1)(x_2) = \mathcal{X}_{E^1}(x_1) \mathcal{X}_{E^2}(x_2) \alpha$ in $L^p(\mu^1; L^p(\mu^2))$ clearly lie in the image of J.

In order to prove the integral formula, we exploit the fact that the integral with respect to μ^2 is a linear continuous map from $L^1(\mu^2)$ to Y. If $f \in L^1(\mu)$, then $Jf \in L^1(\mu^1; L^1(\mu^2))$, and hence (see theorem 5.11)

$$x_1 \longmapsto \int_{S^2} Jf(x_1) \, d\mu^2$$

is a function in $L^1(\mu^1)$. On noting that in addition $Jf_k \to Jf$ in $L^1(\mu^1; L^1(\mu^2))$ as $k \to \infty$, if the f_k are chosen as above, we obtain with the help of 5.11 that as $k \to \infty$

$$\int_{S^1} \left(\int_{S^2} Jf(x_1) \, d\mu^2 \right) d\mu^1(x_1) = \int_{S^2} \left(\int_{S^1} Jf \, d\mu^1 \right) d\mu^2$$
$$\longleftarrow \int_{S^2} \left(\int_{S^1} Jf_k \, d\mu^1 \right) d\mu^2 = \int_{S^1} \left(\int_{S^2} Jf_k(x_1) \, d\mu^2 \right) d\mu^1(x_1)$$
$$= \int_{S^1 \times S^2} f_k \, d\mu \longrightarrow \int_{S^1 \times S^2} f \, d\mu.$$

$\square$

A6.11 Remark on the case $p = \infty$. With the above assumptions, let $f \in L^\infty(\mu^1 \times \mu^2; Y)$. Then $f \in L^q(\mu^1 \times \mu^2; Y)$ for every $1 \leq q < \infty$, so that the result shown in A6.10 yields that

$$Jf \in \bigcap_{1 \leq q < \infty} L^q(\mu^1; L^q(\mu^2; Y)).$$

Moreover, it follows easily from E3.4 and A6.9 that

$$\|f\|_{L^\infty(\mu^1 \times \mu^2)} = \|g\|_{L^\infty(\mu^1)},$$

where $g(x_1) := \|f(x_1, \cdot)\|_{L^\infty(\mu^2)} = \|(Jf)(x_1)\|_{L^\infty(\mu^2)}$.

However, in general Jf is not (!) an element of $L^\infty(\mu^1; L^\infty(\mu^2; Y))$, as can be seen from the example $\mu^1 = \mu^2 = \mathrm{L}^1 \llcorner [0,1]$, $Y = \mathbb{R}$, $f = \mathcal{X}_E$, $E := \{(x_1, x_2) ; \; x_1 < x_2\}$. In this case the function

$$x_1 \mapsto \mathcal{X}_{[x_1, 1]} \in L^\infty(\mu^2; Y)$$

is not μ^1-measurable.

7 Uniform boundedness principle

A fundamental result for linear continuous maps is the uniform boundedness principle. It states that the pointwise boundedness of a family of operators already implies their boundedness in the operator norm. This principle rests upon the following theorem.

7.1 Baire's category theorem. Let X be a nonempty complete metric space and let

$$X = \bigcup_{k \in \mathbb{N}} A_k, \quad \text{with closed sets } A_k \subset X.$$

Then there exists a $k_0 \in \mathbb{N}$ with $\mathring{A}_{k_0} \neq \emptyset$.
Remark: Recall that $\mathring{A}_k = \text{intr}_X(A_k)$.

Proof. Assume that $\mathring{A}_k = \emptyset$ for all k. Then

$$U \subset X \text{ open, nonempty, } k \in \mathbb{N}$$
$$\implies \quad U \setminus A_k \text{ open, nonempty}$$
$$\implies \quad \text{there exists a ball } \overline{B_\varepsilon(x)} \subset U \setminus A_k \text{ with } \varepsilon \leq \tfrac{1}{k}.$$

Hence we can inductively choose balls $B_{\varepsilon_k}(x_k)$ such that

$$\overline{B_{\varepsilon_k}(x_k)} \subset B_{\varepsilon_{k-1}}(x_{k-1}) \setminus A_k \quad \text{and} \quad \varepsilon_k \leq \tfrac{1}{k}.$$

Consequently, we see that $x_l \in B_{\varepsilon_k}(x_k)$ for $l \geq k$ and $\varepsilon_k \to 0$ as $k \to \infty$ and the balls $B_{\varepsilon_k}(x_k)$ are nested, and we conclude that $(x_l)_{l \in \mathbb{N}}$ is a Cauchy sequence. Hence there exists the limit

$$x := \lim_{l \to \infty} x_l \in X$$

and $x \in \overline{B_{\varepsilon_k}(x_k)}$ for all k. Since $\overline{B_{\varepsilon_k}(x_k)} \cap A_k = \emptyset$, we have that

$$x \notin \bigcup_{k \in \mathbb{N}} A_k = X,$$

a contradiction. □

As is evident from the example $X = \mathbb{Q}$, the completeness assumption in 7.1 is essential. With the help of 7.1 we can now show the following:

7.2 Theorem (Uniform boundedness principle). Let X be a nonempty complete metric space and let Y be a normed space. Let $\mathcal{F} \subset C^0(X;Y)$ be a set of functions with

$$\sup_{f \in \mathcal{F}} \|f(x)\|_Y < \infty \quad \text{for every } x \in X. \tag{7-5}$$

Then there exist an $x_0 \in X$ and an $\varepsilon_0 > 0$, such that

$$\sup_{x \in B_{\varepsilon_0}(x_0)} \sup_{f \in \mathcal{F}} \|f(x)\|_Y < \infty. \tag{7-6}$$

Proof. For $f \in \mathcal{F}$ and $k \in \mathbb{N}$ it holds that $\{x \in X; \|f(x)\|_Y \leq k\}$ is a closed set. Hence the sets

$$\begin{aligned} A_k &:= \bigcap_{f \in \mathcal{F}} \{x \in X; \|f(x)\|_Y \leq k\} \\ &= \{x \in X; \|f(x)\|_Y \leq k \text{ for all } f \in \mathcal{F}\}, \end{aligned}$$

being intersections of closed sets, are closed, and it follows from (7-5) that they form a cover of X. Then theorem 7.1 yields that $\mathring{A}_{k_0} \neq \emptyset$ for some k_0, and hence there exists a $\overline{B_{\varepsilon_0}(x_0)} \subset A_{k_0}$. Noting that

$$\sup_{x \in A_{k_0}} \sup_{f \in \mathcal{F}} \|f(x)\|_Y \leq k_0$$

yields the desired result. □

For linear continuous maps 7.2 is reformulated as the

7.3 Banach-Steinhaus theorem. Let X be a Banach space and let Y be a normed space. Suppose $\mathcal{T} \subset \mathcal{L}(X;Y)$ with

$$\sup_{T \in \mathcal{T}} \|Tx\|_Y < \infty \quad \text{for every } x \in X.$$

Then $\mathcal{T}$ is a bounded set in $\mathcal{L}(X;Y)$, i.e.

$$\sup_{T \in \mathcal{T}} \|T\|_{\mathcal{L}(X;Y)} < \infty.$$

Proof. Setting
$$f_T(x) := \|Tx\|_Y \quad \text{for } T \in \mathcal{T}, \ x \in X$$

defines functions $f_T \in C^0(X;\mathbb{R})$, and $\mathcal{F} := \{f_T \ ; \ T \in \mathcal{T}\}$ satisfies the assumptions in 7.2. Hence, by the conclusions of 7.2, there exist an $x_0 \in X$, an $\varepsilon_0 > 0$, and a constant $C < \infty$ with

$$\|Tx\|_Y \le C \quad \text{for } T \in \mathcal{T} \text{ and } \|x - x_0\|_X \le \varepsilon_0.$$

Then it follows for all $T \in \mathcal{T}$ and all $x \neq 0$ that

$$\|Tx\|_Y = \frac{\|x\|_X}{\varepsilon_0} \left\| T\left(x_0 + \varepsilon_0 \frac{x}{\|x\|_X} \right) - T(x_0) \right\|_Y \le \frac{\|x\|_X}{\varepsilon_0} \cdot 2C,$$

whence $\|T\|_{\mathscr{L}(X;Y)} \le \frac{2C}{\varepsilon_0}$. $\qquad\qquad\square$

In the following theorem, we prove that an even weaker assumption on the operators is sufficient. Here we will make use of the following

7.4 Notation. Let X be a normed space. From now on, for $x \in X$ and $x' \in X'$, we will use the notation

$$\langle x, x' \rangle_X \text{ (or simply: } \langle x, x' \rangle) := x'(x)$$

and call this the **duality product** (or **duality pairing**) of the space X. This notation is motivated by the Hilbert space case. Indeed, if X is a Hilbert space and $R_X : X \to X'$ is the isomorphism from the Riesz representation theorem 6.1, then

$$(x, y)_X = \langle x, R_X y \rangle_X \qquad \text{for all } x, y \in X,$$
$$\langle x, x' \rangle_X = \left(x, R_X^{-1} x' \right)_X \qquad \text{for all } x \in X, \ x' \in X'.$$

Especially when applied to function spaces, the notation introduced here proves to be justified.

Notice: In $\langle \cdot_1, \cdot_2 \rangle_X$ the second variable is the linear map, and the first variable is the argument of this map. This is consistent with the fact that in the weak formulation of differential equations (see e.g. the equation (6-9)), the test function appears on the left.

7.5 Theorem. Let X be a Banach space and let Y be a normed space. In addition, let $\mathcal{T} \subset \mathscr{L}(X;Y)$ such that for all $x \in X$ and $y' \in Y'$

$$\sup_{T \in \mathcal{T}} |\langle Tx, y' \rangle_Y| < \infty.$$

Then $\mathcal{T}$ is a bounded subset of $\mathscr{L}(X;Y)$.

Proof. For $x \in X$ and $T \in \mathcal{T}$ it follows from 6.17(3) that

$$S_{x,T}(y') := \langle Tx, y' \rangle_Y \qquad \text{for } y' \in Y'$$

defines an element $S_{x,T} \in (Y')' = \mathscr{L}(Y'; \mathbb{K})$ with $\|S_{x,T}\|_{(Y')'} = \|Tx\|_Y$. Now

$$\sup_{T \in \mathcal{T}} |S_{x,T}(y')| = \sup_{T \in \mathcal{T}} |\langle Tx, y' \rangle_Y| < \infty \qquad \text{for all } y' \in Y'.$$

Moreover, it follows from 5.3(2) that Y' is a Banach space. Hence we can apply the Banach-Steinhaus theorem 7.3 to the set $\{ S_{x,T} \in \mathscr{L}(Y'; \mathbb{K}) \, ; \, T \in \mathcal{T} \}$ and obtain that

$$\sup_{T \in \mathcal{T}} \|Tx\|_Y = \sup_{T \in \mathcal{T}} \|S_{x,T}\|_{(Y')'} < \infty$$

for every $x \in X$. Thus the desired result follows from theorem 7.3.   □

A further consequence of 7.1 is the open mapping theorem. To this end, we introduce the following

7.6 Definition. Let X and Y be topological spaces. Then $f : X \to Y$ is called **open** if

$$U \text{ is open in } X \quad \Longrightarrow \quad f(U) \text{ is open in } Y.$$

If f is bijective, then f is open if and only if f^{-1} is continuous. If X, Y are normed spaces and $T : X \to Y$ is linear, then

$$T \text{ is open} \quad \Longleftrightarrow \quad \text{There exists a } \delta > 0 \text{ with } B_\delta(0) \subset T(B_1(0)).$$

Proof $\Leftarrow$. Let U be open and let $x \in U$. Choose an $\varepsilon > 0$ with $B_\varepsilon(x) \subset U$. Now $B_\delta(0) \subset T(B_1(0))$ implies that $B_{\varepsilon\delta}(Tx) \subset T(B_\varepsilon(x)) \subset T(U)$, and hence $T(U)$ is open.   □

7.7 Open mapping theorem. Let X and Y be Banach spaces. Then it holds for every operator $T \in \mathscr{L}(X;Y)$ that

$$T \text{ is surjective} \quad \Longleftrightarrow \quad T \text{ is open.}$$

Proof $\Rightarrow$. Since T is surjective,

$$Y = \bigcup_{k \in \mathbb{N}} \overline{T(B_k(0))}.$$

It follows from the Baire category theorem 7.1 that there exist a k_0 and a ball $B_{\varepsilon_0}(y_0)$ in Y with

$$B_{\varepsilon_0}(y_0) \subset \overline{T(B_{k_0}(0))}.$$

This means that for $y \in B_{\varepsilon_0}(0)$ there exist points $x_i \in B_{k_0}(0)$ with $Tx_i \to y_0 + y$ as $i \to \infty$. On choosing an $x_0 \in X$ with $Tx_0 = y_0$, this implies that

$$T\left(\frac{x_i - x_0}{k_0 + \|x_0\|}\right) \longrightarrow \frac{y}{k_0 + \|x_0\|} \quad \text{and} \quad \left\|\frac{x_i - x_0}{k_0 + \|x_0\|}\right\| < 1,$$

which proves that

$$B_\delta(0) \subset \overline{T(B_1(0))} \quad \text{with } \delta := \frac{\varepsilon_0}{k_0 + \|x_0\|}. \tag{7-7}$$

However, our aim is to show such an inclusion without the closure of the set on the right-hand side, for a smaller δ if necessary. To this end, we note that (7-7) implies that

$$y \in B_\delta(0) \quad \Longrightarrow \quad \text{there exists an } x \in B_1(0) \text{ with } y - Tx \in B_{\frac{\delta}{2}}(0)$$

$$\Longrightarrow \quad 2(y - Tx) \in B_\delta(0) \ .$$

Hence for $y \in B_\delta(0)$ we can inductively choose points $y_k \in B_\delta(0)$ and $x_k \in B_1(0)$ such that

$$y_0 = y \quad \text{and} \quad y_{k+1} = 2(y_k - Tx_k) \ .$$

Then

$$2^{-k-1} y_{k+1} = 2^{-k} y_k - T(2^{-k} x_k) \ ,$$

and so

$$T\left(\sum_{k=0}^{m} 2^{-k} x_k\right) = y - 2^{-m-1} y_{m+1} \longrightarrow y \quad \text{as } m \to \infty.$$

Since

$$\sum_{k=0}^{m} \|2^{-k} x_k\| < \sum_{k=0}^{m} 2^{-k} \leq 2 < \infty, \quad \text{we have that} \quad \left(\sum_{k=0}^{m} 2^{-k} x_k\right)_{m \in \mathbb{N}}$$

is a Cauchy sequence in X. As X is complete, there exists

$$x := \sum_{k=0}^{\infty} 2^{-k} x_k \quad \text{in } X \text{ with } \|x\| < 2.$$

The continuity of T then yields that

$$Tx = \lim_{m \to \infty} T\left(\sum_{k=0}^{m} 2^{-k} x_k\right) = y \ .$$

This shows that $B_\delta(0) \subset T(B_2(0))$, or equivalently $B_{\frac{\delta}{2}}(0) \subset T(B_1(0))$. Hence, by 7.6, T is open. $\qquad\square$

Proof $\Leftarrow$. The fact that $B_\delta(0) \subset T(B_1(0))$ for some $\delta > 0$ implies that $B_R(0) \subset T(B_{\frac{R}{\delta}}(0))$ for all $R > 0$. $\qquad\square$

As a consequence, we obtain the following results, the first of which is also called the **bounded inverse theorem.**

7.8 Inverse mapping theorem. If X and Y are Banach spaces and if $T \in \mathscr{L}(X; Y)$, then

$$T \text{ is bijective} \quad \Longrightarrow \quad T^{-1} \in \mathscr{L}(Y; X) \ .$$

Proof. T^{-1} is linear. It follows from 7.7 that T is open, and hence T^{-1} is continuous. $\qquad\square$

7.9 Closed graph theorem. Let X and Y be Banach spaces and let $T : X \to Y$ be linear. Then

$$\text{graph}(T) := \{(x, Tx) \in X \times Y \,;\, x \in X\}$$

is closed in $X \times Y$ if and only if $T \in \mathscr{L}(X;Y)$.

Proof $\Rightarrow$. In the formulation of the theorem, $X \times Y$ is considered as a Banach space, equipped with, for example, the norm $\|(x,y)\| := \|x\| + \|y\|$ (see E4.12). As a closed subspace $Z := \text{graph}(T)$ is a Banach space. Let

$$P_X(x,y) := x \quad \text{and} \quad P_Y(x,y) := y \quad \text{for } (x,y) \in Z.$$

P_X and P_Y are linear and continuous, and $P_X : Z \to X$ is bijective. It follows from the inverse mapping theorem 7.8 that $P_X^{-1} \in \mathscr{L}(X;Z)$. Hence $T = P_Y P_X^{-1} \in \mathscr{L}(X;Y)$. $\qquad\square$

Proof $\Leftarrow$. This follows immediately from the continuity of T. $\qquad\square$

E7 Exercises

E7.1 On the adjoint map. Let X, Y be Banach spaces, and let $A : X \to Y$ and $B : Y' \to X'$ be linear. If it holds for all $x \in X$ and $y' \in Y'$ that

$$\langle Ax, y'\rangle_Y = \langle x, By'\rangle_X \,,$$

then A and B are continuous.

Solution. For $x \in X$ it follows from 6.17(3) that $T_x y' := \langle Ax, y'\rangle_Y$ for $y' \in Y'$ defines a $T_x \in (Y')'$ with $\|T_x\| = \|Ax\|_Y$. Since for all $y' \in Y'$

$$\sup_{\|x\|_X \leq 1} |T_x y'| = \sup_{\|x\|_X \leq 1} |\langle x, By'\rangle_X| \leq \|By'\|_{X'} < \infty,$$

it follows from the Banach-Steinhaus theorem that

$$\sup_{\|x\|_X \leq 1} \|Ax\|_Y = \sup_{\|x\|_X \leq 1} \|T_x\| < \infty,$$

i.e. A is continuous. Moreover, since

$$|\langle x, By'\rangle_X| = |\langle Ax, y'\rangle_Y| \leq \|A\| \cdot \|x\|_X \cdot \|y'\|_{Y'},$$

we have that $\|By'\|_{X'} \leq \|A\| \cdot \|y'\|_{Y'}$, and hence also B is continuous. $\qquad\square$

E7.2 Pointwise convergence in $\mathscr{L}(X;Y)$. Let X be a Banach space, let Y be a normed space and let $(T_n)_{n \in \mathbb{N}}$ be a sequence in $\mathscr{L}(X;Y)$ such that

$$T(x) := \lim_{n \to \infty} T_n(x) \quad \text{exists for all } x \in X.$$

(1) Show that $T \in \mathscr{L}(X;Y)$ and $\|T\| \leq \liminf_{n \to \infty} \|T_n\| < \infty$.

(2) Give an example where $(T_n)_{n \in \mathbb{N}}$ does not converge to T in the operator norm.

Solution (1). Clearly T is linear. For all $x \in X$ we have that $(\|T_n(x)\|_Y)_{n \in \mathbb{N}}$ is a bounded sequence, and hence the Banach-Steinhaus theorem yields that $\{\|T_n\|; \ n \in \mathbb{N}\}$ is bounded. In addition,

$$\|T(x)\|_Y = \left\| \lim_{n \to \infty} T_n x \right\|_Y = \lim_{n \to \infty} \|T_n x\|_Y$$
$$\leq \liminf_{n \to \infty} (\|T_n\| \cdot \|x\|_X) = \left(\liminf_{n \to \infty} \|T_n\| \right) \cdot \|x\|_X .$$

$\square$

Solution (2). Let $X = Y = l^2(\mathbb{K})$ and let $T_n : l^2(\mathbb{K}) \to l^2(\mathbb{K})$ be defined by setting for $x = (x_i)_{i \in \mathbb{N}} \in l^2(\mathbb{K})$

$$(T_n x)_i = \begin{cases} x_i & \text{for } i \leq n, \\ 0 & \text{for } i > n. \end{cases}$$

Then

$$\|x - T_n x\|_{\ell^2} = \left(\sum_{i>n} |x_i|^2 \right)^{\frac{1}{2}} \to 0 \quad \text{as } n \to \infty,$$

but $\|\mathrm{Id} - T_n\| \geq 1$, because $\|e_{n+1} - T_n e_{n+1}\|_{\ell^2} = \|e_{n+1}\|_{\ell^2} = 1$ for unit vectors e_{n+1} as in 2.23. $\square$

E7.3 Equivalent norms. Let $\|\cdot\|_1$ and $\|\cdot\|_2$ be two norms on the $\mathbb{K}$-vector space X and let X be complete with respect to both of these norms. If $\|\cdot\|_2$ is stronger than $\|\cdot\|_1$, then the two norms are equivalent.

Solution. By 2.15, there exists a $C_2 > 0$ with $\|x\|_1 \leq C_2 \|x\|_2$ for all $x \in X$. Denote by X_k the Banach space X with respect to the norm $\|\cdot\|_k$, $k = 1, 2$. As $\|\cdot\|_2$ is stronger than $\|\cdot\|_1$, it holds that $\mathrm{Id} : X_2 \to X_1$ is continuous. It follows from the inverse mapping theorem 7.8 that $\mathrm{Id}^{-1} : X_1 \to X_2$ is continuous, i.e. there exists a $C_1 > 0$ with $\|x\|_2 \leq C_1 \|x\|_1$ for all $x \in X$. Hence the two norms are equivalent. $\square$

E7.4 Sesquilinear forms. Let X, Y be Banach spaces and let $b : X \times Y \to$ $\mathbb{K}$ be sesquilinear such that

$$x \mapsto b(x, y) \text{ is continuous for every } y \in Y,$$
$$y \mapsto b(x, y) \text{ is continuous for every } x \in X.$$

Then there exists a constant $0 \leq C < \infty$ such that

$$|b(x, y)| \leq C\|x\|_X \cdot \|y\|_Y \quad \text{for all } x \in X, \ y \in Y.$$

Solution. For $x \in X$ we have that $f_x(y) := b(x, y)$ defines an $f_x \in C^0(Y; \mathbb{K})$, and $\mathcal{F} := \{f_x ; \|x\|_X \leq 1\}$ satisfies the assumptions in 7.2, since it holds for $y \in Y$

$$\sup_{\|x\|_X \leq 1} |f_x(y)| = \sup_{\|x\|_X \leq 1} |b(x, y)| = \|b(\cdot, y)\|_{X'} < \infty$$

on noting that $b(\cdot, y) \in X'$ by 5.1. Hence it follows from 7.2 that there exist a $y_0 \in Y$, an $\varepsilon_0 > 0$ and a constant C such that

$$|b(x, y)| = |f_x(y)| \leq C \quad \text{for } \|x\|_X \leq 1, \ \|y - y_0\|_Y \leq \varepsilon_0.$$

Then for $\|x\|_X \leq 1$ and $\|y\|_Y \leq 1$ we have that

$$|b(x, y)| = \frac{1}{\varepsilon_0}|b(x, y_0 + \varepsilon_0 y) - b(x, y_0)| \leq \frac{2C}{\varepsilon_0},$$

which yields the desired result. (Compare the proof of 7.3.)   □

8 Weak convergence

In many cases the concept of convergence with respect to the norm turns out to be too restrictive. That is why in this chapter we will introduce a weaker notion of convergence which will enable us to solve minimum problems under far weaker assumptions.

In 4.3 we proved the projection theorem in Hilbert spaces and noted subsequently that the same result cannot be expected to hold in general Banach spaces. The difficulty lies in finding a convergent subsequence within a given minimal sequence, something that is in general not possible with respect to the norm convergence, as balls in infinite-dimensional spaces are not precompact (see 4.10). However, we will see (in 8.10) that closed balls are sequentially compact with respect to weak convergence, at least for the class of reflexive spaces (see 8.8). Here we lose the continuity of the norm, but we nonetheless retain its lower semicontinuity (see 8.3(4)). This property will play a crucial role in the proofs of the existence results 8.15 and 8.17. Hence the class of reflexive spaces, which lies between the class of Hilbert spaces and the class of general Banach spaces, plays a significant role in applications.

In this chapter all the spaces are assumed to be complete, except in 8.12-8.14. In the following, we will always use the notation $\langle x, x' \rangle_X := x'(x)$ for $x \in X$ and $x' \in X'$ from 7.4. We will also write $\langle x, x' \rangle := \langle x, x' \rangle_X$. This simple notation is used in the case when only one Banach space X is involved.

8.1 Definition (weak convergence). Let X be a Banach space.

(1) A sequence $(x_k)_{k \in \mathbb{N}}$ in X **converges weakly** to $x \in X$ (we write $x_k \to x$ weakly in X as $k \to \infty$, or $x_k \rightharpoonup x$ as $k \to \infty$) if for all $x' \in X'$

$$\langle x_k, x' \rangle_X \to \langle x, x' \rangle_X \quad \text{as } k \to \infty.$$

(2) A sequence $(x'_k)_{k \in \mathbb{N}}$ in X' **converges weakly*** to $x' \in X'$ (we write $x'_k \to x'$ weakly* in X' as $k \to \infty$, or $x'_k \overset{*}{\rightharpoonup} x'$ as $k \to \infty$) if for all $x \in X$

$$\langle x, x'_k \rangle_X \to \langle x, x' \rangle_X \quad \text{as } k \to \infty.$$

(3) Analogously to (1) and (2) we define weak and weak* Cauchy sequences.

(4) A set $M \subset X$ (X') is called **weakly sequentially compact** (**weakly* sequentially compact**) if every sequence in M contains a weakly (weakly*) convergent subsequence whose weak (weak*) limit lies in M.

Warning: It is possible to define a corresponding weak (weak*) topology (see 8.7). However, if X is not separable, this topology does not have a countable basis of neighbourhoods. It follows that "covering compact" and "sequentially compact" are not equivalent properties (see the example 8.7(4)).

Note: As a complement to weak convergence, convergence with respect to a norm, i.e. norm convergence, will also be referred to as **strong convergence**. This reduces confusion.

The weak convergence may be interpreted as weak* convergence in the bidual space:

8.2 Embedding into the bidual space.

(1) Defining

$$\langle x', J_X x \rangle_{X'} := \langle x, x' \rangle_X \quad \text{for } x \in X, \ x' \in X'$$

yields an isometric map $J_X \in \mathscr{L}(X; X'')$. Here

$$X'' := (X')' = \mathscr{L}(X'; \mathbb{K})$$

is the **bidual space** of X.

(2) Let $x_k, x \in X$ for $k \in \mathbb{N}$. Then:

$$
\begin{array}{ccc}
x_k \to x \text{ weakly} & & J_X x_k \to J_X x \text{ weakly*} \\
\text{in } X \text{ as } k \to \infty & \Longleftrightarrow & \text{in } X'' \text{ as } k \to \infty.
\end{array}
$$

(3) Let $x_k', x' \in X'$ for $k \in \mathbb{N}$. Then:

$$
\begin{array}{ccc}
x_k' \to x' \text{ weakly} & & x_k' \to x' \text{ weakly*} \\
\text{in } X' \text{ as } k \to \infty & \Longrightarrow & \text{in } X' \text{ as } k \to \infty.
\end{array}
$$

Proof (1). See 6.17(3). □

Proof (2). For $x' \in X'$ we have that $\langle x_k, x' \rangle_X = \langle x', J_X x_k \rangle_{X'}$ and $\langle x, x' \rangle_X = \langle x', J_X x \rangle_{X'}$. □

Proof (3). Because $\langle x, x_k' \rangle_X = \langle x_k', J_X x \rangle_{X'}$ for all $x \in X$. □

8.3 Remarks.

(1) It follows from 6.17(2) that the weak limit of a sequence is unique. For the weak* limit this holds trivially.

(2) Strong convergence (i.e. norm convergence) of a sequence implies weak convergence and weak* convergence.

(3) If $x_k' \to x'$ weakly* in X' as $k \to \infty$, then

$$\|x'\|_{X'} \le \liminf_{k \to \infty} \|x_k'\|_{X'}.$$

(4) If $x_k \to x$ weakly in X as $k \to \infty$, then

$$\|x\|_X \leq \liminf_{k \to \infty} \|x_k\|_X .$$

(5) Weakly convergent sequences and weakly* convergent sequences are bounded.

(6) Let $x_k \to x$ (strongly) in X and $x_k' \to x'$ weakly* in X' as $k \to \infty$. Then

$$\langle x_k , x_k' \rangle_X \to \langle x , x' \rangle_X \quad \text{as } k \to \infty. \tag{8-1}$$

The same holds if $x_k \to x$ weakly in X and $x_k' \to x'$ (strongly) in X'.

Remark: Assertion (4) means that the norm is *lower semicontinuous* with respect to the weak convergence of sequences (see also E8.5). Assertion (6) is often used when considering convergence in function spaces.

Proof (3). For all $x \in X$ we have that as $k \to \infty$

$$|\langle x , x' \rangle_X| \longleftarrow |\langle x , x_k' \rangle_X| \leq \|x_k'\|_{X'} \cdot \|x\|_X ,$$

which implies that

$$|\langle x , x' \rangle_X| \leq \liminf_{k \to \infty} \|x_k'\|_{X'} \cdot \|x\|_X .$$

Therefore, by the definition of the X'-norm,

$$\|x'\|_{X'} = \sup_{\|x\|_X \leq 1} |\langle x , x' \rangle_X| \leq \liminf_{k \to \infty} \|x_k'\|_{X'} .$$

$$\square$$

Proof (4). Analogously to the proof of (3) it holds for all $x' \in X'$ that

$$|\langle x , x' \rangle_X| \leq \|x'\|_{X'} \cdot \liminf_{k \to \infty} \|x_k\|_X .$$

If $x \neq 0$, we can choose x' with $\|x'\|_{X'} = 1$ and $\langle x , x' \rangle_X = \|x\|_X$ (see 6.17(1)) to obtain the desired result. For $x = 0$ the result holds trivially. $\square$

Proof (5). If $x_k' \to x'$ weakly* in X', then

$$\sup_{k \in \mathbb{N}} |\langle x , x_k' \rangle_X| < \infty \quad \text{for all } x \in X,$$

and so it follows from the Banach-Steinhaus theorem (see 7.3) that

$$\sup_{k \in \mathbb{N}} \|x_k'\|_{X'} < \infty .$$

If $x_k \to x$ weakly in X, then $J_X x_k \to J_X x$ weakly* in X'' (with J_X as in 8.2), and so it follows from the above that $J_X x_k$ is bounded in X'', and hence also x_k in X. $\square$

Proof (6). The first claim follows on noting that

$$|\langle x, x'\rangle_X - \langle x_k, x'_k\rangle_X| \leq |\langle x, x'-x'_k\rangle_X| + |\langle x_k - x, x'_k\rangle_X|$$

$$\leq \underbrace{|\langle x, x'-x'_k\rangle_X|}_{\to 0 \text{ as } k\to\infty} + \underbrace{\|x - x_k\|_X}_{\to 0 \text{ as } k\to\infty} \cdot \underbrace{\|x'_k\|_{X'}}_{\text{bounded in } k},$$

since, by (5), the sequence $(x'_k)_{k\in\mathbb{N}}$ is bounded in X'. The second claim follows analogously.   □

We now give some characterizations of weak convergence in function spaces.

8.4 Examples.

(1) Let $1 \leq p < \infty$ with $\frac{1}{p} + \frac{1}{p'} = 1$ (where in the case $p = 1$ we assume that the measure space is σ-finite). Then for $f_k, f \in L^p(\mu)$

$$f_k \to f \quad \text{weakly in } L^p(\mu) \text{ as } k \to \infty$$

$$\Longleftrightarrow \quad \int_S f_k \bar{g}\,d\mu \longrightarrow \int_S f\bar{g}\,d\mu \quad \text{as } k \to \infty \text{ for all } g \in L^{p'}(\mu).$$

(2) Let $S \subset \mathbb{R}^n$ be compact. Then for $f_k, f \in C^0(S)$ (see also E8.4)

$$f_k \to f \quad \text{weakly in } C^0(S) \text{ as } k \to \infty$$

$$\Longleftrightarrow \quad \int_S f_k\,d\lambda \longrightarrow \int_S f\,d\lambda \quad \text{as } k \to \infty \text{ for all } \lambda \in rca(S).$$

(3) Let $\Omega \subset \mathbb{R}^n$ be open, let $m \in \mathbb{N}$ and let $1 \leq p \leq \infty$. Then for $u_k, u \in W^{m,p}(\Omega)$

$$u_k \to u \quad \text{weakly in } W^{m,p}(\Omega) \text{ as } k \to \infty$$

$$\Longleftrightarrow \quad \partial^s u_k \to \partial^s u \quad \text{weakly in } L^p(\Omega) \text{ as } k \to \infty \text{ for all } |s| \leq m.$$

The same result holds for the subspace $W_0^{m,p}(\Omega)$.

Proof (1) *and* (2). Follow directly from Theorem 6.12 and Theorem 6.23, respectively.   □

Proof (3). Let $X = W^{m,p}(\Omega)$ or $X = W_0^{m,p}(\Omega)$. Then

$$(Jv)(x) := (\partial^s v(x))_{|s|\leq m} \in \mathbb{K}^M \quad \text{for } v \in X \text{ and almost all } x \in \Omega$$

defines a linear map $J : X \to L^p(\Omega; \mathbb{K}^M)$, where $M := \binom{n+m}{n}$ is the number of multi-indices s with $|s| \leq m$. In addition, $\|Jv\|_{L^p(\Omega;\mathbb{K}^M)}$ can be bounded from above and from below by $\|v\|_{W^{m,p}(\Omega)}$, and so the completeness of X yields that the subspace $Y := J(X) \subset L^p(\Omega; \mathbb{K}^M)$ is closed. Therefore, J is a bijective continuous linear map between X and $Y = J(X)$ with a continuous inverse $J^{-1} \in \mathscr{L}(Y; X)$.

If $u_k \to u$ weakly in X as $k \to \infty$ and $R \in L^p(\Omega; \mathbb{K}^M)'$, then $T := RJ \in X'$ and

$$R(Ju_k) = T(u_k) \longrightarrow T(u) = R(Ju) \quad \text{as } k \to \infty,$$

that is, $Ju_k \to Ju$ weakly in $L^p(\Omega; \mathbb{K}^M)$. On the other hand, if this is true and $T \in X'$, then $\bar{R} := TJ^{-1} \in Y'$. Applying the Hahn-Banach theorem 6.15 we obtain an extension $R \in L^p(\Omega; \mathbb{K}^M)'$ of $\bar{R}$ and therefore

$$T(u_k) = \bar{R}(Ju_k) = R(Ju_k) \longrightarrow R(Ju) = \bar{R}(Ju) = T(u) \quad \text{as } k \to \infty,$$

that is, $u_k \to u$ weakly in X. Finally, with $v_k^s := \partial^s u_k$ and $v^s := \partial^s u$, it is clear that

$$(v_k^s)_{|s| \le m} \longrightarrow (v^s)_{|s| \le m} \text{ weakly in } L^p(\Omega; \mathbb{K}^M) \text{ as } k \to \infty$$

$$\Longleftrightarrow$$

$$\text{for all } |s| \le m : \left(v_k^s \to v^s \text{ weakly in } L^p(\Omega; \mathbb{K}) \text{ as } k \to \infty \right),$$

a property that is true in general. □

Weak convergence can be interpreted as a generalization of convergence of all coordinates or coordinatewise convergence, as we know it for finite-dimensional spaces. As an analogy of this we replace in the infinite-dimensional case the "coordinates of a point" $x \in X$ by the values $\langle x, x' \rangle$ for $x' \in X'$. This is the idea behind the proof of the following theorem, which is the main functional analysis result of this chapter.

8.5 Theorem. Let X be separable. Then the closed unit ball $\overline{B_1(0)}$ in X' is weakly* sequentially compact.

Remark: This then also holds for every other closed ball $\overline{B_R(x)}$ in X'.

Proof. Let $\{x_n ; \ n \in \mathbb{N}\}$ be dense in X. If $(x'_k)_{k \in \mathbb{N}}$ is a sequence in X' with $\|x'_k\| \le 1$, then $(\langle x_n, x'_k \rangle)_{k \in \mathbb{N}}$ are bounded sequences in $\mathbb{K}$. Applying a diagonalization procedure we produce a subsequence $k \to \infty$ such that for all n

$$\lim_{k \to \infty} \langle x_n, x'_k \rangle \quad \text{exists in } \mathbb{K}.$$

Hence we have that for all $y \in Y := \operatorname{span}\{x_n ; \ n \in \mathbb{N}\}$ the limit

$$x'(y) := \lim_{k \to \infty} \langle y, x'_k \rangle \quad \text{exists in } \mathbb{K},$$

and $x' : Y \to \mathbb{K}$ is linear. It follows from

$$|x'(y)| = \lim_{k \to \infty} |\langle y, x'_k \rangle| \le \|y\|$$

that x' is uniformly continuous on Y and so it can be uniquely extended to a continuous linear map x' on $\overline{Y} = X$ (see E5.3). Therefore, $x' \in X'$ with $\|x'\| \le 1$, and for all $x \in X$ and $y \in Y$

$$|\langle x, x' - x'_k \rangle| \le |\langle x - y, x' - x'_k \rangle| + |\langle y, x' - x'_k \rangle|$$
$$\le 2\|x - y\| + |\langle y, x' - x'_k \rangle|.$$

The second term, for every y, converges to zero as $k \to \infty$, while the first term can be made arbitrarily small because $\overline{Y} = X$. □

8.6 Examples.

(1) If $X = L^1(\mu)$ is separable, then we obtain from 6.12 (see proof below) the following result: If $(f_k)_{k \in \mathbb{N}}$ is bounded in $L^\infty(\mu)$, then there exists a subsequence $(f_{k_i})_{i \in \mathbb{N}}$ and an $f \in L^\infty(\mu)$ such that

$$\int_S f_{k_i} \bar{g} \, d\mu \longrightarrow \int_S f \bar{g} \, d\mu \quad \text{as } i \to \infty \text{ for all } g \in L^1(\mu).$$

Note: $L^1(\mu)$ is separable, for example, if $S \subset \mathbb{R}^n$ is Lebesgue measurable and μ is the Lebesgue measure, or if $S \subset \mathbb{R}^n$ is compact and $\mu \in rca(S)$.

(2) If $X = C^0(S)$ with $S \subset \mathbb{R}^n$ being compact, then 4.18(3) and 6.23 yield the following result: If $(\mu_k)_{k \in \mathbb{N}}$ is bounded in $rca(S)$, then there exist a subsequence $(\mu_{k_i})_{i \in \mathbb{N}}$ and a measure $\mu \in rca(S)$ such that

$$\int_S g \, d\mu_{k_i} \longrightarrow \int_S g \, d\mu \quad \text{as } i \to \infty \text{ for all } g \in C^0(S).$$

Proof (1) *Note.* If μ is the Lebesgue measure on $S \subset \mathbb{R}^n$, then $L^1(\mu)$ is separable (see 4.18(4)). This also holds for $\mu \in rca(S)$, when $S \subset \mathbb{R}^n$ is compact, because every function in $L^1(\mu)$ can be approximated in the L^1-norm by step functions, and, as μ is regular, every μ-measurable set can be approximated in measure by relatively open sets (with respect to S). But every open set is a countable union of semi-open cuboids, with each cuboid having its center on the lattice $2^{-i} \cdot \mathbb{Z}^n$ and side length 2^{1-i} for an $i \in \mathbb{N}$. $\square$

Proof (1). Let $L^1(\mu)$ be separable. On recalling that functions in $L^1(\mu)$ can be approximated by step functions, it follows from 4.17(2) that there exists a subset $\{g_i ; \ i \in \mathbb{N}\}$ of step functions which is dense in $L^1(\mu)$, e.g.

$$g_i := \sum_{j=1}^{m_i} \alpha_{ij} \mathcal{X}_{E_{ij}} \quad \text{with } \mu(E_{ij}) < \infty.$$

Let

$$\tilde{S} := \bigcup_{i,j} E_{ij} \quad \text{and} \quad \tilde{\mu}(E) := \mu(E \cap \tilde{S}) \text{ for } E \in \mathcal{B}.$$

Then $\tilde{\mu}$ is σ-finite, and so 6.12 can be applied to $L^1(\tilde{\mu})$. This yields the desired result, because

$$f \in L^1(\mu) \quad \Longrightarrow \quad f = 0 \ \mu\text{-almost everywhere in } S \setminus \tilde{S}.$$

To see the above, observe that there exists a sequence $(i_k)_{k \in \mathbb{N}}$ in $\mathbb{N}$ such that $\| f - g_{i_k} \|_{L^1(\mu)} \to 0$ as $k \to \infty$, and so

$$\int_{S \setminus \tilde{S}} |f| \, d\mu = \int_{S \setminus \tilde{S}} |f - g_{i_k}| \, d\mu \leq \| f - g_{i_k} \|_{L^1(\mu)} \longrightarrow 0 \quad \text{as } k \to \infty,$$

which implies that $f = 0$ almost everywhere in $S \setminus \tilde{S}$. $\square$

8.7 Weak topology. The following results serve to illustrate the concept of weak sequential compactness. They will not be used in the remainder of this book.

(1) *Weak topology.* Let X be a Banach space. For triples (n, z', ε) with $n \in \mathbb{N}$, $z' = (z'_k)_{k=1,\ldots,n}$, $z'_1, \ldots, z'_n \in X'$ and $\varepsilon > 0$ define

$$U_{n,z',\varepsilon} := \left\{ x \in X \; ; \; |\langle x, z'_k \rangle| < \varepsilon \text{ for } k = 1, \ldots, n \right\},$$

and

$$\mathcal{T}_w := \left\{ A \subset X \; ; \; x \in A \Longrightarrow x + U_{n,z',\varepsilon} \subset A \text{ for some } U_{n,z',\varepsilon} \right\}.$$

Then X equipped with $\mathcal{T}_w$ (called the **weak topology**) is a locally convex topological vector space (as in 5.21), and $\mathcal{T}_w$ is the weakest topology for which all $x' \in X'$ are continuous maps $x' : X \to \mathbb{K}$ with respect to $\mathcal{T}_w$.

(2) *Weak* topology.* Let X be a Banach space. For triples (n, z, ε) with $n \in \mathbb{N}$, $z = (z_k)_{k=1,\ldots,n}$, $z_1, \ldots, z_n \in X$ and $\varepsilon > 0$ define

$$U_{n,z,\varepsilon} := \left\{ x' \in X' \; ; \; |\langle z_k, x' \rangle| < \varepsilon \text{ for } k = 1, \ldots, n \right\},$$

and

$$\mathcal{T}'_w := \left\{ A \subset X' \; ; \; x' \in A \Longrightarrow x' + U_{n,z,\varepsilon} \subset A \text{ for some } U_{n,z,\varepsilon} \right\}.$$

Then X' equipped with $\mathcal{T}'_w$ (called the **weak* topology**) is a locally convex topological vector space (as in 5.21).

Moreover, it holds that: If $\mathcal{T}''_w$ is the weak* topology on $(X')'$ and if J_X is as in 8.2(1), then $\mathcal{T}_w = \{J_X^{-1}(A) \; ; \; A \in \mathcal{T}''_w\}$.

(3) *Alaoglu's theorem.* Let X be a Banach space. Then $\overline{B_1(0)} \subset X'$ (the closed unit ball with respect to the norm on X') is covering compact with respect to the weak* topology on X'.

On the proof: We omit the proof. The result can be shown with the help of Tychonoff's theorem (according to A. N. Tikhonov), see e.g. [Conway].

(4) *Counterexample to compactness theorems.* Theorem 8.5 does not hold without the separability of X, that is: In general "weak* sequential compactness" and "cover compactness with respect to the weak* topology" need to be distinguished.

Example: Let $X = L^\infty(\,]0, 1[\,)$ and for $\varepsilon > 0$ define

$$T_\varepsilon f := \frac{1}{\varepsilon} \int_0^\varepsilon f(x)\, dx \quad \text{for } f \in L^\infty(\,]0, 1[\,).$$

Then $T_\varepsilon \in L^\infty(\,]0, 1[\,)'$ with $\|T_\varepsilon\| = 1$, and the following holds: There exists no null sequence $(\varepsilon_k)_{k\in\mathbb{N}}$ such that $(T_{\varepsilon_k})_{k\in\mathbb{N}}$ is weakly* convergent in $L^\infty(\,]0, 1[\,)'$.

Proof (4) Example. Assume that $(T_{\varepsilon_k})_{k\in\mathbb{N}}$ is weakly* convergent. By choosing a subsequence (which is then also weakly* convergent and which we again denote by $(T_{\varepsilon_k})_{k\in\mathbb{N}}$), we can assume that

$$1 > \frac{\varepsilon_{k+1}}{\varepsilon_k} \to 0 \quad \text{as } k \to \infty.$$

Now consider the function $f \in L^\infty(]0,1[)$ defined by

$$f(x) := (-1)^j \quad \text{for } \varepsilon_{j+1} < x < \varepsilon_j \text{ and } j \in \mathbb{N}.$$

Then

$$T_{\varepsilon_k} f = \frac{1}{\varepsilon_k} \left((\varepsilon_k - \varepsilon_{k+1})(-1)^k + \int_0^{\varepsilon_{k+1}} f(x)\, dx \right),$$

and so

$$\left| T_{\varepsilon_k} f - (-1)^k \right| \le \frac{1}{\varepsilon_k} \left(\varepsilon_{k+1} + \int_0^{\varepsilon_{k+1}} |f(x)|\, dx \right) \le \frac{2\varepsilon_{k+1}}{\varepsilon_k} \longrightarrow 0$$

as $k \to \infty$. This shows that the sequence $(T_{\varepsilon_k} f)_{k\in\mathbb{N}}$ has the two cluster points ± 1. Hence $(T_{\varepsilon_k})_{k\in\mathbb{N}}$ cannot be weakly* convergent. □

Reflexive spaces

In the following we consider the class of reflexive spaces. A reflexive space X is characterized by the fact that the bidual space X'' is isometrically isomorphic to the space X itself, however not (!) with respect to an arbitrary isometry, but precisely with respect to the isometry J_X defined in 8.2(1). The class of reflexive spaces contains all Hilbert spaces (see 8.11(1)).

8.8 Reflexivity. Let X be a Banach space and let J_X be the isometry from 8.2(1). Then we call

$$X \text{ } \textit{reflexive} \quad :\Longleftrightarrow \quad J_X \text{ is surjective} .$$

We have the following results:

(1) If X is reflexive, then weak* and weak sequence convergence in X' coincide.

(2) If X is reflexive, then every closed subspace of X is reflexive.

(3) If $T : X \to Y$ is an isomorphism, then

$$X \text{ reflexive} \quad \Longleftrightarrow \quad Y \text{ reflexive} .$$

(4) It holds that

$$X \text{ reflexive} \quad \Longleftrightarrow \quad X' \text{ reflexive} .$$

Proof (2). Let $Y \subset X$ be a closed subspace. Given a $y'' \in Y''$, let

$$\langle x', x'' \rangle_{X'} := \langle x'|_Y , y'' \rangle_{Y'} \quad \text{for } x' \in X'.$$

Then $x'' \in X''$. Let $x := J_X^{-1} x''$. Now for all $x' \in X'$ with $x' = 0$ on Y we have that

$$\langle x\,,x'\rangle_X = \langle x'\,,x''\rangle_{X'} = \langle x'|_Y\,,y''\rangle_{Y'} = 0\,,$$

which, on recalling 6.16, implies that $x \in Y$. Now let $y' \in Y'$, and let $x' \in X'$ denote an extension of y' as in the Hahn-Banach theorem (see 6.15). Then we conclude that

$$\langle x\,,y'\rangle_Y = \langle x\,,x'\rangle_X = \langle x'|_Y\,,y''\rangle_{Y'} = \langle y'\,,y''\rangle_{Y'}\,,$$

i.e. $y'' = J_Y x$. This shows that J_Y is surjective. $\qquad\square$

Proof (3). The claim is symmetric in X and Y, and so it is sufficient to consider the case where X is reflexive. We need to show the reflexivity of Y. Let $y'' \in Y''$. Then

$$\langle x'\,,x''\rangle_{X'} := \langle x'\circ T^{-1}\,,y''\rangle_{Y'} \qquad \text{for } x' \in X'$$

defines an $x'' \in X''$, and for $y' \in Y'$ (setting $x' := y'\circ T$)

$$\langle y'\,,y''\rangle_{Y'} = \langle y'\circ T\,,x''\rangle_{X'} = \langle J_X^{-1}x''\,,y'\circ T\rangle_X = \langle TJ_X^{-1}x''\,,y'\rangle_Y\,,$$

and so $y'' = J_Y T J_X^{-1} x''$. $\qquad\square$

Proof (4)$\Rightarrow$. If $x''' \in X'''$ then $x'''\circ J_X \in X'$, and it holds for all $x'' \in X''$ that

$$\langle x''\,,x'''\rangle_{X''} = \langle J_X^{-1}x''\,,x'''\circ J_X\rangle_X = \langle x'''\circ J_X\,,x''\rangle_{X'}\,,$$

i.e. $x''' = J_{X'}(x'''\circ J_X)$. $\qquad\square$

Proof (4)$\Leftarrow$. Employing the established implication "$\Rightarrow$" for the Banach space X' yields that X'' is reflexive. As J_X is isometric, $J_X(X)$ is a closed subspace of X'', which according to (2) is also reflexive. Hence (3) implies that X is reflexive. $\qquad\square$

The proof of theorem 8.10 below employs the following:

8.9 Lemma. For every Banach space X,

$$X' \text{ separable} \quad\Longrightarrow\quad X \text{ separable}\,.$$

Observe: The converse is false, as shown by the very important example $X = L^1(\mu)$ (see 6.12 and 4.18(4)).

Proof. Let $\{x_n'\,;\,n \in \mathbb{N}\}$ be dense in X'. Choose $x_n \in X$ with

$$|\langle x_n\,,x_n'\rangle_X| \geq \tfrac{1}{2}\|x_n'\| \quad\text{and}\quad \|x_n\| = 1$$

and define $Y := \operatorname{clos}(\operatorname{span}\{x_n\,;\,n \in \mathbb{N}\})$. Now if $x' \in X'$ with $x' = 0$ on Y, then for all n

$$\|x' - x_n'\| \geq |\langle x_n\,,x' - x_n'\rangle_X| = |\langle x_n\,,x_n'\rangle_X|$$
$$\geq \tfrac{1}{2}\|x_n'\| \geq \tfrac{1}{2}(\|x'\| - \|x_n' - x'\|)$$

and so

$$\|x'\| \le 3 \inf_n \|x' - x_n'\| = 0 \,,$$

since $\{x_n' \,;\, n \in \mathbb{N}\}$ is a dense subset. Hence it follows from 6.16 that $Y = X$.

$\square$

We now prove the main theorem for reflexive spaces.

8.10 Theorem. Let X be a reflexive Banach space. Then the closed unit ball $\overline{B_1(0)} \subset X$ is weakly sequentially compact.
Remark: This then also holds for every other closed ball $\overline{B_R(x)}$.

Proof. Let $(x_k)_{k \in \mathbb{N}}$ be a sequence in $\overline{B_1(0)} \subset X$ and set

$$Y := \overline{\operatorname{span}\{x_k \,;\, k \in \mathbb{N}\}}\,.$$

Then Y is reflexive (see 8.8(2)) and, by definition, separable. It follows that $Y'' = J_Y Y$ is separable, and hence so is Y' (see 8.9). That means that we can apply 8.5 to the space Y' and to the sequence $(J_Y x_k)_{k \in \mathbb{N}}$ in Y''. In particular, there exists a $y'' \in Y''$ such that for a subsequence $k \to \infty$

$$\langle y' , J_Y x_k \rangle_{Y'} \to \langle y' , y'' \rangle_{Y'} \quad \text{for all } y' \in Y'.$$

Setting $x := J_Y^{-1} y'' \in Y$, it follows that

$$\langle x_k , y' \rangle_Y = \langle y' , J_Y x_k \rangle_{Y'} \longrightarrow \langle y' , y'' \rangle_{Y'} = \langle x , y' \rangle_Y \quad \text{as } k \to \infty$$

for all $y' \in Y'$. Since for $x' \in X'$ the map $x'|_Y$ lies in Y', it follows that also $\langle x_k , x' \rangle_X \to \langle x , x' \rangle_X$ as $k \to \infty$, and so $x_k \to x$ weakly in X as $k \to \infty$. $\square$

8.11 Examples of reflexive spaces. Here are several consequences of theorem 8.10.

(1) Every Hilbert space X is reflexive. Together with the Riesz representation theorem 6.1 we obtain: If $(x_k)_{k \in \mathbb{N}}$ is a bounded sequence in X, then there exists a subsequence $(x_{k_i})_{i \in \mathbb{N}}$ and an $x \in X$ such that

$$(y , x_{k_i})_X \to (y , x)_X \quad \text{as } i \to \infty \text{ for all } y \in X.$$

(2) $L^p(\mu)$ for $1 < p < \infty$ is reflexive. It follows from 6.12 that: If $(f_k)_{k \in \mathbb{N}}$ is a bounded sequence in $L^p(\mu)$, then there exists a subsequence $(f_{k_i})_{i \in \mathbb{N}}$ and an $f \in L^p(\mu)$ such that

$$\int_S g f_{k_i} \, d\mu \longrightarrow \int_S g f \, d\mu \quad \text{as } i \to \infty \text{ for all } g \in L^{p'}(\mu).$$

(3) $W^{m,p}(\Omega)$ for $1 < p < \infty$ is reflexive. It holds that: If $(f_k)_{k\in\mathbb{N}}$ is a bounded sequence in $W^{m,p}(\Omega)$, then there exist a subsequence $(f_{k_i})_{i\in\mathbb{N}}$ and an $f \in W^{m,p}(\Omega)$ such that for all $|s| \leq m$

$$\int_\Omega g\partial^s f_{k_i}\, d\mathrm{L}^n \longrightarrow \int_\Omega g\partial^s f\, d\mathrm{L}^n \quad \text{as } i \to \infty \text{ for all } g \in L^{p'}(\Omega).$$

(4) $L^1(\mu)$ and $L^\infty(\mu)$ (with the measure μ being σ-finite) are not reflexive if the underlying σ-algebra $\mathcal{B}$ contains infinitely many disjoint sets with positive measure, i.e. if and only if $L^1(\mu)$ and $L^\infty(\mu)$, respectively, are infinite-dimensional.

(5) $C^0(S)$ and $rca(S)$ are not reflexive if $S \subset \mathbb{R}^n$ is compact and contains more than finitely many points, i.e. if and only if $C^0(S)$ and $rca(S)$, respectively, are infinite-dimensional.

Proof (1). Let $R_X : X \to X'$ be the (conjugate linear) isomorphism from the Riesz representation theorem. Then for $x'' \in X''$ letting

$$\langle y, x'\rangle_X := \overline{\langle R_X y, x''\rangle_{X'}} \quad \text{for } y \in X$$

defines an $x' \in X'$. Set $x := R_X^{-1}x'$. Then for all $y \in X$

$$\langle R_X y, x''\rangle_{X'} = \overline{\langle y, R_X x\rangle_X} = \overline{(y, x)_X} = \langle x, R_X y\rangle_X,$$

i.e. $x'' = J_X x$, which shows that J_X is surjective.

Remark: Hence in the real case, i.e. $\mathbb{K} = \mathbb{R}$, it holds that $J_X^{-1} = R_X^{-1}R'_X$, with $R'_X : X'' \to X'$ denoting the adjoint map (see 5.5(8)) of R_X.   □

Proof (2). The isometries

$$J_p : L^p(\mu) \to L^{p'}(\mu)' \quad \text{and} \quad J_{p'} : L^{p'}(\mu) \to L^p(\mu)'$$

from 6.12 satisfy

$$\overline{\langle f, J_{p'}g\rangle_{L^p(\mu)}} = \langle g, J_p f\rangle_{L^{p'}(\mu)} \quad \text{for all } f \in L^p(\mu),\ g \in L^{p'}(\mu).$$

For $f'' \in L^p(\mu)''$ letting

$$\langle g, g'\rangle_{L^{p'}(\mu)} := \overline{\langle J_{p'}g, f''\rangle_{L^p(\mu)'}} \quad \text{for } g \in L^{p'}(\mu)$$

defines a $g' \in L^{p'}(\mu)'$. Set $f := J_p^{-1}g'$. Then for all $g \in L^{p'}(\mu)$

$$\langle g, g'\rangle_{L^{p'}(\mu)} = \langle g, J_p f\rangle_{L^{p'}(\mu)} = \overline{\langle f, J_{p'}g\rangle_{L^p(\mu)}} = \overline{\langle J_{p'}g, J_{L^p(\mu)}f\rangle_{L^p(\mu)'}},$$

where $J_{L^p(\mu)} : L^p(\mu) \to L^p(\mu)''$ denotes the embedding from 8.2. Consequently,

$$\langle J_{p'}g, f''\rangle_{L^p(\mu)'} = \langle J_{p'}g, J_{L^p(\mu)}f\rangle_{L^p(\mu)'} \quad \text{for all } g \in L^{p'}(\mu).$$

As $J_{p'}$ is surjective, it follows that $f'' = J_{L^p(\mu)}f$, which proves the reflexivity of $L^p(\mu)$.

Remark: Hence in the real case, i.e. $\mathbb{K} = \mathbb{R}$, it holds that $J_{L^p(\mu)}^{-1} = J_p^{-1}J_{p'}'$, with $J_{p'}' : L^p(\mu)'' \to L^{p'}(\mu)'$ denoting the adjoint map (see 5.5(8)) of $J_{p'}$. $\square$

Proof (3). Let $J : W^{m,p}(\Omega) \to L^p(\Omega; \mathbb{K}^M)$ be defined as in the proof of 8.4(3). Then combining (2) and 8.8(2) yields that the closed subspace $J(W^{m,p}(\Omega))$ is reflexive (the proof of (2) is the same for functions with values in $\mathbb{K}^M$). The claim now follows from 8.8(3). $\square$

Proof (4). On noting 8.8(4), 6.12 for $p = 1$ and 8.8(3), it is sufficient to show this for $L^1(\mu)$. Let $F \in L^\infty(\mu)'$. If $J_\infty : L^\infty(\mu) \to L^1(\mu)'$ denotes the isomorphism from 6.12, then setting

$$\langle f', G \rangle_{L^1(\mu)'} := \overline{\langle J_\infty^{-1}f', F \rangle_{L^\infty(\mu)}} \quad \text{for } f' \in L^1(\mu)'$$

defines a $G \in L^1(\mu)''$. If $G = J_{L^1(\mu)}f$ for an $f \in L^1(\mu)$, with $J_{L^1(\mu)}$ denoting the embedding from 8.2, then it holds for all $g \in L^\infty(\mu)$ that

$$\overline{\langle g, F \rangle_{L^\infty(\mu)}} = \langle J_\infty g, G \rangle_{L^1(\mu)'} = \langle J_\infty g, J_{L^1(\mu)}f \rangle_{L^1(\mu)'}$$
$$= \langle f, J_\infty g \rangle_{L^1(\mu)} = \int_S f\bar{g} \, d\mu,$$

that is,

$$\langle g, F \rangle_{L^\infty(\mu)} = \int_S g\bar{f} \, d\mu \quad \text{for all } g \in L^\infty(\mu). \tag{8-2}$$

Under the assumption that $L^1(\mu)$ is infinite-dimensional, we now construct an F which does not satisfy this property. To this end, let $E_k \in \mathcal{B}$ be such that

$$E_k \subset E_{k+1}, \ \mu(E_k) < \mu(E_{k+1}) \quad \text{and} \quad E := \bigcup_{k \in \mathbb{N}} E_k.$$

Consider the subspace

$$Y := \text{clos}\left(\left\{ g \in L^\infty(\mu) \, ; \, g = 0 \text{ on } S \setminus E_k \text{ for some } k \right\}\right) \subset L^\infty(\mu).$$

Then $\mathcal{X}_E \notin Y$, and so 6.16 implies that there exists an $F \in L^\infty(\mu)'$ with $F = 0$ on Y and $F(\mathcal{X}_E) = 1$. Hence,

$$F(\mathcal{X}_{E_k}) = 0 \quad \text{and} \quad F(\mathcal{X}_E) = 1,$$

but for every $f \in L^1(\mu)$ we have that

$$\int_S \mathcal{X}_{E_k}\bar{f} \, d\mu \longrightarrow \int_S \mathcal{X}_E\bar{f} \, d\mu.$$

Therefore, F cannot have the representation (8-2). $\square$

Proof (5). Let $C^0(S)$ be reflexive. Then analogously to the proof of (4), and on using 6.23, there exists for every functional $F \in rca(S)'$ an $f \in C^0(S)$ with

$$\langle \nu, F \rangle_{rca(S)} = \int_S f \, d\nu \quad \text{for all } \nu \in rca(S). \tag{8-3}$$

If S is not finite, then there exist points $x_k \in S$ for $k \in \mathbb{N}$ with $x_k \to x \in S$ as $k \to \infty$ and with $x_k \neq x$ for all k. Consider the Dirac measures δ_{x_k} and δ_x and set $Y := \{\nu \in rca(S) ; \; \nu(\{x\}) = 0\}$. It holds that $Y \subset rca(S)$ is a closed subspace with $\delta_{x_k} \in Y$ and $\delta_x \notin Y$. It follows from 6.16 that there exists an $F \in rca(S)'$ with $F(\delta_{x_k}) = 0$ for all k and $F(\delta_x) = 1$. But for every $f \in C^0(S)$ we have that

$$\int_S f \, d\delta_{x_k} = f(x_k) \longrightarrow f(x) = \int_S f \, d\delta_x .$$

Hence F cannot have the representation (8-3). □

Minkowski's functional

In 4.3 we solved the minimal distance problem for closed convex sets in Hilbert spaces, and we saw in E4.3 that in general this is not possible in Banach spaces. We will now show that in reflexive spaces the distance to such sets is attained (see 8.15). This is based on the fact that convex side constraints for elements of an arbitrary Banach space remain valid for limits of weakly convergent sequences, see theorem 8.13. For closed balls this theorem can be obtained directly from 8.3(4), and for general closed convex sets it follows from the following

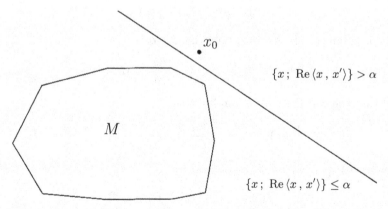

Fig. 8.1. *Separation theorem*

8.12 Separation theorem. Let X be a normed space, let $M \subset X$ be nonempty, closed and convex, and let $x_0 \in X \setminus M$. Then there exist an $x' \in X'$ and an $\alpha \in \mathbb{R}$ with

$$\operatorname{Re} \langle x, x' \rangle \leq \alpha \text{ for } x \in M \quad \text{and} \quad \operatorname{Re} \langle x_0, x' \rangle > \alpha.$$

Remark: It follows that $x' \neq 0$, and hence $\{x \in X; \operatorname{Re} \langle x, x' \rangle = \alpha\}$ is a hyperplane.

Proof. First we consider the case $\mathbb{K} = \mathbb{R}$. We may assume with no loss of generality that

$$0 \in \overset{\circ}{M}.$$

Justification: Choose an $\tilde{x} \in M$ and consider $\tilde{x}_0 := x_0 - \tilde{x}$ and $\widetilde{M} := \overline{B_r(M - \tilde{x})}$ with $0 < r < \operatorname{dist}(x_0, M)$. Then if the theorem is established for $\widetilde{M}$ and $\tilde{x}_0$ with x' and $\tilde{\alpha}$, it follows that the theorem holds for M and x_0 with x' and $\alpha := \tilde{\alpha} + \langle \tilde{x}, x' \rangle$. Consider the **Minkowski functional**

$$p(x) := \inf \left\{ r > 0; \frac{x}{r} \in M \right\} \quad \text{for } x \in X.$$

Since $0 \in \overset{\circ}{M}$, it follows that $0 \leq p(x) < \infty$ for all $x \in X$. Moreover,

$$p \leq 1 \text{ on } M, \quad p(x_0) > 1, \quad p(0) = 0.$$

In addition, we have for $x, y \in X$ that

$$p(ax) = ap(x) \quad \text{for } a \geq 0,$$
$$p(x + y) \leq p(x) + p(y),$$

i.e. p is sublinear. To see this, note that for $\alpha > 0$

$$\frac{x}{r} \in M \quad \Longleftrightarrow \quad \frac{\alpha x}{\alpha r} \in M,$$

and that the convexity of M implies that

$$\frac{x}{r} \in M, \frac{y}{s} \in M \quad \Longrightarrow \quad \frac{x + y}{r + s} = \frac{r}{r + s} \frac{x}{r} + \frac{s}{r + s} \frac{y}{s} \in M.$$

Now let $f : \operatorname{span}\{x_0\} \to \mathbb{R}$ be defined by

$$f(ax_0) := ap(x_0) \quad \text{for } a \in \mathbb{R}.$$

Then

$$f(ax_0) = p(ax_0) \quad \text{for } a \geq 0,$$
$$f(ax_0) \leq 0 \leq p(ax_0) \quad \text{for } a \leq 0.$$

It follows from the Hahn-Banach theorem (see 6.14), applied to the subspace $\operatorname{span}\{x_0\}$, that there exists a linear extension F of f with $F \leq p$ on X. Hence

$$F \le p \le 1 \text{ on } M, \quad F(x_0) = f(x_0) = p(x_0) > 1.$$

On recalling that $\overline{B_\varrho(0)} \subset M$ for some $\varrho > 0$, we note that

$$x \in X \implies \frac{x}{\frac{1}{\varrho}\|x\|} \in M \implies p(x) \le \frac{1}{\varrho}\|x\| \implies F(x) \le \frac{1}{\varrho}\|x\|.$$

Similarly, $-F(x) = F(-x) \le \frac{1}{\varrho}\|x\|$, which implies that $F \in X'$. Hence we have shown the desired result for $x' := F$ and $\alpha = 1$.

In the case $\mathbb{K} = \mathbb{C}$ consider X as an $\mathbb{R}$-vector space $X_{\mathbb{R}}$ and obtain an $F_{\mathbb{R}} \in X'_{\mathbb{R}}$ with the desired properties. Then, as in the proof of 6.15, proceed to the function $F(x) := F_{\mathbb{R}}(x) - iF_{\mathbb{R}}(ix)$. $\qquad\square$

8.13 Theorem. Let X be a normed space and let $M \subset X$ be closed and convex. Then M is **weakly sequentially closed**, i.e. if $x_k, x \in X$ for $k \in \mathbb{N}$, then

$$\begin{matrix} x_k \to x \text{ weakly in } X \text{ as } k \to \infty, \\ x_k \in M \text{ for } k \in \mathbb{N} \end{matrix} \quad \implies \quad x \in M.$$

Proof. If $x \notin M$, then by the separation theorem 8.12 there exist an $x' \in X'$ and an $\alpha \in \mathbb{R}$ such that

$$\text{Re}\,\langle y, x'\rangle \le \alpha \text{ for } y \in M \text{ and } \text{Re}\,\langle x, x'\rangle > \alpha.$$

Now we have that $\text{Re}\,\langle x_k, x'\rangle \le \alpha$, and the weak convergence to x yields that also $\text{Re}\,\langle x, x'\rangle \le \alpha$, a contradiction. $\qquad\square$

The following two results are consequences of this theorem.

8.14 Mazur's lemma. Let $(x_k)_{k \in \mathbb{N}}$ be a sequence in a normed space X that converges weakly to x. Then $x \in \text{clos}\,(\text{conv}\,\{x_k \, ; \, k \in \mathbb{N}\})$.

Proof. $M := \text{conv}\,\{x_k \, ; \, k \in \mathbb{N}\}$ is a convex set, and hence so is $\overline{M}$. Now apply theorem 8.13. $\qquad\square$

8.15 Theorem. Let X be a reflexive Banach space and let $M \subset X$ be nonempty, closed and convex. Then for $x_0 \in X$ there exists an $x \in M$ with

$$\|x - x_0\| = \text{dist}(x_0, M).$$

Proof. Let $(x_k)_{k \in \mathbb{N}}$ be a **minimal sequence**, i.e.

$$x_k \in M \quad \text{and} \quad \|x_k - x_0\| \to \text{dist}(x_0, M) \quad \text{as } k \to \infty.$$

Then $(x_k)_{k \in \mathbb{N}}$ is a bounded sequence, and so it follows from 8.10 that there exists a subsequence $k \to \infty$ such that $x_k \to x$ weakly in X as $k \to \infty$. Hence 8.13 yields $x \in M$. On noting that also $x_k - x_0 \to x - x_0$ weakly in X, it follows from the lower semicontinuity of the norm (see 8.3(4)) that $\|x - x_0\| = \text{dist}(x_0, M)$. $\qquad\square$

Variational methods

Closed convex sets play an important role in existence proofs for elliptic partial differential equations. We now provide applications of theorem 8.13 on closed convex sets to variational problems with side constraints (see 8.17–8.18), where a generalization of the Poincaré inequality 6.7 is needed (see 8.16). The results on partial differential equations will rely on Sobolev spaces, and the theorems required for these spaces will be derived in Appendix A8. Moreover, we always consider open sets $\Omega \subset \mathbb{R}^n$ which are connected.

Remark: An open set $\Omega \subset \mathbb{R}^n$ is connected if and only if it is **path connected**, i.e. if for every two points $x_0, x_1 \in \Omega$ there exists a (continuous) path in Ω from x_0 to x_1, i.e. a continuous map $\gamma : [0,1] \to \Omega$ with $\gamma(0) = x_0$ and $\gamma(1) = x_1$. In the following we will always only make use of this property (see e.g. 10.4). In a general topological space X a subset $A \subset X$ is said to be **connected** if A is not the union of two disjoint, nonempty and relatively in A open sets.

8.16 Generalized Poincaré inequality. Let $\Omega \subset \mathbb{R}^n$ be open, bounded and connected with Lipschitz boundary $\partial\Omega$ (see definition A8.2). Moreover, let $1 < p < \infty$ and let $M \subset W^{1,p}(\Omega)$ be nonempty, closed and convex. Then the following are equivalent for every $u_0 \in M$:

(1) There exists a constant $C_0 < \infty$ such that for all $\xi \in \mathbb{R}$,

$$u_0 + \xi \in M \quad\Longrightarrow\quad |\xi| \le C_0.$$

(2) There exists a constant $C < \infty$ with

$$\|u\|_{L^p(\Omega)} \le C \cdot \left(\|\nabla u\|_{L^p(\Omega)} + 1\right) \quad \text{for all } u \in M.$$

Note: If M, in addition, is a **cone with apex 0**, i.e. if

$$u \in M, \ r \ge 0 \quad\Longrightarrow\quad ru \in M,$$

then the inequality in (2) can be replaced with

$$\|u\|_{L^p(\Omega)} \le C \cdot \|\nabla u\|_{L^p(\Omega)} \quad \text{for all } u \in M.$$

Proof Note. Replace u in (2) with ru and let $r \nearrow \infty$. $\Box$

Proof (2)$\Rightarrow$(1). Let $\xi \in \mathbb{R}$ with $u := u_0 + \xi \in M$. Then $\nabla u = \nabla u_0$, and hence the inequality in (2) for u implies that

$$C \cdot (\|\nabla u_0\|_{L^p} + 1) \ge \|u_0 + \xi\|_{L^p} \ge |\xi| \cdot \|1\|_{L^p} - \|u_0\|_{L^p}.$$

This yields the desired result with a C_0 that depends on C and u_0. $\Box$

Proof (1)$\Rightarrow$(2). Without loss of generality we may assume that $u_0 = 0$. To see this, note that if the desired inequality holds for $\tilde{u} \in \widetilde{M} := M - u_0$ with a constant $\tilde{C}$, then it follows for $u := \tilde{u} + u_0$ that

$$\|u\|_{L^p} \le \|\tilde{u}\|_{L^p} + \|u_0\|_{L^p} \le \tilde{C} \cdot \left(\|\nabla u\|_{L^p} + \|\nabla u_0\|_{L^p} + 1\right) + \|u_0\|_{L^p}.$$

Now let $u_0 = 0$ and assume that the conclusion is false. Then there exist $u_k \in M$, $k \in \mathbb{N}$, with

$$\|\nabla u_k\|_{L^p} + 1 \le \frac{1}{k}\|u_k\|_{L^p}. \tag{8-4}$$

In particular, $\|u_k\|_{L^p} \to \infty$, and so for every given $R > 0$ (for k sufficiently large)

$$\delta_k := \frac{R}{\|u_k\|_{L^p}} \longrightarrow 0 \quad \text{as } k \to \infty.$$

Hence we have that $0 < \delta_k \le 1$ for k sufficiently large, and combining the fact that $0 \in M$ and the convexity of M then yields that $v_k := \delta_k u_k \in M$. Further,

$$\|v_k\|_{L^p} = \delta_k \|u_k\|_{L^p} = R,$$

and the inequality (8-4) yields that

$$\|\nabla v_k\|_{L^p} + \delta_k \le \frac{1}{k}\|v_k\|_{L^p} = \frac{R}{k} \longrightarrow 0 \quad \text{as } k \to \infty.$$

Thus, the v_k are bounded in $W^{1,p}(\Omega)$. Then 8.11(3) implies that there exist a subsequence, again denoted by $(v_k)_{k \in \mathbb{N}}$, and a $v \in W^{1,p}(\Omega)$, such that $v_k \to v$ weakly in $W^{1,p}(\Omega)$ as $k \to \infty$, and so $v \in M$ on recalling 8.13. In particular, $\nabla v_k \to \nabla v$ weakly in $L^p(\Omega)$ (see 8.4(3)). However, the above inequality yields that $\nabla v_k \to 0$ strongly in $L^p(\Omega)$, and hence $\nabla v = 0$. As Ω is connected, it follows that v is (almost everywhere) a constant function (see E8.9). This means that $v = \xi$ almost everywhere in Ω for some $\xi \in \mathbb{R}$, and the assumptions yield that $|\xi| \le C_0$. On the other hand, by Rellich's embedding theorem (see A8.4), the weak convergence in $W^{1,p}(\Omega)$ implies that $v_k \to v$ strongly in $L^p(\Omega)$, and so

$$R = \|v_k\|_{L^p} \longrightarrow \|v\|_{L^p} = |\xi| \cdot \|1\|_{L^p} \le C_0 \|1\|_{L^p}.$$

This yields a contradiction, on initially choosing R sufficiently large. □

In the above result we have considered domains $\Omega \subset \mathbb{R}^n$ with a local Lipschitz boundary $\partial\Omega$. It turns out that the class of such "Lipschitz domains" is mathematically very robust (see, for example, the trace theorem A8.6 or the embedding theorem 10.9, which for Lipschitz domains holds in Sobolev spaces of arbitrary order). And it is the class of domains that is appropriate for applications, as the boundary can have edges and corners (e.g. cubes are allowed, and more general domains with piecewise smooth boundaries, where the pieces meet at nondegenerate angles). We now consider Sobolev functions on Lipschitz domains and solve the

8.17 Elliptic minimum problem. Let $\Omega \subset \mathbb{R}^n$ be open, bounded and connected with Lipschitz boundary (see A8.2). Let $\mathbb{K} = \mathbb{R}$. Then

$$E(u) := \int_\Omega \Big(\frac{1}{2} \sum_{i,j=1}^n \partial_i u \cdot a_{ij} \partial_j u + fu \Big) \, \mathrm{d}L^n \quad \text{for } u \in W^{1,2}(\Omega)$$

defines a map $E : W^{1,2}(\Omega) \to \mathbb{R}$, where we assume that $f \in L^2(\Omega)$ and $a_{ij} \in L^\infty(\Omega)$. In addition, we assume that $(a_{ij})_{i,j=1,\dots,n}$ is *elliptic* (as in (6-8)), i.e. that there exists a positive constant c_0 such that for all $x \in \Omega$

$$\sum_{i,j=1}^n a_{ij}(x)\xi_i\xi_j \geq c_0|\xi|^2 \quad \text{for all } \xi \in \mathbb{R}^n. \tag{8-5}$$

Without loss of generality we may assume *symmetry*, i.e. that

$$a_{ij} = a_{ji} \quad \text{for } i,j = 1,\dots,n. \tag{8-6}$$

(Otherwise replace a_{ij} with $\tilde{a}_{ij} := \frac{1}{2}(a_{ij} + a_{ji})$.) Then for every nonempty, closed and convex subset $M \subset W^{1,2}(\Omega)$ with the property in 8.16 (the property (8-10), below, is stronger) it holds that:

(1) E has an *absolute minimum* u on M, i.e. there exists a $u \in M$ such that
$$E(u) \leq E(v) \quad \text{for all } v \in M. \tag{8-7}$$

(2) The absolute minima u of E on M are precisely the solutions of the following *variational inequality* of E on M:

$$\int_\Omega \Big(\sum_{i,j=1}^n \partial_i(u-v) \cdot a_{ij}\partial_j u + (u-v)f \Big) \, \mathrm{d}L^n \leq 0 \quad \text{for all } v \in M. \tag{8-8}$$

(3) If M is a closed affine subspace, that is, if $M = u_0 + M_0$ for some $u_0 \in M$ and a closed subspace $M_0 \subset W^{1,2}(\Omega)$, then the variational inequality (8-8) for $u \in M$ is equivalent to

$$\int_\Omega \Big(\sum_{i,j=1}^n \partial_i v \cdot a_{ij}\partial_j u + vf \Big) \, \mathrm{d}L^n = 0 \quad \text{for all } v \in M_0. \tag{8-9}$$

(4) If M satisfies

$$v \in M, \ \xi \in \mathbb{R}, \ v + \xi \in M \quad \Longrightarrow \quad \xi = 0, \tag{8-10}$$

then there exists a unique absolute minimum and a unique solution of the variational inequality of E on M.

Proof (1). We begin by showing that there exist positive constants c and C such that

$$E(u) \geq c \int_\Omega |\nabla u|^2 \, \mathrm{dL}^n - C \quad \text{for all } u \in M. \tag{8-11}$$

On noting the elementary **Young's inequality**

$$a \cdot b \leq \delta a^2 + \frac{1}{4\delta} b^2 \quad \text{for } a, b \geq 0 \text{ and } \delta > 0, \tag{8-12}$$

it follows from the ellipticity in (8-5) that

$$E(u) \geq c_0 \int_\Omega |\nabla u|^2 \, \mathrm{dL}^n - \|f\|_{L^2} \|u\|_{L^2}$$

$$\geq c_0 \|\nabla u\|_{L^2}^2 - \delta \|u\|_{L^2}^2 - \frac{1}{4\delta} \|f\|_{L^2}^2 .$$

Letting C_1 denote the constant from the Poincaré inequality 8.16(2),

$$\|u\|_{L^2}^2 \leq 2C_1^2 \|\nabla u\|_{L^2}^2 + 2 ,$$

and so

$$E(u) \geq (c_0 - 2C_1^2 \delta) \|\nabla u\|_{L^2}^2 - C(\delta, f) ,$$

where $C(\delta, f)$ is a quantity depending on δ and f. On choosing δ sufficiently small, we obtain (8-11) with $c = \frac{c_0}{2}$.

It follows from (8-11) that $E(u) \geq -C$ for all $u \in M$, i.e. E is bounded from below on M. Now choose a **minimal sequence** $(u_k)_{k \in \mathbb{N}}$ in M, i.e.

$$E(u_k) \longrightarrow d := \inf_{v \in M} E(v) > -\infty \quad \text{as } k \to \infty.$$

By (8-11), the sequence $(\nabla u_k)_{k \in \mathbb{N}}$ is bounded in $L^2(\Omega)$. Together with the Poincaré inequality 8.16(2) we obtain that $(u_k)_{k \in \mathbb{N}}$ is a bounded sequence in $W^{1,2}(\Omega)$. It follows from 8.11(3) that there exists a $u \in W^{1,2}(\Omega)$ such that $u_k \to u$ weakly in $W^{1,2}(\Omega)$ for a subsequence $k \to \infty$. Since M is closed and convex, it follows from theorem 8.13 that $u \in M$. Moreover, it follows from 8.4(3) that the weak convergence implies that

$$\int_\Omega f(u_k - u) \, \mathrm{dL}^n \longrightarrow 0 \quad \text{and} \quad \sum_{i,j=1}^n \int_\Omega a_{ij} \partial_i u \partial_j (u_k - u) \, \mathrm{dL}^n \longrightarrow 0 .$$

Hence we have that

$$E(u_k) = E(u + u_k - u)$$

$$= E(u) + \underbrace{\sum_{i,j=1}^n \int_\Omega a_{ij} \partial_i u \partial_j (u_k - u) \, \mathrm{dL}^n + \int_\Omega f(u_k - u) \, \mathrm{dL}^n}_{\to 0 \text{ as } k \to \infty}$$

$$+ \underbrace{\int_\Omega \frac{1}{2} \sum_{i,j=1}^n a_{ij} \partial_i (u_k - u) \partial_j (u_k - u) \, \mathrm{dL}^n}_{\geq 0} ,$$

which yields that $E(u) \leq \liminf_{k \to \infty} E(u_k) = d$. On the other hand, $u \in M$ implies that $E(u) \geq \inf_{v \in M} E(v) = d$, and so $E(u) = d$. $\qquad\square$

Proof (2). If u is an absolute minimum and if $v \in M$, then, since M is convex, $(1 - \varepsilon)u + \varepsilon v \in M$ for $0 < \varepsilon \leq 1$, and so

$$
E(u) \leq E\big((1 - \varepsilon)u + \varepsilon v\big) = E\big(u + \varepsilon(v - u)\big)
$$

$$
= E(u) + \varepsilon \int_\Omega \Big(\sum_{i,j=1}^n \partial_i(v - u)a_{ij}\partial_j u + (v - u)f \Big)\, \mathrm{d}L^n
$$

$$
+ \frac{\varepsilon^2}{2} \int_\Omega \underbrace{\sum_{i,j=1}^n \partial_i(v - u)a_{ij}\partial_j(v - u)\, \mathrm{d}L^n}_{\geq 0}\,.
$$

$$(8\text{-}13)$$

Subtracting $E(u)$, dividing by ε and letting $\varepsilon \searrow 0$ then yields the desired variational inequality.

Conversely, if $u \in M$ then the identity in (8-13) (with $\varepsilon = 1$) yields for all $v \in M$ that

$$
E(v) \geq E(u) + \int_\Omega \Big(\sum_{i,j=1}^n \partial_i(v - u)a_{ij}\partial_j u + (v - u)f \Big)\, \mathrm{d}L^n\,.
$$

Now if u is a solution of the variational inequality, then the above integral is nonnegative. Hence u is an absolute minimum of E on M. $\qquad\square$

Proof (3). In (8-8) choose $v = u \pm \tilde{v}$ with $\tilde{v} \in M_0$ (cf. the proof of 4.4(1)). $\qquad\square$

Proof (4). If u_1 and u_2 are two solutions of the variational inequality, then choose $v = u_2$ in the variational inequality for u_1 and $v = u_1$ in the variational inequality for u_2 to obtain

$$
\int_\Omega \Big(\sum_{i,j=1}^n \partial_i(u_1 - u_2) \cdot a_{ij}\partial_j u_1 + (u_1 - u_2)f \Big)\, \mathrm{d}L^n \leq 0\,,
$$

$$
\int_\Omega \Big(\sum_{i,j=1}^n \partial_i(u_2 - u_1) \cdot a_{ij}\partial_j u_2 + (u_2 - u_1)f \Big)\, \mathrm{d}L^n \leq 0\,.
$$

Adding these two inequalities yields that

$$
0 \geq \int_\Omega \sum_{i,j=1}^n \partial_i(u_1 - u_2) \cdot a_{ij}\partial_j(u_1 - u_2)\, \mathrm{d}L^n \geq c_0 \int_\Omega |\nabla(u_1 - u_2)|^2\, \mathrm{d}L^n\,,
$$

and so $\nabla(u_1 - u_2) = 0$ in $L^2(\Omega)$. As in the proof of 8.16 it now follows for some $\xi \in \mathbb{R}$ that $u_1 - u_2 = \xi \in \mathbb{R}$ almost everywhere in Ω, with the assumptions implying that $\xi = 0$. $\qquad\square$

We remark that the techniques used in the proof for the minimum problem in 8.17 carry over to nonquadratic functionals. We now give some important examples of the set M for this minimum problem. Here all of the occurring boundary values are defined with the help of the trace theorem A8.6.

8.18 Examples of minimum problems.

(1) Let

$$M := \left\{ v \in W^{1,2}(\Omega) \; ; \; v = 0 \; H^{n-1}\text{-almost everywhere on } \partial\Omega \right\}.$$

Then it holds: There exists a unique absolute minimum u in 8.17. It satisfies (8-9) with $M_0 = M$. Hence u is the weak solution of the homogeneous **Dirichlet problem** in 6.5(1) (for $h_i = 0$, $b = 0$).
Note: It holds that $M = W_0^{1,2}(\Omega)$. Hence this is a special case of theorem 6.8, which was shown there for general open and bounded sets $\Omega \subset \mathbb{R}^n$.

(2) Let

$$M := \left\{ v \in W^{1,2}(\Omega) \; ; \; \int_\Omega v \, dL^n = 0 \right\}.$$

In addition, we assume that $\int_\Omega f \, dL^n = 0$. Then it holds: There exists a unique absolute minimum u in 8.17. It satisfies the equality (8-9) for all $v \in W^{1,2}(\Omega)$. Hence u is a weak solution of the homogeneous **Neumann problem** in 6.5(2) (for $h_i = 0$, $b = 0$). The solution to this problem is unique up to an additive constant.
Observe: This result differs from theorem 6.6, as there the Neumann problem was solved for $b > 0$.

(3) Let $u_0, \psi \in W^{1,2}(\Omega)$ be given and let $u_0(x) \geq \psi(x)$ for almost all $x \in \Omega$. Define

$$M := \left\{ v \in W^{1,2}(\Omega) \; ; \; v = u_0 \; H^{n-1}\text{-almost everywhere on } \partial\Omega, \right.$$
$$\left. v \geq \psi \; L^n\text{-almost everywhere in } \Omega \right\}.$$

The corresponding minimum problem is called an **obstacle problem**. Then it holds: There exists a unique solution u to the obstacle problem. It satisfies the variational inequality (8-8).
Special case: For the case $n = 1$, see also E8.8.

(4) Let Lebesgue measurable sets $E_1, E_2 \subset \Omega$ with $L^n(E_1) > 0$ and $L^n(E_2) > 0$, and $\psi_1, \psi_2 \in W^{1,2}(\Omega)$ with $\psi_1 \leq \psi_2$ almost everywhere in Ω be given. Define

$$M := \left\{ v \in W^{1,2}(\Omega) \; ; \; v \geq \psi_1 \; L^n\text{-almost everywhere in } E_1, \right.$$
$$\left. v \leq \psi_2 \; L^n\text{-almost everywhere in } E_2 \right\}.$$

The corresponding minimum problem is called a **double obstacle problem**. Then it holds: There exists a solution u to this obstacle problem and it satisfies the variational inequality (8-8).
Remark: The solution need not be unique.

(5) Let $u_0 \in W^{1,2}(\Omega)$ and let $\Gamma \subset \partial\Omega$ be a closed subset with measure $H^{n-1}(\Gamma) > 0$. Define

$$M := \{ v \in W^{1,2}(\Omega) \;;\; v = u_0 \; H^{n-1}\text{-almost everywhere on } \Gamma \} \,.$$

Then it holds: There exists a unique absolute minimum $u \in M$ in 8.17. It satisfies (8-9) with

$$M_0 = \{ v \in W^{1,2}(\Omega) \;;\; v = 0 \; H^{n-1}\text{-almost everywhere on } \Gamma \} .$$

Definition: Then $u \in W^{1,2}(\Omega)$ is called a **weak solution** of the **mixed boundary value problem**

$$-\sum_{i,j=1}^{n} \partial_i(a_{ij}\partial_j u) + f = 0 \quad \text{in } \Omega,$$

$$u = u_0 \quad \text{on } \Gamma,$$

$$\sum_{i,j=1}^{n} \nu_i a_{ij} \partial_j u = 0 \quad \text{on } \partial\Omega \setminus \Gamma,$$

where ν is the outer normal to Ω defined in A8.5(3). The weak solution in $W^{1,2}(\Omega)$ to this boundary value problem is unique.

Proof (1). The continuity of the trace operator yields that $M \subset W^{1,2}(\Omega)$ is a closed subspace (with S as in A8.6 it holds that $M = \mathcal{N}(S)$). Clearly M is nonempty and satisfies (8-10) (from $v \in M$ and $v + \xi \in M$ it follows for the traces that $v = 0$ and $v + \xi = 0$ almost everywhere on $\partial\Omega$, and so $\xi = 0$). Now 8.17 yields the existence of a unique solution u, which satisfies (8-9) with $M_0 = M$.   □

Proof (2). M is a subspace and contains 0 as the only constant function. In addition, M is closed (the embedding from $W^{1,2}(\Omega)$ into $L^1(\Omega)$ is continuous and the side constraint is continuous on $L^1(\Omega)$). Hence M satisfies the property (8-10), and so 8.17 yields the existence of a unique solution u, which satisfies (8-9) with $M_0 = M$.

For arbitrary $v \in W^{1,2}(\Omega)$ it holds that $\widetilde{v} := v - m(v) \in M$, where

$$m(g) := \fint_{\Omega} g \, d\mathrm{L}^n := \frac{1}{\mathrm{L}^n(\Omega)} \int_{\Omega} g \, d\mathrm{L}^n \quad \text{for } g \in L^1(\Omega) \tag{8-14}$$

denotes the **mean** of g on Ω.

On recalling that $m(f) = 0$, we obtain that (8-9) holds for constant functions, and hence it also holds for $v = \widetilde{v} + m(v)$, as claimed.

Now if $\widetilde{u} \in M$ is another function that satisfies (8-9) for all $v \in W^{1,2}(\Omega)$, then

$$\int_{\Omega} \sum_{i,j=1}^{n} \partial_i v \cdot a_{ij}\partial_j(u - \widetilde{u}) \, d\mathrm{L}^n = 0 \quad \text{for all } v \in W^{1,2}(\Omega).$$

Set $v = u - \widetilde{u}$. Then

$$0 = \int_{\Omega} \sum_{i,j=1}^{n} \partial_i(u - \widetilde{u}) \cdot a_{ij}\partial_j(u - \widetilde{u}) \, d\mathrm{L}^n \geq c_0 \int_{\Omega} |\nabla(u - \widetilde{u})|^2 \, d\mathrm{L}^n.$$

Hence we have that $\nabla(u-\widetilde{u}) = 0$ almost everywhere in Ω. As Ω is connected, it follows that there exists a $\xi \in \mathbb{R}$ such that $\widetilde{u} = u + \xi$ almost everywhere in Ω. $\qquad\square$

Proof (3). M is convex and $u_0 \in M$. We show that M is closed. Let $(u_k)_{k\in\mathbb{N}}$ be a sequence in M that converges in $W^{1,2}(\Omega)$ to a $u \in W^{1,2}(\Omega)$. Then it follows from the trace theorem A8.6 that $u_k \to u$ in $L^2(\partial\Omega)$. On noting that $u_k = u_0$ in $L^2(\partial\Omega)$, we also have that $u = u_0$ in $L^2(\partial\Omega)$. In addition, $u_k \to u$ in $L^2(\Omega)$. Hence there exists a subsequence $k \to \infty$ such that $u_k \to u$ almost everywhere in Ω. Now $u_k \geq \psi$ almost everywhere implies that $u \geq \psi$.

Moreover, (8-10) holds. Indeed, it follows from $v \in M$ and $\widetilde{v} := v+\xi \in M$ that $\xi = \widetilde{v} - v = 0$ almost everywhere on $\partial\Omega$, and so $\xi = 0$. By 8.17, there exists a unique solution to the variational inequality. $\qquad\square$

Proof (4). We have that M is convex and that $\psi_1, \psi_2 \in M$. The closedness of M follows as in the proof of (3). In addition, 8.16(1) is satisfied, e.g. with $u_0 = \psi_1$. To see this, note that if $v := \psi_1 + \xi \in M$ with $\xi \in \mathbb{R}$, then $\xi \geq 0$, since $\mathrm{L}^n(E_1) > 0$. Similarly, we have that $\xi \leq \psi_2 - \psi_1$ on E_2, and so it follows from $\mathrm{L}^n(E_2) > 0$ (on applying either the Hölder inequality (see 3.18) or Jensen's inequality (see E4.10)) that

$$\xi \leq \fint_{E_2} |\psi_2 - \psi_1|\,\mathrm{dL}^n \leq \left(\fint_{E_2} |\psi_2 - \psi_1|^2\,\mathrm{dL}^n\right)^{\frac{1}{2}}$$
$$= (\mathrm{L}^n(E_2))^{-\frac{1}{2}} \|\psi_2 - \psi_1\|_{L^2(E_2)} < \infty.$$

By 8.17, there exists a solution to the minimum problem.
On the uniqueness: In general, there exist several solutions. For example, if $\psi_1 = -1$, $\psi_2 = +1$, $f = 0$, then every constant function $u = \xi$ with $\xi \in [-1,1]$ is a solution. This would no longer be the case if, in addition, Dirichlet data were prescribed on $\partial\Omega$ (e.g. as in (3)). $\qquad\square$

Proof (5). M is convex and $u_0 \in M$. The closedness of M follows as in the proof of (3), on restricting the pointwise argument to the subset $\Gamma \subset \partial\Omega$. The same holds for the proof of (8-10), where now we use that $H^{n-1}(\Gamma) > 0$. Then 8.17 yields the existence of a unique solution. On noting that $M_0 := M - u_0$ is a subspace, we conclude that (8-9) holds. $\qquad\square$

E8 Exercises

Throughout these exercises we let $\mathbb{K} = \mathbb{R}$.

E8.1 Weak limit in $L^p(\mu)$. Let μ be a σ-finite measure and let $f_j, f \in L^p(\mu)$ with $1 \leq p \leq \infty$. Then it holds: If $f_j \to f$ weakly in $L^p(\mu)$ and $f_j \to \widetilde{f}$ μ-almost everywhere as $j \to \infty$, then $\widetilde{f} = f$ μ-almost everywhere.

Solution. Let S_m be as in 3.9(4). It follows from Egorov's theorem A3.18 that for $\varepsilon > 0$ there exists a measurable set $E_\varepsilon \subset S_m$ such that $\mu(S_m \setminus E_\varepsilon) \leq \varepsilon$ and $f_j \to \tilde{f}$ uniformly on E_ε as $j \to \infty$. Given $\zeta \in L^\infty(\mu)$, the map

$$g \longmapsto \int_{E_\varepsilon} \zeta g \, d\mu$$

defines a continuous linear functional on $L^p(\mu)$ (for $p < \infty$ this follows from $\mu(E_\varepsilon) < \infty$ and the Hölder inequality), i.e. an element of $L^p(\mu)'$. Hence we have that

$$\int_{E_\varepsilon} \zeta(f_j - f) \, d\mu \longrightarrow 0 \quad \text{as } j \to \infty.$$

Since $f_j \to \tilde{f}$ uniformly on E_ε,

$$\int_{E_\varepsilon} \zeta(\tilde{f} - f) \, d\mu = 0 \quad \text{for all } \zeta \in L^\infty(\mu).$$

Now set $\zeta(x) = \psi(\tilde{f}(x) - f(x))$, where

$$\psi(z) := \begin{cases} \dfrac{z}{|z|} & \text{for } z \neq 0, \\ 0 & \text{for } z = 0. \end{cases}$$

Then $\zeta(\tilde{f} - f) = |\tilde{f} - f|$, and hence we obtain that $\tilde{f} = f$ almost everywhere on E_ε. Letting $\varepsilon \searrow 0$ and $m \nearrow \infty$ yields the desired result. □

E8.2 Weak limit of a product. Let μ be a σ-finite measure and let $1 < p < \infty$. Moreover, let $f_j \to f$ in $L^p(\mu)$ as $j \to \infty$, let $(g_j)_{j \in \mathbb{N}}$ be bounded in $L^{p'}(\mu)$ and let $g_j \to g$ almost everywhere. Then

$$g_j f_j \longrightarrow gf \quad \text{weakly in } L^1(\mu) \text{ as } j \to \infty.$$

In particular,

$$\int_S g_j f_j \, d\mu \longrightarrow \int_S gf \, d\mu \quad \text{as } j \to \infty.$$

Solution. Otherwise it follows from theorem 6.12 that there exists a $\zeta \in L^\infty(\mu)$ such that for a subsequence $j \to \infty$ and a $\delta > 0$ we have that

$$\left| \int_S g_j f_j \zeta \, d\mu - \int_S gf\zeta \, d\mu \right| \geq \delta \quad \text{for all } j. \tag{E8-1}$$

On recalling from 8.11(2) that $L^{p'}(\mu)$ is reflexive for $1 < p' < \infty$, it follows from theorem 8.10 that there exists a $\tilde{g} \in L^{p'}(\mu)$ such that for a further subsequence $g_j \to \tilde{g}$ weakly in $L^{p'}(\mu)$ as $j \to \infty$. Now E8.1 yields that $\tilde{g} = g$, and hence $g_j \to g$ weakly in $L^{p'}(\mu)$. Moreover, $f_j \zeta \to f\zeta$ converges (strongly) in $L^p(\mu)$ as $j \to \infty$. In this situation we can apply 8.3(6):

If $J : L^p(\mu) \rightarrow \left(L^{p'}(\mu)\right)'$ denotes the isomorphism from 6.12, then $J(f_j\zeta) \rightarrow J(f\zeta)$ converges (strongly) in $\left(L^{p'}(\mu)\right)'$ and hence the second result in 8.3(6) yields that $\langle g_j , J(f_j\zeta)\rangle_{L^{p'}} \rightarrow \langle g, J(f\zeta)\rangle_{L^{p'}}$, in contradiction to (E8-1). $\square$

E8.3 Weak limit of a product. Let $\mu(S) < \infty$ and let $1 < p \le \infty$. Assume that $f_j \rightarrow f$ converges weakly in $L^p(\mu)$ as $j \rightarrow \infty$. In addition, let $g_j : S \rightarrow \mathbb{R}$ be measurable and uniformly bounded, and let $g_j \rightarrow g$ almost everywhere as $j \rightarrow \infty$. Then

$$g_j f_j \longrightarrow gf \quad \text{weakly in } L^1(\mu) \text{ as } j \rightarrow \infty.$$

Solution. Since $|g_j - g|^{p'}$ are uniformly bounded and $\mu(S) < \infty$, it follows for a constant C that

$$|g_j - g|^{p'} \le C \in L^1(\mu).$$

Since these functions converge almost everywhere to 0, it follows from Lebesgue's convergence theorem 3.25 that $|g_j - g|^{p'} \rightarrow 0$ in $L^1(\mu)$, and hence $\zeta g_j \rightarrow \zeta g$ (strongly) in $L^{p'}(\mu)$ as $j \rightarrow \infty$ for all $\zeta \in L^\infty(\mu)$. Moreover, the assumptions state that $f_j \rightarrow f$ weakly in $L^p(\mu)$. In this situation we can apply the first result in 8.3(6) (analogously to the solution of E8.2). $\square$

E8.4 Weak convergence in C^0. Let $S \subset \mathbb{R}^n$ be compact and let $f_j, f \in C^0(S)$. Then

$$f_j \longrightarrow f \text{ weakly in } C^0(S) \quad \Longleftrightarrow \quad \begin{array}{l} \sup\limits_{x \in S} \sup\limits_{j \in \mathbb{N}} |f_j(x)| < \infty \text{ and} \\ f_j(x) \longrightarrow f(x) \text{ as } j \rightarrow \infty \\ \text{for all } x \in S. \end{array}$$

as $j \rightarrow \infty$

Remark: It holds that $\sup_{x \in S} \sup_{j \in \mathbb{N}} |f_j(x)| = \sup_{j \in \mathbb{N}} \sup_{x \in S} |f_j(x)|$.

Solution $\Rightarrow$. By 8.3(5), the sequence $(f_j)_{j \in \mathbb{N}}$ is bounded in $C^0(S)$. Moreover, it follows from 6.23 that the weak convergence is equivalent to

$$\int_S f_j \, d\nu \longrightarrow \int_S f \, d\nu \quad \text{as } j \rightarrow \infty \tag{E8-2}$$

for all $\nu \in rca(S)$. Now choose $\nu = \delta_x$ for $x \in S$, where δ_x denotes the Dirac measure at the point x. $\square$

Solution $\Leftarrow$. We have to show (E8-2). Let $\mu \in rca(S)$ be nonnegative. It follows from Egorov's theorem A3.18 that for $\varepsilon > 0$ there exists a measurable set $E_\varepsilon \subset S$ with $\mu(S \setminus E_\varepsilon) \le \varepsilon$ such that $f_j \rightarrow f$ uniformly on E_ε as $j \rightarrow \infty$. On recalling that the functions f_j are uniformly bounded, say $|f_j| \le C$, we have that

$$\left| \int_S (f_j - f) \, d\mu \right| \leq \underbrace{\mu(E_\varepsilon) \sup_{x \in E_\varepsilon} |f_j(x) - f(x)|}_{\substack{\to \, 0 \text{ as } j \to \infty \\ \text{for every } \varepsilon}} + C \cdot \underbrace{\mu(S \setminus E_\varepsilon)}_{\to \, 0 \text{ as } \varepsilon \to 0} .$$

This yields (E8-2) for μ.

Note: The desired result also holds for arbitrary measures in $rca(S; \mathbb{R})$, as they can be decomposed into their real and imaginary parts, and these further into their positive and negative parts (the nonnegative and nonpositive parts, see the Hahn decomposition A6.2).   □

E8.5 Strong convergence in Hilbert spaces. Let X be a Hilbert space. Then it holds for every sequence $(x_k)_{k \in \mathbb{N}}$ in X that:

$$x_k \longrightarrow x \text{ (strongly) in } X \qquad\qquad x_k \longrightarrow x \text{ weakly in } X \text{ and}$$
$$\text{as } k \to \infty \qquad \Longleftrightarrow \qquad \|x_k\|_X \longrightarrow \|x\|_X \text{ as } k \to \infty.$$

Solution ⇐. We have that

$$\|x_k\|_X^2 = \|x\|_X^2 + 2\mathrm{Re}\,(x_k - x \,,\, x)_X + \|x_k - x\|_X^2 .$$

It follows from the Riesz representation theorem that $(x_k - x \,,\, x)_X \to 0$ as $k \to \infty$, and so the convergence $\|x_k\|_X \to \|x\|_X$ yields the desired result.   □

E8.6 Strong convergence in L^p spaces. Prove that the equivalence in E8.5 also holds for the Banach space $X = L^p(\mu)$ with $1 < p < \infty$.

Solution ⇐. Let $f_k, f \in L^p(\mu)$ be such that $f_k \to f$ weakly in $L^p(\mu)$ as $k \to \infty$, which on recalling theorem 6.12 means that

$$\int_S f_k g \, d\mu \longrightarrow \int_S f g \, d\mu \quad \text{for all } g \in L^{p'}(\mu),$$

and such that $\|f_k\|_{L^p} \to \|f\|_{L^p}$ as $k \to \infty$. We employ the elementary inequality

$$|b|^p \geq |a|^p + p \cdot (b - a) \bullet \left(|a|^{p-2} a\right) + c \cdot \left(|b| + |a|\right)^{p-2} |b - a|^2 \qquad \text{(E8-3)}$$

for $a, b \in \mathbb{R}^m, a \neq 0$, with a constant $c > 0$ depending on m and p (proof see below).

Set $a = f(x)$, if $f(x) \neq 0$, and $b = f_k(x)$. With $g(x) := |f(x)|^{p-2} f(x)$ (we consider the real case), if $f(x) \neq 0$, and $g(x) := 0$ otherwise, it follows that

$$\int_S |f_k|^p \, d\mu \geq \int_S |f|^p \, d\mu + p \cdot \mathrm{Re}\left(\int_S (f_k - f) g \, d\mu\right) + c \cdot \delta_k \qquad \text{(E8-4)}$$

with

$$\delta_k := \int_{S_k} (|f_k| + |f|)^{p-2} |f_k - f|^2 \, d\mu \,,$$

where $S_k := \{x \in S \,;\, |f_k(x)| + |f(x)| > 0\}$. On noting that $g \in L^{p'}(\mu)$, it follows from the assumptions that the second term on the right-hand side of (E8-4) converges to 0, and that the left-hand side converges to the first term on the right-hand side. We conclude that $\delta_k \to 0$ as $k \to \infty$. For $p \geq 2$ this yields the desired result, since

$$\delta_k \geq \int_S |f_k - f|^p \, d\mu \,.$$

For $1 < p < 2$ and $\varepsilon > 0$ let

$$E_{\varepsilon,k} := \{ x \in S_k \,;\, |f_k(x) - f(x)| \geq \varepsilon(|f_k(x)| + |f(x)|) \} \,.$$

Then

$$|f_k - f|^p \leq \begin{cases} \varepsilon^{p-2}(|f_k| + |f|)^{p-2}|f_k - f|^2 & \text{on } E_{\varepsilon,k} \,, \\ \varepsilon^p(|f_k| + |f|)^p \leq 2^{p-1}\varepsilon^p(|f_k|^p + |f|^p) & \text{on } S_k \setminus E_{\varepsilon,k} \,, \end{cases}$$

whence

$$\int_S |f_k - f|^p \, d\mu = \int_{S_k} |f_k - f|^p \, d\mu$$

$$\leq 2^{p-1}\varepsilon^p \int_{S_k \setminus E_{\varepsilon,k}} (|f_k|^p + |f|^p) \, d\mu + \varepsilon^{p-2} \int_{E_{\varepsilon,k}} (|f_k| + |f|)^{p-2}|f_k - f|^2 \, d\mu$$

$$\leq 2^{p-1}\varepsilon^p \underbrace{(\|f_k\|_{L^p}^p + \|f\|_{L^p}^p)}_{\text{bounded in } k} + \varepsilon^{p-2}\delta_k$$

for all ε and k, which yields the desired result.

For the proof of (E8-3) let $a_s := (1-s)a + sb$. As (E8-3) depends continuously on b, we may assume that $a_s \neq 0$ for $0 \leq s \leq 1$. Then

$$|a_1|^p - |a_0|^p = p \int_0^1 |a_s|^{p-2} a_s \bullet (a_1 - a_0) \, ds \,,$$

and hence

$$|a_1|^p - |a_0|^p - p|a_0|^{p-2}a_0 \bullet (a_1 - a_0)$$

$$= p \, (a_1 - a_0) \bullet \int_0^1 \int_0^s \frac{d}{dt}(|a_t|^{p-2}a_t) \, dt \, ds$$

$$= p \int_0^1 \int_0^s |a_t|^{p-2} \left(|a_1 - a_0|^2 + (p-2)\left((a_1 - a_0) \bullet \frac{a_t}{|a_t|} \right)^2 \right) dt \, ds$$

$$\geq p \, (1 + \min(p-2, 0)) \cdot \psi(a_0, a_1) \cdot |a_1 - a_0|^2 \,,$$

with

$$\psi(a_0, a_1) := \int_0^1 \int_0^s |a_t|^{p-2} \, dt \, ds \,.$$

Observe that $\psi(a_0, a_1) = (|a_0| + |a_1|)^{p-2} \psi(b_0, b_1)$ with $b_l := (|a_0| + |a_1|)^{-1} a_l$ for $l = 0, 1$. Hence we need to show that

$$\inf\{\psi(b_0, b_1) ; \ |b_0| + |b_1| = 1\} > 0 \,.$$

For $1 < p \leq 2$ we have that $\psi(b_0, b_1) \geq \frac{1}{2}$, because $|(1 - t)b_0 + tb_1| \leq 1$, and for $p > 2$ the value $\psi(b_0, b_1)$ can converge to 0 only if $b_0 \to 0$ and $b_1 \to 0$. □

E8.7 Weak convergence of oscillating functions. Let $I \subset \mathbb{R}$ be an open, bounded interval and let $1 < p < \infty$.

(1) If $g \in L^\infty(\mathbb{R})$ is a *periodic function* with *period* $\kappa > 0$, i.e. $g(x + \kappa) = g(x)$ for almost all x, and if

$$\frac{1}{\kappa} \int_0^\kappa g(x) \, dx = \lambda \,,$$

then the functions $f_n(x) := g(nx)$ converge weakly in $L^p(I)$ to λ as $n \to \infty$.

(2) Let $\alpha, \beta \in \mathbb{R}$, $0 < \theta < 1$, and

$$f_n(x) := \begin{cases} \alpha & \text{for } k < nx < k + \theta, \ k \in \mathbb{Z}, \\ \beta & \text{for } k + \theta < nx < k + 1, \ k \in \mathbb{Z}. \end{cases}$$

Then the functions f_n converge weakly in $L^p(I)$ to the constant function $\theta\alpha + (1 - \theta)\beta$ as $n \to \infty$.

(3) Find functions $f_n, f, g_n, g \in L^\infty(I)$ such that $f_n \to f$, $g_n \to g$ weakly in $L^p(I)$ as $n \to \infty$, but such that $f_n g_n$ does not converge weakly to fg.

Solution (1). Without loss of generality let $\lambda = 0$ (otherwise replace g with $g - \lambda$). Then the assumptions on g yield that

$$h(x) := \int_0^x g(y) \, dy$$

defines a continuous function that is bounded on all of $\mathbb{R}$. If $[a, b] \subset I$, then

$$\int_a^b f_n(x) \, dx = \frac{1}{n} \left(h(nb) - h(na) \right) \longrightarrow 0 \quad \text{as } n \to \infty.$$

Consequently,

$$\int_I f_n(x)\zeta(x) \, dx \longrightarrow 0 \quad \text{as } n \to \infty$$

for all step functions ζ. As these step functions are dense in $L^{p'}(I)$, and as the functions f_n are bounded in $L^p(I)$, we obtain the same result also for all $\zeta \in L^{p'}(I)$ (see E5.4). □

Solution (2). This follows from (1), on noting that

$$\int_0^1 f_1(x)\,\mathrm{d}x = \theta\alpha + (1-\theta)\beta\,.$$

□

Solution (3). Let f_n be as in (2) and define g_n correspondingly for the values $\widetilde{\alpha}, \widetilde{\beta} \in \mathbb{R}$ and the same value θ. Then (2) yields the following weak convergence results in $L^p(I)$:

$$f_n \longrightarrow \theta\alpha + (1-\theta)\beta\,,$$
$$g_n \longrightarrow \theta\widetilde{\alpha} + (1-\theta)\widetilde{\beta}\,,$$
$$f_n g_n \longrightarrow \theta\alpha\widetilde{\alpha} + (1-\theta)\beta\widetilde{\beta}\,.$$

Now the equation

$$\theta\alpha\widetilde{\alpha} + (1-\theta)\beta\widetilde{\beta} = \left(\theta\alpha + (1-\theta)\beta\right)\left(\theta\widetilde{\alpha} + (1-\theta)\widetilde{\beta}\right)$$

is equivalent to $(\alpha - \beta)(\widetilde{\alpha} - \widetilde{\beta}) = 0$, and so for $\alpha \neq \beta$ and $\widetilde{\alpha} \neq \widetilde{\beta}$ we obtain the desired example.

□

E8.8 Variational inequality. Find the solution $u \in W^{1,2}(\Omega)$ of the obstacle problem in 8.18(3) for $n = 1$, $\Omega = \,]-1,1[\, \subset \mathbb{R}$, $u_0 \geq 0$, $\psi = 0$, $f = 1$ and $a = 1$.

Solution. (On recalling E3.6, we use the fact that for $n = 1$ functions in $W^{1,2}(\Omega)$ can be identified with functions in $C^0(\overline{\Omega})$.) Let

$$M = \left\{ v \in W^{1,2}(\Omega)\;;\; v \geq 0 \text{ almost everywhere in } \Omega, \right.$$
$$\left. v(\pm 1) = u_\pm := u_0(\pm 1) \right\}.$$

Then $u \in M \cap C^0([-1,1])$ and

$$\int_{-1}^1 \left((u-v)'u' + (u-v)\right)\mathrm{d}L^1 \leq 0 \quad \text{for all } v \in M.$$

First we consider an interval $]a,b[$ in which $u > 0$. If $\zeta \in C_0^\infty(]a,b[)$, then $u \geq c$ in supp ζ for a $c > 0$, and hence $u + \varepsilon\zeta \in M$ for small $|\varepsilon|$. It follows that

$$0 = \int_a^b (\zeta'u' + \zeta)\,\mathrm{d}L^1 = \int_a^b \zeta'v'\,\mathrm{d}L^1\,,$$

where $v(x) := u(x) - \frac{1}{2}x^2$. This implies (see E8.9) that v is linear in $]a,b[$, and hence there exist $d_0, d_1 \in \mathbb{R}$ such that

$$u(x) = \frac{x^2}{2} + d_1 x + d_0 \quad \text{for } a < x < b.$$

On choosing $]a, b[\subset \{u > 0\}$ maximally, i.e. $u(a) = 0$, if $a > -1$, and $u(b) = 0$, if $b < 1$, the obtained characterization of u implies that we have to distinguish the following cases:

$$a = -1, \ b = 1, \quad \text{and so } u > 0 \text{ in }]-1, 1[,$$
$$a > -1, \ b = 1, \quad \text{and so } u > 0 \text{ in }]a, 1] \text{ with } u(a) = 0,$$
$$a = -1, \ b < 1, \quad \text{and so } u > 0 \text{ in } [-1, b[\text{ with } u(b) = 0.$$

Hence overall we obtain the following two cases for u:

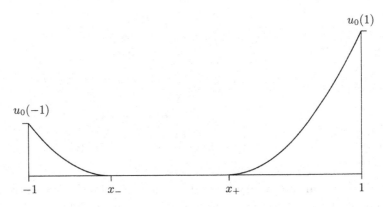

Fig. 8.2. *Solution of the obstacle problem*

(1) $u > 0$ in $]-1, 1[$,
(2) There exist $-1 \le x_- \le x_+ \le 1$ such that $u(x) = 0$ for $x_- \le x \le x_+$ and $u(x) > 0$ otherwise.

In the case (1) the values d_0 and d_1 are determined by the boundary conditions, and we obtain

$$u(x) = \tfrac{1}{2}\left(x^2 - 1 + (u_+ - u_-)x + u_+ + u_-\right)$$

and the necessary condition

$$|u_+ - u_-| \ge 2 \quad \text{or} \quad u_+ + u_- > 1 + \tfrac{1}{4}(u_+ - u_-)^2. \qquad \text{(E8-5)}$$

Correspondingly, in the case (2) we obtain for certain $s_\pm \ge 0$ that

$$u(x) = \tfrac{1}{2}(x - x_+)^2 + s_+(x - x_+)$$
$$\text{for } x \ge x_+ \text{ with } (1 - x_+)s_+ = u_+ - \tfrac{1}{2}(1 - x_+)^2 \ge 0,$$
$$u(x) = \tfrac{1}{2}(x_- - x)^2 + s_-(x_- - x)$$
$$\text{for } x \le x_- \text{ with } (1 + x_-)s_- = u_- - \tfrac{1}{2}(1 + x_-)^2 \ge 0.$$

The uniqueness of the solution means that $x_\pm$ are uniquely determined by $u_\pm$. Hence we further investigate the variational inequality. For $\zeta \in C_0^\infty(]-1,1[)$ with $\zeta \geq 0$ it holds that $u + \zeta \in M$, and so the variational inequality yields that

$$
0 \leq \int_{-1}^1 (\zeta' u' + \zeta) \, dL^1
$$

$$
= \int_{x_+}^1 \zeta'(x)(x - x_+ + s_+) \, dx + \int_{-1}^{x_-} \zeta'(x)(x - x_- - s_-) \, dx + \int_{-1}^1 \zeta \, dL^1
$$

$$
= -\zeta(x_+) s_+ - \zeta(x_-) s_- + \int_{x_-}^{x_+} \zeta \, dL^1 .
$$

If $x_+ < 1$ set $\zeta(x) := \max(0, 1 - \frac{1}{\delta}|x - x_+|)$ and obtain as $\delta \to 0$ that $s_+ \leq 0$. Together with the above inequality for s_+ we obtain that $s_+ = 0$, and similarly for $s_- = 0$. Therefore,

$$
u(x) = \begin{cases} \frac{1}{2}(x_- - x)^2 & \text{for } x \leq x_-, \\ 0 & \text{for } x_- \leq x \leq x_+, \\ \frac{1}{2}(x - x_+)^2 & \text{for } x \geq x_+, \end{cases}
$$

where

$$
u_+ - \tfrac{1}{2}(1 - x_+)^2 = 0 \quad \text{and} \quad u_- - \tfrac{1}{2}(1 + x_-)^2 = 0 .
$$

Apart from $(u_-, u_+) = (0, 2)$ or $(2, 0)$, this case is complementary to the case (E8-5). $\qquad\square$

E8.9 A fundamental lemma. Let $\Omega \subset \mathbb{R}^n$ be open and connected, and suppose that $u \in L^1_{loc}(\Omega)$ satisfies

$$
\int_\Omega u \cdot \partial_i \zeta \, dL^n = 0 \quad \text{for } \zeta \in C_0^\infty(\Omega) \text{ and } i = 1, \ldots, n.
$$

Then u is (almost everywhere) a constant function.

Solution. Let B be a ball with $\overline{B} \subset \Omega$ and let $(\varphi_\varepsilon)_{\varepsilon > 0}$ be a standard Dirac sequence. On setting $\widetilde{\varphi}_\varepsilon(y) := \varphi_\varepsilon(-y)$ we have that $\zeta * \widetilde{\varphi}_\varepsilon \in C_0^\infty(\Omega)$ for $\zeta \in C_0^\infty(B)$ and $\varepsilon < \mathrm{dist}(B, \partial\Omega)$, and so

$$
-\int_\Omega \partial_i(u * \varphi_\varepsilon) \zeta \, dL^n = \int_\Omega (u * \varphi_\varepsilon) \partial_i \zeta \, dL^n = \int_\Omega u \, \partial_i(\zeta * \widetilde{\varphi}_\varepsilon) \, dL^n = 0 .
$$

Hence $\nabla(u * \varphi_\varepsilon) = 0$ in B, which yields that $u * \varphi_\varepsilon$ is constant in B. On recalling that $u * \varphi_\varepsilon \to u$ in $L^1(B)$ as $\varepsilon \to 0$, it follows that u is also a constant almost everywhere in B. As Ω is path connected (see remark above 8.16), this constant does not depend on B. $\qquad\square$

A8 Properties of Sobolev functions

Here we will derive properties of functions in $W^{m,p}(\Omega)$, where we treat bounded sets Ω with Lipschitz boundary $\partial\Omega$ (see definition A8.2). This class of domains, on one hand, allows a functional analytically uniform presentation of the theory of Sobolev spaces, and on the other hand, this class is of major importance in applications, because it contains domains with edges and corners, as they occur in flow domains and also in workpieces.

In applications to boundary value problems on such domains, e.g. on cuboids, often different boundary conditions are prescribed on different sides of the domain (see the mixed boundary value problem in 8.18(5)). For the weak formulation of these boundary value problems we need to prove that functions in $W^{1,p}(\Omega)$ have weak boundary values on $\partial\Omega$ (see A8.6). Then we show (see A8.10) that $W_0^{1,p}(\Omega)$ consists precisely of those functions in $W^{1,p}(\Omega)$ that have weak boundary values 0. This belatedly justifies the weak formulation of the homogeneous Dirichlet problem in 6.5.

We begin with Rellich's embedding theorem A8.1 for $W_0^{m,p}(\Omega)$ and A8.4 for $W^{m,p}(\Omega)$.

A8.1 Rellich's embedding theorem. Let $\Omega \subset \mathbb{R}^n$ be open and bounded, let $1 \le p < \infty$ and let $m \ge 1$. If $u_k \in W_0^{m,p}(\Omega)$ for $k \in \mathbb{N}$ and if $u \in W_0^{m-1,p}(\Omega)$, then

$$
\begin{array}{l}
(u_k)_{k\in\mathbb{N}} \text{ bounded in } W_0^{m,p}(\Omega), \\
u_k \to u \text{ weakly in } W_0^{m-1,p}(\Omega) \\
\text{as } k \to \infty
\end{array}
\implies
\begin{array}{l}
u_k \to u \text{ (strongly) in } W_0^{m-1,p}(\Omega) \\
\text{as } k \to \infty.
\end{array}
$$

Remark: On recalling 8.3(5), it follows if $u_k, u \in W_0^{m,p}(\Omega)$ for $k \in \mathbb{N}$ that

$$
\begin{array}{l}
u_k \to u \text{ weakly in } W_0^{m,p}(\Omega) \\
\text{as } k \to \infty
\end{array}
\implies
\begin{array}{l}
u_k \to u \text{ (strongly) in } W_0^{m-1,p}(\Omega) \\
\text{as } k \to \infty.
\end{array}
$$

Proof. Let $m = 1$. Hence u_k are bounded in $W_0^{1,p}(\Omega)$ and converge weakly in $L^p(\Omega)$ towards u. (For $m > 1$ apply the proof below for $|s| \le m - 1$ to $\partial^s u_k$ in place of u_k. It holds that $\partial^s u_k$ are bounded in $W_0^{1,p}(\Omega)$ and, by 8.4(3), they converge weakly in $L^p(\Omega)$ to $\partial^s u$.)

Extend u_k, u to $\mathbb{R}^n \setminus \Omega$ by 0. Then, by assumption, $u_k \in W^{1,p}(\mathbb{R}^n)$ (see 3.29), with support in $\overline{\Omega}$, and moreover u_k are bounded in $W^{1,p}(\mathbb{R}^n)$ converging by 8.4(1) weakly in $L^p(\mathbb{R}^n)$ towards u.

Now if $(\varphi_\varepsilon)_{\varepsilon>0}$ is a standard Dirac sequence, then $\varphi_\varepsilon * u_k \in C_0^\infty(\mathbb{R}^n)$ and for every $\varepsilon > 0$

$$\varphi_\varepsilon * u_k \to \varphi_\varepsilon * u \quad \text{as } k \to \infty \text{ in } L^p(\mathbb{R}^n). \tag{A8-1}$$

To see this, consider for $x \in \mathbb{R}^n$ the functionals $\Psi_\varepsilon(x) \in L^p(\mathbb{R}^n)'$ defined by

$$\langle v\, , \Psi_\varepsilon(x)\rangle_{L^p} := \int_{\mathbb{R}^n} v(y)\varphi_\varepsilon(x-y)\,dy \quad \text{for } v \in L^p(\mathbb{R}^n).$$

If $x_k \to x$ converges as $k \to \infty$, then $\varphi_\varepsilon(x_k - \cdot) \to \varphi_\varepsilon(x - \cdot)$ converges uniformly on $\mathbb{R}^n$ for $\varepsilon > 0$, hence $\Psi_\varepsilon(x_k) \to \Psi_\varepsilon(x)$ converges in $L^p(\mathbb{R}^n)'$. Since, by assumption, $u_k \to u$ weakly in $L^p(\mathbb{R}^n)$, we obtain, see the second result in 8.3(6),

$$(\varphi_\varepsilon * u_k)(x_k) = \langle u_k\, , \Psi_\varepsilon(x_k)\rangle_{L^p} \longrightarrow \langle u\, , \Psi_\varepsilon(x)\rangle_{L^p} = (\varphi_\varepsilon * u)(x).$$

This shows that $\varphi_\varepsilon * u_k \to \varphi_\varepsilon * u$ locally uniformly on $\mathbb{R}^n$. As $\varphi_\varepsilon * u_k$ and $\varphi_\varepsilon * u$ vanish outside the bounded set $\overline{B}_\varepsilon(\Omega)$, we obtain the result (A8-1). Moreover,

$$\|v - \varphi_\varepsilon * v\|_{L^p} \le \varepsilon\|\nabla v\|_{L^p} \tag{A8-2}$$

for all $v \in W^{1,p}(\mathbb{R}^n)$ with compact support. For the proof of (A8-2) observe that the left- and right-hand sides depend continuously on v with respect to the $W^{1,p}$-norm. Hence on approximating v (e.g. by convolution as in 4.23), it is sufficient to show (A8-2) for $v \in C_0^\infty(\mathbb{R}^n)$. Then

$$(v - \varphi_\varepsilon * v)(x) = \int_{\mathbb{R}^n} \varphi_\varepsilon(y)\big(v(x) - v(x-y)\big)\,dy$$
$$= \int_{\mathbb{R}^n} \varphi_\varepsilon(y)\Big(\int_0^1 \nabla v(x-sy)\bullet y\,ds\Big)\,dy\,,$$

and so it follows from 4.13(1) that

$$\|v - \varphi_\varepsilon * v\|_{L^p} \le \sup_{h\in\mathrm{supp}\,\varphi_\varepsilon} \left\|\int_0^1 \nabla v(\cdot - sh)\bullet h\,ds\right\|_{L^p}$$
$$\le \varepsilon \sup_{|h|\le\varepsilon} \int_0^1 \|\nabla v(\cdot - sh)\|_{L^p}\,ds = \varepsilon\|\nabla v\|_{L^p}\,.$$

Combining (A8-1) and (A8-2) yields that

$$\|u - u_k\|_{L^p} \le \|u - \varphi_\varepsilon * u\|_{L^p} + \underbrace{\|\varphi_\varepsilon * u - \varphi_\varepsilon * u_k\|_{L^p}}_{\substack{\to\, 0 \text{ as } k \to \infty \\ \text{for every } \varepsilon}} + \varepsilon\|\nabla u_k\|_{L^p}\,.$$

Noting that ∇u_k are bounded in $L^p(\mathbb{R}^n)$ and recalling from 4.15(2) that $\varphi_\varepsilon * u \to u$ in $L^p(\mathbb{R}^n)$ as $\varepsilon \to 0$, we obtain the desired result. $\qquad\square$

A8.2 Lipschitz boundary. Let $\Omega \subset \mathbb{R}^n$ be open and bounded. We say that Ω has a **Lipschitz boundary** if $\partial\Omega$ can be covered by finitely many open sets $U^1, \ldots, U^l$ such that $\partial\Omega \cap U^j$ for $j = 1,\ldots,l$ is the graph of a Lipschitz continuous function with $\Omega \cap U^j$ in each case lying on one side of this graph. This means the following: There exists an $l \in \mathbb{N}$ and for

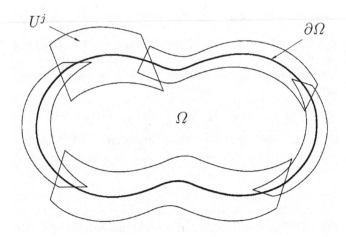

Fig. 8.3. *Cover of the boundary*

$j = 1, \ldots, l$ there exist a Euclidean coordinate system $e_1^j, \ldots, e_n^j$ in $\mathbb{R}^n$, a reference point $y^j \in \mathbb{R}^{n-1}$, numbers $r^j > 0$ and $h^j > 0$ and a Lipschitz continuous function $g^j : \mathbb{R}^{n-1} \to \mathbb{R}$, such that with the notation

$$x_{,n}^j := (x_1^j, \ldots, x_{n-1}^j), \quad \text{where } x = \sum_{i=1}^n x_i^j e_i^j ,$$

it holds that

$$U^j = \left\{ x \in \mathbb{R}^n \ ; \ \left| x_{,n}^j - y^j \right| < r^j \text{ and } \left| x_n^j - g^j(x_{,n}^j) \right| < h^j \right\} ,$$

and for $x \in U^j$

$$
\begin{aligned}
x_n^j = g^j(x_{,n}^j) &\implies x \in \partial\Omega , \\
0 < x_n^j - g^j(x_{,n}^j) < h^j &\implies x \in \Omega , \\
0 > x_n^j - g^j(x_{,n}^j) > -h^j &\implies x \notin \Omega
\end{aligned}
\qquad \text{(A8-3)}
$$

(hence $U^j \cap \Omega = Q^j$, see Fig. 8.4), and

$$\partial\Omega \subset \bigcup_{j=1}^l U^j .$$

Furthermore, we may then add another open set U^0 with $\overline{U^0} \subset \Omega$ such that $U^0, \ldots, U^l$ cover all of $\overline{\Omega}$.

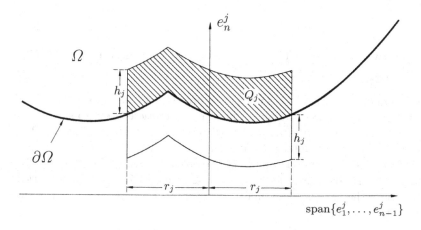

Fig. 8.4. *Local boundary neighbourhood*

A8.3 Localization. Let Ω be as in A8.2. We prove results for Sobolev functions by localizing these functions with respect to the open cover U^j, $j = 0, \ldots, l$ in A8.2. We choose a partition of unity $\eta^0, \ldots, \eta^l$ on $\overline{\Omega}$ with respect to this cover (see 4.20), i.e. $\eta^j \in C^\infty(\mathbb{R}^n)$ with compact support $\mathrm{supp}\,(\eta^j) \subset U^j$ (this means $\eta^j \in C_0^\infty(U^j)$) and

$$0 \le \eta^j \le 1 \text{ in } \mathbb{R}^n \text{ and } \sum_{j=0}^{l} \eta^j = 1 \text{ on } \overline{\Omega}.$$

Now if $u \in W^{m,p}(\Omega)$, then

$$u = \sum_{j=0}^{l} \eta^j u \text{ in } \Omega.$$

In particular, $\eta^0 u \in W^{m,p}(\Omega)$ with compact support in Ω and for $j = 1, \ldots, l$ we have that $\eta^j u \in W^{m,p}(\Omega^j)$, where

$$\Omega^j := \{x \in \mathbb{R}^n ;\ 0 < x_n^j - g^j(x_{,n}^j)\},$$

with $(\eta^j u)(x) = 0$ if $\left| x_{,n}^j - y^j \right| \ge r^j$ or $x_n^j - g^j(x_{,n}^j) \ge h^j$.

A8.4 Rellich's embedding theorem. Let $\Omega \subset \mathbb{R}^n$ be open and bounded with Lipschitz boundary, let $1 \le p < \infty$ and let $m \ge 1$. If $u_k \in W^{m,p}(\Omega)$ for $k \in \mathbb{N}$ and if $u \in W^{m-1,p}(\Omega)$, then

$(u_k)_{k \in \mathbb{N}}$ bounded in $W^{m,p}(\Omega)$

$u_k \to u$ weakly in $W^{m-1,p}(\Omega)$ $\implies$ $u_k \to u$ (strongly) in $W^{m-1,p}(\Omega)$

as $k \to \infty$ $\qquad\qquad\qquad\qquad\qquad$ as $k \to \infty$.

Remark: On recalling 8.3(5), it follows if $u_k, u \in W^{m,p}(\Omega)$ for $k \in \mathbb{N}$ that

$$u_k \to u \text{ weakly in } W^{m,p}(\Omega) \qquad \Longrightarrow \qquad u_k \to u \text{ (strongly) in } W^{m-1,p}(\Omega)$$

$$\text{as } k \to \infty \qquad\qquad\qquad\qquad\qquad \text{as } k \to \infty .$$

Proof. Similarly to A8.1, it is sufficient to prove the theorem for $m = 1$. With the notations as in A8.3, we have that the assumptions are also satisfied by $u_k^j := \eta^j u_k$ and $u^j := \eta^j u$, and we need to show that $u_k^j \to u^j$ in $L^p(\Omega)$ as $k \to \infty$. For $j = 0$ this follows from A8.1.

For $j \geq 1$ this follows on replicating the proof of A8.1. The proofs of (A8-1) and (A8-2) carry over (for the proof of (A8-2) use 4.24 for the approximation) if we replace the integration domain $\mathbb{R}^n$ with Ω^j. Here we have to make sure that in the convolution

$$(\varphi_\varepsilon * v)(x) = \int_{\mathbb{R}^n} \varphi_\varepsilon(x - y) v(y) \, \mathrm{d}y \quad \text{for } v \in W^{1,p}(\Omega^j)$$

the function $y \mapsto \varphi_\varepsilon(x - y)$ has compact support in Ω^j for $x \in \Omega^j$. By the definition of Ω^j this means that

$$x_n^j > g^j(x_{,n}^j), \ \varphi_\varepsilon(x - y) \neq 0 \quad \Longrightarrow \quad y_n^j > g^j(y_{,n}^j).$$

If λ denotes the Lipschitz constant of g^j, then the above holds if

$$\varphi_\varepsilon(z) \neq 0 \quad \Longrightarrow \quad z_n^j < -\lambda |z_{,n}^j| ,$$

i.e. we need to choose the function φ, on which the Dirac sequence $(\varphi_\varepsilon)_{\varepsilon > 0}$ is based, so that

$$\varphi \in C_0^\infty \left(\{ z \in B_1(0) \ ; \ z_n^j < -\lambda |z_{,n}^j| \} \right),$$

which is satisfied, for example, for $\varphi \in C_0^\infty \left(B_\delta \left(-\frac{1}{2} e_n^j \right) \right)$ with $0 < \delta < \frac{1}{2}(1 + \lambda)^{-1}$. This choice has the property that for $x \in \Omega^j$ and $\varphi_\varepsilon(x - y) \neq 0$ the segment connecting x and y lies in Ω^j.

Remark: Another possibility (for $m = 1$) is to extend the functions u_k^j, u^j to functions in $W^{1,p}(\mathbb{R}^n)$ with compact support (see the proof of A8.12), and then apply A8.1. □

The corresponding result for $p = \infty$ plays a special role, because for domains Ω with Lipschitz boundary it holds that $W^{m,\infty}(\Omega) = C^{m-1,1}(\overline{\Omega})$ (see theorem 10.5(2)). The assertion of Rellich's embedding theorem for $p = \infty$ then follows from the Arzelà-Ascoli theorem. The argument for $m = 1$ is as follows: Every sequence bounded in $C^{0,1}(\overline{\Omega})$ contains a subsequence that converges in $C^0(\overline{\Omega})$. But as every cluster point has to coincide with the weak limit, the whole sequence converges strongly in $C^0(\overline{\Omega})$.

Now we want to show that Sobolev functions in $W^{1,p}(\Omega)$ in a weak sense have boundary values in $L^p(\partial\Omega)$. To this end, we first define spaces of functions that are integrable on $\partial\Omega$.

A8.5 Boundary integral. Let Ω be open and bounded with Lipschitz boundary and let Y be a Banach space.

(1) We call $f : \partial\Omega \to Y$ measurable and integrable, respectively, if with the notations as in A8.3 it holds for $j = 1, \dots, l$ that the functions

$$y \longmapsto (\eta^j f)\Big(\sum_{i=1}^{n-1} y_i e_i^j + g^j(y)e_n^j\Big) \quad \text{for } y \in \mathbb{R}^{n-1} \text{ with } |y - y^j| < r^j$$

are measurable and integrable, respectively, with respect to the $(n-1)$-dimensional Lebesgue measure. The boundary integral of f on $\partial\Omega$ is then defined by

$$\int_{\partial\Omega} f \, d\mathrm{H}^{n-1} := \sum_{j=1}^{l} \int_{\partial\Omega} \eta^j f \, d\mathrm{H}^{n-1} \,,$$

where we define, if $\operatorname{supp} h \subset U^j$,

$$\int_{\partial\Omega} h \, d\mathrm{H}^{n-1} := \int_{\mathbb{R}^{n-1}} h\Big(\sum_{i=1}^{n-1} y_i e_i^j + g^j(y)e_n^j\Big) \sqrt{1 + |\nabla g^j(y)|^2} \, d\mathrm{L}^{n-1}(y) \,.$$

Here $\nabla g^j \in L^\infty_{\mathrm{loc}}(\mathbb{R}^{n-1}; \mathbb{R}^{n-1})$, since theorem 10.5(2) implies that the Lipschitz continuous function $g^j : \mathbb{R}^{n-1} \to \mathbb{R}$ lies in $W^{1,\infty}_{\mathrm{loc}}(\mathbb{R}^{n-1})$. Hence the last integral represents a generalization of the surface integral on smooth hypersurfaces as introduced in 3.10(4). Claim: This definition of the integral is independent of the local partition and independent of the representation of the boundary.

(2) For $1 \le p \le \infty$, let

$$L^p(\partial\Omega; Y) := \big\{ f : \partial\Omega \to Y \, ; \, f \text{ is measurable and } \|f\|_{L^p(\partial\Omega)} < \infty \big\} \,,$$

where for $1 \le p < \infty$

$$\|f\|_{L^p(\partial\Omega)} := \Big(\int_{\partial\Omega} |f|^p \, d\mathrm{H}^{n-1}\Big)^{\frac{1}{p}} \,, \quad \text{and} \quad \|f\|_{L^\infty(\partial\Omega)} := \operatorname*{ess\,sup}_{\partial\Omega} |f|$$

with the ess sup-norm defined analogously to 3.15. Then $L^p(\partial\Omega; Y)$ with this norm is a Banach space for $1 \le p \le \infty$, and for $p < \infty$ the set $\big\{ f|_{\partial\Omega} \, ; \, f \in C^\infty(\mathbb{R}^n; Y) \big\}$ is dense in $L^p(\partial\Omega; Y)$.

(3) We define the **outer normal** to Ω at the point $x \in \partial\Omega$ as

$$\nu_\Omega(x) := \big(1 + |\nabla g^j(y)|^2\big)^{-\frac{1}{2}} \Big(\sum_{i=1}^{n-1} \partial_i g^j(y)e_i^j - e_n^j\Big)$$

$$\text{for } x = \sum_{i=1}^{n-1} y_i e_i^j + g^j(y)e_n^j \in U^j \text{ with } |y - y^j| < r^j.$$

It holds that ν_Ω is measurable on $\partial\Omega$ with $|\nu_\Omega| = 1$, and hence $\nu_\Omega \in L^\infty(\partial\Omega; \mathbb{R}^n)$. The definition of ν_Ω is independent of the local representation of the boundary. With the above representation of x, the normal $\nu_\Omega(x)$ is perpendicular to the tangent vectors

$$\tau_k(x) := \partial_{y_k}\left(\sum_{i=1}^{n-1} y_i e_i^j + g^j(y)e_n^j\right) = e_k^j + \partial_k g^j(y)e_n^j \quad \text{for } 1 \leq k \leq n-1.$$

In addition, $\nu_\Omega(x)$ points outward, i.e. $x + \varepsilon\nu_\Omega(x) \notin \Omega$ for $\varepsilon > 0$ sufficiently small, if g is differentiable in y.

Proof (1). In a small open set $U \subset \mathbb{R}^n$ we consider two different representations of $\partial\Omega$ as defined in A8.2, i.e. we consider two coordinate systems $e_1, \ldots, e_n$ and $\tilde{e}_1, \ldots, \tilde{e}_n$, two Lipschitz continuous functions $g : \mathbb{R}^{n-1} \to \mathbb{R}$ and $\tilde{g} : \mathbb{R}^{n-1} \to \mathbb{R}$ and two bounded open sets $V, \tilde{V} \subset \mathbb{R}^{n-1}$, such that with $\Gamma := \partial\Omega \cap U$

$$\left\{\sum_{i=1}^{n-1} y_i e_i + g(y)e_n \; ; \; y \in V\right\} = \left\{\sum_{i=1}^{n-1} \tilde{y}_i\tilde{e}_i + \tilde{g}(\tilde{y})\tilde{e}_n \; ; \; \tilde{y} \in \tilde{V}\right\} = \Gamma.$$

On setting

$$\psi(y) := \sum_{i=1}^{n-1} y_i e_i + g(y)e_n \quad \text{for } y \in \mathbb{R}^{n-1},$$

and similarly for $\tilde{\psi}$, we need to show that for every function $f : \Gamma \to \mathbb{R}$ with $\text{supp}\, f \subset U$ it holds that:

$$f \circ \psi \text{ integrable} \quad \Longleftrightarrow \quad f \circ \tilde{\psi} \text{ integrable}$$

and

$$\int_V f(\psi(y))\sqrt{1 + |\nabla g(y)|^2}\, dy = \int_{\tilde{V}} f(\tilde{\psi}(\tilde{y}))\sqrt{1 + |\nabla\tilde{g}(\tilde{y})|^2}\, d\tilde{y}. \quad (A8\text{-}4)$$

Consider the transformation $\tau := \tilde{\psi}^{-1} \circ \psi$, hence $y \mapsto \tilde{y} = \tau(y)$. Since

$$\left|y^1 - y^2\right| \leq \left|\psi(y^1) - \psi(y^2)\right| \leq \sqrt{1 + \text{Lip}(g)^2}\left|y^1 - y^2\right|,$$

$\psi : V \to \Gamma$ is a Lipschitz continuous map with a Lipschitz continuous inverse $\psi^{-1} : \Gamma \to V$, and the same holds for $\tilde{\psi}$. This implies that $\tau : V \to \tilde{V}$ is bijective and that τ and τ^{-1} are Lipschitz continuous. Hence $f \circ \psi$ is measurable if and only if $f \circ \tilde{\psi}$ is measurable (use 4.27).

In order to prove the integral identity, we first consider the case where $f \in C_0^0(U)$ and $g \in C^1(V)$. Then also $\tilde{g}$ is continuously differentiable. To see this, note that the differentiability of g, and therefore ψ, is equivalent to

$$\psi(y) - \psi(y_0) - P_{\psi(y_0)}(\psi(y) - \psi(y_0)) = o(|\psi(y) - \psi(y_0)|)$$

as $y \to y_0$. Hereby P_{x_0} is the orthogonal projection on the tangent space of Γ in $x_0 := \psi(y_0)$. Now, we have $\psi(y) = \widetilde{\psi}(\widetilde{y})$ if $\widetilde{y} = \tau(y)$, and since τ is continuous, it follows that $y \to y_0$ implies $\widetilde{y} \to \widetilde{y}_0$. Hence

$$\widetilde{\psi}(\widetilde{y}) - \widetilde{\psi}(\widetilde{y}_0) - P_{x_0}(\widetilde{\psi}(\widetilde{y}) - \widetilde{\psi}(\widetilde{y}_0)) = o(|\widetilde{\psi}(\widetilde{y}) - \widetilde{\psi}(\widetilde{y}_0)|)$$

as $\widetilde{y} \to \widetilde{y}_0$. But this is equivalent to the differentiability of $\widetilde{\psi}$ and thus also $\widetilde{g}$. Also the differentiability of τ and τ^{-1} is shown. It follows from the (classical) change-of-variables theorem for C^1-transformations that for every function $\widetilde{f} \in C_0^0(\widetilde{V})$

$$\int_{\widetilde{V}} \widetilde{f} \, d\mathrm{L}^{n-1} = \int_V \widetilde{f} \circ \tau \, |\det D\tau| \, d\mathrm{L}^{n-1} \, .$$

Let $\widetilde{f}(\widetilde{y}) := f(\widetilde{\psi}(\widetilde{y})) \sqrt{1 + |\nabla \widetilde{g}(\widetilde{y})|^2}$. Then we need to show that

$$\sqrt{1 + |\nabla \widetilde{g} \circ \tau|^2} \, |\det D\tau| = \sqrt{1 + |\nabla g|^2} \, .$$

But since $(D\widetilde{\psi}) \circ \tau \, D\tau = D\psi$, this reduces to a purely algebraic result for determinants. Hence in this case the integral identity (A8-4) is proved.

If g is only Lipschitz continuous and $f : \Gamma \to \mathbb{R}$ with $f \in C_0^0(U)$, we shall approximate g by continuously differentiable functions. Let $\operatorname{supp} f \circ \psi \subset V_0$ with an open connected subset V_0 satisfying $\overline{V_0} \subset V$. With $\tau = (\tau_1, \ldots, \tau_{n-1})$ we have for $y \in V$ that

$$\sum_{j=1}^{n-1} \tau_j(y) \widetilde{e}_j + \widetilde{g}(\tau(y)) \widetilde{e}_n = \sum_{i=1}^{n-1} y_i e_i + g(y) e_n \, . \tag{A8-5}$$

In the case that $\widetilde{e}_n \neq e_n$, an $(n-2)$-dimensional subspace of $\mathbb{R}^n$ is given by $\operatorname{span}\{e_1, \ldots, e_{n-1}\} \cap \operatorname{span}\{\widetilde{e}_1, \ldots, \widetilde{e}_{n-1}\}$. As L^{n-1} is invariant under orthogonal transformations, we may assume that $\widetilde{e}_i = e_i$ for $1 < i < n$, hence $\operatorname{span}\{e_1, e_n\} = \operatorname{span}\{\widetilde{e}_1, \widetilde{e}_n\}$. (If $\widetilde{e}_n = e_n$ there is nothing to show due to the invariance.) Then

$$\begin{aligned} \tau_j(y) &= y_j \quad \text{for } 1 < j \leq n-1 \, , \\ \tau_1(y) &= y_1 \widetilde{e}_1 \bullet e_1 + g(y) \widetilde{e}_1 \bullet e_n \, , \\ \widetilde{g}(\tau(y)) &= y_1 \widetilde{e}_n \bullet e_1 + g(y) \widetilde{e}_n \bullet e_n \, . \end{aligned} \tag{A8-6}$$

Now let $g_\varepsilon := \varphi_\varepsilon * g$ for a standard Dirac sequence $(\varphi_\varepsilon)_{\varepsilon > 0}$ and define continuously differentiable functions $\tau_\varepsilon = (\tau_{\varepsilon 1}, \ldots, \tau_{\varepsilon n-1})$ and ψ_ε by

$$\begin{aligned} \tau_{\varepsilon j}(y) &:= y_j \quad \text{for } 1 < j \leq n-1 \, , \\ \tau_{\varepsilon 1}(y) &:= y_1 \widetilde{e}_1 \bullet e_1 + g_\varepsilon(y) \widetilde{e}_1 \bullet e_n \, , \\ \psi_\varepsilon(y) &:= \sum_{i=1}^{n-1} y_i e_i + g_\varepsilon(y) e_n \, . \end{aligned} \tag{A8-7}$$

We want to show that τ_ε is a diffeomorphism. We have shown that τ^{-1} is Lipschitz continuous, wich implies that there exists a constant $c > 0$ such that for $y \in V_0$ and $h > 0$ sufficiently small ($\{e_1, \ldots, e_{n-1}\}$ is the canonical basis of $\mathbb{R}^{n-1}$)

$$c \le \tfrac{1}{h}|\tau(y + h\mathbf{e}_1) - \tau(y)| = \tfrac{1}{h}|\tau_1(y + h\mathbf{e}_1) - \tau_1(y)|$$
$$= \left| \widetilde{e}_1 \bullet e_1 + \tfrac{1}{h}\big(g(y + h\mathbf{e}_1) - g(y)\big)\widetilde{e}_1 \bullet e_n \right|.$$

The term inside the modulus has to have a fixed sign $\sigma \in \{\pm 1\}$ which by the continuity of g is independent of y. It follows that

$$c \le \sigma\widetilde{e}_1 \bullet e_1 + \frac{1}{h}\big(g(y + h\mathbf{e}_1) - g(y)\big) \cdot \sigma\widetilde{e}_1 \bullet e_n.$$

Since $g_\varepsilon = \varphi_\varepsilon * g$ is a convolution of g, it follows that this convex inequality also holds for g_ε, that is, for ε small,

$$c \le \sigma\widetilde{e}_1 \bullet e_1 + \frac{1}{h}\big(g_\varepsilon(y + h\mathbf{e}_1) - g_\varepsilon(y)\big) \cdot \sigma\widetilde{e}_1 \bullet e_n,$$

hence also

$$c \le \sigma\widetilde{e}_1 \bullet e_1 + \partial_1 g_\varepsilon(y) \cdot \sigma\widetilde{e}_1 \bullet e_n.$$

Then it follows for $y \in V_0$

$$\sigma \det D\tau_\varepsilon(y) = \sigma\partial_1\tau_{\varepsilon 1}(y) = \sigma\widetilde{e}_1 \bullet e_1 + \partial_1 g_\varepsilon(y) \cdot \sigma\widetilde{e}_1 \bullet e_n \ge c. \qquad \text{(A8-8)}$$

This implies that τ_ε is a diffeomorphism because τ_ε is defined as in (A8-7). Hence τ_ε^{-1} exists and therefore, with $\widetilde{y} = \tau_\varepsilon(y)$,

$$\widetilde{g}_\varepsilon\big(\tau_\varepsilon(y)\big) := y_1\widetilde{e}_n \bullet e_1 + g_\varepsilon(y)\widetilde{e}_n \bullet e_n = \widetilde{e}_n \bullet \psi_\varepsilon(y),$$
$$\widetilde{\psi}_\varepsilon(\widetilde{y}) := \sum_{j=1}^{n-1} \widetilde{y}_j\widetilde{e}_j + g_\varepsilon(\widetilde{y})\widetilde{e}_n = \psi_\varepsilon(y), \qquad \text{(A8-9)}$$

defines continuously differentiable functions $\widetilde{g}_\varepsilon$ and $\widetilde{\psi}_\varepsilon$.

Now we can show that the integral identity holds. If we define the function $f_\varepsilon := f \circ \widetilde{\psi} \circ \widetilde{\psi}_\varepsilon^{-1}$ on the C^1-surface $\Gamma_\varepsilon := \psi_\varepsilon(V_0)$ we see that for $\varepsilon \to 0$

$$\int_{\widetilde{V}} f(\widetilde{\psi}(\widetilde{y}))\sqrt{1 + |\nabla\widetilde{g}(\widetilde{y})|^2}\, d\widetilde{y} \longleftarrow \int_{\widetilde{V}} \underbrace{f(\widetilde{\psi}(\widetilde{y}))}_{=f_\varepsilon(\widetilde{\psi}_\varepsilon(\widetilde{y}))}\sqrt{1 + |\nabla\widetilde{g}_\varepsilon(\widetilde{y})|^2}\, d\widetilde{y}$$
$$= \int_V \underbrace{f_\varepsilon(\psi_\varepsilon(y))}_{=f(\widetilde{\psi}\circ\tau_\varepsilon(y))}\sqrt{1 + |\nabla g_\varepsilon(y)|^2}\, dy \longrightarrow \int_V \underbrace{f(\widetilde{\psi}\circ\tau(y))}_{=f(\psi(y))}\sqrt{1 + |\nabla g(y)|^2}\, dy.$$

Indeed, the equality is an equation on Γ_ε and follows from the above step for the C^1-case. The first convergence follows from the fact that $\nabla\widetilde{g}_\varepsilon \to \nabla\widetilde{g}$ with respect to the L^p-norm for every $p < \infty$. In fact, the definition (A8-9) of $\widetilde{g}_\varepsilon$ implies

$$(\mathrm{D}\tau_\varepsilon)^T (\nabla \widetilde{g}_\varepsilon) \circ \tau_\varepsilon = (\mathrm{D}\psi_\varepsilon)^T \widetilde{e}_n$$

and by computing the derivative of $\mathrm{D}\tau_\varepsilon$ using (A8-7) and (A8-8) we obtain that $\nabla \widetilde{g}_\varepsilon$ is bounded in L^∞. Since g_ε is bounded in C^0 it follows from the Arzela-Ascoli theorem that for a subsequence $\varepsilon \to 0$ the uniform limit $g_\varepsilon \to \widehat{g}$ exists. Hence by the definition of g_ε

$$\widehat{g}(\tau(y)) \leftarrow \widetilde{g}_\varepsilon(\tau_\varepsilon(y)) = \widetilde{e}_n \bullet \psi_\varepsilon(y) \to \widetilde{e}_n \bullet \psi(y) = \widetilde{g}(\tau(y))$$

for $\varepsilon \to 0$, that is, $\widehat{g} = \widetilde{g}$. This proves the convergence of the gradients of $\widetilde{g}_\varepsilon$ for a subsequence. The second convergence follow from the uniform convergence $\tau_\varepsilon \to \tau$ and from the convergence $\nabla g_\varepsilon \to \nabla g$ with respect to the L^p-norm for every $p < \infty$. Hence the integral identity (A8-4) holds.

Finally, let f be arbitrary. Since $\widehat{f} := f \circ \psi$ has compact support in V, we can approximate $\widehat{f}$ in $L^1(V)$ by functions $\widehat{f}_i \in C_0^\infty(V)$ as $i \to \infty$. Then we can apply the results above to the functions $f_i := \widehat{f}_i \circ \psi^{-1}$, i.e.,

$$\int_V f_i \circ \psi(y) \sqrt{1 + |\nabla g(y)|^2}\, \mathrm{d}y = \int_{\widetilde{V}} f_i \circ \widetilde{\psi}(\widetilde{y}) \sqrt{1 + |\nabla \widetilde{g}(\widetilde{y})|^2}\, \mathrm{d}\widetilde{y}.$$

Moreover, we have that

$$\int_{\widetilde{V}} \left| f_i \circ \widetilde{\psi}(\widetilde{y}) - f_j \circ \widetilde{\psi}(\widetilde{y}) \right| \sqrt{1 + |\nabla \widetilde{g}(\widetilde{y})|^2}\, \mathrm{d}\widetilde{y}$$

$$= \int_V |f_i \circ \psi(y) - f_j \circ \psi(y)| \sqrt{1 + |\nabla g(y)|^2}\, \mathrm{d}y$$

$$\leq C \left\| \widehat{f}_i - \widehat{f}_j \right\|_{L^1(V)} \longrightarrow 0 \quad \text{as } i, j \to \infty.$$

Hence the functions $f_i \circ \widetilde{\psi}$ converge as $i \to \infty$ to a limit in $L^1(\widetilde{V})$. But as $\widehat{f}_i(y) \to \widehat{f}(y)$ for almost all $y \in V$ for a subsequence $i \to \infty$, it follows that also $f_i \circ \widetilde{\psi}(\widetilde{y}) \to f \circ \widetilde{\psi}(\widetilde{y})$ for almost all $\widetilde{y} \in \widetilde{V}$, because τ maps null sets into null sets (see 4.27). This implies that the above limit in $L^1(\widetilde{V})$ must be the function $f \circ \psi$. Hence we obtain the desired integral formula in the general case as well. □

Proof (2). On choosing $f = \mathcal{X}_E$ in (1) for Borel sets $E \subset \partial\Omega$, we obtain that

$$E \longmapsto \mu(E) := \int_{\partial\Omega} \mathcal{X}_E \, \mathrm{d}\mathrm{H}^{n-1}$$

is the $(n-1)$-dimensional Hausdorff measure on $\partial\Omega$ (also denoted by $\mathrm{H}^{n-1}\llcorner\partial\Omega$). Then $L^p(\partial\Omega)$ coincides with the space $L^p(\mu)$ for $\mu = \mathrm{H}^{n-1}\llcorner\partial\Omega$ from Chapter 3. □

Proof (3). In the proof (1) consider the approximation g_ε of g. On noting that $(D\widetilde{\psi}_\varepsilon) \circ \tau_\varepsilon \, D\tau_\varepsilon = D\psi_\varepsilon$, it follows that $(D\widetilde{\psi}) \circ \tau \, D\tau = D\psi$ almost everywhere, and so

$$(D\psi)^T = (D\tau)^T (D\widetilde{\psi})^T \circ \tau \quad \text{almost everywhere.} \tag{A8-10}$$

We have that $\nu = \nu_\Omega$, with respect to g, is uniquely defined by

$$(D\psi)^T \nu = 0, \ |\nu| = 1, \ \nu \bullet e_n < 0.$$

Similarly, $\widetilde{\nu}$ is uniquely defined with respect to $\widetilde{g}$. It follows from (A8-10) and (A8-3) that $\widetilde{\nu} \circ \tau = \nu$ almost everywhere. □

A8.6 Trace theorem. Let $\Omega \subset \mathbb{R}^n$ be open and bounded with Lipschitz boundary and let $1 \le p \le \infty$. Then there exists a unique continuous linear map

$$S : W^{1,p}(\Omega) \longrightarrow L^p(\partial\Omega) \quad (\textit{\textbf{trace operator}})$$

such that

$$Su = u|_{\partial\Omega} \quad \text{for } u \in W^{1,p}(\Omega) \cap C^0(\overline{\Omega}).$$

We call Su the **trace** or the **weak boundary values** of u on $\partial\Omega$.
Notation: In general we write $u(x)$ in place of $(Su)(x)$ for $x \in \partial\Omega$.

Proof. In the case $p = \infty$ it follows from theorem 10.5 that $W^{1,\infty}(\Omega)$ is embedded in $C^{0,1}(\overline{\Omega})$, and so the claim holds trivially. Now let $p < \infty$ and $u \in W^{1,p}(\Omega)$. With the notations as in A8.3, we have that $v := \eta^j u \in W^{1,p}(\Omega^j)$ and for some $\delta > 0$ it holds that

$$v(x) = 0 \quad \text{for } \left| x_{,n}^j - y^j \right| \ge r^j - \delta \text{ and for } x_n^j - g^j(x_{,n}^j) \ge h^j - \delta.$$

For $0 < s < h_j$ we define the functions $v_s : \mathbb{R}^{n-1} \to \mathbb{R}$ via

$$v_s(y) := v(y, g^j(y) + s), \quad \text{where } (y, h) := \sum_{i=1}^{n-1} y_i e_i^j + h e_n^j.$$

Being a Lipschitz transformation, $(y, h) \mapsto (y, g^j(y) + h)$ maps measurable functions into measurable functions (recall 4.27), and so it follows from Fubini's theorem that the v_s are measurable functions for almost all s. In addition, $v_s = 0$ for $s \ge h^j - \delta$. Now the essential observation is that for almost all $s_1, s_2 > 0$ and then for almost all $y \in \mathbb{R}^{n-1}$ we have

$$v_{s_2}(y) - v_{s_1}(y) = v(y, g^j(y) + s_2) - v(y, g^j(y) + s_1)$$
$$= \int_{g^j(y)+s_1}^{g^j(y)+s_2} \partial_{e_n^j} v(y, h) \, \mathrm{d}h. \tag{A8-11}$$

In order to prove this, we approximate v by functions $w_k \in W^{1,p}(\Omega^j) \cap C^\infty(\Omega^j)$ using theorem 4.24. The identity (A8-11) holds for w_k, and setting $D := \mathrm{B}_{r^j}(y^j)$ we have that

$$\int_0^{h^j} \int_D \left| v\big(y, g^j(y) + s\big) - w_k\big(y, g^j(y) + s\big) \right| dy\, ds$$

$$= \int_{\Omega^j} |v - w_k|\, dL^n \longrightarrow 0$$

as $k \to \infty$, and

$$\int_0^{h^j} \int_D \int_{g^j(y)}^{g^j(y)+s} \left| \partial_{e_n^j} v(y, h) - \partial_{e_n^j} w_k(y, h) \right| dh\, dy\, ds$$

$$\leq h^j \int_{\Omega^j} \left| \partial_{e_n^j} (v - w_k) \right| dL^n \longrightarrow 0$$

as $k \to \infty$. Hence the integrands converge for a subsequence $k \to \infty$ for almost all (y, s). This proves (A8-11). Then the Hölder inequality implies for $s_1 < s_2$ that

$$\int_D |v_{s_2} - v_{s_1}|^p\, dL^{n-1} \leq \int_D |s_2 - s_1|^{p-1} \int_{g^j(y)+s_1}^{g^j(y)+s_2} \left| \partial_{e_n^j} v(y, h) \right|^p dh\, dy$$

$$\leq |s_2 - s_1|^{p-1} \int_{D^j(s_1, s_2)} |\nabla v|^p\, dL^n$$

with $D^j(s_1, s_2) := \{x \in \Omega^j \; ; \; s_1 < x_n^j - g^j(x_{,n}^j) < s_2\}$, and hence

$$\left\| v_{s_2} - v_{s_1} \right\|_{L^p(D)} \leq |s_2 - s_1|^{1 - \frac{1}{p}} \left\| \nabla v \right\|_{L^p(D^j(s_1, s_2))}. \tag{A8-12}$$

Since the norm on the right-hand side converges to 0 as $s_1, s_2 \to 0$, the functions v_s form a Cauchy sequence in $L^p(\mathbb{R}^{n-1})$ as $s \to 0$, and hence

$$v_s \to v_0 \quad \text{in } L^p(\mathbb{R}^{n-1}) \text{ as } s \to 0$$

for some $v_0 \in L^p(\mathbb{R}^{n-1})$. Now let

$$S^j v(y, g^j(y)) := v_0(y). \tag{A8-13}$$

That is, the weak boundary values are defined as the limit of the function values on hypersurfaces which are a translation of $\partial\Omega$. It follows from A8.5 that $S^j v \in L^p(\partial\Omega)$ with the bound $\left\| S^j v \right\|_{L^p(\partial\Omega)} \leq C_j \|v_0\|_{L^p(D)}$. Then on choosing a fixed s^j with $h^j - \delta < s^j < h^j$, so that we then have $v_{s^j} = 0$, we obtain from (A8-12), by setting $[s_1, s_2] = [s, s^j]$, that

$$\left\| S^j v \right\|_{L^p(\partial\Omega)} \leq C_j \|v_0\|_{L^p(D)} = C_j \|v_{s^j} - v_0\|_{L^p(D)}$$

$$= C_j \lim_{s \searrow 0} \|v_{s^j} - v_s\|_{L^p(D)} \leq C_j \cdot (s^j)^{1 - \frac{1}{p}} \|\nabla v\|_{L^p(\Omega^j)}.$$

In addition, $\|\nabla v\|_{L^p(\Omega^j)} \leq C(\eta^j) \cdot \|u\|_{W^{1,p}(\Omega^j)}$. For $u \in W^{1,p}(\Omega)$ we now define

$$Su := \sum_{j=1}^{l} S^j(\eta^j u). \qquad\qquad \text{(A8-14)}$$

In particular, we have that $Su = u|_{\partial\Omega}$ if u is continuous on $\overline{\Omega}$. This proves the existence of S. The uniqueness of S follows by establishing that $W^{1,p}(\Omega) \cap C^0(\overline{\Omega})$ is dense in $W^{1,p}(\Omega)$, which will be done in A8.7. □

A8.7 Lemma. Let $\Omega \subset \mathbb{R}^n$ be open and bounded with Lipschitz boundary and let $1 \le p < \infty$ and $m \ge 0$. Then

$$\{\, u|_{\Omega} \; ; \; u \in C_0^\infty(\mathbb{R}^n) \,\} \qquad \text{is dense in } W^{m,p}(\Omega).$$

Proof. Following A8.3, we partition u as

$$u = \sum_{j=0}^{l} \eta^j u.$$

For the part $\eta^0 u$ choose a standard Dirac sequence $(\varphi_\varepsilon)_{\varepsilon>0}$. Since $\eta^0 \in C_0^\infty(\Omega)$, it follows that $\varphi_\varepsilon * (\eta^0 u) \in C_0^\infty(\Omega)$ for ε sufficiently small, and hence $\varphi_\varepsilon * (\eta^0 u) \to \eta^0 u$ in $W^{m,p}(\Omega)$ as $\varepsilon \to 0$. For $j \ge 1$ let Ω^j and $e_1^j, \ldots, e_n^j$ be as in A8.3. For $\delta > 0$ define

$$v_\delta(x) := (\eta^j u)(x + \delta e_n^j) \quad \text{for } x \in \Omega_\delta^j,$$
$$\Omega_\delta^j := \left\{\, x \in \mathbb{R}^n \; ; \; |x_{,n}^j - y^j| < r^j \text{ and } -\delta < x_n^j - g^j(x_{,n}^j) < h^j \,\right\}.$$

Then $v_{\delta,\varepsilon} := \varphi_\varepsilon * \left(\mathcal{X}_{\Omega_\delta^j} v_\delta\right) \in C_0^\infty(\mathbb{R}^n)$ and, on recalling 4.23, it holds on Ω that $v_{\delta,\varepsilon} = \varphi_\varepsilon * v_\delta \in W^{m,p}(\Omega)$ for ε sufficiently small (so that $(1+\mathrm{Lip}(g^j))\cdot\varepsilon < \delta$) with

$$\varphi_\varepsilon * v_\delta \to \eta^j u \quad \text{in } W^{m,p}(\Omega),$$

when first $\varepsilon \searrow 0$ and then $\delta \searrow 0$. This shows that $\eta^j u$ can be approximated in the $W^{m,p}(\Omega)$-norm by functions in $C_0^\infty(\mathbb{R}^n)$, and hence overall also u. □

We now prove some frequently used results on weak boundary values, beginning with integration by parts for Sobolev functions.

A8.8 Weak Gauß's theorem (Weak divergence theorem). Let $\Omega \subset \mathbb{R}^n$ be open and bounded with Lipschitz boundary.

(1) If $u \in W^{1,1}(\Omega)$, then for $i = 1, \ldots, n$

$$\int_\Omega \partial_i u \, \mathrm{dL}^n = \int_{\partial\Omega} u\nu_i \, \mathrm{dH}^{n-1},$$

where ν is the outer normal to $\partial\Omega$ as defined in A8.5.

(2) Let $1 \leq p \leq \infty$. If $u \in W^{1,p}(\Omega)$ and $v \in W^{1,p'}(\Omega)$ with $\frac{1}{p} + \frac{1}{p'} = 1$, then for $i = 1, \ldots, n$

$$\int_\Omega (u\partial_i v + v\partial_i u) \, \mathrm{dL}^n = \int_{\partial\Omega} uv\nu_i \, \mathrm{dH}^{n-1}.$$

Proof (2). It follows from 4.25 that $uv \in W^{1,1}(\Omega)$ with $\partial_i(uv) = u\partial_i v + v\partial_i u$. On recalling A8.7, for $1 < p < \infty$ we approximate u and v by functions in $C^\infty(\mathbb{R}^n)$ and obtain (with S denoting the operator from A8.6) that

$$S(uv) = S(u) \cdot S(v) \quad \text{in } L^1(\partial\Omega). \tag{A8-15}$$

For $p = 1$ we have that $p' = \infty$, and so after modification on a null set, v is in $C^{0,1}(\overline{\Omega})$ (see theorem 10.5). Hence the boundary values of v are well defined and are attained continuously. Now (A8-15) follows from the proof of A8.6. Thus, (2) is reduced to (1). $\qquad\square$

Proof (1). On recalling A8.7 and A8.6, we may assume that $u \in C_0^\infty(\mathbb{R}^n)$. Following A8.3, we partition u into $\eta^j u$, $j = 0, \ldots, l$. For $\eta^0 u \in C_0^\infty(\Omega)$ the boundary integral vanishes and the formula follows from integration by parts in the i-th coordinate direction. For $j \geq 1$ the function $\eta^j u$ is defined on the local set Ω^j. Hence on applying an orthogonal transformation to the canonical Euclidean coordinate system, we need to prove the desired result for functions $u \in C_0^\infty(\mathbb{R}^n)$ and the domain

$$\Omega = \{ (y, h) \in \mathbb{R}^n \; ; \; h > g(y) \}$$

with a Lipschitz continuous function $g : \mathbb{R}^{n-1} \to \mathbb{R}$. By A8.5(3), the normal ν is then defined by

$$\nu(y, g(y)) := \frac{(\nabla g(y), -1)}{\sqrt{1 + |\nabla g(y)|^2}} \quad \text{for } y \in \mathbb{R}^{n-1}.$$

Hence we need to show that

$$\int_\Omega \nabla u(x) \, \mathrm{d}x = \int_{\mathbb{R}^{n-1}} (u\nu)(y, g(y)) \sqrt{1 + |\nabla g(y)|^2} \, \mathrm{d}y$$
$$= \int_{\mathbb{R}^{n-1}} u(y, g(y))(\nabla g(y), -1) \, \mathrm{d}y. \tag{A8-16}$$

When g is continuously differentiable, this is the classical Gauß's theorem, which can be shown for instance as follows: Let $v(y, s) := u(y, g(y) + s)$. Then

$$\partial_n v(y, s) = \partial_n u(y, g(y) + s),$$
$$\partial_i v(y, s) = \partial_i u(y, g(y) + s) + \partial_i g(y)\partial_n u(y, g(y) + s) \quad \text{for } i < n,$$

and hence

$$\int_\Omega \nabla u(x)\,dx = \int_{\mathbb{R}^{n-1}} \int_0^\infty \nabla u\big(y, g(y) + s\big)\,ds\,dy$$

$$= \int_{\mathbb{R}^{n-1}} \int_0^\infty \big(\nabla v - \partial_n v \cdot (\nabla g, 0)\big)(y, s)\,ds\,dy$$

$$= \sum_{i=1}^{n-1} \bigg(\int_0^\infty \Big(\int_{\mathbb{R}^{n-1}} \partial_i v(y, s)\,dy\Big)\,ds\bigg) e_i$$

$$- \int_{\mathbb{R}^{n-1}} \bigg(\int_0^\infty \partial_n v(y, s)\,ds\bigg)\big(\nabla g(y), -1\big)\,dy\,.$$

Integration by parts with respect to y_i yields for $i < n$, since the support of $v(\bullet, s)$ is compact, that

$$\int_{\mathbb{R}^{n-1}} \partial_i v(y, s)\,ds = 0\,,$$

and integration by parts with respect to s gives

$$\int_0^\infty \partial_n v(y, s)\,ds = -v(y, 0) = -u\big(y, g(y)\big)\,.$$

Now we use convolution to approximate the Lipschitz continuous function g by continuously differentiable functions g_k. Letting $\Omega_k := \{(y, h) \in \mathbb{R}^n\,;\, h > g_k(y)\}$ we have that $\mathcal{X}_{\Omega_k} \to \mathcal{X}_\Omega$ as $k \to \infty$ in $L^1(\mathbb{R}^n) \cap B_R(0)$ for every R and $u(\bullet, g_k) \to u(\bullet, g)$ uniformly, because $g_k \to g$ locally uniformly, and also (recall 4.15)

$$\nabla g_k \to \nabla g \quad \text{in } L^p(B_R(0)) \quad \text{for every } p < \infty \text{ and every } R.$$

Hence in (A8-16) we can pass to the limit for g_k. This yields the desired result. □

The following result is a generalization of E3.7 to the n-dimensional case.

A8.9 Lemma. Let $g : \mathbb{R}^{n-1} \to \mathbb{R}$ be Lipschitz continuous, let

$$\Omega_\pm := \{(y, h) \in \mathbb{R}^n\,;\, \pm(h - g(y)) > 0\}\,,$$

and let $u : \mathbb{R}^n \to \mathbb{R}$ with $u|_{\Omega_+} \in W^{1,1}(\Omega_+)$ and $u|_{\Omega_-} \in W^{1,1}(\Omega_-)$. Then, on denoting by $S_\pm$ the trace operators with respect to the domains $\Omega_\pm$ from A8.6,

$$u \in W^{1,1}(\mathbb{R}^n) \quad \Longleftrightarrow \quad S_+\big(u|_{\Omega_+}\big) = S_-\big(u|_{\Omega_-}\big)\,.$$

Corollary: Concerning the removability of singularities in Sobolev spaces we have the following result: If $N \subset \mathbb{R}^{n-1}$ is a closed Lebesgue null set and $A := \{(y, g(y))\,;\, y \in N\}$ with g as above, then for every open set $\Omega \subset \mathbb{R}^n$

$$u \in W^{1,1}(\Omega \setminus A) \quad \Longleftrightarrow \quad u \in W^{1,1}(\Omega)\,.$$

Proof $\Rightarrow$. Setting $u_s(y) := u(y, g(y) + s)$ for $s \in \mathbb{R}$, it holds that (see (A8-12) with $p = 1$)

$$\int_{\mathbb{R}^{n-1}} |u_\varepsilon - u_{-\varepsilon}| \, \mathrm{dL}^{n-1} \leq \int_{\mathbb{R}^{n-1}} \int_{g(y)-\varepsilon}^{g(y)+\varepsilon} |\nabla u(y, h)| \, \mathrm{d}h \, \mathrm{d}y \longrightarrow 0$$

as $\varepsilon \searrow 0$, and so $S_+(u|_{\Omega_+}) = S_-(u|_{\Omega_-})$ by the definition of the trace operator in (A8-13). $\qquad \square$

Proof $\Leftarrow$. Define $u_+ = u|_{\Omega_+}$ and $u_- = u|_{\Omega_-}$. Let $\nu_\pm$ denote the outer normal to $\Omega_\pm$. Then it follows from A8.8(2) for $\zeta \in C_0^\infty(\mathbb{R}^n)$ that

$$\int_{\mathbb{R}^n} (u\nabla\zeta + \zeta\nabla u) \, \mathrm{dL}^n = \int_{\Omega_+} (u\nabla\zeta + \zeta\nabla u) \, \mathrm{dL}^n + \int_{\Omega_-} (u\nabla\zeta + \zeta\nabla u) \, \mathrm{dL}^n$$

$$= \int_{\partial\Omega_+} \zeta S_+(u_+)\nu_+ \, \mathrm{dH}^{n-1} + \int_{\partial\Omega_-} \zeta S_-(u_-)\nu_- \, \mathrm{dH}^{n-1}$$

$$= \int_{\mathrm{graph}(g)} \zeta \cdot \underbrace{(S_+(u_+)\nu_+ + S_-(u_-)\nu_-)}_{=0} \, \mathrm{dH}^{n-1} = 0 \,,$$

because $\nu_- = -\nu_+$ and $S_+(u_+) = S_-(u_-)$. $\qquad \square$

We now show that functions in $W_0^{1,p}(\Omega)$ have weak boundary values 0.

A8.10 Lemma. Let $\Omega \subset \mathbb{R}^n$ be open and bounded with Lipschitz boundary and let $1 \leq p < \infty$. Let S be the trace operator from A8.6. Then

$$W_0^{1,p}(\Omega) = \{u \in W^{1,p}(\Omega) ;\ Su = 0\} \,.$$

Proof $\subset$. Every function $u \in W_0^{1,p}(\Omega)$ can be approximated by $C_0^\infty(\Omega)$-functions u_i as $i \to \infty$. The properties of the trace operator then imply that $0 = Su_i \to Su$ in $L^p(\partial\Omega)$. $\qquad \square$

Proof $\supset$. Let $u \in W^{1,p}(\Omega)$ with $Su = 0$. Choosing η^j as in A8.3, it follows (see (A8-15)) that $S(\eta^j u) = \eta^j S(u) = 0$ on $\partial\Omega$ for $j = 1, \ldots, l$. Now define for $j = 1, \ldots, l$

$$v_j(x) := \begin{cases} (\eta^j u)(x) & \text{for } x \in \Omega^j, \\ 0 & \text{otherwise.} \end{cases}$$

Then A8.9 implies that $v_j \in W^{1,p}(\mathbb{R}^n)$, and hence for $\delta > 0$ also $v_{j\delta} \in W^{1,p}(\mathbb{R}^n)$, where

$$v_{j\delta}(x) := v_j(x - \delta e_n^j) \,,$$

and $v_{j\delta} \to v_j$ in $W^{1,p}(\mathbb{R}^n)$ as $\delta \to 0$. Consequently,

$$u_\delta := \eta^0 u + \sum_{j=1}^l v_{j\delta} \longrightarrow u \quad \text{in } W^{1,p}(\Omega) \text{ as } \delta \to 0.$$

Since u_δ has compact support in Ω, it can be approximated in $W^{1,p}(\Omega)$ with the help of convolution by functions in $C_0^\infty(\Omega)$. $\qquad \square$

A8.11 Remark. Results for Sobolev functions on domains with Lipschitz boundary can also be proved by locally straightening the boundary. In the local situation at the boundary, i.e. $\Omega = \Omega_+$ with the notations as in A8.9, this means that we consider

$$\widetilde{\Omega} := \{(y, h) \in \mathbb{R}^n ;\ h > 0\},$$
$$\widetilde{u}(y, h) := u(y, g(y) + h) \quad \text{for } (y, h) \in \widetilde{\Omega}.$$

It holds that: If $1 \le p \le \infty$ and $u \in W^{1,p}(\Omega)$, then $\widetilde{u} \in W^{1,p}(\widetilde{\Omega})$ with the **chain rule**

$$\begin{aligned}
\partial_n \widetilde{u}(y, h) &= \partial_n u(u, g(y) + h), \\
\partial_i \widetilde{u}(y, h) &= \partial_i u(y, g(y) + h) + \partial_i g(y) \partial_n u(y, g(y) + h)
\end{aligned} \tag{A8-17}$$

for $i < n$.

Proof. Let $\tau(y, h) := (y, g(y) + h)$. For $v \in L^p(\Omega)$ with $p < \infty$ it follows from Fubini's theorem that $v \circ \tau \in L^p(\widetilde{\Omega})$, with

$$\begin{aligned}
\int_\Omega |v|^p \, dL^n &= \int_{\mathbb{R}^{n-1}} \int_{g(y)}^\infty |v(y, h)|^p \, dh \, dy \\
&= \int_{\mathbb{R}^{n-1}} \int_0^\infty |v(y, g(y) + h)|^p \, dh \, dy = \int_{\widetilde{\Omega}} |v \circ \tau|^p \, dL^n.
\end{aligned} \tag{A8-18}$$

Hence we have that $\|v\|_{L^p(\Omega)} = \|v \circ \tau\|_{L^p(\widetilde{\Omega})}$ for $1 \le p \le \infty$. This shows that the right-hand sides in (A8-17) lie in $L^p(\widetilde{\Omega})$, and so (A8-17) (by the definition of the weak derivatives) only needs to be shown locally in $\widetilde{\Omega}$ for the case $p = 1$.

We approximate g by $g_\varepsilon := \varphi_\varepsilon * g$ with a standard Dirac sequence $(\varphi_\varepsilon)_{\varepsilon > 0}$. On setting $\tau_\varepsilon(y, h) := (y, g_\varepsilon(y) + h)$, let

$$\widetilde{u}_\varepsilon := u \circ \tau_\varepsilon \quad \text{on } \widetilde{\Omega}_\varepsilon := \tau_\varepsilon^{-1}(\Omega).$$

By 4.26, we have $\widetilde{u}_\varepsilon \in W^{1,1}(\widetilde{\Omega}_\varepsilon)$ and the chain rule (A8-17) holds for $\widetilde{u}_\varepsilon$. We note that $g_\varepsilon \to g$ locally uniformly as $\varepsilon \to 0$ and $\nabla g_\varepsilon \to \nabla g$ in $L_{loc}^q(\mathbb{R}^{n-1})$ for every $q < \infty$, and so $\nabla g_\varepsilon \to \nabla g$ almost everywhere for a subsequence $\varepsilon \to 0$. Moreover, the ∇g_ε are bounded in $L_{loc}^\infty(\mathbb{R}^{n-1})$. If we can show that for $v \in L_{loc}^1(\Omega)$ and for every $D \subset\subset \Omega$

$$v \circ \tau_\varepsilon \to v \circ \tau \quad \text{as } \varepsilon \to 0 \text{ in } L^1(\tau^{-1}(D)), \tag{A8-19}$$

then this implies the convergence of $u \circ \tau_\varepsilon$ and $(\partial_i u) \circ \tau_\varepsilon$, and we can pass to the limit in the chain rule (A8-17).

Now it follows from (A8-18) that (A8-19) is equivalent to

$$v \circ \tau_\varepsilon \circ \tau^{-1} \to v \quad \text{as } \varepsilon \to 0 \text{ in } L^1(D). \tag{A8-20}$$

Here, we approximate v in the L^1-norm by continuous functions v_k. These functions satisfy (A8-20) and therefore we have (cf. the proof of 4.15(1))

$$\left\| v \circ \tau_\varepsilon \circ \tau^{-1} - v \right\|_{L^1(D)} \leq \left\| v \circ \tau_\varepsilon \circ \tau^{-1} - v_k \circ \tau_\varepsilon \circ \tau^{-1} \right\|_{L^1(D)}$$
$$+ \left\| v_k \circ \tau_\varepsilon \circ \tau^{-1} - v_k \right\|_{L^1(D)} + \left\| v_k - v \right\|_{L^1(D)}$$
$$\leq \left(C(\tau_\varepsilon \circ \tau^{-1}) + 1 \right) \left\| v_k - v \right\|_{L^1(D)} + \left\| v_k \circ \tau_\varepsilon \circ \tau^{-1} - v_k \right\|_{L^1(D)},$$

where $C(\tau_\varepsilon \circ \tau^{-1})$ converges to 1 as $\varepsilon \to 0$. Thus (A8-20) also holds for v. $\square$

A further consequence of A8.9 is:

A8.12 Extension theorem. Let $\Omega \subset \mathbb{R}^n$ be open and bounded with Lipschitz boundary and let $1 \leq p \leq \infty$. Then, for $\delta > 0$, there exists an extension operator

$$E : W^{1,p}(\Omega) \longrightarrow W_0^{1,p}(\mathrm{B}_\delta(\Omega)),$$

i.e. E is linear, continuous, and such that $(Eu)\big|_\Omega = u$ for all $u \in W^{1,p}(\Omega)$.

Proof. We treat E similarly to the operator S in (A8-14). Hence it is sufficient to consider the local situation near the boundary (cf. the proof of A8.8(1)). Let $\Omega = \Omega_+$ with $\Omega_\pm$ as in A8.9. Choose a cut-off function $\eta \in C^\infty(\mathbb{R}^n)$ with $\eta = 1$ in $\mathrm{B}_{\frac{\delta}{2}}(\Omega)$ and $\eta = 0$ in $\mathbb{R}^n \setminus \mathrm{B}_\delta(\Omega)$. Then, define $Eu := \eta \widetilde{u}$ with

$$\widetilde{u}(y, h) := \begin{cases} u(y, h) & \text{for } h > g(y), \\ u(y, 2g(y) - h) & \text{for } h < g(y). \end{cases}$$

(For $p = \infty$ it follows from theorem 10.5 that this defines a $C^{0,1}$-extension of u.) For $p < \infty$ it follows similarly to the proof of A8.11 that $\widetilde{u} \in W^{1,p}(\Omega_-)$, with

$$\|\widetilde{u}\|_{L^p(\Omega_-)} = \|u\|_{L^p(\Omega_+)},$$
$$\|\nabla \widetilde{u}\|_{L^p(\Omega_-)} \leq \left(2 + \mathrm{Lip}(g) \right) \|\nabla u\|_{L^p(\Omega_+)}.$$

Consequently, $Eu \in W^{1,p}(\Omega_-)$ with

$$\|Eu\|_{W^{1,p}(\Omega_-)} \leq C \|u\|_{W^{1,p}(\Omega_+)}.$$

Then by the definition of the trace operator in (A8-13) it holds that for a sequence $\varepsilon \searrow 0$ and for almost all y

$$S_-(Eu)(y, g(y)) \longleftarrow Eu(y, g(y) - \varepsilon)$$
$$= u(y, g(y) + \varepsilon) \longrightarrow S_+(u)(y, g(y))$$

Now A8.9 yields that $Eu \in W^{1,p}(\mathbb{R}^n)$. $\square$

The following theorem implies that sets of the form

$$M := \{u \in W^{1,2}(\Omega);\ \varphi(u) = g \text{ on } \partial\Omega\} \qquad \text{(A8-21)}$$

are weakly sequentially closed in $W^{1,2}(\Omega)$, if $\varphi : \mathbb{R} \to \mathbb{R}$ is continuous and $g : \partial\Omega \to \mathbb{R}$ is measurable.

A8.13 Embedding theorem onto the boundary. If $\Omega \subset \mathbb{R}^n$ is open and bounded with Lipschitz boundary, then for $1 \le p < \infty$ and $u_k, u \in W^{1,p}(\Omega)$ it holds that:

$$u_k \to u \text{ weakly in } W^{1,p}(\Omega) \qquad \Longrightarrow \qquad u_k \to u \text{ (strongly) in } L^p(\partial\Omega)$$
$$\text{as } k \to \infty \qquad\qquad\qquad \text{as } k \to \infty.$$

Proof. Without loss of generality let $u = 0$. If $\eta \in C^\infty(\mathbb{R}^n)$, then also $\eta u_k \to 0$ weakly in $W^{1,p}(\Omega)$, and so it follows from A8.3 and A8.1 that we only need to consider the local situation on the boundary. Hence let $\Omega = \Omega_+$ as in A8.9 and let the supports of u_k, u be contained in a bounded set of $\mathbb{R}^n$. On recalling (A8-12), the functions $u_{ks}(y) := u_k(y, g(y) + s)$ satisfy for almost all ε, s with $0 < \varepsilon < s$ the bound

$$\int_{\mathbb{R}^{n-1}} |u_{ks} - u_{k\varepsilon}|^p \, d\mathrm{L}^{n-1} \le |s - \varepsilon|^{p-1} \int_{E_{\varepsilon,s}} |\nabla u_k|^p \, d\mathrm{L}^n ,$$

where $E_{\varepsilon,s} := \{(y, h) \in \mathbb{R}^n;\ \varepsilon < h - g(y) < s\}$. Let $\delta > 0$. Then for almost all s with $0 < \varepsilon \le s \le \delta$, on setting $C = 2^{p-1}$ (see (3-13)), we have that

$$\int_{\mathbb{R}^{n-1}} |u_{k\varepsilon}|^p \, d\mathrm{L}^{n-1} \le C \int_{\mathbb{R}^{n-1}} |u_{ks}|^p \, d\mathrm{L}^{n-1} + C\delta^{p-1} \int_{E_{0,\delta}} |\nabla u_k|^p \, d\mathrm{L}^n .$$

On letting $\varepsilon \to 0$, we have that $u_{k\varepsilon} \to u_{k0}$ in $L^p(\mathbb{R}^{n-1})$, where u_{k0} are the weak boundary values of u_k. Then integrating this inequality over $s \in [\frac{\delta}{2}, \delta]$ and dividing by $\frac{\delta}{2}$ yields that

$$\int_{\mathbb{R}^{n-1}} |u_{k0}|^p \, d\mathrm{L}^{n-1} \le \frac{2C}{\delta} \int_{E_{\frac{\delta}{2},\delta}} |u_k|^p \, d\mathrm{L}^n + C\delta^{p-1} \int_{E_{0,\delta}} |\nabla u_k|^p \, d\mathrm{L}^n .$$

It follows from Rellich's embedding theorem A8.4 that the first term on the right-hand side converges to 0 for every δ. If $p > 1$ then the second term converges to 0 as $\delta \to 0$, since the functions ∇u_k are bounded in $L^p(\Omega_+)$. In the case $p = 1$ it follows from the following theorem that the integral in the second term converges to 0 uniformly in k as $\delta \to 0$, because the ∇u_k converge weakly to 0 in $L^1(\Omega_+; \mathbb{R}^n)$ and because they have supports in a bounded set. This yields the desired result. □

A8.14 Weak sequential compactness in $L^1(\mu)$. Let $(S, \mathcal{B}, \mu)$ be a measure space and let $M \subset L^1(\mu; \mathbb{R}^m)$. Then every sequence in M contains a subsequence that converges weakly in $L^1(\mu; \mathbb{R}^m)$ if and only if

(1) M is bounded in $L^1(\mu; \mathbb{R}^m)$.

(2) It holds that

$$\sup_{f \in M} \int_E |f| \, d\mu \longrightarrow 0 \quad \text{as } \mu(E) \to 0.$$

(3) There exist sets $S_k \in \mathcal{B}$, for $k \in \mathbb{N}$, with $\mu(S_k) < \infty$, such that

$$\sup_{f \in M} \int_{S \setminus S_k} |f| \, d\mu \longrightarrow 0 \quad \text{as } k \to \infty.$$

Remark: If $\mu(S) < \infty$, condition (3) is not necessary, choose $S_k = S$.

Proof $\Rightarrow$. (1) follows via an indirect argument from 8.3(5).

Assume that (2) is false. Hence there exist a $c > 0$ and measurable sets E_n as well as $f_n \in M$ for $n \in \mathbb{N}$ such that

$$\mu(E_n) \to 0 \text{ as } n \to \infty \quad \text{and} \quad \int_{E_n} |f_n| \, d\mu \geq c \text{ for all } n.$$

From this it follows that there exist $\widetilde{E}_n \in \mathcal{B}$ with $\mu(\widetilde{E}_n) \to 0$ as $n \to \infty$ and

$$\left| \int_{\widetilde{E}_n} f_n \, d\mu \right| \geq \frac{c}{2m}. \tag{A8-22}$$

To see this, let $A_j^\pm := \{x \in S; \ \pm f_n(x) \bullet e_j > 0\}$ for $j = 1, \ldots, m$. Then

$$\int_{E_n} |f_n| \, d\mu \leq \sum_{j=1}^m \left(\int_{E_n \cap A_j^+} |f_n \bullet e_j| \, d\mu + \int_{E_n \cap A_j^-} |f_n \bullet e_j| \, d\mu \right)$$

$$= \sum_{j=1}^m \left(\left| \int_{E_n \cap A_j^+} f_n \bullet e_j \, d\mu \right| + \left| \int_{E_n \cap A_j^-} f_n \bullet e_j \, d\mu \right| \right),$$

which means that for some j (which depends on n) we have that

$$\left| \int_{E_n \cap A_j^+} f_n \bullet e_j \, d\mu \right| \geq \frac{c}{2m} \quad \text{or} \quad \left| \int_{E_n \cap A_j^-} f_n \bullet e_j \, d\mu \right| \geq \frac{c}{2m}.$$

Let $\widetilde{E}_n := E_n \cap A_j^+$ in the first case, and $\widetilde{E}_n := E_n \cap A_j^-$ in the second case. Then

$$\frac{c}{2m} \leq \left| \int_{\widetilde{E}_n} f_n \bullet e_j \, d\mu \right| = \left| e_j \bullet \int_{\widetilde{E}_n} f_n \, d\mu \right| \leq \left| \int_{\widetilde{E}_n} f_n \, d\mu \right|,$$

and $\mu(\widetilde{E}_n) \leq \mu(E_n) \to 0$ as $n \to \infty$. This proves (A8-22). It follows from the assumption on M that there exists a subsequence $n \to \infty$ (there is no extra notation for the subsequence) such that for all μ-measurable E the limit

$$\lim_{n \to \infty} \lambda_n(E) \quad \text{exists, with} \quad \lambda_n(E) := \int_E f_n \, d\mu. \qquad (A8\text{-}23)$$

Since for every n we have $\lambda_n(E) \to 0$ as $\mu_n(E) \to 0$, the following theorem A8.15 yields a contradiction to $\left| \lambda_n(\widetilde{E}_n) \right| \geq \frac{c}{2m}$.

Now assume that (3) is false, i.e. there exists a $c > 0$ such that for all $E \in \mathcal{B}$ with $\mu(E) < \infty$

$$\int_{S \setminus E} |f| \, d\mu \geq c \quad \text{for an } f \in M. \qquad (A8\text{-}24)$$

Moreover, for all $f \in L^1(\mu; \mathbb{R}^m)$ and $\varepsilon > 0$,

$$\int_{S \setminus E} |f| \, d\mu \leq \varepsilon \quad \text{for an } E \in \mathcal{B} \text{ with } \mu(E) < \infty, \qquad (A8\text{-}25)$$

because there exists a step function g with $\|f - g\|_{L^1} \leq \varepsilon$, and then $E := \{x \in S; \ g(x) \neq 0\}$ has finite measure.

On combining (A8-25) and (A8-24) we inductively choose $f_n \in M$ and $E_n \in \mathcal{B}$ with $\mu(E_n) < \infty$ and $E_n \subset E_{n+1}$ such that

$$\int_{S \setminus E_{n+1}} |f_n| \, d\mu \leq \frac{1}{n} \quad \text{and} \quad \int_{S \setminus E_{n+1}} |f_{n+1}| \, d\mu \geq c.$$

Then it holds for $n \geq \frac{2}{c}$ that

$$\int_{E_{n+1} \setminus E_n} |f_n| \, d\mu = \int_{S \setminus E_n} |f_n| \, d\mu - \int_{S \setminus E_{n+1}} |f_n| \, d\mu \geq \frac{c}{2}.$$

Next, as in the proof of (A8-22), there exist measurable sets $\widetilde{E}_n \subset E_{n+1} \setminus E_n$ such that

$$\left| \int_{\widetilde{E}_n} f_n \, d\mu \right| \geq \frac{c}{4m},$$

and for a subsequence $n \to \infty$ the corresponding λ_n satisfy the above property (A8-23). We now consider the measure space $(\widetilde{S}, \widetilde{\mathcal{B}}, \widetilde{\mu})$ with

$$\widetilde{S} := \bigcup_{n \in \mathbb{N}} E_n, \quad \widetilde{\mathcal{B}} := \{E \cap \widetilde{S}; \ E \in \mathcal{B}\},$$

$$\widetilde{\mu}(E) := \sum_{j \in \mathbb{N}} 2^{-j} \frac{\mu(E \cap E_j \setminus E_{j-1})}{1 + \mu(E_j \setminus E_{j-1})},$$

where $E_0 := \emptyset$. Since $\widetilde{\mu}(E) \to 0$ implies $\mu(E \cap E_j \setminus E_{j-1}) \to 0$ for all j, and since, for fixed n, for $E \subset \widetilde{S} \setminus E_j$ it holds that

$$|\lambda_n(E)| \leq \int_{\widetilde{S} \setminus E_j} |f_n| \, d\mu \longrightarrow 0 \quad \text{as } j \to \infty,$$

we obtain that $|\lambda_n(E)| \to 0$ for $E \in \widetilde{\mathcal{B}}$ with $\widetilde{\mu}(E) \to 0$ for fixed n. Combining the following theorem A8.15 applied to the measure $\widetilde{\mu}$ and the facts that

$$\left| \lambda_n(\widetilde{E}_n) \right| \geq \frac{c}{4m} \quad \text{and} \quad \widetilde{\mu}(\widetilde{E}_n) \leq 2^{-n} \to 0 \text{ as } n \to \infty,$$

we arrive at a contradiction. $\qquad\qquad\qquad\qquad\qquad\qquad\qquad\qquad\qquad\qquad\square$

Proof $\Leftarrow$ for regular measures. Let $S \subset \mathbb{R}^n$ be compact and let μ be a non-negative measure in $rca(S)$. We may assume that $m = 1$. For every sequence $(f_n)_{n \in \mathbb{N}}$ in M it follows from (1) and A3.17(2) that

$$\lambda_n(E) := \int_E f_n \, d\mu$$

defines a bounded sequence $(\lambda_n)_{n \in \mathbb{N}}$ in $rca(S)$. By 8.6(2), there exists a $\lambda \in rca(S)$ such that for a subsequence $n \to \infty$,

$$\int_S g \, d\lambda_n \to \int_S g \, d\lambda \quad \text{for all } g \in C^0(S). \tag{A8-26}$$

If E is a μ-null set, then, on recalling that μ is regular, for $\varepsilon > 0$ there exists a relatively in S open set U with $E \subset U$ and $\mu(U) \leq \varepsilon$. Moreover, as λ is regular, there exist finitely many disjoint closed sets $K_j \subset U$ such that

$$|\lambda|(U) \leq \varepsilon + \sum_j |\lambda(K_j)|.$$

For $\delta > 0$ choose $g_j \in C^0(S)$ with $\mathcal{X}_{K_j} \leq g_j \leq \mathcal{X}_{B_\delta(K_j)}$. Then it follows that

$$|\lambda|(U) \leq \varepsilon + \sum_j |\lambda|(B_\delta(K_j) \setminus K_j) + \sum_j \left| \int_S g_j \, d\lambda \right|$$

and

$$\sum_j \left| \int_S g_j \, d\lambda \right| \longleftarrow \sum_j \left| \int_S g_j \, d\lambda_n \right| \quad (\text{as } n \to \infty)$$

$$= \sum_j \left| \int_S g_j f_n \, d\mu \right| \leq \int_S \Big(\sum_j g_j \Big) |f_n| \, d\mu \leq \int_U |f_n| \, d\mu,$$

where we observe that the last inequality holds for δ sufficiently small, because then the sets $B_\delta(K_j)$ are disjoint subsets of U. Letting $\delta \searrow 0$ and noting assumption (2) we get

$$|\lambda|(U) \leq \varepsilon + \sup_{f \in M} \int_U |f| \, \mathrm{d}\mu \longrightarrow 0 \quad \text{as } \varepsilon \to 0.$$

This shows that E is also a $|\lambda|$-null set. Hence we can apply the Radon-Nikodým theorem 6.11 and obtain that there exists an $f \in L^1(\mu)$ with

$$\lambda(E) = \int_E f \, \mathrm{d}\mu$$

for all μ-measurable sets E. It follows from (A8-26) that

$$\int_S g f_n \, \mathrm{d}\mu \longrightarrow \int_S g f \, \mathrm{d}\mu \quad \text{as } n \to \infty \qquad \text{(A8-27)}$$

for all $g \in C^0(S)$. On recalling 6.12, we need to show that this also holds for all $g \in L^\infty(\mu)$. First let $g = \mathcal{X}_E$ with a μ-measurable set E. For $\varepsilon > 0$ choose K closed and U relatively open in S such that $K \subset E \subset U$ with $\mu(U \setminus K) \leq \varepsilon$ and $\widetilde{g} \in C^0(S)$ with $\mathcal{X}_K \leq \widetilde{g} \leq \mathcal{X}_U$. Then

$$\left| \int_E f_n \, \mathrm{d}\mu - \int_E f \, \mathrm{d}\mu \right| \leq \left| \int_S \widetilde{g}(f_n - f) \, \mathrm{d}\mu \right| + \sup_{n'} \int_{U \setminus K} (|f_{n'}| + |f|) \, \mathrm{d}\mu,$$

where, thanks to (2), the second term converges to 0 as $\varepsilon \to 0$. Since the first term converges to 0 as $n \to \infty$ by (A8-27), we obtain

$$\int_E f_n \, \mathrm{d}\mu \longrightarrow \int_E f \, \mathrm{d}\mu \quad \text{as } n \to \infty.$$

Recalling that the characteristic functions span a dense subspace of $L^\infty(\mu)$ then yields that (A8-27) also holds for all $g \in L^\infty(\mu)$. □

Proof ⟸ *for bounded measures.* The idea is to use a separable analogue of $C^0(S)$ in the above proof. As before, let $m = 1$. Let $(f_n)_{n \in \mathbb{N}}$ be a sequence in M, and let

$$g_n = \sum_{j=1}^{k_n} \alpha_{nj} \mathcal{X}_{E_{nj}} \quad \text{with } \mu(E_{nj}) < \infty$$

be step functions with $\|f_n - g_n\|_{L^1} \leq \frac{1}{n}$. On noting that for every $n_0 \in \mathbb{N}$ it holds that

$$\int_E |g_n| \, \mathrm{d}\mu \leq \max_{i \leq n_0} \int_E |g_i| \, \mathrm{d}\mu + \frac{1}{n_0} + \sup_{f \in M} \int_E |f| \, \mathrm{d}\mu,$$

we have that $\{g_n ; n \in \mathbb{N}\}$ also satisfies the assumption (2), and it is sufficient to show that $(g_n)_{n \in \mathbb{N}}$ contains a weakly convergent subsequence.

Now the algebra $\mathcal{B}_0$ induced by the set $\{E_{nj} ; j \leq k_n, n \in \mathbb{N}\}$ is countable. Hence it follows from (1) that with the help of a diagonalization procedure we obtain a subsequence such that (without special notation for the subsequence)

$$\lambda(E) := \lim_{n\to\infty} \int_E g_n \, d\mu$$

exists for all $E \in \mathcal{B}_0$. It holds that λ is additive on $\mathcal{B}_0$. Let $\mathcal{B}_1$ be the smallest σ-algebra that contains $\mathcal{B}_0$ and all μ-null sets, and define $\mu_1 := \mu \llcorner \mathcal{B}_1$. Then $(S, \mathcal{B}_1, \mu_1)$ is a finite measure space. Since $\mu_1(S) < \infty$, we can show that λ admits a σ-additive extension to $\mathcal{B}_1$. To see this, let $(E_k)_{k\in\mathbb{N}}$ be a shrinking sequence of sets in $\mathcal{B}_1$ for which the above limit exists, and let

$$E := \bigcap_{k\in\mathbb{N}} E_k \,.$$

Then

$$\left| \int_E (g_n - g_l) \, d\mu \right| \le \underbrace{\left| \int_{E_k} (g_n - g_l) \, d\mu \right|}_{\substack{\to\, 0 \text{ as } n,l\to\infty \\ \text{for any } k}} + 2 \underbrace{\sup_j \int_{E_k\setminus E} |g_j| \, d\mu}_{\substack{\to\, 0 \text{ as } k\to\infty, \\ \text{recall } (2)}} \,,$$

which shows that the above limit defines λ on all of $\mathcal{B}_1$. On noting that, in addition,

$$|\lambda(E_k \setminus E)| = \lim_{n\to\infty} \left| \int_{E_k\setminus E} g_n \, d\mu \right| \le \sup_n \int_{E_k\setminus E} |g_n| \, d\mu \longrightarrow 0 \quad \text{as } k\to\infty,$$

we see that λ is even σ-additive on $\mathcal{B}_1$ and that $\lambda(E) = 0$ if $\mu(E) = 0$. Hence it follows from the Radon-Nikodým theorem that there exists an $f \in L^1(\mu_1)$ with

$$\lambda(E) = \int_E f \, d\mu_1 \quad \text{for all } E \in \mathcal{B}_1.$$

As the characteristic functions span a dense subspace of $L^\infty(\mu_1)$, this means, on recalling 6.12, that $g_n \to f$ weakly in $L^1(\mu_1)$. Now $L^1(\mu_1) \subset L^1(\mu)$ implies that $g_n \to f$ weakly also in $L^1(\mu)$. $\qquad\square$

Proof $\Leftarrow$ the general case. As before, let $m = 1$. Let $(f_n)_{n\in\mathbb{N}}$ be a sequence in M and let S_k for $k \in \mathbb{N}$ be the sets from (3), which we can choose such that $S_k \subset S_{k+1}$. We apply the result just shown to the sets

$$M_k := \{\mathcal{X}_{S_k} f ; \ f \in M\}$$

(with the measure μ being restricted to S_k). Hence a diagonalization procedure yields a subsequence $n \to \infty$ and $h_k \in L^1(\mu)$ with $h_k = 0$ on $S \setminus S_k$, such that

$$\int_{S_k} f_n \, g \, d\mu \longrightarrow \int_{S_k} h_k \, g \, d\mu \quad \text{as } n\to\infty \text{ for all } g \in L^\infty(\mu).$$

Then $h_{k+1} = h_k$ almost everywhere on S_k, and on setting $\widetilde{S} := \bigcup_{k\in\mathbb{N}} S_k$ we have that

$$h(x) := \begin{cases} h_k(x) & \text{for } x \in S_k, \ k \in \mathbb{N}, \\ 0 & \text{for } x \in S \setminus \widetilde{S}, \end{cases}$$

defines a μ-measurable function. Now it holds for $k < l$ and for all $g \in L^\infty(\mu)$ that as $n \to \infty$

$$\left| \int_S (h_l - h_k) g \, d\mu \right| = \left| \int_{S_l} h_l \, \mathcal{X}_{S \setminus S_k} \, g \, d\mu \right|$$

$$\longleftarrow \left| \int_{S_l} f_n \, \mathcal{X}_{S \setminus S_k} \, g \, d\mu \right| \le \delta_k \|g\|_{L^\infty},$$

where

$$\delta_k := \sup_{f \in M} \int_{S \setminus S_k} |f| \, d\mu.$$

It follows that $\|h_l - h_k\|_{L^1} \le \delta_k \to 0$ as $k \to \infty$, on recalling (3). Hence $h \in L^1(\mu)$ and for $g \in L^\infty(\mu)$

$$\int_{\widetilde{S}} (h - f_n) \, g \, d\mu \le \|g\|_{L^\infty} \cdot \underbrace{\left(\int_{\widetilde{S} \setminus S_k} |h| \, d\mu + \delta_k \right)}_{\substack{\to 0 \text{ as } k \to \infty}} + \underbrace{\left| \int_{S_k} (h_k - f_n) \, g \, d\mu \right|}_{\substack{\to 0 \text{ as } n \to \infty \\ \text{for any } k}}.$$

This shows that $f_n \to h$ weakly in $L^1(\mu)$ as $n \to \infty$. To see this, note that if $\widetilde{\mu}$ is the measure μ restricted to $\widetilde{S}$, then $\widetilde{\mu}$ is σ-finite and

$$Jf(x) := \begin{cases} f(x) & \text{for } x \in \widetilde{S}, \\ 0 & \text{for } x \in S \setminus \widetilde{S}, \end{cases}$$

defines an embedding $J : L^1(\widetilde{\mu}) \to L^1(\mu)$. Hence for $F \in L^1(\mu)'$ we have that $\widetilde{F} := F \circ J \in L^1(\widetilde{\mu})'$, which by 6.12 can be represented by means of $g \in L^\infty(\widetilde{\mu})$. Consequently,

$$F(h - f_n) = \widetilde{F}(h - f_n) = \int_{\widetilde{S}} (h - f_n) g \, d\mu \longrightarrow 0 \quad \text{as } n \to \infty.$$

$\square$

A8.15 Theorem (Vitali-Hahn-Saks). Let $(S, \mathcal{B}, \mu)$ be a measure space and let $\lambda_n : \mathcal{B} \to \mathbb{K}$ be σ-additive for $n \in \mathbb{N}$. Suppose that

$$\forall \, n \in \mathbb{N} : \ \Big(\, |\lambda_n(E)| \to 0 \text{ as } \mu(E) \to 0 \, \Big),$$

and that the limit

$$\lim_{n \to \infty} \lambda_n(E) \in \mathbb{K} \text{ exists for all } E \in \mathcal{B}.$$

Then

$$\sup_{n \in \mathbb{N}} |\lambda_n(E)| \to 0 \text{ as } \mu(E) \to 0.$$

Proof. The set
$$\mathcal{M} := \{E \in \mathcal{B};\ \mu(E) < \infty\},$$

equipped with the distance

$$d(E_1, E_2) := \int_S |\mathcal{X}_{E_1} - \mathcal{X}_{E_2}|\,\mathrm{d}\mu,$$

is a complete metric space if the equivalence relation

$$E_1 = E_2 \text{ in } \mathcal{M} \quad :\Longleftrightarrow \quad \mathcal{X}_{E_1} = \mathcal{X}_{E_2} \ \mu\text{-almost everywhere}$$

is used in $\mathcal{M}$. The completeness follows from the fact that the limit of characteristic functions in $L^1(\mu)$ is again a characteristic function (this follows from A3.11). The assumptions yield that the λ_n are continuous on $\mathcal{M}$. Indeed, $d(E_k, E) \to 0$ as $k \to \infty$ implies that $\mu(E_k \setminus E) \to 0$ and $\mu(E \setminus E_k) \to 0$, and so

$$|\lambda_n(E_k) - \lambda_n(E)| = |\lambda_n(E_k \setminus E) - \lambda_n(E \setminus E_k)|$$
$$\leq |\lambda_n(E_k \setminus E)| + |\lambda_n(E \setminus E_k)| \to 0.$$

Hence for $\varepsilon > 0$ and $k \in \mathbb{N}$ the sets

$$\mathcal{A}_k^\varepsilon := \big\{\, E \in \mathcal{M};\ |\lambda_k(E) - \lambda_j(E)| \leq \varepsilon \text{ for all } j \geq k \,\big\}$$

are closed subsets in $\mathcal{M}$ and the assumptions of the theorem imply that

$$\bigcup_{k \in \mathbb{N}} \mathcal{A}_k^\varepsilon = \mathcal{M}$$

for all $\varepsilon > 0$. It follows from the Baire category theorem 7.1 that at least one $\mathcal{A}_k^\varepsilon$ has a nonempty interior, i.e. there exist $k_\varepsilon \in \mathbb{N}$, $A_\varepsilon \in \mathcal{M}$, $\delta_\varepsilon > 0$ with

$$d(E, A_\varepsilon) \leq \delta_\varepsilon \quad \Longrightarrow \quad |\lambda_{k_\varepsilon}(E) - \lambda_j(E)| \leq \varepsilon \text{ for all } j \geq k_\varepsilon.$$

Now for $E \in \mathcal{M}$ arbitrary and $E_1 := A_\varepsilon \cup E$, $E_2 := A_\varepsilon \setminus E$

$$E = E_1 \setminus E_2, \quad d(E_1, A_\varepsilon) \leq \mu(E), \quad d(E_2, A_\varepsilon) \leq \mu(E).$$

If $\mu(E) \leq \delta_\varepsilon$ it then follows for $j \geq k_\varepsilon$ that

$$|\lambda_j(E)| \leq |\lambda_{k_\varepsilon}(E)| + |(\lambda_{k_\varepsilon}(E_1) - \lambda_j(E_1)) - (\lambda_{k_\varepsilon}(E_2) - \lambda_j(E_2))|$$
$$\leq |\lambda_{k_\varepsilon}(E)| + 2\varepsilon,$$

and so

$$\sup_{j \in \mathbb{N}} |\lambda_j(E)| \leq 2\varepsilon + \underbrace{\max_{j \leq k_\varepsilon} |\lambda_j(E)|}_{\substack{\to\, 0 \text{ as } \mu(E) \to 0 \\ \text{for any } \varepsilon > 0.}}$$

This proves the desired result. □

9 Finite-dimensional approximation

In this chapter we consider certain finite-dimensional subspaces of Banach spaces. This plays an important role in applications, where we regard elements in such subspaces as approximations of elements in the entire Banach space X. Clearly we require the approximating subspaces to be finite-dimensional, because in numerical computations only a prescribed finite number of coordinates can be stored.

Hence in what follows points $x \in X$ will always be approximated by a countable sequence of points $(x_n)_{n \in \mathbb{N}}$. It is for this reason that in applications the weak and weak* sequential compactness introduced in Chapter 8 plays a far more important role than compactness with respect to the weak and weak* topology, respectively.

An important optimization problem is to characterize a function from a function space X approximately by finitely many numerical values. This can be achieved by suitably exhausting X by finite-dimensional subspaces X_n, $n \in \mathbb{N}$. Another important problem is to numerically solve linear equations between Banach spaces. This concerns for instance the numerical solution of a boundary value problem for linear partial differential equations (see 6.5). Here it is once again necessary to approximate a Banach space, e.g. the space $X = W^{1,2}(\Omega)$, by suitable finite-dimensional subspaces X_n, $n \in \mathbb{N}$, and then to find an approximative solution in these subspaces (see the Ritz-Galerkin method in 9.23–9.25). In all applications the subspaces X_n are chosen so that they are appropriate for the problem at hand, i.e. they should be easy to handle and, on the other hand, they should also retain as much of the structure of the infinite-dimensional problem as possible.

To approximate a space by countably many finite-dimensional subspaces is only possible for separable normed spaces:

9.1 Lemma. Let X be an infinite-dimensional normed space. Then the following are equivalent:

(1) X is separable.

(2) There exist finite-dimensional subspaces $X_n \subset X$ for $n \in \mathbb{N}$ such that $X_n \subset X_{n+1}$ for all n and

$$\bigcup_{n \in \mathbb{N}} X_n \quad \text{is dense in } X.$$

(3) There exist finite-dimensional subspaces $E_n \subset X$ for $n \in \mathbb{N}$ such that $E_n \cap E_m = \{0\}$ for $n \neq m$ and

$$\bigoplus_{k \in \mathbb{N}} E_k := \bigcup_{n \in \mathbb{N}} \Big(E_1 \oplus \cdots \oplus E_n \Big) \quad \text{is dense in } X.$$

(4) There exists a linearly independent set $\{e_k \, ; \, k \in \mathbb{N}\} \subset X$ such that

$$\operatorname{span}\{e_k \, ; \, k \in \mathbb{N}\} \quad \text{is dense in } X.$$

Proof (1)$\Rightarrow$(2). Choose a countable set $\{x_n \, ; \, n \in \mathbb{N}\}$ that is dense in X. Define $X_n := \operatorname{span}\{x_1, \ldots, x_n\}$. $\qquad\square$

Proof (2)$\Rightarrow$(3). Let $E_1 := X_1$ and for $n \in \mathbb{N}$ choose subspaces E_{n+1} with $X_{n+1} = X_n \oplus E_{n+1}$. $\qquad\square$

Proof (3)$\Rightarrow$(4). Let $d_n := \dim E_n$ and let $\{e_{n,j} \, ; \, j = 1, \ldots, d_n\}$ be a basis of E_n. Then we have that

$$X_n := E_1 \oplus \cdots \oplus E_n = \operatorname{span}\{e_{i,j} \, ; \, 1 \leq i \leq n, \, 1 \leq j \leq d_i\}$$

and hence $M := \{e_{i,j} \, ; \, i \in \mathbb{N}, \, 1 \leq j \leq d_i\}$ is a desired linearly independent set, since

$$\operatorname{span}(M) = \operatorname{span}\Big(\bigcup_{n \in \mathbb{N}} X_n \Big)$$

is a dense set in X. $\qquad\square$

Proof (4)$\Rightarrow$(1). For $n \in \mathbb{N}$ it holds that

$$A_n := \Big\{ \sum_{k=1}^{n} \alpha_k e_k \, ; \, \alpha_k \in Q \text{ for } 1 \leq k \leq n \Big\},$$

with Q as in the proof of 4.17(4), is countable with $\operatorname{clos}(A_n) = \operatorname{span}\{e_k \, ; \, 1 \leq k \leq n\}$. The desired result then follows from 4.17(1). $\qquad\square$

On recalling 4.18(4) we can apply these results, for example, to the spaces $L^p(\Omega)$, $1 \leq p < \infty$, with $\Omega \subset \mathbb{R}^n$ open. Moreover, by 4.15(3), we have $C_0^\infty(\Omega)$ is dense in $L^p(\Omega)$, and so 4.17(2) yields that it is also separable with respect to the L^p-norm. Hence we can apply 9.1 to the space $C_0^\infty(\Omega)$ equipped with the L^p-norm. The denseness results in 9.1(2)-9.1(4) for this space then also hold with respect to the space $L^p(\Omega)$. This implies that the finite-dimensional spaces with respect to $L^p(\Omega)$ in 9.1 can be chosen as subspaces of $C_0^\infty(\Omega)$. The same argument can equally be applied to other pairs of function spaces.

As a further consequence of 9.1 we now give a constructive proof of the Hahn-Banach theorem 6.15 for separable spaces X, i.e. for the extension of a functional $y' \in Y'$ onto X for a subspace $Y \subset X$. Recalling that the extension of y' to the closure $\overline{Y}$ is already uniquely defined by y' (see E5.3), we restrict our attention to closed subspaces $Y \subset X$.

9.2 On the Hahn-Banach theorem. Let X be an infinite-dimensional separable normed $\mathbb{R}$-vector space and let $Y \subset X$ be a closed subspace. Then:

(1) If $E \subset X$ is a finite-dimensional subspace with $Y \cap E = \{0\}$, then $Y \oplus E$ is also a closed subspace.

(2) There exists a linearly independent set $\{e_k \,;\; k \in \mathbb{N}\}$ such that denoting $E_n := \mathrm{span}\{e_k \,;\; k \leq n\}$

$$Y \oplus \bigcup_{n \in \mathbb{N}} E_n \quad \text{is dense in } X.$$

(3) Let $y' \in Y'$ and let e_n, E_n for $n \in \mathbb{N}$ be as in (2). Then there exist numbers $c_n \in \mathbb{R}$ for $n \in \mathbb{N}$ such that the following holds: The inductively defined maps

$$x'_n(y + \alpha e_n) := x'_{n-1}(y) + \alpha c_n \quad \text{for } y \in Y \oplus E_{n-1}, \; \alpha \in \mathbb{R},$$

where $x'_0 := y'$ and $E_0 := \{0\}$, yield a map $x' : X \to \mathbb{R}$ defined by

$$x'(y) := \lim_{n \to \infty} x'_n(y_n) \quad \text{for } y = \lim_{n \to \infty} y_n \text{ with } y_n \in Y \oplus E_n .$$

Moreover, it holds: $x' \in X'$ is an extension of y' with $\|x'\|_{X'} = \|y'\|_{Y'}$.

Proof (1). Let $\widetilde{X}$ be the vector space X with the equivalence relation

$$x_1 = x_2 \text{ in } \widetilde{X} \quad :\Longleftrightarrow \quad x_1 - x_2 \in Y.$$

As Y is closed, we have that $\widetilde{X}$ with $\|x\|_{\widetilde{X}} := \mathrm{dist}(x, Y)$ is a normed space. Since $Y \cap E = \{0\}$, the dimension of E in $\widetilde{X}$ is the same as in X. Now let $y_k \in Y$, $z_k \in E$ for $k \in \mathbb{N}$ with $y_k + z_k \to x \in X$ as $k \to \infty$. It follows that $\|z_k - x\|_{\widetilde{X}} \to 0$ as $k \to \infty$, and so $(z_k)_{k \in \mathbb{N}}$ is a Cauchy sequence in $\widetilde{X}$. Hence, by 4.9, there exists a $z \in E$ with $\|z_k - z\|_{\widetilde{X}} \to 0$ as $k \to \infty$, and so $\|x - z\|_{\widetilde{X}} = 0$, i.e. there exists a $y \in Y$ with $x - z = y$. It follows that $x = y + z \in Y \oplus E$. $\square$

Proof (2). Apply 9.1(4) to $\widetilde{X}$. $\square$

Proof (3). Let $R := \|y'\|_{Y'}$. We need to inductively find c_n such that on letting $Y_n := Y \oplus E_n$ it holds that

$$x'_n(y + \alpha e_n) \leq R \, \|y + \alpha e_n\|_X \quad \text{for } y \in Y_{n-1}, \; \alpha \in \mathbb{R} .$$

Similarly to the proof of 6.14 this means that we need to choose c_n such that

$$\sup_{y \in Y_{n-1}} \left(x'_{n-1}(y) - R\|y - e_n\|_X \right) \leq c_n \leq \inf_{\widetilde{y} \in Y_{n-1}} \left(R\|\widetilde{y} + e_n\|_X - x'_{n-1}(\widetilde{y}) \right),$$

which is possible, because the induction hypothesis implies that

$$x'_{n-1}(y) + x'_{n-1}(\tilde{y}) \le R\,\|y + \tilde{y}\|_X \le R\big(\|y - e_n\|_X + \|\tilde{y} + e_n\|_X\big)\,.$$

As $x'_n = x'_{n-1}$ on Y_{n-1} for all $n \in \mathbb{N}$, we have that this defines a linear map $x' : \bigcup_{n \in \mathbb{N}} Y_n \to \mathbb{R}$ with $x'(y) = x'_n(y)$ for $y \in Y_n$, which implies that $|x'(y)| \le R\|y\|_X$ for all such y. It now follows from (2) (recall E5.3) that there exists a unique extension of x' to X.   □

The result 9.1(2) means that for $x \in X$ there exist points $x_n \in X_n$ such that $x_n \to x$ as $n \to \infty$. The result 9.1(4) implies that for $x \in X$ and $n \in \mathbb{N}$ there exist numbers $\alpha_{n,k}$ for $k = 1, \ldots, n$ such that

$$\left\| x - \sum_{k=1}^{n} \alpha_{n,k} e_k \right\|_X \longrightarrow 0 \quad \text{as } n \to \infty.$$

If the coefficients $\alpha_{n,k}$ can be chosen independently of n, we speak of a Schauder basis:

9.3 Definition (Schauder basis). Let X be a normed space. A sequence $(e_k)_{k \in \mathbb{N}}$ in X is called a *Schauder basis* of X if:

For $x \in X$ there exist unique $\alpha_k \in \mathbb{K}$ for $k \in \mathbb{N}$

such that $\displaystyle\sum_{k=1}^{n} \alpha_k e_k \longrightarrow x$ in X as $n \to \infty$.

The fact that the coefficients α_k of x are unique, and hence depend linearly on x, implies that we can define linear maps $e'_k : X \to \mathbb{K}$ by setting

$$e'_k(x) := \alpha_k \text{ for } k \in \mathbb{N}, \quad \text{if} \quad \sum_{k=1}^{n} \alpha_k e_k \longrightarrow x \text{ as } n \to \infty.$$

Observe: In this definition we can also write

$$x = \sum_{k=1}^{\infty} \alpha_k e_k := \lim_{n \to \infty} \sum_{k=1}^{n} \alpha_k e_k \quad \text{in } X.$$

However, this does not imply anything about the absolute convergence of this series. In particular, no infinite rearrangment of terms in this series is allowed (!). This means that together with the set $\{e_k \,;\, k \in \mathbb{N}\}$ a specific order of the elements of this set is prescribed. Hence a Schauder basis is defined by the sequence $(e_k)_{k \in \mathbb{N}}$, and not by a set.

We now show that the maps e'_k are continuous.

9.4 Theorem (Dual basis). Let $(e_k)_{k \in \mathbb{N}}$ be a Schauder basis of the Banach space X and let e'_k be the corresponding maps from 9.3. Then:

(1) $\{e_k \; ; \; k \in \mathbb{N}\}$ satisfies 9.1(4) with $\dim \operatorname{span}\{e_k \; ; \; 1 \le k \le n\} = n$ for $n \in \mathbb{N}$.

(2) $e_k' \in X'$ for $k \in \mathbb{N}$.

(3) The sequence $(e_k')_{k\in\mathbb{N}}$ in X' is a **dual basis** of $(e_k)_{k\in\mathbb{N}}$, i.e.

$$\langle e_l, e_k' \rangle_X = \delta_{k,l} \quad \text{for all } k, l.$$

(4) The dual basis is unique.

Proof (1). Follows from the uniqueness of the coefficients in 9.3. $\qquad\square$

Proof (3). For $n \ge l$

$$e_l = \sum_{k=1}^{n} \delta_{k,l} e_k \,.$$

The uniqueness of the coefficients yields that $e_k'(e_l) = \delta_{k,l}$. $\qquad\square$

Proof (2). Consider the set

$$Y := \Big\{ \alpha = (\alpha_k)_{k\in\mathbb{N}} \in \mathbb{K}^{\mathbb{N}} \; ; \; \lim_{n\to\infty} \sum_{k=1}^{n} \alpha_k e_k \text{ exists in } X \Big\}$$

with

$$\|\alpha\|_Y := \sup_n \Big\| \sum_{k=1}^{n} \alpha_k e_k \Big\|_X$$

and define a map $T : Y \to X$ by setting

$$T(\alpha) := \lim_{n\to\infty} \sum_{k=1}^{n} \alpha_k e_k \,.$$

Clearly, Y is a $\mathbb{K}$-vector space, $\|\cdot\|_Y$ is a norm and T is linear. The fact that $(e_k)_{k\in\mathbb{N}}$ is a Schauder basis is equivalent to the bijectivity of T, with

$$T^{-1}(x) = (e_k'(x))_{k\in\mathbb{N}} \,.$$

Moreover, we have that $T \in \mathscr{L}(Y; X)$, because

$$\|T(\alpha)\|_X = \lim_{n\to\infty} \Big\| \sum_{k=1}^{n} \alpha_k e_k \Big\|_X \le \|\alpha\|_Y \,. \tag{9-28}$$

If Y is complete, which we will prove below, then by the inverse mapping theorem $T^{-1} \in \mathscr{L}(X; Y)$, i.e. there exists a $C < \infty$ with

$$\|(e_k'(x))_{k\in\mathbb{N}}\|_Y \le C\|x\|_X \quad \text{for } x \in X.$$

Then on combining

$$|e_n'(x)| \cdot \|e_n\|_X = \|e_n'(x)e_n\|_X$$

$$= \left\| \sum_{k=1}^{n} e_k'(x)e_k - \sum_{k=1}^{n-1} e_k'(x)e_k \right\|_X \tag{9-29}$$

$$\leq 2 \left\| (e_k'(x))_{k\in\mathbb{N}} \right\|_Y \leq 2\, C \, \|x\|_X \,,$$

and $e_n \neq 0$ (otherwise $e_n'(x)$ would not be unique, or see (1)), it follows that $e_n' \in X'$, with the products $\|e_n'\|_{X'} \cdot \|e_n\|_X$ being bounded.   □

Proof (4). The fact that the dual basis of a Schauder basis is unique can be easily seen. Namely, if $e_k'^{1}, e_k'^{2} \in X'$ with $\left\langle e_l, e_k'^{m} \right\rangle_X = \delta_{k,l}$ for all $k,l \in \mathbb{N}$, it follows that

$$\left\langle \textstyle\sum_{l=1}^{n} \alpha_l e_l,\, e_k'^{1} - e_k'^{2} \right\rangle_X = 0$$

for all such linear combinations. As these are dense in X, we obtain that $\left\langle x,\, e_k'^{1} - e_k'^{2} \right\rangle_X = 0$ for all $x \in X$, i.e. $e_k'^{1} = e_k'^{2}$.   □

Proof (2) *completeness*. It remains to show that Y is complete. If $(\alpha^i)_{i\in\mathbb{N}}$ is a Cauchy sequence in Y, then for all $n \in \mathbb{N}$ (similarly to (9-29) and (9-28))

$$|\alpha_n^i - \alpha_n^j| \leq \frac{2}{\|e_n\|_X} \|\alpha^i - \alpha^j\|_Y$$

and

$$\|T(\alpha^i) - T(\alpha^j)\|_X = \|T(\alpha^i - \alpha^j)\|_X \leq \|\alpha^i - \alpha^j\|_Y \,,$$

i.e. the $(\alpha_n^i)_{i\in\mathbb{N}}$ are Cauchy sequences in $\mathbb{K}$ for all $n \in \mathbb{N}$, and $(T(\alpha^i))_{i\in\mathbb{N}}$ is a Cauchy sequence in X. Hence, there exist the limits

$$\alpha_n := \lim_{i\to\infty} \alpha_n^i \in \mathbb{K} \quad \text{and} \quad x := \lim_{i\to\infty} T(\alpha^i) \in X.$$

Now we show that $\alpha := (\alpha_n)_{n\in\mathbb{N}} \in Y$ and $T(\alpha) = x$. First we see

$$\left\| x - \sum_{k=1}^{n} \alpha_k e_k \right\|_X \leq \underbrace{\|x - T(\alpha^i)\|_X}_{\to\, 0 \text{ as } i\to\infty} + \underbrace{\left\| T(\alpha^i) - \sum_{k=1}^{n} \alpha_k^i e_k \right\|_X}_{\to\, 0 \text{ as } n\to\infty \text{ and any } i}$$

$$+ \underbrace{\left\| \sum_{k=1}^{n} (\alpha_k^i - \alpha_k^j)e_k \right\|_X}_{\substack{\leq\, \|\alpha^i - \alpha^j\|_Y \,\to\, 0 \\ \text{as } i,j\to\infty}} + \underbrace{\left\| \sum_{k=1}^{n} (\alpha_k^j - \alpha_k)e_k \right\|_X}_{\substack{\to\, 0 \text{ as } j\to\infty \\ \text{and any } n}} .$$

Hence the left-hand side converges to 0 as $n \to \infty$. This follows by first letting $j \to \infty$ to arrive at

$$\left\| x - \sum_{k=1}^{n} \alpha_k e_k \right\|_X \leq \left\| x - T(\alpha^i) \right\|_X + \left\| T(\alpha^i) - \sum_{k=1}^{n} \alpha_k^i e_k \right\|_X$$
$$+ \limsup_{j \to \infty} \left\| \alpha^i - \alpha^j \right\|_Y,$$

and then $n \to \infty$ to obtain

$$\limsup_{n \to \infty} \left\| x - \sum_{k=1}^{n} \alpha_k e_k \right\|_X \leq \left\| x - T(\alpha^i) \right\|_X + \limsup_{j \to \infty} \left\| \alpha^i - \alpha^j \right\|_Y,$$

where now the right-hand side converges to 0 if $i \to \infty$. We also have that $\alpha^i \to \alpha$ in Y, on noting that

$$\left\| \sum_{k=1}^{n} (\alpha_k - \alpha_k^i) e_k \right\|_X \leq \left\| \sum_{k=1}^{n} (\alpha_k - \alpha_k^j) e_k \right\|_X + \left\| \alpha^j - \alpha^i \right\|_Y,$$

which on letting $j \to \infty$ implies that

$$\left\| \alpha - \alpha^i \right\|_Y = \sup_n \left\| \sum_{k=1}^{n} (\alpha_k - \alpha_k^i) e_k \right\|_X \leq \underbrace{\limsup_{j \to \infty} \left\| \alpha^j - \alpha^i \right\|_Y}_{\to 0 \text{ as } i \to \infty}.$$

$\square$

Let X^1, X^2 be Banach spaces and let $(e_k^m)_{k \in \mathbb{N}}$ be a Schauder basis of X^m, $m = 1, 2$. Now if $S \in \mathscr{L}(X^1; X^2)$, then Se_k^1 have representations with respect to $(e_l^2)_{l \in \mathbb{N}}$, i.e. there exist uniquely defined numbers $a_{k,l} \in \mathbb{K}$ with

$$Se_k^1 = \sum_{l=1}^{\infty} a_{k,l} e_l^2.$$

It follows that

$$x = \sum_{k=1}^{\infty} \alpha_k e_k^1 \quad \Longrightarrow \quad Sx = \sum_{k=1}^{\infty} \sum_{l=1}^{\infty} \alpha_k a_{k,l} e_l^2.$$

This means that the operator S is uniquely determined by the **infinite matrix** $(a_{k,l})_{k,l \in \mathbb{N}}$. Clearly, not every such matrix defines an operator in $\mathscr{L}(X^1; X^2)$. The numbers $a_{k,l}$ need to satisfy certain conditions as $k, l \to \infty$, which guarantee that the above infinite sums converge (for a special case see 12.11).

Orthogonal systems

In Hilbert spaces a special role is played by bases whose elements are perpendicular to each other.

9.5 Definition. Let X be a pre-Hilbert space. A sequence $(e_k)_{k \in N}$, $N \subset \mathbb{N}$, in X is called an **orthogonal system** if

$$(e_k, e_l)_X = 0 \text{ for } k \neq l \quad \text{and} \quad e_k \neq 0 \text{ for all } k, \tag{9-30}$$

and an **orthonormal system** if

$$(e_k, e_l)_X = \delta_{k,l} \text{ for all } k, l. \tag{9-31}$$

Remark: In contrast to 9.3, this definition may also be formulated for the countable set $\{e_k; k \in N\}$. All of the following infinite sums can be arbitrarily rearranged.

9.6 Bessel's inequality. Let $(e_k)_{k=1,\ldots,n}$ be a (finite) orthonormal system in the pre-Hilbert space X. Then for every $x \in X$

$$0 \leq \|x\|^2 - \sum_{k=1}^{n} |(x, e_k)_X|^2$$

$$= \left\| x - \sum_{k=1}^{n} (x, e_k)_X e_k \right\|^2 = \operatorname{dist}(x, \operatorname{span}\{e_1, \ldots, e_n\})^2.$$

Proof. For $\alpha_1, \ldots, \alpha_n \in \mathbb{K}$

$$\left\| x - \sum_{k=1}^{n} \alpha_k e_k \right\|^2 = \|x\|^2 - \sum_{k=1}^{n} (x, e_k)_X \overline{\alpha_k} - \sum_{k=1}^{n} \alpha_k (e_k, x)_X + \sum_{k=1}^{n} |\alpha_k|^2$$

$$= \|x\|^2 - \sum_{k=1}^{n} |(x, e_k)_X|^2 + \sum_{k=1}^{n} |(x, e_k)_X - \alpha_k|^2,$$

i.e. the left-hand side is minimal when $\alpha_k = (x, e_k)_X$ for $k = 1, \ldots, n$.   □

9.7 Orthonormal basis. Let $(e_k)_{k \in \mathbb{N}}$ be an orthonormal system in the pre-Hilbert space X. Then $(e_k)_{k \in \mathbb{N}}$ is called an **orthonormal basis** if one of the following equivalent conditions is satisfied:

(1) $\operatorname{span}\{e_k; k \in \mathbb{N}\}$ is dense in X.

(2) $(e_k)_{k \in \mathbb{N}}$ is a Schauder basis of X.

(3) *Representation.*

$$x = \sum_{k=1}^{\infty} (x, e_k)_X e_k \quad \text{for all } x \in X.$$

(4) Parseval's identity.

$$(x\,,\,y)_X = \sum_{k=1}^{\infty} (x\,,\,e_k)_X\,\overline{(y\,,\,e_k)_X} \quad \text{for all } x, y \in X.$$

(5) Completeness relation.

$$\|x\|^2 = \sum_{k=1}^{\infty} |(x\,,\,e_k)_X|^2 \quad \text{for all } x \in X.$$

Proof (1)$\Rightarrow$(3). Let $x \in X$ and let

$$x_n = \sum_{k=1}^{m_n} \alpha_{n,k} e_k \longrightarrow x \quad \text{in } X \text{ as } n \to \infty.$$

For n and m with $m \geq m_n$ it holds, by Bessel's inequality, that

$$\|x - x_n\| \geq \mathrm{dist}(x, \mathrm{span}\{e_1, \ldots, e_{m_n}\}) \geq \mathrm{dist}(x, \mathrm{span}\{e_1, \ldots, e_m\})$$
$$= \left\| x - \sum_{k=1}^{m} (x\,,\,e_k)_X\, e_k \right\|.$$

$\square$

Proof (3)$\Rightarrow$(2). We need to show that the coefficients are unique. If

$$0 = \sum_{k=1}^{\infty} \alpha_k e_k\,,$$

then the continuity of the scalar product yields for all $l \in \mathbb{N}$ that

$$0 = \left(\sum_{k=1}^{\infty} \alpha_k e_k\,,\,e_l \right)_X = \sum_{k=1}^{\infty} \alpha_k \underbrace{(e_k\,,\,e_l)_X}_{=\delta_{k,l}} = \alpha_l\,.$$

$\square$

Proof (2)$\Rightarrow$(1). Follows from 9.4(1). $\square$

Proof (3)$\Rightarrow$(4). The continuity of the scalar product implies that

$$(x\,,\,y)_X = \lim_{n \to \infty} \left(\sum_{k=1}^{n} (x\,,\,e_k)_X\, e_k\,,\,\sum_{l=1}^{n} (y\,,\,e_l)_X\, e_l \right)_X$$
$$= \lim_{n \to \infty} \sum_{k,l=1}^{n} (x\,,\,e_k)_X\,\overline{(y\,,\,e_l)_X}\,\underbrace{(e_k\,,\,e_l)_X}_{=\delta_{k,l}} = \lim_{n \to \infty} \sum_{k=1}^{n} (x\,,\,e_k)_X\,\overline{(y\,,\,e_k)_X}\,.$$

$\square$

Proof (4)$\Rightarrow$(5). Set $y = x$. $\square$

Proof (5)$\Rightarrow$(3). It follows from Bessel's inequality that

$$\left\| x - \sum_{k=1}^{n} (x, e_k)_X \, e_k \right\|^2 = \|x\|^2 - \sum_{k=1}^{n} |(x, e_k)_X|^2 \longrightarrow 0 \qquad \text{as } n \to \infty.$$

$\square$

9.8 Theorem. For every infinite-dimensional Hilbert space X over $\mathbb{K}$ the following are equivalent:

(1) X is separable.

(2) X has an orthonormal basis.

Note: If one of these conditions is satisfied, then X is isometrically isomorphic to $\ell^2(\mathbb{K})$. (Every finite-dimensional Hilbert space X is isometrically isomorphic to $\mathbb{K}^{\dim X}$.)

Proof (2)$\Rightarrow$(1). Follows from 9.7(1) and 9.1. $\square$

Proof (1)$\Rightarrow$(2). Let $X_n := \text{span}\{e_1, \dots, e_n\}$ with $(e_k)_{k \in \mathbb{N}}$ from 9.1(4). (Or choose subspaces X_n as in 9.1(2) with $\dim X_n = n$ and then $e_n \in X_n \setminus X_{n-1}$, where $X_0 := \{0\}$.) Then we inductively define $\widehat{e}_n$ for $n \in \mathbb{N}$ via

$$\widetilde{e}_n := e_n - \sum_{1 \le k < n} (e_n, \widehat{e}_k)_X \, \widehat{e}_k \quad \in X_n \setminus X_{n-1},$$

$$\widehat{e}_n := \frac{\widetilde{e}_n}{\|\widetilde{e}_n\|_X} \qquad (\text{hence } \|\widehat{e}_n\|_X = 1).$$

Then $\widehat{e}_n \in X_n \cap X_{n-1}^\perp$, and so $(\widehat{e}_n)_{n \in \mathbb{N}}$ is an orthonormal system. Since

$$\text{span}\{\widehat{e}_k \, ; \, k = 1, \dots, n\} = X_n,$$

it follows that $\text{span}\{\widehat{e}_k \, ; \, k \in \mathbb{N}\}$ is dense in X. Hence 9.7(1) yields that $(\widehat{e}_k)_{k \in \mathbb{N}}$ is an orthonormal basis. This procedure to compute (or to construct) an orthonormal basis is called the **Gram-Schmidt process**. $\square$

Proof of Note. Let $(e_k)_{k \in \mathbb{N}}$ be an orthonormal basis. It follows from 9.7(5) that the map $J : X \to \ell^2(\mathbb{K})$ with $Jx := ((x, e_k)_X)_{k \in \mathbb{N}}$ is well defined. In addition, 9.7(4) implies that $(Jx, Jy)_{\ell^2} = (x, y)_X$ for $x, y \in X$. Moreover, J is bijective with $J^{-1}((\alpha_k)_{k \in \mathbb{N}}) = \sum_{k \in \mathbb{N}} \alpha_k e_k$. (In the finite-dimensional case consider the index set $\{1, \dots, \dim X\}$ in place of $\mathbb{N}$.) $\square$

We now give the standard example of an orthonormal basis in a function space.

9.9 Example. Consider $L^2(]-\pi,\pi[\,;\mathbb{K})$ with $\mathbb{K}=\mathbb{R}$ or $\mathbb{K}=\mathbb{C}$. Then

$$e_k(x) := \frac{1}{\sqrt{2\pi}}\,\mathrm{e}^{\mathrm{i}kx} \quad \text{for } k \in \mathbb{Z}$$

defines an orthonormal basis $(e_k)_{k\in\mathbb{Z}}$ of $L^2(]-\pi,\pi[\,;\mathbb{C})$. Moreover,

$$\widetilde{e}_k(x) := \begin{cases} \dfrac{1}{\sqrt{\pi}}\sin(kx) & \text{for } k > 0, \\[2mm] \dfrac{1}{\sqrt{2\pi}} & \text{for } k = 0, \\[2mm] \dfrac{1}{\sqrt{\pi}}\cos(kx) & \text{for } k < 0 \end{cases}$$

defines an orthonormal basis $(\widetilde{e}_k)_{k\in\mathbb{Z}}$ of $L^2(]-\pi,\pi[\,;\mathbb{R})$.

Proof. First let us show that we are dealing with orthonormal systems. We have

$$(e_k\,,\,e_l)_{L^2} = \frac{1}{2\pi}\int_{-\pi}^{\pi} \mathrm{e}^{\mathrm{i}kx}\mathrm{e}^{-\mathrm{i}lx}\,\mathrm{d}x$$

$$= \begin{cases} \dfrac{1}{2\pi}\displaystyle\int_{-\pi}^{\pi} 1\,\mathrm{d}x = 1 & \text{for } k = l, \\[3mm] \dfrac{1}{2\pi\mathrm{i}(k-l)}\displaystyle\int_{-\pi}^{\pi}\frac{\mathrm{d}}{\mathrm{d}x}(\mathrm{e}^{\mathrm{i}(k-l)x})\,\mathrm{d}x = 0 & \text{for } k \neq l. \end{cases}$$

Regarding $\widetilde{e}_k$ as elements of $L^2(]-\pi,\pi[\,;\mathbb{C})$, we have that

$$\widetilde{e}_k = \begin{cases} \frac{1}{\mathrm{i}\sqrt{2}}(e_k - e_{-k}) & \text{for } k > 0, \\[2mm] e_0 & \text{for } k = 0, \\[2mm] \frac{1}{\sqrt{2}}(e_k + e_{-k}) & \text{for } k < 0, \end{cases}$$

which immediately implies that $(\widetilde{e}_k\,,\,\widetilde{e}_l)_{L^2} = 0$ for $|k| \neq |l|$. In addition, it holds for $k > 0$ that

$$\|\widetilde{e}_{\pm k}\|_{L^2}^2 = \frac{1}{2}\|e_k \mp e_{-k}\|_{L^2}^2$$

$$= \frac{1}{2}\big(\|e_k\|_{L^2}^2 \mp 2\mathrm{Re}\,(e_k\,,\,e_{-k})_{L^2} + \|e_{-k}\|_{L^2}^2\big) = 1$$

and

$$(\widetilde{e}_k\,,\,\widetilde{e}_{-k})_{L^2} = \frac{1}{2\mathrm{i}}\,(e_k - e_{-k}\,,\,e_k + e_{-k})_{L^2}$$

$$= \frac{1}{2\mathrm{i}}\big(\|e_k\|_{L^2}^2 + (e_k\,,\,e_{-k})_{L^2} - (e_{-k}\,,\,e_k)_{L^2} - \|e_{-k}\|_{L^2}^2\big) = 0,$$

and hence $(\widetilde{e}_k)_{k\in\mathbb{Z}}$ is also an orthonormal system.

On recalling 9.7(1) we need to show that span$\{e_k \; ; \; k \in \mathbb{Z}\}$ is dense in $L^2(] - \pi, \pi[; \mathbb{C})$. As $C_0^\infty(] - \pi, \pi[; \mathbb{C})$ is dense in $L^2(] - \pi, \pi[; \mathbb{C})$ (see 4.15(3)), we only need to show that smooth functions can be approximated in the L^2-norm by finite linear combinations of the e_k. We give two possible proofs.

1st possibility: Let $f \in C^{0,1}([- \pi, \pi]; \mathbb{C})$ and set

$$P_n f := \sum_{|k| \le n} (f, e_k)_{L^2} \, e_k \, .$$

Then Bessel's inequality 9.6 yields (first for the finite sum over all k with $|k| \le n$, and then for the limit as $n \to \infty$) that

$$\sum_{k \in \mathbb{Z}} |(f, e_k)_{L^2}|^2 \le \|f\|_{L^2}^2 < \infty \, ,$$

and so for $m > n$

$$\|P_m f - P_n f\|_{L^2}^2 \le \sum_{|k| > n} |(f, e_k)_{L^2}|^2 \longrightarrow 0 \quad \text{as } n \to \infty,$$

i.e. the limit $\widetilde{f} := \lim_{n \to \infty} P_n f$ exists in $L^2(] - \pi, \pi[; \mathbb{C})$, which implies for a subsequence $n \to \infty$ that $P_n f(x) \to \widetilde{f}(x)$ for almost all x. Then we obtain $f = \widetilde{f}$ with the help of lemma 9.11 below on the pointwise convergence of Fourier series.

2nd possibility: Let $f \in C^0([- \pi, \pi]; \mathbb{C})$ with $f(-\pi) = f(\pi)$. Define $g \in C^0(\mathbb{R}^2; \mathbb{C})$ by

$$g(re^{i\theta}) := \eta(r) f(\theta) \quad \text{for } r \ge 0 \text{ and } -\pi \le \theta \le \pi,$$

where η is a continuous function on $[0, \infty[$ with $\eta(0) = 0$ and $\eta(1) = 1$. We need to show that g can be uniformly approximated in $\overline{B_1(0)}$ by polynomials in two variables, as this implies that the restriction to $\partial B_1(0)$, i.e. the representation $f(\theta) = g(e^{i\theta})$, yields the desired approximation. Hence we need to prove the following theorem. □

9.10 Weierstraß approximation theorem. Every function $g \in C^0(\mathbb{R}^n)$ can be uniformly approximated by polynomials on any ball $\overline{B_R(0)}$.

Note: Here we have $C^0(\mathbb{R}^n) = C^0(\mathbb{R}^n; \mathbb{K})$ by the definition 3.2. However, the proof carries over unchanged to functions in $C^0(\mathbb{R}^n; Y)$ with an arbitrary Banach space Y. We then obtain polynomials with "coefficients" in Y.

Proof. We may assume that $R = \frac{1}{4}$ and $g \in C_0^0(B_{\frac{1}{2}}(0))$, because otherwise we can consider $\widetilde{g}(x) := \eta(x) g(4Rx)$, where $\eta \in C_0^0(B_{\frac{1}{2}}(0))$ with $\eta = 1$ on $\overline{B_{\frac{1}{4}}(0)}$. Define

$$\varphi_m(x) := \frac{1}{c_m} \prod_{i=1}^{n} (1 - x_i^2)^m, \quad \text{if } |x_i| \le 1 \text{ for } i = 1, \ldots, n,$$

and $\varphi_m(x) := 0$ otherwise, where

$$c_m := \left(\int_{-1}^{1} (1 - t^2)^m \, dt \right)^n.$$

Then $(\varphi_m)_{m \in \mathbb{N}}$ is a general Dirac sequence (see 4.14(1)) and hence

$$\| \varphi_m * g - g \|_{C^0} \le \sup_{x \in \mathbb{R}^n} \int_{\mathbb{R}^n} |g(x - y) - g(x)| \varphi_m(y) \, dy$$

$$\le \underbrace{\sup_{x \in \mathbb{R}^n, \, |y| \le \delta} |g(x - y) - g(x)|}_{\longrightarrow \, 0 \text{ as } \delta \to 0} + \underbrace{2 \cdot \|g\|_{C^0} \int_{\mathbb{R}^n \setminus B_\delta(0)} \varphi_m \, dL^n}_{\substack{\longrightarrow \, 0 \text{ as } m \to \infty \\ \text{for any } \delta.}}$$

Since $g = 0$ outside of $\overline{B_{\frac{1}{2}}(0)}$, for $x \in \overline{B_{\frac{1}{2}}(0)}$ we have that

$$(\varphi_m * g)(x) = \int_{\mathbb{R}^n} \varphi_m(x - y) g(y) \, dy = \frac{1}{c_m} \int_{\mathbb{R}^n} \prod_{i=1}^{n} (1 - (x_i - y_i)^2)^m g(y) \, dy,$$

which clearly is a polynomial in the variables $x_1, \ldots, x_n$. $\qquad \square$

9.11 Lemma. Given $f \in L^2(]-\pi, \pi[; \mathbb{C})$ let

$$P_n f := \sum_{|k| \le n} (f, e_k)_{L^2} e_k$$

be the **Fourier sum** of f with e_k as in 9.9. If $-\pi < x < \pi$ and $\xi \in \mathbb{C}$ with

$$\limsup_{r \searrow 0} \left| \frac{f(x + r) + f(x - r) - 2\xi}{r} \right| < \infty,$$

then $\xi = \lim_{n \to \infty} P_n f(x)$.

Special case: The assumption is satisfied for $\xi = f(x)$ if f is Lipschitz continuous.

Proof. Noting that $(\xi, e_k)_{L^2} = 0$ for $k \ne 0$ and extending f by periodicity, i.e. setting $f(x + 2\pi k) := f(x)$ for $|x| < \pi$ and $k \in \mathbb{Z} \setminus \{0\}$, we have that

$$P_n f(x) - \xi = \sum_{|k| \le n} (f - \xi, e_k)_{L^2} e_k(x)$$

$$= \frac{1}{2\pi} \int_{-\pi}^{\pi} (f(x - y) - \xi) \cdot \sum_{|k| \le n} e^{iky} \, dy.$$

The expression (see E9.8)

$$\sum_{|k|\leq n} e^{iky} = \frac{\sin((n+\frac{1}{2})y)}{\sin(\frac{1}{2}y)} \qquad \text{is symmetric in } y.$$

Hence the changes of variables $y = 2z$ and $y = -2z$ yield

$$P_n f(x) - \xi = \frac{1}{2\pi} \int_{-\frac{\pi}{2}}^{\frac{\pi}{2}} \frac{f(x+2z) + f(x-2z) - 2\xi}{\sin z} \cdot \sin((2n+1)z) \, dz$$

$$= -(g_x, e_{2n+1})_{L^2} + (g_x, e_{-2n-1})_{L^2},$$

where

$$g_x(y) := \frac{\sqrt{2\pi}}{4\pi i} \cdot \chi_{]-\frac{\pi}{2},\frac{\pi}{2}[}(y) \frac{f(x+2y) + f(x-2y) - 2\xi}{\sin y}.$$

As $g_x \in L^2(]-\pi, \pi[; \mathbb{C})$, it follows from Bessel's inequality that $(g_x, e_{2n+1})_{L^2}$ and $(g_x, e_{-2n-1})_{L^2}$ converge to 0 as $n \to \infty$. $\qquad \square$

Projection

Thus, we have shown that for $f \in L^2(]-\pi, \pi[; \mathbb{C})$ the Fourier sum $P_n f$ (see 9.11) converges in this space to f (recall 9.7). Moreover, for Lipschitz continuous functions the Fourier sum converges pointwise. It is often of interest, in particular for numerical error estimations, how well $P_n f$ approximates the given function f. For Sobolev functions this error can be bounded in the L^2-norm as follows.

9.12 Lemma. Let $m \geq 1$ and let $f \in W^{m,2}(]-\pi, \pi[; \mathbb{C})$ satisfy the *periodic boundary conditions*

$$f^{(j)}(-\pi) = f^{(j)}(+\pi) \qquad \text{for } 0 \leq j \leq m-1.$$

Then it holds for the Fourier sums $P_n f$ in 9.11 that

$$\|f - P_n f\|_{L^2} \leq (\tfrac{1}{n+1})^m \|f^{(m)}\|_{L^2}.$$

Definition: The functions $f^{(j)}$ denote the weak derivatives of f.

Proof. It holds that $\frac{d}{dx} e_k(x) = ik \, e_k(x)$, and so integration by parts yields for $j < m$ (as $f^{(j)} \cdot \overline{e_k}$ is in $W^{1,2}(]-\pi, \pi[; \mathbb{C})$ by the product rule 4.25 and hence also in $AC([-\pi, \pi]; \mathbb{C})$ by E3.6) that

$$\left(f^{(j)}, e_k\right)_{L^2} = -\frac{1}{ik} \int_{-\pi}^{\pi} f^{(j)}(x) \frac{d}{dx} \overline{e_k(x)} \, dx$$

$$= -\frac{e^{-ik\pi}}{ik\sqrt{2\pi}} \left(f^{(j)}(\pi) - f^{(j)}(-\pi)\right) + \frac{1}{ik} \int_{-\pi}^{\pi} f^{(j+1)}(x) \cdot \overline{e_k(x)} \, dx$$

$$= \frac{1}{ik} \left(f^{(j+1)}, e_k\right)_{L^2}.$$

We obtain via induction that

$$(f, e_k)_{L^2} = \left(\tfrac{1}{ik}\right)^m \cdot \left(f^{(m)}, e_k\right)_{L^2}.$$

Noting

$$(f - P_n f, e_k)_{L^2} = \begin{cases} 0 & \text{for } |k| \le n, \\ (f, e_k)_{L^2} & \text{for } |k| > n, \end{cases}$$

and recalling 9.7(5) yields that

$$\|f - P_n f\|_{L^2}^2 = \sum_{|k| \ge n+1} |(f, e_k)_{L^2}|^2 \le \frac{1}{(n+1)^{2m}} \sum_{|k| \ge n+1} \left| \left(f^{(m)}, e_k\right)_{L^2} \right|^2$$

$$\le \frac{1}{(n+1)^{2m}} \left\| f^{(m)} \right\|_{L^2}^2.$$

$\square$

The Fourier sum $P_n f$ for a given function $f \in X := L^2(] - \pi, \pi[; \mathbb{C})$ is an explicit rule for computing an approximation in the subspace $X_n := \text{span}\{e_k ; |k| \le n\}$. Moreover, $P_n f = f$ means that already $f \in X_n$. This property we can also require for a general separable Banach space X, i.e. one has certain subspaces $X_n \subset X$, and for $x \in X$ an approximating $x_n \in X_n$ defined by $x_n = P_n x$, where $P_n : X \to X_n$ is a linear map, and in addition it should hold that $x_n = x$ if $x \in X_n$, that is, $P_n = \text{Id}$ on X_n. Maps with this property are called projections. Note that this definition uses only the vector space structure of X. In the following, i.e. starting with 9.14, we will then consider continuous projections.

9.13 Linear projections. Let Y be a subspace of the vector space X. A linear map $P : X \to X$ is called a *(linear) projection onto Y* if (cf. 5.5(3))

$$P^2 = P \quad \text{and} \quad \mathcal{R}(P) = Y.$$

It holds that:

(1) P is a projection onto Y if and only if

$$P : X \to Y \quad \text{and} \quad P = \text{Id on } Y.$$

(2) If $P : X \to X$ is a projection, then

$$X = \mathcal{N}(P) \oplus \mathcal{R}(P).$$

(3) If P is a projection then so is $\text{Id} - P$, with

$$\mathcal{N}(\text{Id} - P) = \mathcal{R}(P) \quad \text{and} \quad \mathcal{R}(\text{Id} - P) = \mathcal{N}(P).$$

(4) For every subspace Y of X there exists a projection onto Y.
Observe: The projections in (4) are in general not (!) continuous (see also theorem 9.15).

Proof (2). For $x \in X$ we have that

$$x = \underbrace{(x - Px)}_{\in \mathcal{N}(P)} + \underbrace{Px}_{\in \mathcal{R}(P)} \; .$$

If $x \in \mathcal{N}(P) \cap \mathcal{R}(P)$, then $Px = 0$ and $x = Px$, and so $x = 0$. □

Proof (3). It holds that

$$(\mathrm{Id} - P)^2 = \mathrm{Id} - 2P + P^2 = \mathrm{Id} - 2P + P = \mathrm{Id} - P$$

and

$$x \in \mathcal{N}(\mathrm{Id} - P) \quad \Longleftrightarrow \quad x - Px = 0 \quad \Longleftrightarrow \quad x \in \mathcal{R}(P),$$

and so $\mathcal{N}(\mathrm{Id} - P) = \mathcal{R}(P)$, and hence also $\mathcal{N}(P) = \mathcal{N}(\mathrm{Id} - (\mathrm{Id} - P)) = \mathcal{R}(\mathrm{Id} - P)$. □

Proof (4). Similarly to the proof of the Hahn-Banach theorem 6.14, set

$$\mathcal{M} := \big\{ (Z, P) \,;\; Y \subset Z \subset X, \; Z \text{ subspace},$$
$$P : Z \to Y \text{ linear}, \; P = \mathrm{Id} \text{ on } Y \big\}$$

with the order as in 6.14. As in the proof of 6.14, it follows that $\mathcal{M}$ has a maximal element (Z, P). If we assume that there exists a $z_0 \in X \setminus Z$, then

$$Z_0 := Z \oplus \mathrm{span}\{z_0\}, \quad P_0(z + \alpha z_0) := P(z) \text{ for } z \in Z, \; \alpha \in \mathbb{K},$$

defines a $(Z_0, P_0) \in \mathcal{M}$ with $(Z, P) < (Z_0, P_0)$ (i.e. $(Z, P) \leq (Z_0, P_0)$ and $(Z, P) \neq (Z_0, P_0)$), a contradiction. □

9.14 Continuous projections. Let X be a normed space. Then (cf. 5.5(3))

$$\mathscr{P}(X) := \{P \in \mathscr{L}(X); \; P^2 = P\}$$

denotes the set of *continuous (linear) projections*. If $P \in \mathscr{P}(X)$, then:

(1) $\mathcal{N}(P)$ and $\mathcal{R}(P)$ are closed,
(2) $\|P\| \geq 1$ or $P = 0$.

Proof (1). It follows from P being continuous that $\mathcal{N}(P) = P^{-1}(\{0\})$ is closed as the preimage of a closed set. Combining the facts that $\mathrm{Id} - P$ is continuous and a projection, recall 9.13(3), yields that $\mathcal{R}(P) = \mathcal{N}(\mathrm{Id} - P)$ is closed. □

Proof (2). We use the fact that $\mathscr{L}(X)$ is a Banach algebra (see 5.3(3)), and conclude that $\|P\| = \|P^2\| \leq \|P\|^2$. Therefore $\|P\| = 0$ or $\|P\| \geq 1$. □

The result in 9.13(4) does not hold for continuous projections in Banach spaces, i.e. in general it is not possible to continuously project onto every closed subspace. However, the following theorem is valid.

9.15 Closed complement theorem. Let X be a Banach space. Let Y be a closed subspace and let Z be a subspace with $X = Z \oplus Y$ (i.e. $Z \cap Y = \{0\}$ and $\mathrm{span}(Z \cup Y) = X$). Then the following are equivalent:

(1) There exists a continuous projection P onto Y with $Z = \mathcal{N}(P)$.
(2) Z is closed.

Remark: The theorem can also be formulated as follows: If Y is a closed subspace of a Banach space X, then Y has a **closed complement** (i.e. there exists a closed subspace Z with $X = Z \oplus Y$) if and only if there exists a continuous projection onto Y.

Proof (1)$\Rightarrow$(2). $\mathcal{N}(P)$ is closed. $\qquad\qquad\qquad\qquad\qquad\qquad\qquad$ □

Proof (2)$\Rightarrow$(1). Consider the Banach space $\widetilde{X} := Z \times Y$ with the norm

$$\|(z, y)\|_{\widetilde{X}} := \|z\|_X + \|y\|_X \quad \text{for } z \in Z,\ y \in Y$$

and define $T(z, y) := z + y$. It follows from $X = Z \oplus Y$ that $T : \widetilde{X} \to X$ is linear and bijective. Define $P_Z : X \to Z$ and $P_Y : X \to Y$ by

$$T^{-1}x = (P_Z x, P_Y x) \quad \text{for } x \in X.$$

Then P_Z, P_Y are linear. As $T^{-1}(y) = (0, y)$ for $y \in Y$, we have that $P_Y = \mathrm{Id}$ on Y, i.e. P_Y is a projection onto Y. On noting that $\|P_Y x\|_X \le \|T^{-1}x\|_{\widetilde{X}}$, it holds that P_Y is continuous if T^{-1} is continuous. Now $\|T(z, y)\|_X \le \|(z, y)\|_{\widetilde{X}}$ yields that T is continuous, and hence the continuity of T^{-1} follows from the inverse mapping theorem. $\qquad\qquad\qquad\qquad\qquad\qquad\qquad$ □

We now exhibit two classes of subspaces that have a closed complement. These are finite-dimensional subspaces in arbitrary Banach spaces and arbitrary closed subspaces in Hilbert spaces. In general, there exist closed subspaces of Banach spaces that do not have a closed complement (see [Murray]).

9.16 Theorem. Let X be a normed vector space, let E be an n-dimensional subspace with basis $\{e_i\ ;\ i = 1, \ldots, n\}$ and let Y be a closed subspace with $Y \cap E = \{0\}$. Then:

(1) There exist $e'_1, \ldots, e'_n$ in X' with $e'_j = 0$ on Y and $\langle e_i,\, e'_j \rangle = \delta_{i,j}$.
Remark: Compare also with E9.3.
(2) There exists a continuous projection P onto E with $Y \subset \mathcal{N}(P)$.

Proof (1). It follows from 9.2(1) that $Y_j := \mathrm{span}\{e_k\ ;\ k \ne j\} \oplus Y$ are closed subspaces. Moreover, $e_j \notin Y_j$. Hence, by 6.16, there exist $e'_j \in X'$ such that $e'_j = 0$ on Y_j and $\langle e_j,\, e'_j \rangle = 1$. $\qquad\qquad\qquad\qquad\qquad\qquad$ □

Proof (2). Define $Px := \sum_{j=1}^n \langle x,\, e'_j \rangle e_j$. $\qquad\qquad\qquad\qquad\qquad\qquad$ □

9.17 Lemma. Let Y be a closed subspace of a Hilbert space X and let P be the orthogonal projection onto Y from 4.3. Then:

(1) $P \in \mathscr{P}(X)$.

(2) $\mathscr{R}(P) = Y$ and $\mathscr{N}(P) = Y^\perp := \{x \in X ; \ (x, y)_X = 0 \text{ for all } y \in Y\}$.

(3) $X = Y^\perp \perp Y$.

(4) If $Z \subset X$ is a subspace with $X = Z \perp Y$, then $Z = Y^\perp$. Hence we call $Y^\perp$ the **orthogonal complement** of Y.

Proof. By recalling 4.4(2), we have that P in 4.3 is characterized by

$$(x - Px, y)_X = 0 \quad \text{for all } y \in Y. \tag{9-32}$$

This immediately implies, as already shown in 4.4(2), that P is linear. Moreover P is continuous, since choosing $y = Px$ in (9-32) yields that

$$\|Px\|^2 = (x, Px)_X \leq \|x\| \|Px\|,$$

and so $\|Px\| \leq \|x\|$. In addition it immediately follows from (9-32) that P is a projection onto Y in the sense of 9.13. To see this, let $x \in Y$. Choosing $y = x - Px \in Y$ in (9-32) then yields that $x - Px = 0$, which implies that $P = \mathrm{Id}$ on Y. Furthermore, the property (9-32) states that

$$x \in \mathscr{N}(P) \iff Px = 0 \iff (x, y)_X = 0 \text{ for all } y \in Y \iff x \in Y^\perp.$$

Hence we have shown assertions (1) and (2). Assertion (3) then follows from 9.13(2). In order to show (4) we first observe that $Z \subset Y^\perp$, by the definition in 2.3(2). But if $x \in Y^\perp$ with the representation $x = z + y$, $z \in Z$, $y \in Y$, then also $x - z \in Y^\perp$, and so $0 = (x - z, y)_X = \|y\|^2$, i.e. $x = z \in Z$. $\quad\square$

There exist alternative characterizations of linear orthogonal projections to those given in 4.3 and 4.4. In 9.18(3) they are identified as precisely the self-adjoint projections (see the definition 12.2).

9.18 Lemma. Let X be a Hilbert space and let $P : X \to X$ be linear. Then the following are equivalent:

(1) P is an orthogonal projection onto $\mathscr{R}(P)$, i.e. (as in 4.3)

$$\|x - Px\| \leq \|x - Py\| \quad \text{for all } x, y \in X.$$

(2) $(x - Px, Py)_X = 0$ for all $x, y \in X$.

(3) $P^2 = P$ and $(Px, y)_X = (x, Py)_X$ for all $x, y \in X$.

(4) $P \in \mathscr{P}(X)$ with $\|P\| \leq 1$ (9.14(2) then implies that $P = 0$ or $\|P\| = 1$).

Proof (1)$\Leftrightarrow$(2). See the proofs of 4.3 and 4.4. $\quad\square$

Proof (2)⇒(3). We have for $x, y \in X$ that

$$
\begin{aligned}
0 &= (x - Px,\, Py)_X - \overline{(y - Py,\, Px)_X} \\
&= (x,\, Py)_X - (Px,\, Py)_X - \overline{(y,\, Px)_X} + \overline{(Py,\, Px)_X} \\
&= (x,\, Py)_X - (Px,\, y)_X \,.
\end{aligned}
$$

Applying this identity we see that for $x \in X$

$$
\left(P^2 x - Px,\, y\right)_X = (P(Px - x),\, y)_X = (Px - x,\, Py)_X = 0
$$

for all y, which means that $P^2 x = Px$.   □

Proof (3)⇒(4). Set $y = Px$ in (3) and obtain

$$
\|Px\|^2 = \left(x,\, P^2 x\right)_X = (x,\, Px)_X \leq \|x\| \|Px\|,
$$

and hence $\|Px\| \leq \|x\|$, i.e. $P \in \mathscr{L}(X)$ with $\|P\| \leq 1$ and $P^2 = P$.   □

Proof (4)⇒(2). Let $x \in X$ and $y \in \mathscr{R}(P)$, and let $z := x - Px$. It follows from $P^2 = P$ that $Py = y$ and $Pz = 0$. Hence, for $\varepsilon > 0$ and $|\alpha| = 1$ we have that

$$
\begin{aligned}
\|y\|^2 &= \|P(\varepsilon z + \alpha y)\|^2 \leq \|\varepsilon z + \alpha y\|^2 \quad (\text{recall } \|P\| \leq 1) \\
&= \varepsilon^2 \|z\|^2 + \varepsilon \, 2\mathrm{Re}\,(z,\, \alpha y)_X + \|y\|^2,
\end{aligned}
$$

and so

$$
0 \leq \varepsilon \|z\|^2 + 2\mathrm{Re}\,(z,\, \alpha y)_X \longrightarrow 2\mathrm{Re}(\overline{\alpha}\,(z,\, y)_X) \quad \text{as } \varepsilon \searrow 0.
$$

As this holds for all $\alpha \in \mathbb{K}$ with $|\alpha| = 1$, it follows that

$$
0 = (z,\, y)_X = (x - Px,\, y)_X \,.
$$

This yields the desired result (2).   □

9.19 Remark. Let X be a Banach space. If X_n are finite-dimensional subspaces as in 9.1(2), then 9.16(2) yields the existence of $P_n \in \mathscr{P}(X)$ with $X_n = \mathscr{R}(P_n)$. Then a stronger property than 9.1(2) is the **approximation property**:

(P1) $P_n x \to x$ as $n \to \infty$ for all $x \in X$.

On recalling the Banach-Steinhaus theorem (see 7.3), it follows from (P1) that

$$
C := \sup_n \|P_n\| < \infty.
$$

In addition, we may require that the projections P_n satisfy the following **commutativity relation**:

(P2) $P_n P_m = P_{\min(n,m)}$ for all n, m.

This also contains once again the projection property of the P_n. A sequence $(P_n)_{n \in \mathbb{N}}$ in $\mathscr{P}(X)$ with (P1) and (P2) corresponds, via

$$Q_n := P_n - P_{n-1} \text{ (where } P_0 := 0) \quad \text{or} \quad P_n := \sum_{i=1}^{n} Q_i,$$

to a sequence $(Q_n)_{n \in \mathbb{N}}$ in $\mathscr{P}(X)$ with the properties

(Q1) $\sum_{i=1}^{n} Q_i x \to x$ as $n \to \infty$ for all $x \in X$,
(Q2) $Q_n Q_m = \delta_{n,m} Q_n$ for all n, m.

The subspaces $E_n := \mathscr{R}(Q_n)$ then satisfy 9.1(3), while 9.1(2) is satisfied with $X_n = E_1 \oplus \cdots \oplus E_n$.

Proof. Defining the Q_n through the P_n as above yields that

$$Q_n^2 = (P_n - P_{n-1})(P_n - P_{n-1}) = P_n - P_n P_{n-1} - P_{n-1} P_n + P_{n-1}$$
$$= P_n - P_{n-1} = Q_n,$$

and for $m > n$

$$Q_m Q_n = (P_m - P_{m-1})(P_n - P_{n-1})$$
$$= P_m P_n - P_m P_{n-1} - P_{m-1} P_n + P_{m-1} P_{n-1}$$
$$= P_n - P_{n-1} - P_n + P_{n-1} = 0,$$

and similarly $Q_n Q_m = 0$, which proves (Q2). Defining the P_n through the Q_n as above yields that $P_n \in \mathscr{P}(X)$ (see E9.7) and

$$P_n P_m = \sum_{\substack{1 \le k \le n \\ 1 \le l \le m}} Q_k Q_l = \sum_{k=1}^{n} \underbrace{\left(\sum_{l=1}^{m} \delta_{k,l} \right)}_{\substack{= 1 \text{ for } k \le m \\ = 0 \text{ for } k > m}} Q_k = \sum_{k=1}^{\min(n,m)} Q_k,$$

which proves (P2). $\qquad\qquad\square$

9.20 Examples. The properties in 9.19 are satisfied in the following cases:

(1) In the sequence space $X = \ell^2(\mathbb{K})$ define for $x = (x_i)_{i \in \mathbb{N}} \in \ell^2(\mathbb{K})$

$$Q_n x := x_n e_n \quad \text{and} \quad P_n x := \sum_{i=1}^{n} x_i e_i.$$

(2) If X is a Hilbert space and $X = \overline{\bigcup_{n \in \mathbb{N}} X_n}$ with $\dim X_n < \infty$, $X_n \subset X_{n+1}$, then let P_n be the orthogonal projection onto X_n and, on setting

$$X_{n+1} = X_n \perp E_n, \quad \text{i.e. } E_n := X_n^\perp \cap X_{n+1},$$

let Q_n be the orthogonal projection onto E_n (see E9.9).

(3) If in (2) we have in particular that $X_n = \operatorname{span}\{e_i \; ; \; 1 \le i \le n\}$ for an orthonormal basis $(e_i)_{i \in \mathbb{N}}$, then

$$Q_n x = (x, e_n)_X \, e_n \quad \text{and} \quad P_n x = \sum_{i=1}^n (x, e_i)_X \, e_i.$$

(4) If $(e_i)_{i \in \mathbb{N}}$ is a Schauder basis of a Banach space X and if $(e'_k)_{k \in \mathbb{N}}$ is the corresponding dual basis, then let (see also E9.10)

$$Q_n x := \langle x, e'_n \rangle \, e_n \quad \text{and} \quad P_n x := \sum_{i=1}^n \langle x, e'_i \rangle \, e_i.$$

We now provide some explicit examples in function spaces, where we restrict ourselves to functions in one variable.

9.21 Piecewise constant approximation. For the following examples we partition the unit interval $[0, 1]$ into points

$$0 = x_{n,0} < x_{n,1} < \ldots < x_{n,m_n} = 1 \quad \text{and} \quad M_n := \{x_{n,i} \; ; \; i = 0, \ldots, m_n\}$$

for $n \in \mathbb{N} \cup \{0\}$ (where $m_0 = 1$, i.e. for $n = 0$ we consider only the whole interval) and require that the **mesh of the partition** (or the **maximal step size**)

$$h_n := \max_{1 \le i \le m_n} |x_{n,i} - x_{n,i-1}| \longrightarrow 0 \quad \text{as } n \to \infty.$$

The commutativity relation (P2) will be guaranteed for the following projections by the requirement that

$$M_n \subset M_{n+1} \quad \text{for all } n,$$

i.e. the partition on level $n + 1$ is a refinement of the partition on level n. Furthermore, let $A_{n,i} := \,]x_{n,i-1}, x_{n,i}[$ and $h_{n,i} := x_{n,i} - x_{n,i-1}$. The standard example is $x_{n,i} := i \, 2^{-n}$, $m_n = 2^n$. The space of **piecewise constant functions** with respect to the partition on level $n \ge 0$ is

$$X_n := \left\{ \sum_{i=1}^{m_n} \alpha_i \chi_{A_{n,i}} \; ; \; \alpha_i \in \mathbb{K} \text{ for } i = 1, \ldots, m_n \right\},$$

with $\dim X_n = m_n$. For $f \in L^1(]0, 1[)$ we define

$$P_n f := \sum_{i=1}^{m_n} \left(\frac{1}{h_{n,i}} \int_{A_{n,i}} f(s) \, ds \right) \chi_{A_{n,i}}.$$

The coefficients of $P_n f$ are the average values of f over the intervals $A_{n,i}$. It holds that $\mathscr{R}(P_n) = X_n$. Then:

(1) For $X = L^p(]0,1[)$, $1 \le p < \infty$, the P_n satisfy the properties in 9.19 with $X_n = \mathscr{R}(P_n)$ and $\|P_n\| \le 1$.

(2) For $f \in W^{1,p}(]0,1[)$, $1 \le p \le \infty$, it holds that

$$\|f - P_n f\|_{L^p} \le h_n \|f'\|_{L^p}.$$

(3) For the standard partition $x_{n,i} = i2^{-n}$ for $n \ge 0, 0 \le i \le 2^n$, it holds that $Q_0 = P_0$ and for $n \ge 1$

$$Q_n f = P_n f - P_{n-1} f$$

$$= 2^{n-1} \sum_{i=1}^{2^{n-1}} \left(\int_{A_{n,2i-1}} f(s)\,ds - \int_{A_{n,2i}} f(s)\,ds \right) \cdot \left(\mathcal{X}_{A_{n,2i-1}} - \mathcal{X}_{A_{n,2i}} \right).$$

(4) This yields that $E_0 = \operatorname{span}\{e_0\}$ and $E_n = \operatorname{span}\{e_{ni}\,;\, 1 \le i \le 2^{n-1}\}$ for $n \ge 1$, where

$$e_0 := \mathcal{X}_{]0,1[}, \quad e_{ni} := \mathcal{X}_{A_{n,2i-1}} - \mathcal{X}_{A_{n,2i}}, \quad i = 1, \dots, 2^{n-1}, \quad n \ge 1.$$

These functions, in any order in which the first index is monotonically increasing, form a Schauder basis of $L^p(]0,1[)$, $1 \le p < \infty$.

Proof (1). Using the Hölder inequality (see also Jensen's inequality E4.10) we obtain that

$$\int_0^1 |P_n f(t)|^p \,dt = \sum_{i=1}^{m_n} h_{n,i} \left| \frac{1}{h_{n,i}} \int_{A_{n,i}} f(s)\,ds \right|^p$$

$$\le \sum_{i=1}^{m_n} \int_{A_{n,i}} |f(s)|^p \,ds = \int_0^1 |f(s)|^p \,ds,$$

and hence $P_n \in \mathscr{L}(X)$ with $\|P_n\| \le 1$. Moreover, $P_n f \to f$ in $L^p(]0,1[)$ for all $f \in L^p(]0,1[)$, because for $f \in C^0([0,1])$ we have that

$$\|f - P_n f\|_{L^p} \le \|f - P_n f\|_{\sup}$$

$$\le \sup_{|s-t| \le h_n} |f(s) - f(t)| \longrightarrow 0 \quad \text{as } n \to \infty.$$

The claim then follows on recalling from 4.15(3) that $C^0([0,1])$ is dense in $L^p(]0,1[)$, on noting that the $\|P_n\|$ are bounded, and on recalling E5.5.

It remains to show that the commutativity relation (P2) is satisfied (which also includes the projection property). We have that

$$P_n P_m f = \sum_{j=1}^{m_m} \sum_{i=1}^{m_n} \frac{1}{h_{n,i}} \frac{1}{h_{m,j}} \int_{A_{m,j}} f(s)\,ds \cdot \int_{A_{n,i}} \mathcal{X}_{A_{m,j}}(s)\,ds \cdot \mathcal{X}_{A_{n,i}}.$$

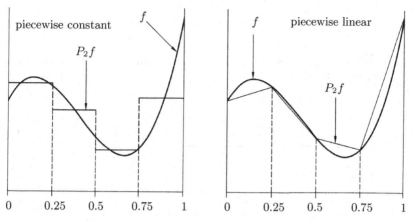

Fig. 9.1. *Piecewise constant and piecewise linear approximations*

The assumption on the interval partitions imply that for $n \geq m$ we have

$$\int_{A_{n,i}} \mathcal{X}_{A_{m,j}}(s)\,\mathrm{d}s = \begin{cases} h_{n,i} & \text{if } A_{n,i} \cap A_{m,j} \neq \emptyset, \\ 0 & \text{otherwise,} \end{cases}$$

and so

$$\sum_i \frac{1}{h_{n,i}} \int_{A_{n,i}} \mathcal{X}_{A_{m,j}}(s)\,\mathrm{d}s \cdot \mathcal{X}_{A_{n,i}} = \sum_{i\,:\,A_{n,i}\cap A_{m,j}\neq\emptyset} \mathcal{X}_{A_{n,i}} = \mathcal{X}_{A_{m,j}}$$

and hence $P_n P_m f = P_m f$. For $n \leq m$ we obtain similarly that

$$\int_{A_{n,i}} \mathcal{X}_{A_{m,j}}(s)\,\mathrm{d}s = \begin{cases} h_{m,j} & \text{if } A_{n,i} \cap A_{m,j} \neq \emptyset, \\ 0 & \text{otherwise,} \end{cases}$$

and hence

$$\sum_j \frac{1}{h_{m,j}} \int_{A_{m,j}} f(s)\,\mathrm{d}s \int_{A_{n,i}} \mathcal{X}_{A_{m,j}}(s)\,\mathrm{d}s$$

$$= \sum_{j\,:\,A_{n,i}\cap A_{m,j}\neq\emptyset} \int_{A_{m,j}} f(s)\,\mathrm{d}s = \int_{A_{n,i}} f(s)\,\mathrm{d}s,$$

which yields that $P_n P_m f = P_n f$. $\square$

Proof (2). By E3.6, for almost all $t \in A_{n,i}$

$$(f - P_n f)(t) = \frac{1}{h_{n,i}} \int_{A_{n,i}} \big(f(t) - f(s)\big)\,\mathrm{d}s = \frac{1}{h_{n,i}} \int_{A_{n,i}} \int_s^t f'(r)\,\mathrm{d}r\,\mathrm{d}s,$$

and so

$$|(f - P_n f)(t)| \leq \int_{A_{n,i}} |f'(r)|\, dr \leq h_n \cdot \frac{1}{h_{n,i}} \int_{A_{n,i}} |f'(s)|\, ds.$$

This yields the desired result for $p = \infty$ and, similarly to the beginning of the proof of (1), for $p < \infty$. $\qquad \square$

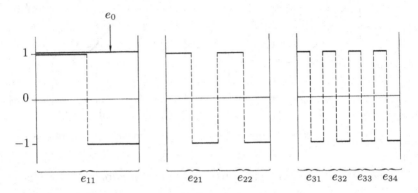

Fig. 9.2. *Piecewise constant basis functions*

Proof (3). This follows from

$$P_n f = \sum_{i=1}^{2^{n-1}} \left(2^n \int_{A_{n,2i-1}} f(s)\, ds \cdot \mathcal{X}_{A_{n,2i-1}} + 2^n \int_{A_{n,2i}} f(s)\, ds \cdot \mathcal{X}_{A_{n,2i}} \right)$$

for $n \geq 1$. $\qquad \square$

Proof (4). The identities for E_n follow from (3), which on recalling (Q1) also implies that

$$f = \alpha_0 e_0 + \sum_{n=1}^{\infty} \sum_{i=1}^{2^{n-1}} \alpha_{n,i}\, e_{ni}$$

where

$$\alpha_0 := \int_0^1 f(s)\, ds \quad \text{and} \quad \alpha_{n,i} = 2^{n-1} \left(\int_{A_{n,2i-1}} f(s)\, ds - \int_{A_{n,2i}} f(s)\, ds \right).$$

We then obtain the desired result (similarly to the proof of E9.10), if we can show for $n \geq 1$ and $\alpha_i \in \mathbb{K}$, $i = 1, \ldots, 2^{n-1}$, that

$$\left\| \sum_{i=1}^{2^{n-1}} |\alpha_i e_{ni}| \right\|_{L^p} \leq \left\| \sum_{i=1}^{2^{n-1}} \alpha_i e_{ni} \right\|_{L^p}.$$

This follows from the fact that the supports of the functions e_{ni}, with $i = 1, \ldots, 2^{n-1}$ are, apart from their endpoints, disjoint. $\qquad \square$

9.22 Continuous piecewise linear approximation. We consider partitions of $[0,1]$ as in 9.21. For $f \in C^0([0,1])$ let

$$P_n f(s) := \frac{(x_{n,i} - s)f(x_{n,i-1}) + (s - x_{n,i-1})f(x_{n,i})}{x_{n,i} - x_{n,i-1}}$$

$$\text{for } x_{n,i-1} \le s \le x_{n,i} \text{ and } i = 1, \dots, m_n.$$

The space of **continuous piecewise linear functions** with respect to the partition on level $n \ge 0$ is

$$X_n := \{ f \in C^0([0,1]) \, ; \, f = P_n f \}.$$

Then:

(1) For $X = C^0([0,1])$ the P_n satisfy the properties in 9.19 with $X_n = \mathscr{R}(P_n)$ and $\|P_n\| \le 1$. Moreover,

$$P_n f = \sum_{i=0}^{m_n} f(x_{n,i}) \, \widehat{e}_{n,i} \in C^{0,1}([0,1])$$

with $P_n f(x_{n,i}) = f(x_{n,i})$, where the **hat functions** are defined by

$$\widehat{e}_{n,i}(s) = \begin{cases} \dfrac{1}{h_{n,i}}(s - x_{n,i-1}) & \text{for } i > 0, \ x_{n,i-1} \le s \le x_{n,i}, \\[2mm] \dfrac{1}{h_{n,i+1}}(x_{n,i+1} - s) & \text{for } i < m_n, \ x_{n,i} \le s \le x_{n,i+1}, \\[2mm] 0 & \text{otherwise}, \end{cases}$$

and hence $\dim X_n = m_n + 1$.

(2) For $f \in W^{m,p}(]0,1[)$ with $1 \le p \le \infty$ it holds for $0 \le l < m \in \{1,2\}$ that

$$\left\| (f - P_n f)^{(l)} \right\|_{L^p} \le 2 h_n^{m-l} \left\| f^{(m)} \right\|_{L^p}.$$

(3) We have that $Q_0 = P_0$ and for $n \ge 1$ that

$$Q_n f = \sum_{i \, : \, x_{n,i} \notin M_{n-1}} \left(f(x_{n,i}) - (P_{n-1}f)(x_{n,i}) \right) \widehat{e}_{n,i}.$$

(4) This yields that $E_0 = \operatorname{span}\{\widehat{e}_{0,0}, \widehat{e}_{0,1}\}$ and for $n \ge 1$

$$E_n = \operatorname{span}\{\widehat{e}_{n,i} \, ; \, x_{n,i} \notin M_{n-1}\}.$$

These basis functions, in any order in which the first index is monotonically increasing, form a Schauder basis of $C^0([0,1])$. For the standard partition $x_{n,i} = i2^{-n}$ these are the functions

$$e_{01}(x) := x, \quad e_{02}(x) := 1 - x,$$
$$e_{ni}(x) := \max(0, \, 1 - 2^n \left| x - (2i-1)2^{-n} \right|) \quad \text{for } i = 1, \dots, 2^{n-1}, \ n \ge 1.$$

Proof (1). It holds that $P_n f \to f$ as $n \to \infty$ because

$$\|f - P_n f\|_{C^0} \leq \sup_{|s-t| \leq h_n} |f(t) - f(s)|.$$

$\square$

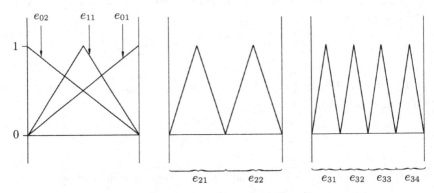

Fig. 9.3. *Piecewise linear basis functions*

Proof (2). It follows from E3.6 that every $f \in W^{1,p}(]0,1[)$ has a unique continuous representative for which $P_n f$ is well defined and lies in the space $W^{1,\infty}(]0,1[)$, where for $x_{n,i-1} < s < x_{n,i}$ we have that

$$(P_n f)'(s) = \frac{1}{h_{n,i}} \big(f(x_{n,i}) - f(x_{n,i-1})\big) = \big(\widetilde{P_n} f'\big)(s),$$

if the $\widetilde{P_n}$ denote the projections from 9.21. Consequently,

$$(f - P_n f)'(s) = \big(f' - \widetilde{P_n} f'\big)(s).$$

For $p < \infty$, on applying the Hölder inequality (see also Jensen's inequality), we obtain that

$$\int_{A_{n,i}} \big|(f - P_n f)(t)\big|^p \, dt = \int_{A_{n,i}} \left| \int_{x_{n,i}}^{t} \big(f' - \widetilde{P_n} f'\big)(s) \, ds \right|^p dt$$

$$\leq h_{n,i}^p \int_{A_{n,i}} \big|(f' - \widetilde{P_n} f')(t)\big|^p dt,$$

and so

$$\|f - P_n f\|_{L^p} \leq h_n \left\|f' - \widetilde{P_n} f'\right\|_{L^p},$$

which also holds for $p = \infty$. Hence the desired results follow from 9.21. $\square$

Proof (3). For affine linear functions g it holds that $P_n g = g$. Now let $n \geq 1$ and let g be the function that for a given i agrees with $P_{n-1} f$ on $[x_{n,i-1}, x_{n,i}]$. Then for $t \in [x_{n,i-1}, x_{n,i}]$

$$(Q_n f)(t) = (P_n f - g)(t) = P_n(f - g)(t)$$

$$= \sum_{j=0}^{m_n} (f - g)(x_{n,j}) \widehat{e}_{n,j}(t) = \sum_{j=0}^{m_n} (f - P_{n-1}f)(x_{n,j}) \widehat{e}_{n,j}(t),$$

because

$$\widehat{e}_{n,j}(t) = 0 \quad \text{for } j \notin \{i-1, i\},$$

$$g(x_{n,j}) = (P_{n-1}f)(x_{n,j}) \quad \text{for } j \in \{i-1, i\}.$$

Noting that $(f - P_{n-1})(x_{n,j}) = 0$ for $x_{n,j} \in M_{n-1}$ yields the desired result. □

Proof (4). This follows as in 9.21(4), upon observing that

$$\left\| \sum_i |\alpha_i \widehat{e}_{n,i}| \right\|_{C^0} = \max_i |\alpha_i| = \left\| \sum_i \alpha_i \widehat{e}_{n,i} \right\|_{C^0}.$$

□

The constructions in 9.21-9.22 can be generalized to partitions of sets in $\mathbb{R}^m$, where m-dimensional simplices are needed for the partitions in 9.22. However, 9.22(2) no longer holds in the multidimensional case, because then $W^{1,p}$-functions in general no longer have a continuous representative (see also theorem 10.13).

Ritz-Galerkin approximation

We conclude with the finite-dimensional approximation of the solutions of the boundary value problems in 6.5.

9.23 Ritz-Galerkin approximation. Let $u \in X := W^{1,2}(\Omega)$ be the solution of the homogeneous Neumann problem from 6.6, or let $u \in X := W_0^{1,2}(\Omega)$ be the solution of the homogeneous Dirichlet problem from 6.8, respectively. Choose finite-dimensional subspaces X_N, $N \in \mathbb{N}$, as in 9.1(2) (X is separable, by 4.18(6) and 4.17(2)). Then the following is true: There exists a unique $u_N \in X_N$ with

$$\int_\Omega \left(\sum_i \partial_i \zeta \left(\sum_j a_{ij} \partial_j u_N + h_i \right) + \zeta(b u_N + f) \right) dL^n = 0 \quad \text{for } \zeta \in X_N.$$

Proof. Analogously to the proofs of 6.6 and 6.8, the bilinear form a satisfies the assumptions of the Lax-Milgram theorem 6.2 on the subspace X_N, which by 4.9 is a Hilbert space. □

9.24 Remark. If $d_N = \dim X_N$ and if $\{\varphi_k^{(N)}\; ;\; k = 1, \ldots, d_N\}$ is a basis of X_N, then the solution u_N in 9.23 has a unique representation

$$u_N = \sum_{k=1}^{d_N} u_{N,k}\, \varphi_k^{(N)} \quad \text{with } u_{N,k} \in \mathbb{R}$$

and is given as the solution of the system of linear equations

$$\sum_{l=1}^{d_N} a_{kl}^{(N)} u_{N,l} + c_k^{(N)} = 0 \quad \text{for } k = 1, \ldots, d_N,$$

where

$$a_{kl}^{(N)} := \int_\Omega \left(\sum_{i,j} a_{ij}\, \partial_i \varphi_k^{(N)} \partial_j \varphi_l^{(N)} + b\varphi_k^{(N)} \varphi_l^{(N)} \right) \mathrm{d}\mathrm{L}^n,$$

$$c_k^{(N)} := \int_\Omega \left(\sum_i \partial_i \varphi_k^{(N)} h_i + \varphi_k^{(N)} f \right) \mathrm{d}\mathrm{L}^n.$$

In a numerical computation these integrals may have to be computed approximatively. In general it is important for a numerical computation of the solution that the spaces X_N, and in particular the basis functions $\varphi_k^{(N)}$, are chosen so that the coefficients $a_{kl}^{(N)}$ and $c_k^{(N)}$ can be determined effectively and so that the solution $(u_{N,l})_{l=1,\ldots,d_N}$ can be computed effectively. Here by "effective" we mean a suitable balance between "sufficiently fast" and "sufficiently accurate". Ensuring effectivity is a challenging mathematical task.

That the finite dimensional solution to the boundary value problem is an approximation is shown by the following fundamental error estimate.

9.25 Céa's lemma. For u and u_N from 9.23 it holds that

$$\|u - u_N\|_{W^{1,2}} \le \frac{C}{c} \inf_{v \in X_N} \|u - v\|_{W^{1,2}} \longrightarrow 0 \quad \text{as } N \to \infty,$$

where c and C are the constants in the proof of 6.6 and 6.8, respectively.

Proof. With the notation as in the proof of 6.6

$$a(v,u) \;\; = F(v) \quad \text{for all } v \in X,$$
$$a(v, u_N) = F(v) \quad \text{for all } v \in X_N.$$

Set $v = w - u_N$ with $w \in X_N$. Then

$$a(w - u_N, u - u_N) = 0 \quad \text{for all } w \in X_N,$$

and hence

$$c\|u - u_N\|_{W^{1,2}}^2 \le a(u - u_N, u - u_N) = a(u - w, u - u_N)$$
$$\le C\|u - w\|_{W^{1,2}} \cdot \|u - u_N\|_{W^{1,2}}.$$

It follows that

$$\|u - u_N\|_{W^{1,2}} \le \frac{C}{c}\mathrm{dist}(u, X_N) \longrightarrow 0 \quad \text{as } N \to \infty,$$

thanks to 9.1(2). $\qquad\square$

The inequality in 9.25 means that the error between u and the approximate solution u_N can be bounded by the error that arises due to the choice of the space X_N. This error is called the **discretization error**, because X_N is chosen, for example, as in 9.22. If the solution satisfies $u \in W^{2,2}(\Omega)$ (see also A12.3), then it follows from 9.22(2) that the discretization error can be bounded by the maximal step size h_N. (For the corresponding result in higher space dimensions see e.g. [Braess].)

E9 Exercises

E9.1 Hamel basis. No infinite-dimensional Banach space X has a countable Hamel basis. Here a set $B \subset X$ is called a **Hamel basis** if every element of X can be uniquely represented as a finite (!) linear combination of elements from B.

Solution. Let $B = \{e_i;\ i \in \mathbb{N}\}$ be a Hamel basis. Set $X_n := \mathrm{span}\{e_i;\ i \le n\}$, then $X = \bigcup_{n\in\mathbb{N}} X_n$. On recalling from 4.9 that the X_n are closed, it follows from the Baire category theorem that $\overset{\circ}{X}_{n_0} \ne \emptyset$ for some $n_0 \in \mathbb{N}$. Hence it holds for an $x_0 \in X_{n_0}$ and an $\varepsilon_0 > 0$ that

$$x \in X,\ \|x\|_X < \varepsilon_0 \quad\Longrightarrow\quad x_0 + x \in X_{n_0}.$$

Since X_{n_0} is a subspace and $x_0 \in X_{n_0}$, it follows that $\mathrm{B}_{\varepsilon_0}(0) \subset X_{n_0}$. This implies that $X \subset X_{n_0}$. Hence X must be finite-dimensional. $\qquad\square$

E9.2 Discontinuous linear maps. Let X be a normed $\mathbb{K}$-vector space. Then X is finite-dimensional if and only if every linear map from X to $\mathbb{K}$ is continuous.

Solution $\Rightarrow$. Let $n := \dim X$ and let $\{e_1, \ldots, e_n\}$ be a basis of X. If $T : X \to \mathbb{K}$ is linear, then it holds for $\alpha_1, \ldots, \alpha_n \in \mathbb{K}$ that

$$\left| T\left(\sum_{i=1}^{n} \alpha_i e_i\right)\right| \le \sum_{i=1}^{n} |\alpha_i| \cdot |T(e_i)| \le \sum_{i=1}^{n} |T(e_i)| \cdot \max_{i=1,\ldots,n} |\alpha_i|.$$

Hence T is continuous with respect to the norm in (4-18) (and on recalling 4.8 also with respect to any given norm on X). $\qquad\square$

Solution ⇐. We give an indirect proof: If X is not finite-dimensional, then there exist linearly independent $e_i \in X$, $i \in \mathbb{N}$, e.g. defined inductively by

$$e_{i+1} \in X \setminus \text{span}\{e_1, \ldots, e_i\}.$$

Without loss of generality let $\|e_i\|_X = 1$. Choose a subspace $Y \subset X$ with (see 9.13(4))

$$X = Y \oplus \text{span}\{e_i; \ i \in \mathbb{N}\}.$$

Then every $x \in X$ has a unique representation

$$x = y + \sum_{i \in N_x} \alpha_i(x) e_i, \quad N_x \subset \mathbb{N} \text{ finite.} \tag{E9-1}$$

Then

$$Tx := \sum_{i \in N_x} i \alpha_i(x)$$

defines a linear map $T : X \to \mathbb{K}$ that is not continuous, because $Te_i = i \to \infty$ as $i \to \infty$, but $\|e_i\|_X = 1$.   □

This solution implies the following:

Theorem: If all norms in a normed vector space X are pairwise equivalent, then X is finite-dimensional.
Note: See also E7.3.

Proof. Let e_i, $i \in \mathbb{N}$, be as above. In X consider the norm

$$\|x\| := \|y\|_X + \sum_{i \in N_x} i |\alpha_i(x)| \quad \text{for } x \text{ as in (E9-1)}.$$

On noting that $\|e_i\|_X = 1$ and $\|e_i\| = i$ it follows that the two norms $\|\cdot\|$ and $\|\cdot\|_X$ are not equivalent.   □

E9.3 Dual basis. Let X be a normed vector space. Given linearly independent $x_i' \in X'$ for $i = 1, \ldots, n$, there exist $x_i \in X$ with

$$\langle x_i, x_j' \rangle = \delta_{i,j} \quad \text{for } i, j = 1, \ldots, n.$$

The vectors x_i for $i = 1, \ldots, n$ are then also linearly independent.

Solution. We prove this by induction on n. For $n = 1$ the claim is trivial. For the induction step for $n > 1$ consider the subspace

$$N := \bigcap_{j < n} \mathcal{N}(x_j').$$

For every $x \in X$ we then have that

$$x - \sum_{i<n} \langle x, x_i' \rangle x_i \in N,$$

because for $j < n$

$$\left\langle x - \sum_{i<n} \langle x, x_i' \rangle x_i, x_j' \right\rangle = \langle x, x_j' \rangle - \sum_{i<n} \langle x, x_i' \rangle \underbrace{\langle x_i, x_j' \rangle}_{= \delta_{i,j}} = 0.$$

Assuming that $x_n' = 0$ on N then yields for all x that

$$\langle x, x_n' \rangle = \left\langle \sum_{i<n} \langle x, x_i' \rangle x_i, x_n' \right\rangle$$

$$= \sum_{i<n} \langle x, x_i' \rangle \langle x_i, x_n' \rangle = \left\langle x, \sum_{i<n} \langle x_i, x_n' \rangle x_i' \right\rangle,$$

which is a contradiction to x_n' being linearly independent of $\{x_1', \ldots, x_{n-1}'\}$. Hence there exists an $x_n \in N$ with $\langle x_n, x_n' \rangle = 1$. The definition of N yields that $\langle x_n, x_j' \rangle = 0$ for $j < n$. $\qquad\square$

E9.4 Orthogonal system. Let $B_1(0)$ be the unit ball in $\mathbb{C}$ and define $e_k(z) := z^k$ for $k \in \mathbb{N} \cup \{0\}$. Show that $(e_k)_{k\geq 0}$ is an orthogonal system in $L^2(B_1(0); \mathbb{C})$ and compute $\|e_k\|_{L^2}$.

Solution. For $k, l \geq 0$ it holds that (see 9.9)

$$\int_{B_1(0)} e_k \overline{e_l} \, dL^2 = \int_0^1 \int_0^{2\pi} r^{1+k+l} e^{i(k-l)\theta} \, d\theta \, dr$$

$$= 2\pi \delta_{k,l} \int_0^1 r^{1+k+l} \, dr = \frac{2\pi}{2+k+l} \delta_{k,l}.$$

$\qquad\square$

E9.5 Weak convergence of unit vectors. If $(e_k)_{k\in\mathbb{N}}$ is an orthonormal system in the Hilbert space X, then

$$e_k \to 0 \text{ weakly in } X \text{ as } k \to \infty.$$

Solution. By the Riesz representation theorem, we need to show that for all $x \in X$ it holds that: $(x, e_k)_X \to 0$ as $k \to \infty$. Now Bessel's inequality yields that

$$\sum_{k\in\mathbb{N}} |(x, e_k)_X|^2 \leq \|x\|^2 < \infty,$$

and hence $((x, e_k)_X)_{k\in\mathbb{N}}$ must be a null sequence. $\qquad\square$

E9.6 On the convergence of the Fourier coefficients. Show that for $f \in L^2(]-\pi, \pi[; \mathbb{C})$

$$\int_{-\pi}^{\pi} f(x)\, \mathrm{e}^{\mathrm{i}kx}\, \mathrm{d}x \longrightarrow 0 \quad \text{as } k \in \mathbb{Z},\ |k| \to \infty.$$

E9.7 Projections in a Banach space. Let X be a Banach space and let

$$Q_1, \ldots, Q_n \in \mathscr{P}(X) \quad \text{with} \quad Q_i Q_j = 0 \text{ for } i \neq j.$$

Then $P := Q_1 + \cdots + Q_n \in \mathscr{P}(X)$ with

$$\mathscr{R}(P) = \mathscr{R}(Q_1) \oplus \cdots \oplus \mathscr{R}(Q_n) \quad \text{and} \quad \mathscr{N}(P) = \mathscr{N}(Q_1) \cap \cdots \cap \mathscr{N}(Q_n).$$

E9.8 Projections in $L^2(]-\pi, \pi[)$. Show that

$$(S_n f)(x) := \frac{1}{2\pi} \int_{-\pi}^{\pi} \frac{\sin\big((n + \frac{1}{2})(x - y)\big)}{\sin(\frac{1}{2}(x - y))} f(y)\, \mathrm{d}y$$

for $n \in \mathbb{N} \cup \{0\}$ defines maps $S_n \in \mathscr{P}\big(L^2(]-\pi, \pi[)\big)$ with $S_n S_m = S_{\min(n,m)}$ for $n, m \geq 0$.

Solution. We have that

$$\sin\Big(\frac{x}{2}\Big) \cdot \sum_{k=-n}^{n} \mathrm{e}^{\mathrm{i}kx} = \frac{1}{2\mathrm{i}}\big(\mathrm{e}^{\mathrm{i}\frac{x}{2}} - \mathrm{e}^{-\mathrm{i}\frac{x}{2}}\big) \sum_{k=-n}^{n} \mathrm{e}^{\mathrm{i}kx}$$

$$= \frac{1}{2\mathrm{i}} \sum_{k=-n}^{n} \big(\mathrm{e}^{\mathrm{i}(k+\frac{1}{2})x} - \mathrm{e}^{\mathrm{i}(k-\frac{1}{2})x}\big) \quad \text{(this is a \textbf{telescoping sum})}$$

$$= \frac{1}{2\mathrm{i}}\big(\mathrm{e}^{\mathrm{i}(n+\frac{1}{2})x} - \mathrm{e}^{-\mathrm{i}(n+\frac{1}{2})x}\big) = \sin\Big(\big(n + \frac{1}{2}\big)x\Big).$$

Consequently,

$$S_n f(x) = \frac{1}{2\pi} \sum_{k=-n}^{n} \int_{-\pi}^{\pi} \mathrm{e}^{\mathrm{i}k(x-y)} f(y)\, \mathrm{d}y = \sum_{k=-n}^{n} (f, e_k)_{L^2(]-\pi,\pi[)}\, e_k(x),$$

where $(e_k)_{k \in \mathbb{Z}}$ denotes the orthonormal basis from 9.9. Hence S_n is the orthogonal projection onto $\mathrm{span}\{e_{-n}, \ldots, e_n\}$. The identity for the S_n then follows from the orthonormality of the e_k. $\qquad\square$

E9.9 Projections in a Hilbert space. Prove the properties of the projections in 9.20(2).

Solution. For $x \in X$ there exist $x_n \in X_n$ with $x_n \to x$ as $n \to \infty$. On recalling 4.3 this implies that $\|x - P_n x\|_X \leq \|x - x_n\|_X \to 0$ as $n \to \infty$, i.e. (P1) is satisfied. For the proof of (P2) let $m > n$. For $x \in X$ it holds that

$P_n x \in X_n \subset X_m$, and so $P_m P_n x = P_n x$. Moreover, it follows from 4.4(2) that $P_n x$ is characterized by

$$(y, x - P_n x)_X = 0 \quad \text{for } y \in X_n, \tag{E9-2}$$

and similarly $P_m x$ by $(y, x - P_m x)_X = 0$ for $y \in X_n \subset X_m$. Replace x by $P_m x$ in (E9-2) and obtain that

$$(y, x - P_n P_m x)_X = 0 \quad \text{for } y \in X_n. \tag{E9-3}$$

Comparing (E9-2) and (E9-3) yields that $P_n P_m x = P_n x$ (recalling 9.18(3) the same can also be derived directly from $P_m P_n x = P_n x$).   □

E9.10 Schauder basis. Let $(Q_n)_{n \in \mathbb{N}}$ be a sequence of projections in $\mathscr{P}(X)$ as in 9.19 and, on setting $d_n := \dim \mathscr{R}(Q_n)$, let $\{e_{n,i} \; ; \; i = 1, \dots, d_n\}$ be arbitrary bases of $\mathscr{R}(Q_n)$. Find a sufficient condition for

$$e_{1,1}, \dots, e_{1,d_1}, e_{2,1}, e_{2,2}, \dots, e_{2,d_2}, e_{3,1}, \dots \text{ etc.}$$

to be a Schauder basis in this order.

Solution. The desired condition is formulated in (E9-5). Let $E_n := \mathscr{R}(Q_n)$. There exist linear coefficient functions $\alpha_{n,i} : E_n \to \mathbb{K}$ with

$$x = \sum_{i=1}^{d_n} \alpha_{n,i}(x) e_{n,i} \quad \text{for } x \in E_n,$$

and hence it follows from (Q1) for all $x \in X$ that

$$x = \sum_{n=1}^{\infty} Q_n(x) = \sum_{n=1}^{\infty} \sum_{i=1}^{d_n} \alpha_{n,i}(Q_n(x)) e_{n,i}. \tag{E9-4}$$

Moreover, if $\beta_{n,i} \in \mathbb{K}$ with

$$0 = \sum_{n=1}^{\infty} \sum_{i=1}^{d_n} \beta_{n,i} e_{n,i},$$

then the continuity of Q_m and (Q2) imply that

$$0 = Q_m(0) = \sum_{n=1}^{\infty} Q_m \underbrace{\left(\sum_{i=1}^{d_n} \beta_{n,i} e_{n,i} \right)}_{\in E_n} = \sum_{n=1}^{\infty} \underbrace{Q_m Q_n}_{= \delta_{m,n} Q_m} \left(\sum_{i=1}^{d_n} \beta_{n,i} e_{n,i} \right)$$

$$= Q_m \left(\sum_{i=1}^{d_m} \beta_{m,i} e_{m,i} \right) = \sum_{i=1}^{d_m} \beta_{m,i} e_{m,i},$$

and so $\beta_{m,i} = 0$, because $e_{m,i}$, $i = 1, \ldots, d_m$, are linearly independent. This proves the uniqueness of the coefficients in (E9-4).

As a sufficient condition we now assume that there exists a constant C, such that

$$\left\| \sum_{i=1}^{j} \alpha_i e_{n,i} \right\| \leq C \cdot \left\| \sum_{i=1}^{d_n} \alpha_i e_{n,i} \right\| \quad \text{for } n \in \mathbb{N} \text{ and } j \in \{1, \ldots, d_n\}. \quad \text{(E9-5)}$$

Then it holds for all $m \in \mathbb{N}$ and $j \in \{1, \ldots, d_m\}$ that

$$\left\| \sum_{i=j}^{d_m} \alpha_{m,i} (Q_m(x)) e_{m,i} + \sum_{n=m+1}^{\infty} \sum_{i=1}^{d_n} \alpha_{n,i} (Q_n(x)) e_{n,i} \right\|$$

$$= \left\| \sum_{n=m}^{\infty} Q_n(x) - \sum_{i=1}^{j-1} \alpha_{m,i} (Q_m(x)) e_{m,i} \right\|$$

$$\leq \left\| \sum_{n=m}^{\infty} Q_n(x) \right\| + C \cdot \underbrace{\left\| \sum_{i=1}^{d_m} \alpha_{m,i} (Q_m(x)) e_{m,i} \right\|}_{= \| Q_m(x) \|} \longrightarrow 0$$

as $m \to \infty$, on recalling (Q1).   □

10 Compact operators

In this chapter we consider the properties of compact linear operators between Banach spaces. The space $\mathscr{K}(X;Y)$ of (linear) compact operators from X to Y was already defined in 5.5(2). Because here we are always concerned with linear operators, for convenience we simply speak of compact operators.

All the spaces in this chapter are assumed to be Banach spaces. We begin with a discussion of the elementary properties of compact operators and then give the most important examples of such operators. These include compact embeddings between function spaces and compact integral operators.

10.1 Compact operators. Let X and Y be Banach spaces over $\mathbb{K}$. Then a linear map $T : X \to Y$ is called a ***compact (linear) operator*** if one of the following equivalent properties is satisfied:

(1) $\overline{T(\mathrm{B}_1(0))} \subset Y$ is compact (see the definition 5.5(2)).

(2) $T(\mathrm{B}_1(0)) \subset Y$ is precompact.

(3) $M \subset X$ is bounded $\implies$ $T(M) \subset Y$ is precompact.

(4) For every bounded sequence $(x_n)_{n \in \mathbb{N}}$ in X, the sequence $(Tx_n)_{n \in \mathbb{N}}$ contains a subsequence that is convergent in Y.

It follows from (2) that $T(\mathrm{B}_1(0))$ is bounded (see 4.7(2)), and so $T \in \mathscr{L}(X;Y)$, by 5.1. Hence it holds for the set defined in 5.5(2) that

$$\mathscr{K}(X;Y) := \{T : X \to Y \, ; \, T \text{ is a compact linear operator}\}$$
$$= \{T \in \mathscr{L}(X;Y) \, ; \, T \text{ satisfies (4)}\} \, .$$

Moreover, let $\mathscr{K}(X) := \mathscr{K}(X;X)$.

Note: The fact that compact maps (with the property (1)) are continuous only holds for linear maps. General nonlinear maps which satisfy (1) need not be continuous.

Proof (1)$\Leftrightarrow$(2). This follows from 4.7(5), as Y is complete. $\qquad\qquad$ □

Proof (2)$\Leftrightarrow$(3). The linearity of T implies that for every $R > 0$ statement (2) is equivalent to the precompactness of $T(\mathrm{B}_R(0))$. Because every bounded set M is contained in a ball $\mathrm{B}_R(0)$, it then follows that the smaller set $T(M)$ is also precompact. $\qquad\qquad$ □

Proof (1)⇒(4). If $x_n \in X$ for $n \in \mathbb{N}$ with $\|x_n\|_X < R$, then $\frac{1}{R}Tx_n = T(\frac{1}{R}x_n)$ are elements of the compact, and hence (by 4.6) also sequentially compact, set $\overline{T(B_1(0))}$. □

Proof (4)⇒(1). Let $y_n \in \overline{T(B_1(0))}$ for $n \in \mathbb{N}$. Then there exist $x_n \in B_1(0)$ with $\|y_n - Tx_n\|_Y \le \frac{1}{n}$. It follows from (4) that there exists a $y \in Y$ such that $Tx_n \to y$ for a subsequence $n \to \infty$, and hence also $y_n \to y$. This shows that $\overline{T(B_1(0))}$ is sequentially compact, and so, by 4.6, is also compact. □

We now prove some basic results.

10.2 Lemma.

(1) If X is a reflexive space, then it holds for every linear map $T : X \to Y$ that

$$T \in \mathscr{K}(X;Y) \quad \Longleftrightarrow \quad T \text{ is } \boldsymbol{completely\ continuous}, \text{ i.e.}$$

if $x_n \to x$ converges weakly in X as $n \to \infty$, then $Tx_n \to Tx$ converges strongly in Y.

(2) $\mathscr{K}(X;Y)$ is a closed subspace of $\mathscr{L}(X;Y)$.

(3) If $T \in \mathscr{L}(X;Y)$ with $\dim \mathscr{R}(T) < \infty$, then $T \in \mathscr{K}(X;Y)$.

(4) If Y is a Hilbert space and $T \in \mathscr{L}(X;Y)$, then

$$T \in \mathscr{K}(X;Y) \quad \Longleftrightarrow \quad \begin{array}{l} \text{there exist } T_n \in \mathscr{L}(X;Y) \text{ with } \dim \mathscr{R}(T_n) < \infty, \\ \text{such that } \|T - T_n\| \to 0 \text{ as } n \to \infty. \end{array}$$

(5) For projections $P \in \mathscr{P}(X)$ it holds that

$$P \in \mathscr{K}(X) \quad \Longleftrightarrow \quad \dim \mathscr{R}(P) < \infty.$$

Proof (1)⇒. (In the proof of this implication the reflexivity of X is not needed.) Let $x_n \to x$ weakly as $n \to \infty$. By 8.3(5), the sequence $(x_n)_{n\in\mathbb{N}}$ is bounded, and so 10.1(4) yields the existence of a $y \in Y$ such that $Tx_n \to y$ strongly in Y for a subsequence $n \to \infty$. For $y' \in Y'$ the map $z \mapsto \langle Tz, y' \rangle$ defines an element in X'. Therefore,

$$\langle Tx_n, y' \rangle \to \langle Tx, y' \rangle \quad \text{as } n \to \infty.$$

This yields that $Tx_n \to Tx$ weakly in Y. As strong convergence implies weak convergence, one must have $y = Tx$. Hence $Tx_n \to Tx$ converges strongly for a subsequence $n \to \infty$. On noting that all of the above argumentation can be applied to every subsequence of $(x_n)_{n\in\mathbb{N}}$, it follows that the whole (!) sequence $(Tx_n)_{n\in\mathbb{N}}$ has only one cluster point Tx, i.e. it converges strongly to Tx. □

Proof (1)⇐. Being completely continuous implies that T is continuous, and so $T \in \mathscr{L}(X;Y)$. Moreover, it follows from theorem 8.10 that bounded sequences in reflexive spaces contain weakly convergent subsequences. □

Proof (2). In order to see that $\mathscr{K}(X;Y)$ is a subspace, let $T_1, T_2 \in \mathscr{K}(X;Y)$, let $\alpha \in \mathbb{K}$ and let $(x_n)_{n \in \mathbb{N}}$ be a bounded sequence in X. Then there exists a subsequence $(T_1 x_{n_k})_{k \in \mathbb{N}}$ that is convergent in Y. Similarly, we may then choose a convergent subsequence $\left(T_2 x_{n_{k_l}}\right)_{l \in \mathbb{N}}$. This implies that $\left((\alpha T_1 + T_2)(x_{n_{k_l}})\right)_{l \in \mathbb{N}}$ converges in Y, which shows that $\alpha T_1 + T_2 \in \mathscr{K}(X;Y)$.

To prove that $\mathscr{K}(X;Y)$ is closed, assume that $T_n \in \mathscr{K}(X;Y)$ converges in $\mathscr{L}(X;Y)$ as $n \to \infty$ to $T \in \mathscr{L}(X;Y)$. For $\varepsilon > 0$ first choose n_ε with $\|T - T_{n_\varepsilon}\| \le \varepsilon$ and then (recall 10.1(2)) balls $B_\varepsilon(y_i)$, $i = 1, \ldots, m_\varepsilon$, such that

$$T_{n_\varepsilon}(B_1(0)) \subset \bigcup_{i=1}^{m_\varepsilon} B_\varepsilon(y_i), \quad \text{which implies:} \quad T(B_1(0)) \subset \bigcup_{i=1}^{m_\varepsilon} B_{2\varepsilon}(y_i).$$

Hence $T(B_1(0))$ is precompact, and so T is compact. □

Proof (3). We have that $Z := \mathscr{R}(T) \subset Y$ is finite-dimensional, and so it follows from 4.9 that with the Y-norm it is a Banach space. On setting $R := \|T\|$ we have that

$$T(B_1(0)) \subset K_R := \{y \in Z \,;\, \|y\|_Y \le R\} \subset Z.$$

By 4.10, we have that $\overline{K_R} \subset Z$ is compact, and hence combining 4.7(5) and 4.7(1) yields that $\overline{T(B_1(0))}$ is compact. □

Proof (4)$\Leftarrow$. We have from (3) that $T_n \in \mathscr{K}(X;Y)$. Then (2) yields that $T \in \mathscr{K}(X;Y)$. □

Proof (4)$\Rightarrow$. Let $\varepsilon > 0$. It follows from 10.1(2) that we can choose balls $B_\varepsilon(y_i)$, $i = 1, \ldots, m_\varepsilon$, with

$$T(B_1(0)) \subset \bigcup_{i=1}^{m_\varepsilon} B_\varepsilon(y_i).$$

Set $Y_\varepsilon := \operatorname{span}\{y_1, \ldots, y_{m_\varepsilon}\}$ and let P_ε denote the orthogonal projection onto Y_ε. Then we have from 9.18 that $\operatorname{Id} - P_\varepsilon$ is also an orthogonal projection (equivalence of 9.18(1) and 9.18(2)), with $\|\operatorname{Id} - P_\varepsilon\| \le 1$ (equivalence of 9.18(1) and 9.18(4)). Now $T_\varepsilon := P_\varepsilon T$ maps to Y_ε, and for $x \in B_1(0)$ it holds that $Tx \in B_\varepsilon(y_i)$ for some i and that

$$(T - T_\varepsilon)(x) = (\operatorname{Id} - P_\varepsilon)Tx = (\operatorname{Id} - P_\varepsilon)(Tx - y_i),$$

and hence $\|(T - T_\varepsilon)(x)\|_Y \le \varepsilon$. □

Proof (5)$\Leftarrow$. Follows from (3). □

Proof (5)$\Rightarrow$. It holds that $B_1(0) \cap \mathscr{R}(P) \subset P(B_1(0))$ is precompact, and so it follows from 4.10 that $\mathscr{R}(P)$ is finite-dimensional. $\qquad\square$

In applications compact operators often occur as a composition of a continuous map and an embedding which is compact (we prove this in 10.3). The compact part of such a composition is often a canonical embedding. That is, if X, Y are Banach spaces and if X as a vector space is contained in Y, then we ask whether the map $\mathrm{Id} : X \to Y$ is injective, continuous and compact, respectively. We will answer this question completely for the function spaces $C^{k,\alpha}(\overline{\Omega})$ and $W^{m,p}(\Omega)$ (see 10.6 – 10.13) and we call the corresponding theorems embedding theorems.

10.3 Lemma. For $T_1 \in \mathscr{L}(X; Y)$ and $T_2 \in \mathscr{L}(Y; Z)$ it holds that:

$$T_1 \text{ or } T_2 \text{ is compact} \quad \Longrightarrow \quad T_2 T_1 \text{ is compact.}$$

Proof. Let $(x_n)_{n \in \mathbb{N}}$ be a bounded sequence in X. As T_1 is continuous, the sequence $(T_1 x_n)_{n \in \mathbb{N}}$ is bounded in Y. If T_2 is compact, it follows that there exists a convergent subsequence $(T_2 T_1 x_{n_k})_{k \in \mathbb{N}}$. If T_1 is compact, there exists a convergent subsequence $(T_1 x_{n_k})_{k \in \mathbb{N}}$, and the continuity of T_2 then yields that also $(T_2 T_1 x_{n_k})_{k \in \mathbb{N}}$ converges. $\qquad\square$

Embedding theorems

The embedding theorem 10.6 for Hölder spaces depends on the Arzelà-Ascoli theorem and the first result in theorem 10.5. For the latter we need the following

10.4 Lemma. Let $\Omega \subset \mathbb{R}^n$ be open and bounded with Lipschitz boundary. If Ω, in addition, is path connected (see the remark preceding 8.16), then for any two points $x_0, x_1 \in \Omega$ there exists a smooth curve γ in Ω which connects x_0 and x_1 and whose length $L(\gamma)$ can be bounded by $|x_1 - x_0|$, i.e. there exists a $\gamma \in C^\infty([0,1]; \Omega)$ with $\gamma(0) = x_0$, $\gamma(1) = x_1$, such that, with a constant C_Ω depending only on Ω,

$$L(\gamma) := \int_0^1 |\gamma'(t)|\, dt \leq \sup_{0 \leq t \leq 1} |\gamma'(t)| \leq C_\Omega \cdot |x_1 - x_0|.$$

Proof. It is sufficient to find a $\gamma \in C^{0,1}([0,1]; \Omega)$ with $\gamma(0) = x_0$, $\gamma(1) = x_1$ and with Lipschitz constant $\mathrm{Lip}(\gamma) \leq C \cdot |x_1 - x_0|$. To see this, observe that we can then let $\gamma(t) := x_0$ for $t < 0$ and $\gamma(t) := x_1$ for $t > 1$ and set $\gamma_\varepsilon := \varphi_\varepsilon * \gamma$, with a standard Dirac sequence $(\varphi_\varepsilon)_{\varepsilon > 0}$. On noting that $\|\gamma'_\varepsilon\|_{\sup} \leq \mathrm{Lip}(\gamma_\varepsilon) \leq \mathrm{Lip}(\gamma)$, it follows that for $\varepsilon > 0$ sufficiently small γ_ε has all the desired properties on $[-\varepsilon, 1+\varepsilon]$, and hence we only need to map $[0,1]$ affine linearly to $[-\varepsilon, 1+\varepsilon]$.

We consider a cover $\left(U^j\right)_{j=1,\ldots,k}$ of $\partial\Omega$ as in A8.2 and choose points $z^j \in U^j \cap \Omega$. Then we choose an open set D with $\overline{D} \subset \Omega$, such that $z^1, \ldots, z^k \in \overline{D}$, and such that $\overline{\Omega}$ is covered by $D, U^1, \ldots, U^k$. Moreover, we cover $\overline{D}$ with finitely many balls $U^j := B_\varrho(z^j) \subset \Omega$ with $j = k+1, \ldots, l$.

For general points x_0 and x_1 we can then define a γ as a composition of subpaths, such that for these subpaths only the following three cases can occur. Altogether, the number of subpaths is bounded by the given cover.

If $x_0, x_1 \in U^j$ for some $j > k$, then define $\gamma(t) := (1-t)x_0 + tx_1$.

If $x_0, x_1 \in U^j$ for some $j \leq k$, then define

$$\gamma(t) := \tau\big((1-t)\tau^{-1}(x_0) + t\tau^{-1}(x_1)\big),$$

where with the notations from A8.2 we set

$$\tau(y) := \sum_{i=1}^{n-1} y_i e_i^j + \big(y_n + g^j(y_{,n})\big)e_n^j.$$

This defines a Lipschitz continuous path γ in Ω from x_0 to x_1 with

$$\mathrm{Lip}(\gamma) \leq \mathrm{Lip}(\tau) \cdot \big|\tau^{-1}(x_1) - \tau^{-1}(x_0)\big| \leq \mathrm{Lip}(\tau) \cdot \mathrm{Lip}(\tau^{-1}) \cdot |x_1 - x_0|.$$

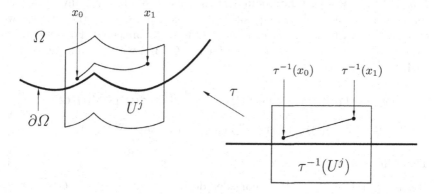

Fig. 10.1. *Construction of curves close to the boundary*

As a third case, let x_0 and x_1 be such that for no $j \in \{1, \ldots, l\}$ they lie in the same set U^j of the above cover of $\overline{\Omega}$. Then there exists a constant $c > 0$, which depends only on the cover, such that

$$|x_0 - x_1| \geq c.$$

This follows from the fact that for every j and for points $x \in \overline{\Omega} \cap U^j$ that are sufficiently close to ∂U^j it must hold that $x \in U^k$ for some $k \neq j$.

We thus have to connect x_0 and x_1 by a curve with a bounded Lipschitz constant. We make use of the fact that Ω is connected, and so path connected. Hence for $j, k \in \{1, \ldots, l\}$ there exists a $\gamma_{j,k} \in C^1([0, 1]; \Omega)$ with $\gamma_{j,k}(0) = z^j$ and $\gamma_{j,k}(1) = z^k$. Now let $x_0 \in U^{j_0}$ and $x_1 \in U^{j_1}$ with $j_0 \neq j_1$. First we connect x_0 with z^{j_0} inside U^{j_0} (as in the first two cases above) with a path such that the Lipschitz constant can be bounded by $C \cdot \left| z^{j_0} - x_0 \right| \leq C \cdot \operatorname{diam} U^{j_0}$. Then we connect z^{j_0} with z^{j_1} by γ_{j_0, j_1}, and finally z^{j_1} with x_1 inside U^{j_1}. A reparametrization of the concatenated paths to the interval $[0, 1]$ then yields the desired result. $\square$

10.5 Theorem. Let $\Omega \subset \mathbb{R}^n$ be open and bounded with Lipschitz boundary. Then it holds for $k \geq 0$ that:

(1) The embedding
$$\operatorname{Id} : C^{k+1}(\overline{\Omega}) \to C^{k,1}(\overline{\Omega})$$
is well defined and continuous.

(2) The embedding
$$\operatorname{Id} : C^{k,1}(\overline{\Omega}) \to W^{k+1,\infty}(\Omega)$$
is well defined and an isomorphism, in the sense that for $u \in W^{k+1,\infty}(\Omega)$ there exists a unique $\widetilde{u} \in C^{k,1}(\overline{\Omega})$ such that $\widetilde{u} = u$ almost everywhere in Ω (i.e. $\widetilde{u} = u$ in $W^{k+1,\infty}(\Omega)$).

Proof. As Ω has a Lipschitz boundary, it consists of finitely many connected components, which all lie at positive distance from one another. Hence we may assume without loss of generality that Ω is connected. For two points $x_0, x_1 \in \Omega$ let γ be as in 10.4. Then for $v \in C^1(\Omega)$, with the notations as in 10.4, we have that

$$|v(x_1) - v(x_0)| = \left| \int_0^1 (v \circ \gamma)'(t) \, dt \right| \leq \int_0^1 |\nabla v(\gamma(t))| \cdot |\gamma'(t)| \, dt \qquad (10\text{-}6)$$
$$\leq \sup_{0 \leq t \leq 1} |\nabla v(\gamma(t))| \cdot L(\gamma) \leq C_\Omega \cdot |x_1 - x_0| \cdot \sup_{0 \leq t \leq 1} |\nabla v(\gamma(t))|.$$

This will be used in the following parts of the proof. $\square$

Proof (1). For $u \in C^{k+1}(\overline{\Omega})$ consider derivatives $v := \partial^s u \in C^1(\overline{\Omega})$ with $|s| = k$. It follows from (10-6) that the Lipschitz constant of v can be bounded by the C^1-norm of v. The fact that this holds for all s of order k yields that $\|u\|_{C^{k,1}(\overline{\Omega})} \leq C \cdot \|u\|_{C^{k+1}(\overline{\Omega})}$ with a constant C. $\square$

Proof (2) well definedness. First let $k = 0$. Let $u \in C^{0,1}(\overline{\Omega})$. If $\mathbf{e}_i$ denotes the i-th unit vector, and if $\zeta \in C_0^\infty(\Omega)$, then as $h \to 0$,

$$\left| \int_\Omega u(x) \partial_i \zeta(x) \, dx \right| \longleftarrow \left| \int_\Omega u(x) \frac{\zeta(x + h\mathbf{e}_i) - \zeta(x)}{h} \, dx \right|$$
$$= \left| \int_\Omega \frac{u(x - h\mathbf{e}_i) - u(x)}{h} \zeta(x) \, dx \right| \leq \operatorname{Lip}(u) \int_\Omega |\zeta(x)| \, dx.$$

This implies (see E6.7) that $u \in W^{1,\infty}(\Omega)$ with $\|\partial_i u\|_{L^\infty} \leq \mathrm{Lip}(u)$ for $i = 1, \ldots, n$. For $k > 0$ apply this result to the derivatives $\partial^s u$ with $|s| = k$. $\square$

Proof (2) surjectivity. First let $k = 0$. Let $u \in W^{1,\infty}(\Omega)$. Consider $u_\varepsilon := \varphi_\varepsilon * (\mathcal{X}_\Omega u)$ for a standard Dirac sequence $(\varphi_\varepsilon)_{\varepsilon > 0}$. Then it follows from (10-6) (with the notations as there) that

$$|u_\varepsilon(x_1) - u_\varepsilon(x_0)| \leq C_\Omega \cdot |x_1 - x_0| \cdot \sup_{0 \leq t \leq 1} |\nabla u_\varepsilon(\gamma(t))|,$$

and, if ε is sufficiently small, for all $x = \gamma(t)$ with $0 \leq t \leq 1$ we have that

$$|\nabla u_\varepsilon(x)| = |\nabla(\varphi_\varepsilon * u)(x)| = |(\varphi_\varepsilon * \nabla u)(x)| \leq \|\nabla u\|_{L^\infty(\Omega)}.$$

This implies

$$\frac{|u_\varepsilon(x_1) - u_\varepsilon(x_0)|}{|x_1 - x_0|} \leq C_\Omega \cdot \|\nabla u\|_{L^\infty(\Omega)}. \qquad (10\text{-}7)$$

Recalling from 4.15(2) that $u_\varepsilon \to u$ in $L^p(\Omega)$ for every $p < \infty$, there exists a subsequence $\varepsilon \to 0$ such that $u_\varepsilon \to u$ almost everywhere in Ω. Hence it follows from (10-7) that for almost all $x_0, x_1 \in \Omega$ (say, $x_0, x_1 \in \Omega \setminus N$),

$$\frac{|u(x_1) - u(x_0)|}{|x_1 - x_0|} \leq C_\Omega \cdot \|\nabla u\|_{L^\infty(\Omega)}, \qquad (10\text{-}8)$$

i.e. u is Lipschitz continuous outside of the null set N. Since $\overline{\Omega \setminus N} = \overline{\Omega}$, it follows from E4.18 that we can modify u on this null set so that $u \in C^{0,1}(\overline{\Omega})$. (After this modification u remains the same (!) element in $L^\infty(\Omega)$.) Since then $\|u\|_{C^0} = \|u\|_{L^\infty}$, we have shown that $\|u\|_{C^{0,1}} \leq C \cdot \|u\|_{W^{1,\infty}}$.

If $u \in W^{k+1,\infty}(\Omega)$ with $k > 0$, then we can apply the above to the weak derivatives $v_s := \partial^s u$ for $|s| \leq k$. In particular, upon modification on a null set we have that $v_s \in C^{0,1}(\overline{\Omega})$ with the above estimate in (10-8),

$$\mathrm{Lip}(v_s, \overline{\Omega}) \leq C_\Omega \cdot \|\nabla v_s\|_{L^\infty(\Omega)} \leq C_\Omega \cdot \|u\|_{W^{k+1,\infty}(\Omega)},$$

since for the weak derivatives with $|s| \leq k$ it holds that $\partial_i v_s = \partial_i \partial^s u = \partial^{s+\mathbf{e}_i} u \in L^\infty(\Omega)$. Hence we obtain the desired result. $\square$

10.6 Embedding theorem in Hölder spaces. Let $\Omega \subset \mathbb{R}^n$ be open and bounded and let $k_1, k_2 \geq 0$ and $0 \leq \alpha_1, \alpha_2 \leq 1$, with

$$k_1 + \alpha_1 > k_2 + \alpha_2.$$

In the case $k_1 > 0$ we assume in addition that Ω has a Lipschitz boundary (see also E10.1). Then the embedding

$$\mathrm{Id} : C^{k_1, \alpha_1}(\overline{\Omega}) \to C^{k_2, \alpha_2}(\overline{\Omega})$$

is compact. Here $C^{k,0}(\overline{\Omega}) := C^k(\overline{\Omega})$ for $k \geq 0$.

Remark: For $k_1 = k_2 = 0$ the set $\overline{\Omega}$ can be replaced with an arbitrary compact set $S \subset \mathbb{R}^n$.

Proof. Let $(u_i)_{i \in \mathbb{N}}$ be a bounded sequence in $C^{k_1, \alpha_1}(\overline{\Omega})$. We need to show that a subsequence converges in $C^{k_2, \alpha_2}(\overline{\Omega})$.

First let $k_2 = k_1 = 0$, and so $0 \leq \alpha_2 < \alpha_1 \leq 1$. By the Arzelà-Ascoli theorem, there exist a $u \in C^0(\overline{\Omega})$ and a subsequence $i \to \infty$ such that u_i converges to u uniformly on $\overline{\Omega}$. Consider only this subsequence and $x, y \in \Omega$ with $x \neq y$. For $|y - x| \leq \delta$ it then holds that

$$\frac{|(u - u_i)(y) - (u - u_i)(x)|}{|y - x|^{\alpha_2}} = \lim_{j \to \infty} \frac{|(u_j - u_i)(y) - (u_j - u_i)(x)|}{|y - x|^{\alpha_2}}$$

$$\leq \delta^{\alpha_1 - \alpha_2} \sup_j \|u_j - u_i\|_{C^{0,\alpha_1}} \leq 2\delta^{\alpha_1 - \alpha_2} \sup_j \|u_j\|_{C^{0,\alpha_1}},$$

while for $|y - x| \geq \delta$ we have that

$$\frac{|(u - u_i)(y) - (u - u_i)(x)|}{|y - x|^{\alpha_2}} \leq 2\delta^{-\alpha_2} \|u - u_i\|_{C^0}.$$

Overall, there is a constant C such that

$$\sup_{\substack{x, y \in \overline{\Omega} \\ x \neq y}} \frac{|(u - u_i)(y) - (u - u_i)(x)|}{|y - x|^{\alpha_2}} \leq \underbrace{C\delta^{\alpha_1 - \alpha_2}}_{\to 0 \text{ as } \delta \to 0} + 2\delta^{-\alpha_2} \underbrace{\|u - u_i\|_{C^0}}_{\to 0 \text{ as } i \to \infty},$$

i.e. the Hölder constant for the exponent α_2 of $u - u_i$ converges to 0 as $i \to \infty$.

Now we consider the case $k_2 = k_1 \geq 1$, and so once again $0 \leq \alpha_2 < \alpha_1 \leq 1$. Then $(\partial^s u_i)_{i \in \mathbb{N}}$ for $|s| < k_1$ are bounded sequences in $C^1(\overline{\Omega})$, and hence, by 10.5(1), also in $C^{0,1}(\overline{\Omega})$, and for $|s| = k_1$ they are bounded sequences in $C^{0,\alpha_1}(\overline{\Omega})$. Applying the result shown above for the sequence $(\partial^s u_i)_{i \in \mathbb{N}}$ in $C^{0,\alpha_1}(\overline{\Omega})$ we can choose successively for s with $|s| \leq k_1$ subsequences so that they converge in $C^{0,\alpha_2}(\overline{\Omega})$. Finally, one obtains a subsequence (which we again denote by $(u_i)_{i \in \mathbb{N}}$) which converges for all (!) s with $|s| \leq k_1$

$$\partial^s u_i \to v_s \quad \text{as } i \to \infty \text{ in } C^{0,\alpha_2}(\overline{\Omega})$$

with certain functions $v_s \in C^{0,\alpha_2}(\overline{\Omega})$. In particular, we obtain that $(u_i)_{i \in \mathbb{N}}$ is a Cauchy sequence in $C^{k_1}(\overline{\Omega})$. As this space is complete we necessarily have that $u := v_0 \in C^{k_1}(\overline{\Omega})$ with $\partial^s u = v_s$, i.e. u_i converges to u in $C^{k_1, \alpha_2}(\overline{\Omega})$.

Finally, let $k_1 > k_2$. By the results shown above, in the case $\alpha_2 < 1$ the embedding from $C^{k_2,1}(\overline{\Omega})$ to $C^{k_2, \alpha_2}(\overline{\Omega})$ is compact, and in the case $\alpha_1 > 0$ the embedding from $C^{k_1, \alpha_1}(\overline{\Omega})$ to $C^{k_1}(\overline{\Omega})$ is compact. In addition, we have from 10.5(1) that the embedding from $C^{k_1}(\overline{\Omega})$ to $C^{k_1-1,1}(\overline{\Omega})$ is continuous. Hence it remains to consider the map from $C^{k_1-1,1}(\overline{\Omega})$ to $C^{k_2,1}(\overline{\Omega})$, which in the case $k_1 = k_2 + 1$ is the identity. In this case we have that $1 + \alpha_1 > \alpha_2$, and so $\alpha_2 < 1$ or $\alpha_1 > 0$, which means that the desired result follows from 10.3.

In the case $k_1 > k_2 + 1$ (e.g. when $\alpha_1 = 0$ and $\alpha_2 = 1$) it follows from the above result that the map from $C^{k_1-1,1}(\overline{\Omega})$ to $C^{k_1-1}(\overline{\Omega})$ is compact. Since

$k_1 - 1 \geq k_2 + 1$, the map from $C^{k_1-1}(\overline{\Omega})$ to $C^{k_2+1}(\overline{\Omega})$ is obviously continuous and the map from $C^{k_2+1}(\overline{\Omega})$ to $C^{k_2,1}(\overline{\Omega})$ is continuous thanks to 10.5(1). The desired result now follows on using 10.3. □

We now want to prove embedding theorems for Sobolev spaces. To this end, we consider on $\overline{B_1(0)} \subset \mathbb{R}^n$ the function $x \mapsto |x|^\varrho$ with real ϱ and investigate to which Sobolev space $W^{m,p}(B_1(0))$, respectively, to which Hölder space $C^{k,\alpha}(\overline{B_1(0)})$ it belongs. The answer will motivate the formulation of the embedding theorems 10.9 and 10.13.

10.7 Sobolev number. Let $f_\varrho(x) := |x|^\varrho$ for $x \in \mathbb{R}^n \setminus \{0\}$, where $\varrho \in \mathbb{R}$. Then it holds that:

(1) f_ϱ is real analytic on $\mathbb{R}^n \setminus \{0\}$ and for $m \geq 0$ there exist positive numbers c_m, C_m, which depend also on n and ϱ, such that

$$c_m \left| \binom{\varrho}{m} \right| \cdot |x|^{\varrho-m} \leq \sum_{|s|=m} |\partial^s f_\varrho(x)| \leq C_m |x|^{\varrho-m}.$$

(2) For $k \geq 0$ and $0 < \alpha \leq 1$ it holds in the case $\varrho \notin \mathbb{N} \cup \{0\}$ that:

$$f_\varrho \in C^{k,\alpha}(\overline{B_1(0)}) \quad \Longleftrightarrow \quad \varrho \geq k + \alpha.$$

(3) For $m \geq 0$ and $1 \leq p < \infty$ it holds in the case $\varrho \notin \mathbb{N} \cup \{0\}$ that:

$$f_\varrho \in W^{m,p}(B_1(0)) \quad \Longleftrightarrow \quad \varrho > m - \frac{n}{p}.$$

Remark: If we consider the exponent ϱ as a measure of the regularity of the function f_ϱ, then it is natural to associate the following **characteristic number** (which we also call the **Sobolev number** or **regularity number**) with the Hölder spaces and Sobolev spaces (where $C^{k,0}(\overline{\Omega}) := C^k(\overline{\Omega})$):

$$\begin{array}{lll} k + \alpha & \text{for } C^{k,\alpha}(\overline{\Omega}) & \text{if } k \geq 0,\ 0 \leq \alpha \leq 1, \\ m - \dfrac{n}{p} & \text{for } W^{m,p}(\Omega) & \text{if } m \geq 0,\ 1 \leq p \leq \infty. \end{array} \tag{10-9}$$

The fact that this Sobolev number does indeed characterize the regularity of the functions in these spaces is a consequence of the following embedding theorems.

Proof (1). The lower bound holds because on setting $e_x := \frac{x}{|x|}$ we have that

$$\pm \binom{\varrho}{m} |x|^{\varrho-m} = \frac{\pm 1}{m!} \partial_{e_x}^m f_\varrho(x) = \pm \sum_{|s|=m} \frac{\partial^s f_\varrho(x)}{s!} e_x^s \leq \sum_{|s|=m} |\partial^s f_\varrho(x)|$$

(with $c_m = 1$), and the upper bound follows from the fact that for all s

$$\partial^s f_\varrho(x) = p_s(x)|x|^{\varrho - 2|s|} \tag{10-10}$$

with homogeneous polynomials p_s of degree $|s|$ or $p_s = 0$. This follows by induction on s, on noting that

$$\partial_i \partial^s f_\varrho(x) = \left(|x|^2 \partial_i p_s(x) + (\varrho - 2|s|)x_i p_s(x)\right) \cdot |x|^{\varrho - 2(|s|+1)},$$

which yields the recurrence formula

$$p_{s+e_i}(x) := |x|^2 \partial_i p_s(x) + (\varrho - 2|s|)x_i p_s(x). \tag{10-11}$$

$\square$

Proof (2). If $\varrho \geq k + \alpha$, then (1) yields that $|\partial^s f_\varrho(x)| \to 0$ as $|x| \to 0$ for $|s| \leq k$, because $\varrho > k$. Hence $f_\varrho \in C^k\left(\overline{B_1(0)}\right)$. If $|s| = k$, then it holds for $0 < |x_0| \leq |x_1| \leq 1$ in the case $|x_1 - x_0| \geq \frac{1}{2}|x_1|$ that

$$|\partial^s f_\varrho(x_1) - \partial^s f_\varrho(x_0)| \leq C_k \cdot \left(|x_0|^{\varrho - k} + |x_1|^{\varrho - k}\right)$$
$$\leq 2^{1+\varrho-k}C_k \cdot |x_1 - x_0|^{\varrho-k} \leq 2^{1+2(\varrho-k)}C_k \cdot |x_1 - x_0|^\alpha.$$

In the case $|x_1 - x_0| \leq \frac{1}{2}|x_1|$ let $x_t := (1-t)x_0 + tx_1$ for $0 \leq t \leq 1$. Then $|x_t| \geq |x_1| - |x_1 - x_0| \geq |x_1 - x_0|$ and so

$$|\partial^s f_\varrho(x_1) - \partial^s f_\varrho(x_0)| \leq \int_0^1 |\nabla \partial^s f_\varrho(x_t)| \, dt \cdot |x_1 - x_0|$$
$$\leq C_{k+1} \int_0^1 |x_t|^{\varrho-k-1} \, dt \cdot |x_1 - x_0|$$
$$\leq C_{k+1} \int_0^1 |x_t|^{\alpha-1} \, dt \cdot |x_1 - x_0| \leq C_{k+1}|x_1 - x_0|^\alpha.$$

Therefore, $f_\varrho \in C^{k,\alpha}\left(\overline{B_1(0)}\right)$. Conversely, if this holds then (1) yields for $0 < |x| \leq 1$ that

$$\infty > \|f_\varrho\|_{C^k} \geq c(n,k) \sum_{|s|=k} |\partial^s f_\varrho(x)| \geq c(n,k) \cdot c_k \cdot \left|\binom{\varrho}{k}\right| \cdot |x|^{\varrho-k},$$

and so $\varrho > k$, because $\varrho \notin \mathbb{N} \cup \{0\}$. As before this means that (1) implies that $\partial^s f_\varrho(x) \to 0$ as $|x| \to 0$ for all $|s| \leq k$. Hence it follows from (1) that for $0 < |x| \leq 1$

$$\infty > \|f_\varrho\|_{C^{k,\alpha}} \geq c(n,k) \sum_{|s|=k} \frac{|\partial^s f_\varrho(x)|}{|x|^\alpha} \geq c(n,k) \cdot c_k \cdot \left|\binom{\varrho}{k}\right| \cdot |x|^{\varrho-k-\alpha},$$

and so $\varrho \geq k + \alpha$.

$\square$

Proof (3). Let $\varrho \notin \mathbb{N} \cup \{0\}$. It follows from (1) that

$$\left\| D^l f_\varrho \right\|^p_{L^p(B_1(0)\setminus\{0\})} \quad \text{for } l \geq 0 \text{ and } 1 \leq p < \infty$$

is bounded from above and below by

$$\int_{B_1(0)} |x|^{p(\varrho-l)} \, dx = C(n) \int_0^1 r^{n-1+p(\varrho-l)} \, dr \, .$$

Hence, $f_\varrho \in W^{m,p}(B_1(0) \setminus \{0\})$ if and only if the integral on the right-hand side is finite for all $0 \leq l \leq m$. This holds if and only if $n + p(\varrho - m) > 0$. The fact that this then yields $f_\varrho \in W^{m,p}(B_1(0))$ follows upon observing that for $|s| < m$ and $\zeta \in C_0^\infty(B_1(0))$ with $0 < \varepsilon < 1$

$$\int_{B_1(0)\setminus B_\varepsilon(0)} \partial_i \zeta \partial^s f_\varrho \, dL^n$$

$$= -\int_{\partial B_\varepsilon(0)} \nu_i \zeta \partial^s f_\varrho \, dH^{n-1} - \int_{B_1(0)\setminus B_\varepsilon(0)} \zeta \partial^{s+e_i} f_\varrho \, dL^n \, ,$$

where, by (1), the first integral on the right-hand side can be bounded by

$$C(n)\|\zeta\|_{\sup} \cdot \varepsilon^{n-1+\varrho-|s|} \to 0 \quad \text{as } \varepsilon \to 0 \, ,$$

since

$$n - 1 + \varrho - |s| \geq n + \varrho - m > n\left(1 - \frac{1}{p}\right) \geq 0 \, . \qquad \square$$

The Sobolev embedding theorem 10.9 rests on the following theorem and for the compactness result makes use of Rellich's embedding theorem (see A8.1 and A8.4).

10.8 Theorem (Sobolev). Let $1 \leq p, q < \infty$ with

$$1 - \frac{n}{p} = -\frac{n}{q} \, . \tag{10-12}$$

Let $u \in W^{1,1}_{\text{loc}}(\mathbb{R}^n)$ with $u \in L^s(\mathbb{R}^n)$ for an $s \in [1, \infty[$ and with $\nabla u \in L^p(\mathbb{R}^n; \mathbb{K}^n)$. Then $u \in L^q(\mathbb{R}^n)$, with

$$\|u\|_{L^q(\mathbb{R}^n)} \leq q \cdot \frac{n-1}{n} \|\nabla u\|_{L^p(\mathbb{R}^n)} \, . \tag{10-13}$$

In particular: The assumptions on u are satisfied for $u \in W^{1,p}(\mathbb{R}^n)$.
Remark: Since $q < \infty$ we must have $p < n$, and so $n \geq 2$. For the case $q = \infty$ see E10.7. For $n = 1$ it holds that $\|u\|_{L^\infty(\mathbb{R})} \leq \|\nabla u\|_{L^1(\mathbb{R})}$ for u as in the assumptions of the theorem (see also E3.6).

Proof. It is sufficient to establish the desired result for functions $u \in L^s(\mathbb{R}^n) \cap C^\infty(\mathbb{R}^n)$. To see this take u as in the assertion and set $u_\varepsilon := \varphi_\varepsilon * u \in C^\infty(\mathbb{R}^n)$ for a standard Dirac sequence $(\varphi_\varepsilon)_{\varepsilon > 0}$. Then $u_\varepsilon \to u$ in $L^s(\mathbb{R}^n)$ and $\nabla u_\varepsilon = \varphi_\varepsilon * \nabla u \to \nabla u$ in $L^p(\mathbb{R}^n; \mathbb{K}^n)$. If the claim has been shown for smooth functions, then for $\varepsilon, \delta > 0$ we have

$$\|u_\varepsilon\|_{L^q} \leq q \cdot \frac{n-1}{n} \|\nabla u_\varepsilon\|_{L^p},$$

$$\|u_\varepsilon - u_\delta\|_{L^q} \leq q \cdot \frac{n-1}{n} \|\nabla (u_\varepsilon - u_\delta)\|_{L^p}.$$

Hence the u_ε as $\varepsilon \searrow 0$ form a Cauchy sequence in $L^q(\mathbb{R}^n)$, which yields that $u_\varepsilon \to \tilde{u}$ in $L^q(\mathbb{R}^n)$ as $\varepsilon \searrow 0$ for some $\tilde{u} \in L^q(\mathbb{R}^n)$. It follows that

$$\|\tilde{u}\|_{L^q} \leq q \cdot \frac{n-1}{n} \|\nabla u\|_{L^p}.$$

Combining the above L^s-convergence and the L^q-convergence yields the existence of a subsequence $\varepsilon_k \searrow 0$ such that $u_{\varepsilon_k} \to u$ and $u_{\varepsilon_k} \to \tilde{u}$ as $k \to \infty$ almost everywhere in $\mathbb{R}^n$. Consequently, $\tilde{u} = u$ almost everywhere in $\mathbb{R}^n$ and we obtain the desired result.

Now let $u \in L^s(\mathbb{R}^n) \cap C^\infty(\mathbb{R}^n)$. In all of the following we will only make use of the fact that $u \in L^s(\mathbb{R}^n) \cap C^1(\mathbb{R}^n)$. First we consider the case

$$p = 1, \quad \text{and so } q = \frac{n}{n-1} \quad \text{(recall that } n \geq 2\text{)}.$$

For $i \in \{1, \ldots, n\}$ it follows from Fubini's theorem that $\xi \mapsto u(x', \xi)$ for almost all $x_1, \ldots, x_{i-1}, x_{i+1}, \ldots, x_n \in \mathbb{R}$ is an element of $L^s(\mathbb{R})$, where we use the notation

$$(x', \xi) := (x_1, \ldots, x_{i-1}, \xi, x_{i+1}, \ldots, x_n).$$

Hence we have that $u(x', z_k) \to 0$ for a sequence $z_k \to \infty$ as $k \to \infty$. It follows for $x_i \in \mathbb{R}$ and sufficiently large k that

$$|u(x)| \leq \int_{x_i}^{z_k} |\partial_i u(x', \xi)| \, d\xi + |u(x', z_k)|,$$

and so

$$|u(x)| \leq \int_{\mathbb{R}} |\partial_i u(x', \xi)| \, d\xi.$$

For ease of exposition we will write this from now on in the compact notation

$$|u(x)| \leq \int_{\mathbb{R}} |\partial_i u| \, d\xi_i.$$

(Observe that the above already proves the remark for the case $n = 1$.) Upon multiplying these n inequalities we obtain that

$$|u(x)|^{\frac{n}{n-1}} \le \prod_{i=1}^{n} \Big(\int_{\mathbb{R}} |\partial_i u|\, d\xi_i\Big)^{\frac{1}{n-1}}.$$

Integration over x_1 yields

$$\int_{\mathbb{R}} |u|^{\frac{n}{n-1}}\, d\xi_1 \le \Big(\int_{\mathbb{R}} |\partial_1 u|\, d\xi_1\Big)^{\frac{1}{n-1}} \cdot \int_{\mathbb{R}} \prod_{i=2}^{n} \Big(\int_{\mathbb{R}} |\partial_i u|\, d\xi_i\Big)^{\frac{1}{n-1}}\, d\xi_1$$

and applying the generalized Hölder inequality we obtain that this is

$$\le \Big(\int_{\mathbb{R}} |\partial_1 u|\, d\xi_1\Big)^{\frac{1}{n-1}} \cdot \prod_{i=2}^{n} \Big(\int_{\mathbb{R}^2} |\partial_i u|\, d(\xi_1,\xi_i)\Big)^{\frac{1}{n-1}}.$$

Now we integrate over x_2 and obtain in the case $n = 2$ the desired result. In the case $n \ge 3$ it follows once again with the help of the Hölder inequality that

$$\int_{\mathbb{R}} \int_{\mathbb{R}} |u|^{\frac{n}{n-1}}\, d\xi_1\, d\xi_2$$

$$\le \Big(\int_{\mathbb{R}^2} |\partial_2 u|\, d(\xi_1,\xi_2)\Big)^{\frac{1}{n-1}}$$

$$\cdot \int_{\mathbb{R}} \Big(\int_{\mathbb{R}} |\partial_1 u|\, d\xi_1\Big)^{\frac{1}{n-1}} \prod_{i=3}^{n} \Big(\int_{\mathbb{R}^2} |\partial_i u|\, d(\xi_1,\xi_i)\Big)^{\frac{1}{n-1}}\, d\xi_2$$

$$\le \Big(\int_{\mathbb{R}^2} |\partial_2 u|\, d(\xi_1,\xi_2)\Big)^{\frac{1}{n-1}}$$

$$\cdot \Big(\int_{\mathbb{R}^2} |\partial_1 u|\, d(\xi_1,\xi_2)\Big)^{\frac{1}{n-1}} \cdot \prod_{i=3}^{n} \Big(\int_{\mathbb{R}^3} |\partial_i u|\, d(\xi_1,\xi_2,\xi_i)\Big)^{\frac{1}{n-1}}.$$

Continuing this procedure inductively we obtain for $j = 1,\dots,n$ that

$$\int_{\mathbb{R}^j} |u|^{\frac{n}{n-1}}\, d(\xi_1,\dots,\xi_j)$$

$$\le \prod_{i=1}^{j} \Big(\int_{\mathbb{R}^j} |\partial_i u|\, d(\xi_1,\dots,\xi_j)\Big)^{\frac{1}{n-1}}$$

$$\cdot \prod_{i=j+1}^{n} \Big(\int_{\mathbb{R}^{j+1}} |\partial_i u|\, d(\xi_1,\dots,\xi_j,\xi_i)\Big)^{\frac{1}{n-1}},$$

and hence for $j = n$ that

$$\int_{\mathbb{R}^n} |u|^{\frac{n}{n-1}}\, dL^n \le \prod_{i=1}^{n} \Big(\int_{\mathbb{R}^n} |\partial_i u|\, dL^n\Big)^{\frac{1}{n-1}} \le \Big(\int_{\mathbb{R}^n} |\nabla u|\, dL^n\Big)^{\frac{n}{n-1}},$$

i.e. the desired result

$$\|u\|_{L^{\frac{n}{n-1}}(\mathbb{R}^n)} \leq \|\nabla u\|_{L^1(\mathbb{R}^n)} . \tag{10-14}$$

For $p > 1$ we want to apply this result to $v = |u|^{\frac{q(n-1)}{n}}$, where on letting p' denote the dual exponent to p it holds that

$$\frac{n-1}{n} - \frac{1}{p'} = \frac{1}{n} + \frac{1}{p} = \frac{1}{q}, \quad \text{and so} \quad \frac{q(n-1)}{n} = 1 + \frac{q}{p'} > 1.$$

In order to avoid unnecessary difficulties, we consider for $\varepsilon > 0$ the functions

$$v_\varepsilon(x) := \psi_\varepsilon(|u(x)|)^{\frac{q(n-1)}{n}},$$

where $\psi_\varepsilon : [0, \infty[\to [0, \infty[$ is continuously differentiable, with

$$\psi_\varepsilon(z) \leq z, \quad \psi'_\varepsilon(z) \leq 1, \quad \psi_\varepsilon(z) \nearrow z \text{ as } \varepsilon \searrow 0.$$

As $u \in C^1(\mathbb{R}^n)$, we also have that $v_\varepsilon \in C^1(\mathbb{R}^n)$, with

$$|\nabla v_\varepsilon| \leq \frac{q(n-1)}{n} w_\varepsilon \cdot |\nabla u|, \quad \text{where } w_\varepsilon := \psi_\varepsilon(|u|)^{\frac{q}{p'}}.$$

For $\varrho > 1$ we choose in particular

$$\psi_\varepsilon(z) := \left(\varepsilon^\varrho + \left(\frac{z}{1+\varepsilon z}\right)^\varrho\right)^{\frac{1}{\varrho}} - \varepsilon,$$

which means that there exists a constant C_ε depending on ε such that

$$\psi_\varepsilon(z) \leq C_\varepsilon \cdot \min(1, z^\varrho).$$

It follows that

$$w_\varepsilon \in L^{p'}(\mathbb{R}^n) \text{ and } v_\varepsilon \in L^1(\mathbb{R}^n), \quad \text{if } \varrho q \frac{n-1}{n} \geq s.$$

The Hölder inequality then yields that $\nabla v_\varepsilon \in L^1(\mathbb{R}^n; \mathbb{R}^n)$. It follows from inequality (10-14) that $v_\varepsilon \in L^{\frac{n}{n-1}}(\mathbb{R}^n)$, i.e. $\psi_\varepsilon(|u|) \in L^q(\mathbb{R}^n)$, with

$$\left(\int_{\mathbb{R}^n} \psi_\varepsilon(|u|)^q \, d\mathrm{L}^n\right)^{\frac{n-1}{n}} = \left(\int_{\mathbb{R}^n} v_\varepsilon^{\frac{n}{n-1}} \, d\mathrm{L}^n\right)^{\frac{n-1}{n}} \leq \int_{\mathbb{R}^n} |\nabla v_\varepsilon| \, d\mathrm{L}^n$$

$$\leq \frac{q(n-1)}{n} \int_{\mathbb{R}^n} w_\varepsilon \cdot |\nabla u| \, d\mathrm{L}^n$$

$$\leq \frac{q(n-1)}{n} \left(\int_{\mathbb{R}^n} \psi_\varepsilon(|u|)^q \, d\mathrm{L}^n\right)^{\frac{1}{p'}} \|\nabla u\|_{L^p},$$

and hence

$$\left(\int_{\mathbb{R}^n} \psi_\varepsilon(|u|)^q \, d\mathrm{L}^n\right)^{\frac{1}{q}} \leq \frac{q(n-1)}{n} \|\nabla u\|_{L^p}.$$

Letting $\varepsilon \searrow 0$ we obtain the desired result from the monotone convergence theorem. $\square$

10.9 Embedding theorem in Sobolev spaces. Let $\Omega \subset \mathbb{R}^n$ be open and bounded with Lipschitz boundary. Further, let $m_1 \geq 0$, $m_2 \geq 0$ be integers, and let $1 \leq p_1 < \infty$ and $1 \leq p_2 < \infty$. Then the following holds:

(1) If

$$m_1 - \frac{n}{p_1} \geq m_2 - \frac{n}{p_2}, \quad \text{and} \quad m_1 \geq m_2, \tag{10-15}$$

then the embedding

$$\text{Id} : W^{m_1,p_1}(\Omega) \to W^{m_2,p_2}(\Omega)$$

exists and is continuous. Here $W^{0,p}(\Omega) = L^p(\Omega)$. The following estimate holds: There exists a constant C, which depends on n, Ω, m_1, p_1, m_2, p_2, such that for $u \in W^{m_1,p_1}(\Omega)$

$$\|u\|_{W^{m_2,p_2}(\Omega)} \leq C\|u\|_{W^{m_1,p_1}(\Omega)}. \tag{10-16}$$

(2) If

$$m_1 - \frac{n}{p_1} > m_2 - \frac{n}{p_2}, \quad \text{and} \quad m_1 > m_2, $$

then the embedding

$$\text{Id} : W^{m_1,p_1}(\Omega) \to W^{m_2,p_2}(\Omega)$$

exists and is continuous and compact.

(3) For arbitrary open, bounded sets $\Omega \subset \mathbb{R}^n$ assertions (1) and (2) hold with the spaces $W^{m_i,p_i}(\Omega)$ replaced by $W_0^{m_i,p_i}(\Omega)$. Here $W_0^{0,p}(\Omega) = L^p(\Omega)$.

Proof (1). We also prove the corresponding result in (3), i.e. we let $\Omega \subset \mathbb{R}^n$ be open and bounded. For $m_1 = m_2$ the claim follows from the Hölder inequality. For $m_1 = m_2 + 1$ we have that

$$1 - \frac{n}{p_1} \geq -\frac{n}{p_2}.$$

Let $u \in W_0^{m_1,p_1}(\Omega)$. For $|s| \leq m_2$ it holds that $v := \partial^s u \in W_0^{1,p_1}(\Omega)$. As Ω is bounded, it follows from the Hölder inequality that then v is also an element of $W_0^{1,p}(\Omega)$ for $1 \leq p \leq p_1$. Extending v by 0 on $\mathbb{R}^n \setminus \Omega$ yields that $v \in W^{1,p}(\mathbb{R}^n)$ (see 3.29). If $n = 1$, choose $p = 1$ and obtain from the remark in 10.8 that with $\varrho := \mathrm{L}^n(\Omega)$

$$\|v\|_{L^{p_2}(\Omega)} \leq \varrho^{\frac{1}{p_2}}\|v\|_{L^\infty(\mathbb{R})} \leq \varrho^{\frac{1}{p_2}}\|\nabla v\|_{L^1(\mathbb{R})} = \varrho^{\frac{1}{p_2}}\|\nabla v\|_{L^1(\Omega)}$$

and in the case $p_1 > 1$, with p_1' denoting the dual exponent to p_1, that

$$\|\nabla v\|_{L^1(\Omega)} \leq \varrho^{\frac{1}{p_1'}}\|\nabla v\|_{L^{p_1}(\Omega)}.$$

If $n \geq 2$, choose $1 \leq p \leq p_1 < \infty$ and $1 \leq p_2 \leq q < \infty$ with

$$1 - \frac{n}{p_1} \geq 1 - \frac{n}{p} = -\frac{n}{q} \geq -\frac{n}{p_2},$$

e.g. $q = \max\left(\frac{n}{n-1}, p_2\right)$, and obtain from 10.8 that $v \in L^{p_2}(\Omega)$, with

$$\|v\|_{L^{p_2}(\Omega)} \leq \varrho^{\frac{1}{p_2} - \frac{1}{q}} \|v\|_{L^q(\mathbb{R}^n)} \leq \varrho^{\frac{1}{p_2} - \frac{1}{q}} \cdot q \frac{n-1}{n} \|\nabla v\|_{L^p(\mathbb{R}^n)}$$

and

$$\|\nabla v\|_{L^p(\mathbb{R}^n)} = \|\nabla v\|_{L^p(\Omega)} \leq \varrho^{\frac{1}{p} - \frac{1}{p_1}} \|\nabla v\|_{L^{p_1}(\Omega)}.$$

If Ω has a Lipschitz boundary, and if $u \in W^{m_1, p_1}(\Omega)$, then we have that $v := \partial^s u \in W^{1, p_1}(\Omega)$ for $|s| \leq m_2$. Then let $\tilde{v} := E(v)$, where $E : W^{1, p_1}(\Omega) \to W_0^{1, p_1}(\tilde{\Omega})$ with $\tilde{\Omega} = B_1(\Omega)$ is the extension operator from A8.12. Similarly to the above we then obtain the bound

$$\|\tilde{v}\|_{L^{p_2}(\tilde{\Omega})} \leq \tilde{C} \cdot \|\nabla \tilde{v}\|_{L^{p_1}(\tilde{\Omega})},$$

and hence, since $\tilde{v} = v$ on Ω,

$$\|v\|_{L^{p_2}(\Omega)} \leq \|\tilde{v}\|_{L^{p_2}(\tilde{\Omega})} \leq \tilde{C} \cdot \|\tilde{v}\|_{W^{1, p_1}(\tilde{\Omega})} \leq \tilde{C} \cdot \|E\| \cdot \|v\|_{W^{1, p_1}(\Omega)}.$$

Now we consider the case $m_1 = m_2 + k$ with $k \geq 2$. Then let $\tilde{m}_i := m_2 + i$ for $i = 0, \ldots, k$. Choose $1 \leq \tilde{p}_i < \infty$ with $\tilde{p}_0 = p_2$ and $\tilde{p}_k = p_1$, such that

$$\tilde{m}_i - \frac{n}{\tilde{p}_i} \geq \tilde{m}_{i-1} - \frac{n}{\tilde{p}_{i-1}} \qquad \text{for } i = 1, \ldots, k, \qquad (10\text{-}17)$$

e.g. $\tilde{p}_i$ for $1 \leq i < k$ with $\frac{1}{\tilde{p}_i} = \min\left(1, \frac{1}{n} + \frac{1}{\tilde{p}_{i-1}}\right)$. Now apply the above proof successively for $i = k, \ldots, 1$. $\qquad \square$

Proof (2). Once again we also prove the corresponding result in (3). For $m_1 = m_2 + 1$ choose $p_2 < p < \infty$ with

$$1 - \frac{n}{p_1} \geq -\frac{n}{p} > -\frac{n}{p_2}.$$

Let $(u_k)_{k \in \mathbb{N}}$ be a bounded sequence in $W^{m_1, p_1}(\Omega)$ (for (3) in $W_0^{m_1, p_1}(\Omega)$). For $|s| \leq m_2$ it then holds that $v_k := \partial^s u_k$ are bounded in $W^{1, p_1}(\Omega)$ (or $W_0^{1, p_1}(\Omega)$). By (1), the sequence $(v_k)_{k \in \mathbb{N}}$ is bounded in $L^p(\Omega)$. Since $L^p(\Omega)$ is reflexive, theorem 8.10 yields the existence of a subsequence $(v_{k_i})_{i \in \mathbb{N}}$, which can be chosen as the same subsequence for all $|s| \leq m_2$, that converges weakly in $L^p(\Omega)$ to $v \in L^p(\Omega)$. As Ω is bounded, $v_{k_i} \to v$ converges weakly in $L^1(\Omega)$ as $i \to \infty$ and $(v_{k_i})_{i \in \mathbb{N}}$ is bounded in $W^{1,1}(\Omega)$ (or $W_0^{1,1}(\Omega)$). Hence it follows from Rellich's embedding theorem (A8.1 and A8.4) that $v_{k_i} \to v$ strongly in $L^1(\Omega)$. Noting that $1 \leq p_2 < p$ then yields the strong convergence also in $L^{p_2}(\Omega)$ (see E10.11).

For $m_1 = m_2 + k$ with $k \geq 2$ we again choose $\tilde{m}_i, \tilde{p}_i$ as in the proof of (1), where now (10-17) needs to be a strict inequality for an $i_0 \in \{1, \ldots, k\}$. Then for i_0 we can apply the above proof, and for $i \neq i_0$ the result (1). $\qquad \square$

Now we consider the embedding of Sobolev spaces into Hölder spaces. The proof of theorem 10.13 rests on two results: a bound on the supremum norm and a bound on the Hölder constant.

10.10 Theorem. Let $\Omega \subset \mathbb{R}^n$ be open and bounded and let $1 < p < \infty$, with

$$1 - \frac{n}{p} > 0 \quad (\text{and so } p > n \geq 1).$$

For every function $u \in W_0^{1,p}(\Omega)$ it then holds that $u \in L^\infty(\Omega)$ with

$$\|u\|_{L^\infty(\Omega)} \leq C(n, p, \operatorname{diam} \Omega) \|\nabla u\|_{L^p(\Omega)}.$$

Proof. Analogously to the proof of 10.8, it is sufficient to establish the desired result for functions $u \in C_0^\infty(\Omega)$. Further, let $R := \operatorname{diam} \Omega$, so that $\Omega \subset B_R(x_0)$ for all $x_0 \in \Omega$. Then it holds for all $\xi \in \partial B_1(0)$ that

$$|u(x_0)| = \left| \int_0^R \frac{d}{dr} (u(x_0 + r\xi)) \, dr \right| \leq \int_0^R |\nabla u(x_0 + r\xi)| \, dr.$$

Integrating this inequality over ξ with respect to the surface measure H^{n-1} and denoting the surface area of the unit sphere by $\sigma_n := H^{n-1}(\partial B_1(0))$ we get

$$\sigma_n |u(x_0)| \leq \int_0^R \int_{\partial B_1(0)} |\nabla u(x_0 + r\xi)| \, dH^{n-1}(\xi) \, dr.$$

A transformation to Euclidean coordinates shows that the right-hand side is

$$= \int_{B_R(x_0)} \frac{|\nabla u(x)|}{|x - x_0|^{n-1}} \, dx,$$

and the Hölder inequality yields that this can be bounded by

$$\leq \left(\int_{B_R(x_0)} \frac{dx}{|x - x_0|^{p'(n-1)}} \right)^{\frac{1}{p'}} \cdot \|\nabla u\|_{L^p(\Omega)}.$$

The first factor is independent of x_0 and finite if $p'(n-1) < n$, i.e. if $p' < n'$ (where n' is the dual exponent to n), which is equivalent to $p > n$. But this was part of the assumption. $\qquad \square$

10.11 Theorem (Morrey). Let $\Omega \subset \mathbb{R}^n$ be open, let $0 < \alpha \leq 1$ and let $u \in W_0^{1,1}(\Omega)$ satisfy

$$\int_{B_r(x_0) \cap \Omega} |\nabla u| \, dL^n \leq M \cdot r^{n-1+\alpha} \tag{10-18}$$

for all $x_0 \in \Omega$ and $r > 0$. Then for almost all $x_1, x_2 \in \Omega$,

$$\frac{|u(x_1) - u(x_2)|}{|x_1 - x_2|^\alpha} \leq C(n, \alpha) \cdot M. \tag{10-19}$$

Note: A p-version of the result is given in 10.12(1).

Proof. We may assume that $u \in W^{1,1}(\mathbb{R}^n)$, because u can be extended by 0 on $\mathbb{R}^n \setminus \Omega$ to yield a function in $W^{1,1}(\mathbb{R}^n)$ (see 3.29). Then for every ball $B_r(x_0)$ with $x_0 \in \mathbb{R}^n$ we have that

$$\int_{B_r(x_0)} |\nabla u| \, \mathrm{d}L^n \leq M(2r)^{n-1+\alpha}, \tag{10-20}$$

on noting that in the case $B_r(x_0) \cap \Omega = \emptyset$ this is trivially true, and that otherwise there exists an $x_1 \in B_r(x_0) \cap \Omega$ and then $B_r(x_0) \subset B_{2r}(x_1)$, and for this latter ball we can apply (10-18).

We begin by proving the bound on the Hölder constant for the case where u is a C^1-function. Given two points $x_1, x_2 \in \mathbb{R}^n$, let

$$x_0 := \frac{1}{2}(x_1 + x_2) \quad \text{and} \quad \varrho := \frac{1}{2}|x_2 - x_1|.$$

Denoting the volume of the n-dimensional unit ball by κ_n, we have that

$$\begin{aligned}
\kappa_n \varrho^n |u(x_1) - u(x_2)| &= \int_{B_\varrho(x_0)} |u(x_1) - u(x_2)| \, \mathrm{d}x \\
&\leq \int_{B_\varrho(x_0)} |u(x_1) - u(x)| \, \mathrm{d}x + \int_{B_\varrho(x_0)} |u(x_2) - u(x)| \, \mathrm{d}x.
\end{aligned} \tag{10-21}$$

Because of symmetry we only need to bound the first integral. Now it holds for $x \in B_\varrho(x_0)$ that

$$\begin{aligned}
|u(x) - u(x_1)| &= \left| \int_0^1 \frac{\mathrm{d}}{\mathrm{d}t}\big(u(x_1 + t(x - x_1))\big) \, \mathrm{d}t \right| \\
&\leq |x - x_1| \int_0^1 |\nabla u(x_1 + t(x - x_1))| \, \mathrm{d}t.
\end{aligned}$$

Since $|x - x_1| \leq 2\varrho$, integration over x yields

$$\int_{B_\varrho(x_0)} |u(x) - u(x_1)| \, \mathrm{d}x \leq 2\varrho \int_0^1 \int_{B_\varrho(x_0)} |\nabla u(x_1 + t(x - x_1))| \, \mathrm{d}x \, \mathrm{d}t.$$

With the transformation of variables $y(x) := x_1 + t(x - x_1)$ this is

$$\begin{aligned}
&= 2\varrho \int_0^1 t^{-n} \int_{B_{t\varrho}(x_1 + t(x_0 - x_1))} |\nabla u(y)| \, \mathrm{d}y \, \mathrm{d}t \\
&\leq 2\varrho \int_0^1 t^{-n} M(2t\varrho)^{n-1+\alpha} \, \mathrm{d}t = \frac{M}{\alpha}(2\varrho)^{n+\alpha},
\end{aligned}$$

where we used (10-20). Hence it follows from (10-21) that

$$|u(x_1) - u(x_2)| \leq \frac{2^{n+1}M}{\alpha\kappa_n}(2\varrho)^\alpha = \frac{2^{n+1}M}{\alpha\kappa_n}|x_1 - x_2|^\alpha. \tag{10-22}$$

For an arbitrary $u \in W^{1,1}(\mathbb{R}^n)$ one can consider the convolution with a standard Dirac sequence $(\varphi_\varepsilon)_{\varepsilon>0}$. Then the functions $u_\varepsilon := \varphi_\varepsilon * u$ are in $C^\infty(\mathbb{R}^n)$ and satisfy (10-20). Indeed,

$$\nabla u_\varepsilon(x) = \int_{\mathbb{R}^n} \varphi_\varepsilon(y) \nabla u(x-y)\, dy$$

and so, using (10-20) for u, we have

$$\int_{B_r(x_0)} |\nabla u_\varepsilon(x)|\, dx \le \int_{\mathbb{R}^n} \left(\int_{B_r(x_0)} |\nabla u(x-y)|\, dx \right) \varphi_\varepsilon(y)\, dy$$

$$= \int_{\mathbb{R}^n} \left(\int_{B_r(x_0-y)} |\nabla u(x)|\, dx \right) \varphi_\varepsilon(y)\, dy$$

$$\le M(2r)^{n-1+\alpha} \int_{\mathbb{R}^n} \varphi_\varepsilon(y)\, dy = M(2r)^{n-1+\alpha}.$$

Hence we obtain (10-22) for u_ε, and noting that $u_\varepsilon \to u$ almost everywhere for a subsequence as $\varepsilon \to 0$ then yields the desired result. $\square$

10.12 Remarks. The inequality (10-19) states that u is Hölder continuous outside of a null set N. But then the function u restricted to $\Omega \setminus N$ can be uniquely extended to a $C^{0,\alpha}$-function on $\overline{\Omega}$. Hence the given function $u \in W_0^{1,1}(\Omega)$ has a unique Hölder continuous representative. Moreover, it holds that:

(1) Theorem 10.11 can also be applied in the general case where $u \in W_0^{1,p}(\Omega)$ with $1 \le p < \infty$. If u then satisfies for $0 < \alpha \le 1$ the inequality

$$\|\nabla u\|_{L^p(B_r(x_0)\cap\Omega)} \le M \cdot r^{\frac{n}{p}-1+\alpha} \tag{10-23}$$

for all $x_0 \in \Omega$ and $r > 0$, then the conclusion of 10.11 holds true.

(2) If $u \in W_0^{1,p}(\Omega)$ with $1 - \frac{n}{p} > 0$, then (1) holds with $\alpha := 1 - \frac{n}{p}$.

(3) Theorem 10.11 also holds for $\Omega = \mathbb{R}^n$ and $u \in W_{\text{loc}}^{1,1}(\mathbb{R}^n)$.

Proof (1). The Hölder inequality yields that

$$\int_{B_r(x_0)\cap\Omega} |\nabla u|\, d\mathrm{L}^n \le C(n) r^{\frac{n}{p'}} \left(\int_{B_r(x_0)\cap\Omega} |\nabla u|^p\, d\mathrm{L}^n \right)^{\frac{1}{p}} \le C(n) M r^{n-1+\alpha}.$$

$\square$

10.13 Embedding theorem of Sobolev spaces into Hölder spaces. Let $\Omega \subset \mathbb{R}^n$ be open and bounded with Lipschitz boundary. Moreover, let $m \ge 1$ be an integer and let $1 \le p < \infty$. In addition, let $k \ge 0$ be an integer and let $0 \le \alpha \le 1$. Then the following holds:

(1) If

$$m - \frac{n}{p} = k + \alpha \quad \text{and} \quad 0 < \alpha < 1 \text{ (and so } \alpha \neq 0,1), \qquad (10\text{-}24)$$

then the embedding

$$\text{Id} : W^{m,p}(\Omega) \to C^{k,\alpha}(\overline{\Omega})$$

exists and is continuous. In particular, for $u \in W^{m,p}(\Omega)$ there exists a unique continuous function that agrees almost everywhere with u (and which we again denote by u) such that

$$\|u\|_{C^{k,\alpha}(\overline{\Omega})} \leq C(\Omega, n, m, p, k, \alpha) \|u\|_{W^{m,p}(\Omega)} . \qquad (10\text{-}25)$$

(2) If

$$m - \frac{n}{p} > k + \alpha,$$

then the embedding

$$\text{Id} : W^{m,p}(\Omega) \to C^{k,\alpha}(\overline{\Omega})$$

exists and is continuous and compact. Here $C^{k,0}(\overline{\Omega}) := C^k(\overline{\Omega})$ for $k \geq 0$.

(3) For arbitrary open, bounded sets $\Omega \subset \mathbb{R}^n$ assertions (1) and (2) hold with the space $W^{m,p}(\Omega)$ replaced by $W_0^{m,p}(\Omega)$.

Proof (1). We also prove the corresponding result in (3). We may assume that $k = 0$. Otherwise apply the following argument to all functions $\partial^s u \in W^{m-k,p}(\Omega)$ (or $W_0^{m-k,p}(\Omega)$) for $|s| \leq k$, on noting that $m - k \geq 1$.

Next we reduce the proof to the case $m = 1$. If $m > 1$, we may choose $1 \leq q < \infty$ such that

$$\alpha - 1 = -\frac{n}{q}, \quad \text{and so} \quad m - \frac{n}{p} = \alpha = 1 - \frac{n}{q}.$$

It then follows from 10.9(1) that the embedding from $W^{m,p}(\Omega)$ into $W^{1,q}(\Omega)$ is continuous (use 10.9(3) for the embedding from $W_0^{m,p}(\Omega)$ into $W_0^{1,q}(\Omega)$). Thus we have to consider only functions in $W^{1,q}(\Omega)$ (or $W_0^{1,q}(\Omega)$).

Hence we consider only the case where in the statement of the theorem $k = 0$ and $m = 1$, i.e.

$$1 - \frac{n}{p} = \alpha.$$

For the case in (3), the desired result follows upon combining theorem 10.10 and theorem 10.11 (see 10.12(2)). Otherwise we consider the continuous extension operator $E : W^{1,p}(\Omega) \to W_0^{1,p}(B_1(\Omega))$ from A8.12 and then apply the theorems 10.10 and 10.11 to the functions Eu. $\qquad \square$

Proof (2). We also prove the corresponding result in (3). Choose $\widetilde{m} \leq m$ and $1 < \widetilde{p} < \infty$, as well as $\widetilde{k} \geq 0$ and $0 < \widetilde{\alpha} < 1$, such that

$$m - \frac{n}{p} \geq \widetilde{m} - \frac{n}{\widetilde{p}} = \widetilde{k} + \widetilde{\alpha} > k + \alpha,$$

where we can set $\widetilde{m} = m$ and $\widetilde{p} = p$ if $\frac{n}{p}$ is not an integer. Then, by 10.9(1) and (1), the embeddings from $W^{m,p}(\Omega)$ into $W^{\widetilde{m},\widetilde{p}}(\Omega)$ and from $W^{\widetilde{m},\widetilde{p}}(\Omega)$ into $C^{\widetilde{k},\widetilde{\alpha}}(\Omega)$ are continuous, respectively (for (3) we argue correspondingly with 10.9(3)). Finally, by 10.6, the embedding from $C^{\widetilde{k},\widetilde{\alpha}}(\overline{\Omega})$ into $C^{k,\alpha}(\overline{\Omega})$ is compact. $\qquad\square$

Laplace operator

We now present a typical application of the embedding theorems for the Laplace operator. This is essential for the treatment of the corresponding eigenvalue problem (see 12.16).

10.14 Inverse Laplace operator. We consider the homogeneous Dirichlet problem from 6.5(1) with the assumptions stated there and with

$$h_i = 0, \quad b \geq 0.$$

For $u \in W_0^{1,2}(\Omega)$ and $f \in L^2(\Omega)$ let $A(u)$ and $J(f)$ be the functionals in $W_0^{1,2}(\Omega)'$ defined by

$$\langle v, A(u)\rangle_{W_0^{1,2}} := \int_\Omega \Big(\sum_{i,j=1}^n \partial_i v \cdot a_{ij}\partial_j u + vbu\Big)\, \mathrm{dL}^n,$$

$$\langle v, J(f)\rangle_{W_0^{1,2}} := \int_\Omega vf\, \mathrm{dL}^n$$

for $v \in W_0^{1,2}(\Omega)$. Then it holds that:

(1) $J : L^2(\Omega) \to W_0^{1,2}(\Omega)'$ is continuous and injective.

(2) $A : W_0^{1,2}(\Omega) \to W_0^{1,2}(\Omega)'$ is an isomorphism. We call A the **weak differential operator** corresponding to the boundary value problem 6.5(1). For $a_{ij} = \delta_{i,j}$ and $b = 0$ this is the **weak Laplace operator** with respect to homogeneous Dirichlet boundary conditions.

(3) $A^{-1}J : L^2(\Omega) \to L^2(\Omega)$ is compact.

(4) $A^{-1}J : W_0^{1,2}(\Omega) \to W_0^{1,2}(\Omega)$ is compact, and for domains Ω with Lipschitz boundary the operator $A^{-1}J : W^{1,2}(\Omega) \to W_0^{1,2}(\Omega)$ is also compact.

(5) $JA^{-1} : W_0^{1,2}(\Omega)' \to W_0^{1,2}(\Omega)'$ is compact.

Proof (1),(2). We have that $\langle v, Au \rangle = a(v, u)$, where a is defined as in (6-11). The fact that J and A are well defined and continuous follows as in the proof of 6.6. It follows from 4.22 that J is injective. A is injective thanks of the coercivity of a, as shown in the proof of 6.8. Recalling 6.3(1) with $X := W_0^{1,2}(\Omega)$ yields that for $u' \in X'$ there exists a unique $u \in X$ such that

$$\langle v, Au \rangle = a(v, u) = \langle v, u' \rangle \quad \text{for all } v \in X,$$

where $\|u\|_X$ can be bounded by $\|u'\|_{X'}$. $\quad\square$

Proof (3). We recall from (1) and (2) that $J : L^2(\Omega) \to W_0^{1,2}(\Omega)'$ and $A^{-1} : W_0^{1,2}(\Omega)' \to W_0^{1,2}(\Omega)$, respectively, are continuous. The embedding $\mathrm{Id} : W_0^{1,2}(\Omega) \to L^2(\Omega)$ is compact, by 10.1(4) and A8.1. The desired result then follows from 10.3.

Remark: If Ω has a Lipschitz boundary, then it follows from 10.9 that $\mathrm{Id} : W^{1,2}(\Omega) \to L^2(\Omega)$ is also compact. $\quad\square$

Proof (4),(5). We can argue with the above maps in the order Id, J, A^{-1} and A^{-1}, Id, J, respectively. $\quad\square$

Integral operators

As a second class of compact maps we now investigate some integral operators. Such operators occur, for example, when boundary value problems are reformulated as integral equations with the help of a Green's function (see 10.18). First we prove the compactness of Hilbert-Schmidt operators and of integral operators with a weakly singular kernel.

10.15 Hilbert-Schmidt integral operator. We have defined in 5.12 an integral operator $T : L^p(\Omega_2) \to L^q(\Omega_1)$, which we claim is compact.

Proof. We recall from 5.12 that T is continuous with $\|T\| \le \|K\|$. In order to prove the compactness of T we extend K by 0 outside $\Omega_1 \times \Omega_2$, i.e. $K(x,y) := 0$ if $x \notin \Omega_1$ or $y \notin \Omega_2$. Then it follows for $h \in \mathbb{R}^{n_1}$ and $f \in L^p(\Omega_2)$ with $\|f\|_{L^p(\Omega_2)} \le 1$, in the same way as in the proof of 5.12, that

$$\int_{\mathbb{R}^{n_1}} |Tf(x+h) - Tf(x)|^q \, dx$$
$$\le \int_{\mathbb{R}^{n_1}} \left(\int_{\mathbb{R}^{n_2}} |K(x+h,y) - K(x,y)|^{p'} \, dy \right)^{\frac{q}{p'}} dx \tag{10-26}$$

and

$$\int_{\mathbb{R}^{n_1} \setminus B_R(0)} |Tf(x)|^q \, dx \le \int_{\mathbb{R}^{n_1} \setminus B_R(0)} \left(\int_{\mathbb{R}^{n_2}} |K(x,y)|^{p'} \, dy \right)^{\frac{q}{p'}} dx. \tag{10-27}$$

The right-hand side in (10-27) converges to 0 as $R \to \infty$, since $\|K\| < \infty$. If, in addition, the right-hand side in (10-26) converges to 0 as $h \to 0$, then the compactness of T follows from the Riesz compactness criterion in theorem 4.16. To show this let $K^h(x,y) := K(x+h, y)$. We need to consider $\|K^h - K\|$, where here the norm of the kernel is defined by integrating over all of $\mathbb{R}^{n_1} \times \mathbb{R}^{n_2}$. We begin by approximating K by bounded kernels with compact support

$$K_R(x,y) := \begin{cases} K(x,y) & \text{if } |x| \le R, \ |y| \le R, \ |K(x,y)| \le R, \\ 0 & \text{otherwise.} \end{cases}$$

Then, on setting $E_R := \left\{ (x,y) \in \mathbb{R}^{n_1} \times \mathbb{R}^{n_2} \ ; \ K(x,y) \ne K_R(x,y) \right\}$, we have that

$$\left| K^h - K \right| \le \left| (K_R)^h - K_R \right| + \left| (\mathcal{X}_{E_R} K)^h \right| + \left| \mathcal{X}_{E_R} K \right|,$$

which yields that

$$\left\| K^h - K \right\| \le C \left(\left\| (K_R)^h - K_R \right\| + \left\| \mathcal{X}_{E_R} K \right\| \right).$$

Noting that $E_{R'} \subset E_R$ for $R' > R$ and that $\bigcap_{R>0} E_R$ is a null set we see that the second term on the right-hand side converges to 0 as $R \to \infty$ (analogously to (10-27) consider the monotone convergence of $(1 - \mathcal{X}_{E_R})|K|$). Since K_R is bounded with compact support, the first term in the case $\frac{q}{p'} \ge 1$ obeys the inequality

$$\left\| (K_R)^h - K_R \right\|^q \le C(R, \tfrac{q}{p'}) \int_{\mathbb{R}^{n_1}} \int_{\mathbb{R}^{n_2}} \left| (K_R)^h - K_R \right|^{p'} (x,y) \, dy \, dx \,,$$

while in the case $r := \frac{p'}{q} > 1$ the Hölder inequality with exponent r gives

$$\left\| (K_R)^h - K_R \right\|^{p'} = \left(\int_{\mathbb{R}^{n_1}} \left(\int_{\mathbb{R}^{n_2}} \left| (K_R)^h - K_R \right|^{p'} (x,y) \, dy \right)^{\frac{1}{r}} dx \right)^r$$

$$\le C(R, r) \int_{\mathbb{R}^{n_1}} \int_{\mathbb{R}^{n_2}} \left| (K_R)^h - K_R \right|^{p'} (x,y) \, dy \, dx \,.$$

Now we use the fact that $(K_R)^h \to K_R$ in $L^{p'}(\mathbb{R}^{n_1} \times \mathbb{R}^{n_2})$ as $h \to 0$, recall 4.15(1).

In the Hilbert space case $p = 2$, $q = 2$ the compactness can also be shown as follows: Choose an orthonormal basis $(e_n)_{n \in \mathbb{N}}$ of $L^2(\Omega_2)$ (see 9.8). Then, by the completeness relation 9.7(5),

$$\|K\|^2 = \int_{\Omega_1} \left\| \overline{K(x, \cdot)} \right\|_{L^2(\Omega_2)}^2 dx = \int_{\Omega_1} \sum_{n \in \mathbb{N}} \left| \left(\overline{K(x, \cdot)}, e_n \right)_{L^2(\Omega_2)} \right|^2 dx$$

$$= \int_{\Omega_1} \sum_{n \in \mathbb{N}} |T e_n(x)|^2 \, dx = \sum_{n \in \mathbb{N}} \|T e_n\|_{L^2(\Omega_1)}^2 \,.$$

We define the continuous projections P_n by

$$P_n f := \sum_{k=1}^{n} (f, e_k)_{L^2(\Omega_2)} \, e_k \, .$$

Then using 9.7(3) and the continuity of T, we see that

$$\|Tf - TP_n f\|_{L^2(\Omega_1)} = \left\| T\left(\sum_{k>n} (f, e_k)_{L^2(\Omega_2)} e_k \right) \right\|_{L^2(\Omega_1)}$$

$$= \left\| \sum_{k>n} (f, e_k)_{L^2(\Omega_2)} Te_k \right\|_{L^2(\Omega_1)} \le \sum_{k>n} \left| (f, e_k)_{L^2(\Omega_2)} \right| \|Te_k\|_{L^2(\Omega_1)} \, .$$

On applying the Cauchy-Schwarz inequality in $\ell^2(\mathbb{R})$ we find that this is

$$\le \underbrace{\left(\sum_{k>n} \left| (f, e_k)_{L^2(\Omega_2)} \right|^2 \right)^{\frac{1}{2}}}_{\le \|f\|_{L^2(\Omega_2)}} \cdot \underbrace{\left(\sum_{k>n} \|Te_k\|_{L^2(\Omega_1)}^2 \right)^{\frac{1}{2}}}_{\to 0 \text{ as } n \to \infty} \, .$$

Hence, $TP_n \to T$ in $\mathscr{L}\big(L^2(\Omega_2); L^2(\Omega_1)\big)$ as $n \to \infty$. Since $\mathscr{R}(P_n)$, and hence also $\mathscr{R}(TP_n) = T\big(\mathscr{R}(P_n)\big)$, are finite-dimensional, it follows from 10.2(4) that $T \in \mathscr{K}\big(L^2(\Omega_2); L^2(\Omega_1)\big)$. □

We now discuss operators with **weakly singular integral kernels**, i.e. kernel functions $(x, y) \mapsto K(x, y)$ that for x fixed are locally integrable in y.

10.16 Schur integral operators. Let $\Omega \subset \mathbb{R}^n$ be open and bounded (!). Let $K : (\overline{\Omega} \times \overline{\Omega}) \setminus D \to \mathbb{K}$ be continuous, where $D := \{(x, x) \, ; \, x \in \overline{\Omega}\}$ is the diagonal of $\overline{\Omega} \times \overline{\Omega}$. Assume that

$$|K(x, y)| \le \frac{C}{|x - y|^\alpha} \quad \text{with } \alpha < n.$$

Then it holds that:

(1) The definition

$$(Tf)(x) := \int_\Omega K(x, y) f(y) \, \mathrm{d}y$$

yields a map $T \in \mathscr{K}\big(C^0(\overline{\Omega})\big)$.

(2) The composition of operators of Schur type is again a Schur operator. In particular, the iterated operators T^m are integral operators of the above type, with exponent

$$\alpha_m = \begin{cases} n - m(n-\alpha) & \text{if } 1 \le m < \frac{n}{n-\alpha}, \\ \varepsilon & \text{for every } \varepsilon > 0, \text{ if } m = \frac{n}{n-\alpha}, \\ 0 & \text{if } m > \frac{n}{n-\alpha}. \end{cases}$$

(3) If $1 \le p < \infty$ with $\alpha < \frac{n}{p'}$, then T is a Hilbert-Schmidt operator on $L^p(\Omega)$ and $T \in \mathscr{K}\left(L^p(\Omega); C^0(\overline{\Omega})\right)$.

Proof (1) *and* (3). We can always ensure that $\alpha < \frac{n}{p'}$, on choosing p sufficiently large. Moreover, the boundedness of Ω yields that the embedding from $C^0(\overline{\Omega})$ into $L^p(\Omega)$ is continuous for all p. Hence it follows from 10.3 that we only need to show the compactness of $T : L^p(\Omega) \to C^0(\overline{\Omega})$. We have that $Tf(x)$ exists for all x and

$$|Tf(x)| \le C \cdot \left(\int_\Omega \frac{dy}{|x-y|^{\alpha p'}} \right)^{\frac{1}{p'}} \|f\|_{L^p(\Omega)} .$$

Since $\alpha p' < n$ and Ω is bounded, the integral on the right-hand side is bounded uniformly in x. Hence the functions Tf with $\|f\|_{L^p(\Omega)} \le 1$ are uniformly bounded. It follows from the Arzelà-Ascoli theorem that it is sufficient to show that they are also equicontinuous, since then 10.1(2) is satisfied. It holds that

$$|Tf(x_1) - Tf(x_2)| \le \|f\|_{L^p(\Omega)} \cdot \left(\int_\Omega |K(x_1,y) - K(x_2,y)|^{p'} \, dy \right)^{\frac{1}{p'}}$$

and the integral on the right-hand side can be bounded for every $\delta > 0$ by

$$\le \int_{\Omega \setminus B_\delta(x_1)} |K(x_1,y) - K(x_2,y)|^{p'} \, dy$$
$$+ C \cdot \int_{B_\delta(x_1)} \left(\frac{1}{|y-x_1|^{\alpha p'}} + \frac{1}{|y-x_2|^{\alpha p'}} \right) dy .$$

For $|x_1 - x_2| \le \frac{\delta}{2}$ the first term is

$$\le C \sup \left\{ |K(x_1,y) - K(x_2,y)|^{p'} ; \ (x_1,y),(x_2,y) \notin B_{\frac{\delta}{4}}(D) \right\}$$
$$\longrightarrow 0 \quad \text{as } |x_1 - x_2| \to 0 \text{ and for every } \delta ,$$

thanks to the continuity of K away from the diagonal D, and the second term is

$$\le C \int_{B_{2\delta}(0)} \frac{dy}{|y|^{\alpha p'}} \le C\delta^{n-\alpha p'} \longrightarrow 0 \quad \text{as } \delta \to 0.$$

Here we assume the usual **convention on constants**, which states that constants that occur in a chain of inequalities may all be denoted by C, even though the constant will in general change after each step. In addition, this convention states that large positive constants are denoted by C, while small positive constants are denoted by c.

The bound above proves the equicontinuity of the functions Tf with $\|f\|_{L^p(\Omega)} \le 1$, and hence we have shown that $T \in \mathcal{K}\left(L^p(\Omega); C^0(\overline{\Omega})\right)$.   □

Proof (2). Now let T_1, T_2 be two such integral operators with kernels K_1, K_2 and exponents $\alpha_1 < n$ and $\alpha_2 < n$. By Fubini's theorem, for $f \in C^0(\overline{\Omega})$ we have that

$$T_1 T_2 f(x) = \int_\Omega K_1(x, z)\left(\int_\Omega K_2(z, y) f(y)\,dy\right) dz$$

$$= \int_\Omega \underbrace{\left(\int_\Omega K_1(x, z) K_2(z, y)\,dz\right)}_{=: \, K(x,y)} f(y)\,dy \, ,$$

if we can show that for each fixed x the function

$$y \longmapsto \tilde{K}(x, y) := \int_\Omega |K_1(x, z) K_2(z, y)|\,dz$$

is in $L^1(\Omega)$. To this end, we show that for $x \ne y$ (with the usual convention on constants)

$$|K(x, y)| \le \tilde{K}(x, y) \le C \int_\Omega \frac{dz}{|z - x|^{\alpha_1} |z - y|^{\alpha_2}}$$

$$\le \begin{cases} \dfrac{C}{|x - y|^{\alpha_1 + \alpha_2 - n}} & \text{if } \alpha_1 + \alpha_2 > n, \\[2ex] C_R \log \dfrac{R}{|x - y|} \le \dfrac{C_{R,\varepsilon}}{|x - y|^\varepsilon} & \text{if } \alpha_1 + \alpha_2 = n \\ & \text{for large } R \text{ and every } \varepsilon > 0, \\[2ex] C & \text{if } \alpha_1 + \alpha_2 < n, \end{cases}$$

where in the last case K is bounded. In order to prove these bounds, we replace z by $\frac{x+y}{2} - |x - y|z$ and set

$$e := \frac{x - y}{2|x - y|}, \qquad \Omega_{x,y} := \{z \in \mathbb{R}^n; \, \frac{x + y}{2} - |x - y|z \in \Omega\}.$$

Then

$$\int_\Omega \frac{dz}{|z - x|^{\alpha_1} |z - y|^{\alpha_2}} = |x - y|^{n - \alpha_1 - \alpha_2} \int_{\Omega_{x,y}} \frac{dz}{|z + e|^{\alpha_1} |z - e|^{\alpha_2}} \qquad (10\text{-}28)$$

and

$$\frac{1}{|z+e|^{\alpha_1}|z-e|^{\alpha_2}} \leq \begin{cases} 2^{\alpha_1}|z-e|^{-\alpha_2} & \text{for } |z-e| \leq \frac{1}{2}, \\ 2^{\alpha_2}|z+e|^{-\alpha_1} & \text{for } |z+e| \leq \frac{1}{2}, \\ \left(|z|-\frac{1}{2}\right)^{-\alpha_1-\alpha_2} & \text{for } |z| \geq 1, \\ 2^{\alpha_1+\alpha_2} & \text{otherwise.} \end{cases} \tag{10-29}$$

We distinguish between the three stated cases.

For $\alpha_1 + \alpha_2 > n$ it follows that

$$\int_{\Omega_{x,y}} \frac{dz}{|z+e|^{\alpha_1}|z-e|^{\alpha_2}} \leq \int_{\mathbb{R}^n} \frac{dz}{|z+e|^{\alpha_1}|z-e|^{\alpha_2}}. \tag{10-30}$$

Since $\alpha_1 < n$, $\alpha_2 < n$ and $\alpha_1 + \alpha_2 > n$, the integral on the right-hand side exists and its value is independent of e and depends only on n, α_1, α_2. To see this, let $e_1, e_2 \in \partial B_{\frac{1}{2}}(0)$ and choose a linear orthogonal transformation which maps e_1 to e_2. It follows from the transformation (change-of-variables) theorem that the integrals for e_1 and e_2 are equal. This proves that the last integral in (10-30) depends only on n, α_1, α_2.

For $\alpha_1 + \alpha_2 = n$ we choose a radius R with $\Omega \subset B_{\frac{R}{2}}(0)$. Then it follows that $|z-x|^{-\alpha_1} \leq C_{R,\varepsilon}|z-x|^{-\alpha_1-\varepsilon}$ for $z, x \in \Omega$ for every fixed $\varepsilon > 0$. Hence for ε sufficiently small we can apply the first case to $\alpha_1 + \varepsilon$ and α_2. This is the second estimate. It follows that $\Omega \subset B_R\left(\frac{x+y}{2}\right)$ for $x, y \in \Omega$, hence (10-28) implies

$$\int_\Omega \frac{dz}{|z-x|^{\alpha_1}|z-y|^{\alpha_2}} \leq \int_{B_{\frac{R}{|x-y|}}(0)} \frac{dz}{|z+e|^{\alpha_1}|z-e|^{\alpha_2}}$$

$$\leq C_R \cdot \left(1 + \int_{B_{\frac{R}{|x-y|}}(0)\setminus B_1(0)} \frac{dz}{|z|^{\alpha_1+\alpha_2}}\right)$$

$$\leq C_R \cdot \left(1 + \log \frac{R}{|x-y|}\right),$$

hence the desired first estimate.

For the case $\alpha_1 + \alpha_2 < n$ we decompose the integral over Ω into integrals over $B_\delta(x)$, $B_\delta(y)$ and $\Omega\setminus(B_\delta(x)\cup B_\delta(y))$, where $\delta := \frac{3}{4}|x-y|$. On noting that in the latter set it holds that $|z-x| \geq c\left|z-\frac{x+y}{2}\right|$ and $|z-y| \geq c\left|z-\frac{x+y}{2}\right|$ with a small constant c, we obtain that

$$\int_\Omega \frac{dz}{|z-x|^{\alpha_1}|z-y|^{\alpha_2}} \leq C\delta^{-\alpha_2}\int_{B_\delta(x)} \frac{dz}{|z-x|^{\alpha_1}}$$

$$+ C\delta^{-\alpha_1}\int_{B_\delta(y)} \frac{dz}{|z-y|^{\alpha_2}} + C\int_\Omega \frac{dz}{\left|z-\frac{x+y}{2}\right|^{\alpha_1+\alpha_2}}$$

$$\leq C \cdot \left(\delta^{n-\alpha_1-\alpha_2} + 1\right).$$

This ends the three cases.

It remains to show that K is continuous outside of the diagonal D. For $(x_2, y_2) \to (x_1, y_1)$ with $x_1 \neq y_1$ we have that

$$|K(x_2, y_2) - K(x_1, y_1)|$$
$$\leq C \int_\Omega \frac{|K_1(x_2, z) - K_1(x_1, z)|}{|z - y_2|^{\alpha_2}} \, dz + C \int_\Omega \frac{|K_2(z, y_2) - K_2(z, y_1)|}{|z - x_1|^{\alpha_1}} \, dz.$$

We decompose the first integral (the second integral can be bounded correspondingly) into the parts over $\Omega \setminus B_\delta(x_1)$ and $B_\delta(x_1)$. The former part is

$$\leq C \underbrace{\sup_{|z - x_1| \geq \delta} |K_1(x_2, z) - K_1(x_1, z)|}_{\to \, 0 \text{ as } x_2 \to x_1 \text{ for every } \delta} \cdot \underbrace{\int_\Omega \frac{dz}{|z - y_2|^{\alpha_2}}}_{\text{bounded in } y_2}.$$

Since $|z - y_2| \geq \frac{1}{2}|x_1 - y_1| > 0$ for $z \in B_\delta(x_1)$ if y_2 is close to $y_1 \neq x_1$ and if δ is sufficiently small, the second part is

$$\leq \frac{C}{|x_1 - y_1|^{\alpha_2}} \underbrace{\int_{B_\delta(x_1)} \left(\frac{1}{|z - x_2|^{\alpha_1}} + \frac{1}{|z - x_1|^{\alpha_1}} \right) dz}_{\leq \, C\delta^{n - \alpha_1} \to 0 \text{ as } \delta \to 0}.$$

In the case $\alpha_1 + \alpha_2 < n$ it holds that $K(x_2, y_2) \to K(x_1, y_1)$ even if $x_1 = y_1$, because the part of the integral over $\Omega \setminus B_\delta(x_1)$ converges to 0 as before, while the integral over $B_\delta(x_1)$ is

$$\leq C \int_{B_\delta(x_1)} \left(\frac{1}{|z - x_2|^{\alpha_1}} + \frac{1}{|z - x_1|^{\alpha_1}} \right) \frac{dz}{|z - y_2|^{\alpha_2}}$$
$$\longrightarrow C \int_{B_\delta(x_1)} \frac{2 \, dz}{|z - x_1|^{\alpha_1 + \alpha_2}} \qquad \text{as } x_2 \to x_1, y_2 \to y_1 = x_1$$
$$\leq C \, \delta^{n - \alpha_1 - \alpha_2} \longrightarrow 0 \qquad \text{as } \delta \to 0.$$

This proves the result on the composition of T_1 with T_2.   □

The fundamental solution

For integral kernels K as in 10.16 with $\alpha = n$ the induced T is no longer compact, and even the existence of the operator T is no longer guaranteed. That is because the function $y \mapsto |x - y|^{-n}$ is no longer integrable in a neighbourhood of x. However, such kernels play an essential role in the potential theoretic approach to partial differential equations, as we will see in 10.18.

10.17 Fundamental solution of the Laplace operator. For $x \in \mathbb{R}^n \setminus \{0\}$ let

$$F(x) := \begin{cases} \dfrac{1}{\sigma_n(n-2)}|x|^{2-n} & \text{for } n \geq 3, \\[2mm] \dfrac{1}{2\pi}\log\dfrac{1}{|x|} & \text{for } n = 2, \\[2mm] -\dfrac{1}{2}|x| & \text{for } n = 1, \end{cases}$$

where σ_n denotes the surface area of $\partial B_1(0) \subset \mathbb{R}^n$ ($\sigma_3 = 4\pi$, $\sigma_2 = 2\pi$, $\sigma_1 = 2$, $\sigma_n = n\kappa_n$, with κ_n the volume of $B_1(0) \subset \mathbb{R}^n$).

(1) It holds that $F \in C^\infty(\mathbb{R}^n \setminus \{0\})$ and

$$\partial_i F(x) = -\frac{1}{\sigma_n}\frac{x_i}{|x|^n} \ , \quad \partial_{ij}F(x) = -\frac{1}{\sigma_n|x|^n}\left(\delta_{i,j} - n\frac{x_i}{|x|}\frac{x_j}{|x|}\right) , \quad \Delta F = 0 \ .$$

(2) It holds that $F \in W_{\text{loc}}^{1,1}(\mathbb{R}^n)$ and with the notations as in 5.15 we have that

$$-\Delta[F] = -\sum_{i=1}^n \partial_i[\partial_i F] = [\delta_0] \quad \text{in } \mathscr{D}'(\mathbb{R}^n).$$

Note: F is the fundamental solution for $-\Delta$.

(3) If $f : \mathbb{R}^n \to \mathbb{R}$ is measurable and bounded with compact support, then

$$u(x) := \int_{\mathbb{R}^n} F(x-y)f(y)\,\mathrm{d}y = (F * f)(x)$$

defines a $u \in C^1(\mathbb{R}^n)$ which satisfies

$$-\Delta[u] = -\sum_{i=1}^n \partial_i[\partial_i u] = [f] \quad \text{in } \mathscr{D}'(\mathbb{R}^n),$$

i.e. u is a weak solution of the differential equation $-\Delta u = f$ in $\mathbb{R}^n$.

Proof (1). By direct calculation. □

Proof (2). We have that $F \in W^{1,1}(B_R(0) \setminus \{0\})$ for $R > 0$. Similarly to the end of the proof of 10.7(3) (or on recalling the corollary in A8.9) it then follows that $F \in W^{1,1}(B_R(0))$, where outside of the null set $\{0\}$ the weak derivatives coincide with the classical ones. Hence, $\partial_i[F] = [\partial_i F]$, which yields for $\zeta \in C_0^\infty(B_R(0))$ that as $\varepsilon \searrow 0$

$$\int_{\mathbb{R}^n}(-\Delta\zeta)F\,\mathrm{d}L^n = \int_{\mathbb{R}^n}\nabla\zeta \bullet \nabla F\,\mathrm{d}L^n \longleftarrow \int_{\mathbb{R}^n\setminus B_\varepsilon(0)}\nabla\zeta \bullet \nabla F\,\mathrm{d}L^n$$

$$= -\int_{\partial B_\varepsilon(0)}\zeta\nu_{B_\varepsilon(0)} \bullet \nabla F\,\mathrm{d}H^{n-1} = \frac{1}{\sigma_n}\int_{\partial B_1(0)}\zeta(\varepsilon y)\,\mathrm{d}H^{n-1}(y) \longrightarrow \zeta(0),$$

since $\Delta F = 0$ in $\mathbb{R}^n \setminus \{0\}$. □

Proof (3). Applying 10.16(3) for the kernel $(x, y) \mapsto F(x - y)$ shows that $u \in C^0(\mathbb{R}^n)$. For $\zeta \in C_0^\infty(\mathbb{R}^n)$ it follows, since $F \in W_{\mathrm{loc}}^{1,1}(\mathbb{R}^n)$, that

$$\int_{\mathbb{R}^n} (\partial_i \zeta u + \zeta v_i) \, \mathrm{dL}^n = 0 \quad \text{with} \quad v_i(x) := \int_{\mathbb{R}^n} \partial_i F(x - y) f(y) \, \mathrm{d}y \, .$$
(10-31)

By 10.16(3) it follows that $v_i \in C^0(\mathbb{R}^n)$, whence $u \in C^1(\mathbb{R}^n)$, with $\partial_i u = v_i$. Moreover, it follows from (2) that

$$\int_{\mathbb{R}^n} (-\Delta \zeta(x)) u(x) \, \mathrm{d}x = \int_{\mathbb{R}^n} \left(\int_{\mathbb{R}^n} (-\Delta \zeta(x + y)) F(x) \, \mathrm{d}x \right) f(y) \, \mathrm{d}y$$
$$= \int_{\mathbb{R}^n} \left(-\Delta[F] (\zeta(\cdot + y)) \right) f(y) \, \mathrm{d}y = \int_{\mathbb{R}^n} \zeta(y) f(y) \, \mathrm{d}y \, .$$

$\square$

10.18 Singular integral operators. For motivational purposes we continue the considerations in 10.17. We approximate $\partial_i u = v_i$ in (10-31) for $\varepsilon > 0$ by

$$v_i^\varepsilon(x) := \int_{\mathbb{R}^n \setminus B_\varepsilon(x)} \partial_i F(x - y) f(y) \, \mathrm{d}y \, .$$

If $f \in C_0^0(\mathbb{R}^n)$, then $v_i^\varepsilon \in C^1(\mathbb{R}^n)$, with

$$\partial_j v_i^\varepsilon(x) = \int_{\mathbb{R}^n \setminus B_\varepsilon(x)} \partial_{ji} F(x - y) f(y) \, \mathrm{d}y - w_{ji}^\varepsilon(x) \, , \quad \text{where}$$

$$w_{ji}^\varepsilon(x) := \int_{\partial B_\varepsilon(x)} \nu_{B_\varepsilon(x)}(y) \bullet e_j \partial_i F(x - y) f(y) \, \mathrm{dH}^{n-1}(y)$$

$$= \frac{1}{\sigma_n} \int_{\partial B_1(0)} y_j y_i \, f(x + \varepsilon y) \, \mathrm{dH}^{n-1}(y) \, .$$

We note that as $\varepsilon \searrow 0$

$$w_{ji}^\varepsilon(x) \longrightarrow \frac{1}{\sigma_n} \cdot \int_{\partial B_1(0)} y_j y_i \, \mathrm{dH}^{n-1}(y) \cdot f(x) = \frac{1}{n} \delta_{i,j} f(x) \, . \qquad (10\text{-}32)$$

Hence, if we want to show that u in 10.17(3) belongs to the space $C^2(\mathbb{R}^n)$, then we have to investigate whether the limit

$$(T_{ji} f)(x) := \lim_{\varepsilon \searrow 0} \int_{\mathbb{R}^n \setminus B_\varepsilon(x)} \partial_{ji} F(x - y) \, f(y) \, \mathrm{d}y$$

exists, and whether T_{ji} is well defined as a continuous operator on appropriate function spaces. On recalling the identity for the second derivatives $\partial_{ji} F(x - y)$ of the fundamental solution from 10.17(1), we note that the above kernel $(x, y) \mapsto K(x, y) := \partial_{ji} F(x - y)$ is a *singular integral kernel*, i.e. a kernel as in 10.16 but with $\alpha = n$. However, we recall from 10.17(1) that this kernel has the particular form

$$K(x,y) = \frac{\omega\left(\frac{x-y}{|x-y|}\right)}{|x-y|^n} \quad \text{with} \quad \omega(\xi) := -\frac{1}{\sigma_n}(\delta_{j,i} - n\xi_j\xi_i) \text{ for } |\xi| = 1,$$

where the mean value of $\omega : \partial B_1(0) \to \mathbb{R}$ vanishes (see (10-32)), i.e.

$$\int_{\partial B_1(0)} \omega(\xi)\, d\mathrm{H}^{n-1}(\xi) = 0. \tag{10-33}$$

Now we consider arbitrary kernels K of the above type with the property (10-33) and prove that for certain functions f the limit

$$(Tf)(x) := \lim_{\varepsilon \searrow 0} \int_{\mathbb{R}^n \backslash B_\varepsilon(x)} K(x,y)f(y)\, dy$$

exists. This limit is also referred to as the **Cauchy principal value** of $\int_{\mathbb{R}^n} K(x,y)f(y)\, dy$ at the point x (observe that $y \mapsto K(x,y)f(y)$ in general is not integrable!). Classes of functions on which T can still be shown to be a continuous operator include C^α-spaces (see 10.19) and L^p-spaces (see 10.20). In both cases T is not (!) a compact operator. For ease of presentation we also define

$$\omega(x) := \omega\left(\frac{x}{|x|}\right) \quad \text{for } x \in \mathbb{R}^n \backslash \{0\}. \tag{10-34}$$

10.19 Hölder-Korn-Lichtenstein inequality. Let $\omega : \mathbb{R}^n \backslash \{0\} \to \mathbb{R}$ be a Lipschitz continuous function on $\partial B_1(0)$ which satisfies (10-33) and (10-34). Then for $0 < \alpha < 1$ and $f \in C^{0,\alpha}(\overline{B_R(0)})$ with $f = 0$ on $\partial B_R(0)$ the limit

$$(Tf)(x) := \lim_{\varepsilon \searrow 0} \int_{B_R(0) \backslash B_\varepsilon(x)} \frac{\omega(x-y)}{|x-y|^n} f(y)\, dy$$

exists pointwise for $x \in \mathbb{R}^n$, and for all $\widetilde{R} > 0$ it holds that

$$\|Tf\|_{C^{0,\alpha}(\overline{B_{\widetilde{R}}(0)})} \leq C(n, R, \alpha) \cdot \|\omega\|_{C^{0,1}(\partial B_1(0))} \cdot \|f\|_{C^{0,\alpha}(\overline{B_R(0)})}.$$

Proof. We extend f by 0 on $\mathbb{R}^n \backslash B_R(0)$. As the mean value of ω is equal to 0, for $|x| \leq 2R$ we have

$$\int_{B_R(0) \backslash B_\varepsilon(x)} \frac{\omega(x-y)}{|x-y|^n} f(y)\, dy = \int_{B_{3R}(x) \backslash B_\varepsilon(x)} \frac{\omega(x-y)}{|x-y|^n} (f(y) - f(x))\, dy,$$

because a transformation to polar coordinates yields that

$$\int_{B_{3R}(x) \backslash B_\varepsilon(x)} \frac{\omega(x-y)}{|x-y|^n}\, dy = \int_\varepsilon^{3R} r^{n-1} \int_{\partial B_1(0)} \frac{\omega(\xi)}{r^n}\, d\mathrm{H}^{n-1}(\xi)\, dr = 0.$$

Noting that with a constant C depending on ω it holds that

$$\left| \frac{\omega(x-y)}{|x-y|^n} (f(y) - f(x)) \right| \leq C \cdot |x-y|^{\alpha-n} \|f\|_{C^{0,\alpha}},$$

we see that the integrand is integrable over $B_{3R}(x)$, and hence

$$Tf(x) = \int_{B_{3R}(x)} \frac{\omega(x-y)}{|x-y|^n} (f(y) - f(x))\, dy$$

and

$$|Tf(x)| \leq C \int_{B_{3R}(0)} |y|^{\alpha-n}\, dy \cdot \|f\|_{C^{0,\alpha}} = C(\omega, n, R, \alpha) \|f\|_{C^{0,\alpha}}.$$

For $|x| \geq 2R$,

$$|Tf(x)| \leq C \|f\|_{C^0} \int_{B_R(0)} \frac{dy}{|x-y|^n} \leq \frac{C(\omega, n, R)}{(|x|-R)^n} \|f\|_{C^0}.$$

Similarly, for $x_1, x_2 \in \mathbb{R}^n$ and $\varrho \geq R + \max(|x_1|, |x_2|)$,

$$Tf(x_1) - Tf(x_2)$$

$$= \int_{B_\varrho(x_1)} \frac{\omega(x_1-y)}{|x_1-y|^n} (f(y) - f(x_1))\, dy$$

$$- \int_{B_\varrho(x_2)} \frac{\omega(x_2-y)}{|x_2-y|^n} (f(y) - f(x_2))\, dy$$

$$= \int_{B_\varrho(x_1)} \left(\frac{\omega(x_1-y)}{|x_1-y|^n} (f(y) - f(x_1)) - \frac{\omega(x_2-y)}{|x_2-y|^n} (f(y) - f(x_2)) \right) dy$$

$$+ \int_{\mathbb{R}^n} \frac{\omega(x_2-y)}{|x_2-y|^n} (f(y) - f(x_2)) \left(\mathcal{X}_{B_\varrho(x_1)}(y) - \mathcal{X}_{B_\varrho(x_2)}(y) \right) dy.$$

The second integral can be bounded by

$$\leq C \|f\|_{C^0} \int_{\mathbb{R}^n} \left| \mathcal{X}_{B_\varrho(0)}(y) - \mathcal{X}_{B_\varrho(x_2-x_1)}(y) \right| \frac{dy}{|y|^n}$$

$$= C \|f\|_{C^0} \int_{\mathbb{R}^n} \left| \mathcal{X}_{B_1(0)}(\tilde{y}) - \mathcal{X}_{B_1(\frac{1}{\varrho}(x_2-x_1))}(\tilde{y}) \right| \frac{d\tilde{y}}{|\tilde{y}|^n}$$

(with the variable transformation $y = \varrho\tilde{y}$), which converges to 0 for every x_1 and x_2 as $\varrho \to \infty$. Setting $\delta := |x_2 - x_1|$, the first integral from above can be bounded on $B_{2\delta}(x_1)$ by (we employ the usual convention on constants)

$$\leq C \|f\|_{C^{0,\alpha}} \cdot \int_{B_{2\delta}(x_1)} \left(|y-x_1|^{\alpha-n} + |y-x_2|^{\alpha-n} \right) dy$$

$$\leq C \|f\|_{C^{0,\alpha}} \cdot \int_{B_{3\delta}(0)} |y|^{\alpha-n}\, dy \leq C \|f\|_{C^{0,\alpha}} \cdot \delta^\alpha.$$

On the remaining domain $B_\varrho(x_1) \setminus B_{2\delta}(x_1)$ we write the integrand as

$$\frac{\omega(x_1 - y)}{|x_1 - y|^n}(f(x_2) - f(x_1)) + \left(\frac{\omega(x_1 - y)}{|x_1 - y|^n} - \frac{\omega(x_2 - y)}{|x_2 - y|^n}\right)(f(y) - f(x_2)).$$

Recalling that the mean value of ω is equal to 0 yields that the integral of the first term vanishes. The Lipschitz continuity of ω implies that

$$|\omega(x_1 - y) - \omega(x_2 - y)| \le C \left|\frac{x_1 - y}{|x_1 - y|} - \frac{x_2 - y}{|x_2 - y|}\right|$$

$$= C\frac{||x_2 - y|(x_1 - y) - |x_1 - y|(x_2 - y)|}{|x_1 - y|\,|x_2 - y|} \le C\frac{|x_1 - x_2|}{|x_2 - y|},$$

and we have

$$\left|\frac{1}{|x_1 - y|^n} - \frac{1}{|x_2 - y|^n}\right|$$

$$\le \frac{|x_1 - x_2|}{|x_1 - y|^n\,|x_2 - y|^n}\sum_{i=0}^{n-1}|x_1 - y|^i\,|x_2 - y|^{n-1-i}$$

$$\le n|x_1 - x_2|\left(\frac{1}{|x_2 - y|\,|x_1 - y|^n} + \frac{1}{|x_1 - y|\,|x_2 - y|^n}\right).$$

Together this gives

$$\left|\frac{\omega(x_1 - y)}{|x_1 - y|^n} - \frac{\omega(x_2 - y)}{|x_2 - y|^n}\right|$$

$$\le C \cdot |x_1 - x_2|\left(\frac{1}{|x_2 - y|\,|x_1 - y|^n} + \frac{1}{|x_1 - y|\,|x_2 - y|^n}\right).$$

On noting that $\frac{1}{2}|x_1 - y| \le |x_2 - y| \le 2|x_1 - y|$ for $|y - x_1| \ge 2\delta$, it follows that the remaining integral over $B_\varrho(x_1) \setminus B_{2\delta}(x_1)$ is bounded uniformly in ϱ by

$$\le C\|f\|_{C^{0,\alpha}} \cdot \delta \int_{\mathbb{R}^n \setminus B_{2\delta}(x_1)} |x_1 - y|^{\alpha - n - 1}\,dy$$

$$\le C\|f\|_{C^{0,\alpha}} \cdot \delta \int_{2\delta}^\infty r^{\alpha - 2}\,dr \le C\|f\|_{C^{0,\alpha}} \cdot \delta^\alpha.$$

$\square$

10.20 Calderón-Zygmund inequality. Let $\omega : \mathbb{R}^n \setminus \{0\} \to \mathbb{R}$ on $\partial B_1(0)$ be measurable with respect to the measure H^{n-1} and bounded and such that it satisfies (10-33) and (10-34). Then for $f \in L^p(\mathbb{R}^n)$ with $1 < p < \infty$ and $0 < \varepsilon \le 1$ the integral

$$(T_\varepsilon f)(x) := \int_{\mathbb{R}^n \setminus B_\varepsilon(x)} \frac{\omega(x - y)}{|x - y|^n} f(y)\,dy$$

exists for almost all $x \in \mathbb{R}^n$. This defines operators $T_\varepsilon \in \mathscr{L}(L^p(\mathbb{R}^n))$ and for $f \in L^p(\mathbb{R}^n)$ there exists

$$Tf := \lim_{\varepsilon \searrow 0} T_\varepsilon f \quad \text{in } L^p(\mathbb{R}^n) \text{ with}$$

$$\|Tf\|_{L^p(\mathbb{R}^n)} \le C(n,p) \cdot \|\omega\|_{L^\infty(\partial B_1(0))} \cdot \|f\|_{L^p(\mathbb{R}^n)} \cdot$$

Proof. See Appendix A10.   □

Remark: For $n = 1$ we have that $\omega(-1) = -\omega(+1)$, hence up to a multiplicative constant $\omega(1) = 1$ and $\omega(-1) = -1$. Then

$$(Tf)(x) = \lim_{\varepsilon \searrow 0} \int_{\mathbb{R} \setminus]x-\varepsilon, x+\varepsilon[} \frac{f(y)}{x-y} \, dy$$

is called the **Hilbert transform** of f.

E10 Exercises

E10.1 Counterexample to embedding theorems. Show that theorem 10.6 in the case $k_1 > 0$ does not (!) hold for arbitrary open bounded sets $\Omega \subset \mathbb{R}^n$.

Solution. A characteristic counterexample is the following: Let $e \in \mathbb{R}^n$ with $|e| = 1$ and set

$$\Omega := \bigcup_{k \in \mathbb{N}} B_{r_k}(x_k) \quad \text{with } x_k = \frac{1}{k}e, \ r_k = \frac{1}{4k^2},$$

so that the closed balls $\overline{B_{r_k}(x_k)}$ are pairwise disjoint. Now if $(a_k)_{k \in \mathbb{N}}$ is a sequence that converges in $\mathbb{R}$ to a, then

$$u(x) := \begin{cases} a_k & \text{for } |x - x_k| \le r_k, \ k \in \mathbb{N}, \\ a & \text{for } x = 0, \end{cases}$$

defines a $u \in C^0(\overline{\Omega})$. Since $\nabla u = 0$ in Ω it follows that also $u \in C^1(\overline{\Omega})$ (see definition 3.6). Note that for $0 < \alpha \le 1$

$$\sup_{x \in \overline{\Omega}, \ x \ne 0} \frac{|u(x) - u(0)|}{|x|^\alpha} \ge \sup_k \left(\left(\frac{k}{2}\right)^\alpha |a_k - a| \right),$$

and $a_k = a + (1 + \log k)^{-1}$ yields that u lies in none of the spaces $C^{0,\alpha}(\overline{\Omega})$. Hence the embedding in 10.6 for $(k_1, \alpha_1) = (1, 0)$ and $(k_2, \alpha_2) = (0, \alpha)$ does not even exist for the above Ω.   □

E10.2 Ehrling's lemma. Let X, Y, Z be Banach spaces. Assume $K \in \mathscr{K}(X;Y)$ and let $T \in \mathscr{L}(Y;Z)$ be injective. Then for every $\varepsilon > 0$ there exists a $C_\varepsilon < \infty$, such that for all $x \in X$

$$\|Kx\|_Y \leq \varepsilon\|x\|_X + C_\varepsilon\|TKx\|_Z.$$

Solution. Otherwise for an $\varepsilon > 0$ there exist points $\widetilde{x}_n \in X$ with

$$\|K\widetilde{x}_n\|_Y > \varepsilon\|\widetilde{x}_n\|_X + n\|TK\widetilde{x}_n\|_Z.$$

Then $x_n := \frac{\widetilde{x}_n}{\|\widetilde{x}_n\|_X}$ are bounded in X and

$$\|Kx_n\|_Y > \varepsilon + n\|TKx_n\|_Z. \tag{E10-1}$$

Since K is compact, there exists a subsequence (which we again denote by $(x_n)_{n \in \mathbb{N}}$) such that $Kx_n \to y \in Y$ as $n \to \infty$, and so

$$\|Ty\|_Z \longleftarrow \|TKx_n\|_Z \leq \frac{1}{n}\|Kx_n\|_Y \longrightarrow 0.$$

As T is injective, it follows that $y = 0$ and hence $\|Kx_n\|_Y \to 0$, which contradicts (E10-1). $\qquad\square$

E10.3 Application of Ehrling's lemma. Let $\Omega \subset \mathbb{R}^n$ be open and bounded, let $1 < p < \infty$ and let $m \geq 2$. Show that:

(1) For every $\varepsilon > 0$ there exists a constant C_ε such that for all $u \in W_0^{m,p}(\Omega)$

$$\|u\|_{W_0^{m-1,p}(\Omega)} \leq \varepsilon\|u\|_{W_0^{m,p}(\Omega)} + C_\varepsilon\|u\|_{L^p(\Omega)}.$$

(2) An equivalent norm on $W_0^{m,p}(\Omega)$ is given by

$$\|u\| := \|D^m u\|_{L^p(\Omega)} + \|u\|_{L^p(\Omega)}.$$

Solution (1). This follows from Ehrling's lemma, on noting that the embedding from $W_0^{m,p}(\Omega)$ into $W_0^{m-1,p}(\Omega)$ is compact (either on recalling 8.11(3), 8.10, Rellich's embedding theorem A8.1 and 10.1(4), or on recalling Sobolev's embedding theorem 10.9). $\qquad\square$

Solution (2). We have from (1) that

$$\|u\|_{W^{m-1,p}} \leq \varepsilon\|u\|_{W^{m,p}} + C_\varepsilon\|u\|_{L^p}$$
$$\leq \varepsilon\|D^m u\|_{L^p} + \varepsilon\|u\|_{W^{m-1,p}} + C_\varepsilon\|u\|_{L^p},$$

which for $\varepsilon \leq \frac{1}{2}$ yields the bound

$$\|u\|_{W^{m-1,p}} \leq 2\varepsilon\|D^m u\|_{L^p} + 2C_\varepsilon\|u\|_{L^p}.$$

Consequently,

$$\|u\| \leq \|u\|_{W^{m,p}} \leq \max(1 + 2\varepsilon, 2C_\varepsilon) \cdot \|u\|.$$

$\qquad\square$

E10.4 On Ehrling's lemma. Let $\Omega = B_R(0) \subset \mathbb{R}^n$. Show that: For $\varepsilon > 0$ there exists a constant C_ε such that for all $u \in C^2(\overline{\Omega})$

$$\|\nabla u\|_{C^0(\overline{\Omega})} \le \varepsilon \|D^2 u\|_{C^0(\overline{\Omega})} + C_\varepsilon \|u\|_{C^0(\overline{\Omega})},$$

and obtain an explicit bound for the constant C_ε.

Solution. First let $R = 1$ and $\varepsilon \le 1$. For $x_0 \in \Omega$ with $\nabla u(x_0) \ne 0$ we choose $y_0, y_1 \in \overline{\Omega} \cap \overline{B_\varepsilon(x_0)}$ such that $y_1 - y_0$ points in the direction of $\nabla u(x_0)$ and $|y_1 - y_0| \ge \frac{\varepsilon}{2}$.
Remark: This is possible because $\Omega = B_1(0)$. If $B_\varepsilon(x_0) \subset \Omega$, then we can choose $y_0 = x_0$ and $y_1 = x_0 + \varepsilon \frac{\nabla u(x_0)}{|\nabla u(x_0)|}$.
   Then, setting $y_t := (1 - t) y_0 + t y_1$, it holds that

$$
\begin{aligned}
u(y_1) - u(y_0) &= \int_0^1 \nabla u(y_t) \cdot (y_1 - y_0)\, dt \\
&= \nabla u(x_0) \cdot (y_1 - y_0) \\
&\quad + \int_0^1 \int_0^1 \sum_{i,j=1}^n \partial_{ij} u\big((1-s)x_0 + s y_t\big)(y_t - x_0)_i (y_1 - y_0)_j\, ds\, dt
\end{aligned}
$$

and

$$\nabla u(x_0) \cdot (y_1 - y_0) = |\nabla u(x_0)|\,|y_1 - y_0|.$$

It follows that

$$
\begin{aligned}
|\nabla u(x_0)| &\le \|D^2 u\|_{C^0(\overline{\Omega})} \cdot \sup_{0 \le t \le 1} |y_t - x_0| + \frac{|u(y_1) - u(y_0)|}{|y_1 - y_0|} \\
&\le \varepsilon \|D^2 u\|_{C^0(\overline{\Omega})} + \frac{4}{\varepsilon}\|u\|_{C^0(\overline{\Omega})},
\end{aligned}
$$

and hence the desired bound with $C_\varepsilon = \frac{4}{\varepsilon}$. (For $\varepsilon \ge 1$ the claim follows with $C_\varepsilon = 4$.) If R is arbitrary, then define

$$v(x) := u\big(\tfrac{x}{R}\big).$$

The established bound for v

$$\|\nabla v\|_{C^0(\overline{B_1(0)})} \le \varepsilon \|D^2 v\|_{C^0(\overline{B_1(0)})} + \frac{4}{\min(\varepsilon, 1)}\|v\|_{C^0(\overline{B_1(0)})}$$

transforms to

$$\|\nabla u\|_{C^0(\overline{B_R(0)})} \le \frac{\varepsilon}{R}\|D^2 u\|_{C^0(\overline{B_R(0)})} + \frac{4R}{\min(\varepsilon, 1)}\|u\|_{C^0(\overline{B_R(0)})}.$$

Now replace ε by $R\varepsilon$ and set $C_\varepsilon = 4\big(\min(\varepsilon, \tfrac{1}{R})\big)^{-1}$.   □

E10.5 An a priori estimate. Let $u \in C^2([0,1])$ be a solution of the linear differential equation

$$au'' + bu' + du = 0 \quad \text{in }]0,1[,$$

where $a, b, d \in C^0([0,1])$ and $a \geq c_0$ with a positive constant c_0. Then there exists a constant C, which depends only on the coefficients, such that

$$\|u\|_{C^2} \leq C \cdot \|u\|_{C^0}.$$

Solution. The differential equation implies that

$$c_0 \|u''\|_{C^0} \leq C(\|u'\|_{C^0} + \|u\|_{C^0}) \quad \text{with } C := \|b\|_{C^0} + \|d\|_{C^0},$$

and so

$$c_0(\|u''\|_{C^0} + \|u'\|_{C^0}) \leq (C + c_0)(\|u'\|_{C^0} + \|u\|_{C^0}).$$

It follows from E10.4 that this can be bounded by

$$\leq (C + c_0)\varepsilon \|u''\|_{C^0} + (C + c_0) \cdot (C_\varepsilon + 1)\|u\|_{C^0}.$$

On choosing ε with $(C + c_0)\varepsilon = \frac{c_0}{2}$, we obtain, with a new constant C, that

$$\|u''\|_{C^0} + \|u'\|_{C^0} \leq C\|u\|_{C^0}.$$

$\square$

E10.6 Equivalent norm. Let $\Omega \subset \mathbb{R}^n$ be open and bounded with Lipschitz boundary and let $m \geq 2$. Then an equivalent norm on $C^m(\overline{\Omega})$ is given by

$$\|u\| := \|D^m u\|_{C^0(\overline{\Omega})} + \|u\|_{C^0(\overline{\Omega})}.$$

E10.7 Counterexample to embedding theorems. Let Ω be as in theorem 10.9 and let

$$1 - \tfrac{n}{p} = 0.$$

Then $W^{1,p}(\Omega)$ is not (!) embedded in $L^\infty(\Omega)$, except in the case $n = 1$. *Note:* In theorem 10.9 the case $m_1 = m_2 + 1$, $p_2 = \infty$ is not allowed, while theorem 10.8 does not permit $q = \infty$.

Solution. The case $n = 1$ (we then have $p = 1$) was solved in E3.6. For $n \geq 2$ a counterexample is

$$u(x) := \log|\log|x|| \quad \text{for } 0 < |x| < \tfrac{1}{2}.$$

Let $\Omega := B_{\frac{1}{2}}(0)$. Then $u \in L^s(\Omega)$ for $1 \leq s < \infty$, but u is not bounded. Moreover, $u \in W^{1,n}(\Omega \setminus \{0\})$, because $u \in C^\infty(\Omega \setminus \{0\})$, with

$$\int_\Omega |\nabla u|^n \, \mathrm{dL}^n = \int_\Omega \frac{\mathrm{d}x}{\left(|x||\log|x||\right)^n} = C(n) \int_0^{\frac{1}{2}} \frac{\mathrm{d}r}{r|\log r|^n}$$

$$= \tilde{C}(n) \left[\frac{1}{|\log r|^{n-1}} \right]_{r=0}^{r=\frac{1}{2}} < \infty.$$

It follows that $u \in W^{1,n}(\Omega)$, similarly to the end of the proof of 10.7(3) (or alternatively by using the corollary in A8.9). $\qquad\square$

E10.8 Sobolev spaces on $\mathrm{I\!R}^n$. For $m \geq 1$ and $1 \leq p < \infty$,

$$W^{m,p}(\mathrm{I\!R}^n) = W_0^{m,p}(\mathrm{I\!R}^n).$$

Proof. Recalling that $C^\infty(\mathrm{I\!R}^n) \cap W^{m,p}(\mathrm{I\!R}^n)$ is dense in $W^{m,p}(\mathrm{I\!R}^n)$ (see 4.24), it is sufficient to approximate functions $u \in C^\infty(\mathrm{I\!R}^n) \cap W^{m,p}(\mathrm{I\!R}^n)$ in the $W^{m,p}$-norm by functions in $C_0^\infty(\mathrm{I\!R}^n)$. To this end, choose a function $\eta \in C^\infty(\mathrm{I\!R}^n)$ with

$$\eta(x) = \begin{cases} 1 & \text{for } |x| \leq 1, \\ 0 & \text{for } |x| \geq 2 \end{cases}$$

(see 4.19), and define $\eta_R(x) := \eta(\frac{x}{R})$. Then for all multi-indices s with $|s| \leq m$,

$$\partial^s(u - \eta_R u) = (1 - \eta_R)\partial^s u - \sum_{\substack{0 \leq r \leq s \\ r \neq s}} \binom{s}{r}(\partial^{s-r}\eta_R)\partial^r u.$$

Noting that $1 - \eta_R = 0$ on $B_R(0)$ and that $|\partial^{s-r}\eta_R| \leq C R^{-|s-r|}$ in $\mathrm{I\!R}^n$ yields that

$$\|\partial^s(u - \eta_R u)\|_{L^p(\mathrm{I\!R}^n)}$$
$$\leq \|\partial^s u\|_{L^p(\mathrm{I\!R}^n \setminus B_R(0))} + C \sum_{\substack{0 \leq r \leq s \\ r \neq s}} R^{-|s-r|}\|\partial^r u\|_{L^p(\mathrm{I\!R}^n)},$$

which converges to 0 as $R \to \infty$. $\qquad\square$

E10.9 Embedding theorem. Let $m_1, m_2 \geq 0$ and $1 \leq p_1, p_2 < \infty$ with

$$m_1 - \frac{n}{p_1} = m_2 - \frac{n}{p_2}, \quad \text{where } m_1 \geq m_2.$$

Then the embedding $\mathrm{Id} : W^{m_1,p_1}(\mathrm{I\!R}^n) \to W^{m_2,p_2}(\mathrm{I\!R}^n)$ exists and is continuous.

Observe: In theorem 10.9 this result was shown for bounded open sets $\Omega \subset \mathrm{I\!R}^n$ with Lipschitz boundary. (Theorem 10.9 also holds for an inequality between the Sobolev numbers.) Here we prove the theorem for $\Omega = \mathrm{I\!R}^n$, where it is essential that the two Sobolev numbers are equal, which is also the case in theorem 10.8.

Solution. For $m_1 = m_2$ the result is trivial. For $m_1 = m_2 + 1$ let $u \in W^{m_1, p_1}(\mathbb{R}^n)$. Then $\partial^s u$ is in $W^{1, p_1}(\mathbb{R}^n)$ for all multi-indices s with $|s| \leq m_1 - 1 = m_2$. Sobolev's theorem 10.8 then yields that $\partial^s u \in L^{p_2}(\mathbb{R}^n)$ with

$$\|\partial^s u\|_{L^{p_2}(\mathbb{R}^n)} \leq C(p_2, n)\|\nabla \partial^s u\|_{L^{p_1}(\mathbb{R}^n)} \leq C(p_2, n)\|u\|_{W^{m_1, p_1}(\Omega)}.$$

For $m_1 = m_2 + k$ with $k \geq 2$ define $\widetilde{m}_i$ and $\widetilde{p}_i$ for $i = 0, \ldots, k$ by

$$\widetilde{m}_i := m_2 + i, \quad \widetilde{m}_i - \frac{n}{\widetilde{p}_i} = m_2 - \frac{n}{p_2}, \quad \text{i.e. } \frac{1}{\widetilde{p}_i} = \frac{i}{n} + \frac{1}{p_2}.$$

Then $\widetilde{p}_0 = p_2$ and $\widetilde{p}_i$ is monotonically decreasing in i with $\widetilde{p}_k = p_1$, and hence $1 \leq \widetilde{p}_i < \infty$ for $i = 0, \ldots, k$. The desired result now follows from successive applications of theorem 10.8. $\qquad\square$

E10.10 Poincaré inequalities. Let $1 \leq p, q < \infty$ with $1 - \frac{n}{p} = -\frac{n}{q}$ and let $u \in W^{1,p}(\mathbb{R}^n)$. Then

$$\|u\|_{L^r(\mathbb{R}^n)} \leq C(n, p) \mathrm{L}^n(\{u \neq 0\})^{\frac{1}{r} - \frac{1}{q}} \cdot \|\nabla u\|_{L^p(\mathbb{R}^n)}$$

for $1 \leq r < q$, and

$$\|u\|_{L^p(\mathbb{R}^n)} \leq C(n, p) \mathrm{L}^n(\{u \neq 0\})^{\frac{1}{n}} \cdot \|\nabla u\|_{L^p(\mathbb{R}^n)}.$$

Solution. If $\{u \neq 0\} := \{x \in \mathbb{R}^n \,;\, u(x) \neq 0\}$ has finite Lebesgue measure then it follows from the Hölder inequality for $1 \leq r < q$ that

$$\int_{\mathbb{R}^n} |u|^r \, d\mathrm{L}^n = \int_{\mathbb{R}^n} \mathcal{X}_{\{u \neq 0\}} \cdot |u|^r \, d\mathrm{L}^n \leq \mathrm{L}^n(\{u \neq 0\})^{1 - \frac{r}{q}} \left(\int_{\mathbb{R}^n} |u|^q \, d\mathrm{L}^n \right)^{\frac{r}{q}},$$

and so 10.8 yields the first inequality. Setting $r = p$, and noting that $\frac{1}{p} - \frac{1}{q} = \frac{1}{n}$, we obtain the second inequality. $\qquad\square$

E10.11 Convergence in L^p-spaces. Let $1 \leq p_0 < p_1 < \infty$, and suppose $u_k \in L^{p_0}(\mu) \cap L^{p_1}(\mu)$ for $k \in \mathbb{N}$ and $u \in L^{p_0}(\mu)$. Then it holds for $p_0 \leq p < p_1$ that

$\{u_k \,;\, k \in \mathbb{N}\}$ bounded in $L^{p_1}(\mu)$,	$u_k, u \in L^p(\mu)$,
$u_k \to u$ strongly in $L^{p_0}(\mu)$ $\implies$	$u_k \to u$ strongly in $L^p(\mu)$
as $k \to \infty$	as $k \to \infty$.

Solution. We have for all $\varepsilon > 0$ the elementary inequality

$$a^p \leq \varepsilon \, a^{p_1} + C_\varepsilon a^{p_0} \quad \text{for all } a \geq 0,$$

where C_ε is a constant depending on ε, p, p_1, p_0. It follows that

$$\int_\Omega |u_k - u_l|^p \, d\mu \leq \varepsilon \underbrace{\int_\Omega |u_k - u_l|^{p_1} \, d\mu}_{\text{bounded in } k, l} + C_\varepsilon \underbrace{\int_\Omega |u_k - u_l|^{p_0} \, d\mu}_{\to\, 0 \text{ as } k, l \to \infty},$$

which implies that $\{u_k; \; k \in \mathbb{N}\}$ is a Cauchy sequence in $L^p(\mu)$ as well. Hence there exists a $\tilde{u} \in L^p(\mu)$ with $u_k \to \tilde{u}$ in $L^p(\mu)$. It follows for a subsequence $k \to \infty$ that $u_k \to u$ and $u_k \to \tilde{u}$ μ-almost everywhere, and so $u = \tilde{u}$ in $L^p(\mu)$. $\qquad\square$

E10.12 Compact sets in c_0. Let c_0 be the space of null sequences, equipped with the supremum norm $\|\cdot\|_{\sup}$.

(1) Show that $M \subset c_0$ is precompact if and only if M is bounded and for every $\varepsilon > 0$ there exists an index n_ε such that $|x_n| \leq \varepsilon$ for all $n \geq n_\varepsilon$ and all $x \in M$.

(2) Let $F : c_0 \to c_0$ be defined by $F(x) = \{x_i^3; \; i \in \mathbb{N}\}$. Prove that $F(\mathrm{B}_1(0))$ is not precompact, but $DF(x)(\mathrm{B}_1(0))$ is for every $x \in c_0$.

E10.13 Nuclear operators. Let X, Y be Banach spaces and let $T : X \to Y$ be *nuclear*, i.e. there exist $\lambda_k \in \mathbb{K}$, $x_k' \in X'$, $y_k \in Y$ for $k \in \mathbb{N}$ with

$$\sum_{k=1}^\infty |\lambda_k| < \infty, \quad \|x_k'\|_{X'} = 1, \quad \|y_k\|_Y = 1,$$

such that

$$Tx = \sum_{k=1}^\infty \lambda_k \, \langle x, \, x_k' \rangle_X \, y_k \quad \text{for all } x \in X.$$

Then T is compact.

Solution. The operators

$$T_n x := \sum_{k=1}^n \lambda_k \, \langle x, \, x_k' \rangle_X \, y_k$$

are compact on recalling 10.2(3). Moreover,

$$\|(T - T_n)x\| \leq \left(\sum_{k=n+1}^\infty |\lambda_k| \right) \|x\|,$$

and so $T_n \to T$ in $\mathscr{L}(X; Y)$. Hence 10.2(2) yields that T is compact. $\qquad\square$

E10.14 Compact operator without eigenvalues. Setting

$$Tx := \sum_{k=1}^\infty \frac{x_k}{k+1} e_{k+1} \quad \text{for } x = (x_k)_{k \in \mathbb{N}}$$

defines an operator $T : \ell^2(\mathbb{C}) \to \ell^2(\mathbb{C})$. Show that T is compact, but that T has no eigenvalues (see 11.2(2)).

Solution. Noting that

$$T\big(\mathrm{B}_1(0)\big) \subset \big\{ x \in \ell^2(\mathbb{C}) \; ; \; |x_i| \le \tfrac{1}{i} \text{ for all } i \big\}$$

and recalling E4.13, we have that T is compact. If we assume that $\lambda \in \mathbb{C}$ is an eigenvalue, then $Tx = \lambda x$ for an $x \ne 0$. If $\lambda = 0$, then $Tx = 0$, and so $x = 0$, a contradiction. If $\lambda \ne 0$, it follows that $x_1 = 0$ and $x_{k+1} = \frac{1}{\lambda(k+1)}x_k$ for $k \ge 1$, and so again $x = 0$, a contradiction.   □

E10.15 Bound on the dimension of eigenspaces. Let $\Omega \subset \mathbb{R}^n$, let $K \in L^2(\Omega \times \Omega; \mathbb{C})$ and let $T \in \mathscr{L}\big(L^2(\Omega; \mathbb{C})\big)$ be the Hilbert-Schmidt integral operator defined by

$$(Tf)(x) := \int_\Omega K(x,y)f(y)\,\mathrm{d}y.$$

Show that

$$\dim \mathscr{N}(\mathrm{Id} - T) \le \|K\|^2_{L^2(\Omega \times \Omega)}.$$

Solution. By 10.15, $T \in \mathscr{K}\big(L^2(\Omega; \mathbb{C})\big)$. This implies, on noting that $(\mathrm{Id} - T)(x) = 0$ is equivalent to $x = Tx \in \mathscr{R}(T)$, that $\mathscr{N}(\mathrm{Id} - T) \cap \mathrm{B}_1(0) \subset T\big(\mathrm{B}_1(0)\big)$ is precompact, and hence, by 4.10, that $\mathscr{N}(\mathrm{Id} - T)$ is finite-dimensional. Choose an orthonormal system $f_1, \dots, f_n$ in $\mathscr{N}(\mathrm{Id} - T)$, where $n := \dim \mathscr{N}(\mathrm{Id} - T)$. Then

$$n = \sum_{i=1}^n \|f_i\|^2_{L^2(\Omega)} = \sum_{i=1}^n \|Tf_i\|^2_{L^2(\Omega)} = \int_\Omega \sum_{i=1}^n \left| \int_\Omega K(x,y)f_i(y)\,\mathrm{d}y \right|^2 \mathrm{d}x.$$

Setting $K_x(y) := \overline{K(x,y)}$ and using Bessel's inequality 9.6 we obtain that

$$n = \int_\Omega \sum_{i=1}^n (K_x, f_i)^2_{L^2(\Omega)}\,\mathrm{d}x \le \int_\Omega \|K_x\|^2_{L^2(\Omega)}\,\mathrm{d}x = \|K\|^2_{L^2(\Omega \times \Omega)}.$$

□

E10.16 Norm of Hilbert-Schmidt operators. Under the same assumptions as in E10.15 show that

$$\|T\| = \|K\|_{L^2(\Omega \times \Omega)} \quad \Longleftrightarrow \quad \begin{array}{l} \text{There exist } K_1, K_2 \in L^2(\Omega) \text{ with} \\ K(x,y) = K_1(x)K_2(y) \text{ for almost all } x, y \in \Omega. \end{array}$$

Remark: In this case T is a nuclear operator as in E10.13, with only a single term in the sum.

Solution $\Rightarrow$. Let $K \ne 0$. The assumption yields that for $\varepsilon > 0$ there exist functions $f_\varepsilon \in L^2(\Omega)$ with $\|f_\varepsilon\|_{L^2(\Omega)} = 1$ such that

$$(1 - \varepsilon)\|K\|^2_{L^2(\Omega \times \Omega)} \leq \|Tf_\varepsilon\|^2_{L^2(\Omega)}$$

$$= \int_\Omega \left(\int_\Omega K(z,x)f_\varepsilon(x)\,dx \right)\left(\int_\Omega \overline{K(z,y)}\,\overline{f_\varepsilon(y)}\,dy \right)dz$$

$$= \int_\Omega \int_\Omega f_\varepsilon(x)\overline{f_\varepsilon(y)} \left(\int_\Omega K(z,x)\overline{K(z,y)}\,dz \right)dx\,dy$$

$$\leq \underbrace{\left(\int_\Omega \int_\Omega |f_\varepsilon(x)|^2 |f_\varepsilon(y)|^2\,dx\,dy \right)^{\frac{1}{2}}}_{=1}$$

$$\cdot \left(\int_\Omega \int_\Omega \left| \int_\Omega K(z,x)\overline{K(z,y)}\,dz \right|^2 dx\,dy \right)^{\frac{1}{2}}.$$

Letting $\varepsilon \to 0$ we obtain the inequality

$$\int_\Omega \int_\Omega |K(x,y)|^2\,dx\,dy \leq \left(\int_\Omega \int_\Omega \left| \int_\Omega K(z,x)\overline{K(z,y)}\,dz \right|^2 dx\,dy \right)^{\frac{1}{2}}.$$
(E10-2)

Moreover, the Cauchy-Schwarz inequality yields that for almost all $x, y \in \Omega$ we have that

$$\left| \int_\Omega K(z,x)\overline{K(z,y)}\,dz \right|^2 \leq \int_\Omega |K(z,x)|^2\,dz \cdot \int_\Omega |K(z,y)|^2\,dz. \qquad \text{(E10-3)}$$

Integrating over x and y, we obtain the opposite inequality (E10-2). This implies that in fact equality holds in (E10-2), and therefore for almost all $(x,y) \in \Omega \times \Omega$ also equality holds in (E10-3). On recalling the remark in 2.3(3), the functions $K_x(z) := K(z,x)$ and $K_y(z) := K(z,y)$ are linearly dependent in $L^2(\Omega)$ for almost all $(x,y) \in \Omega \times \Omega$. In other words (see A6.9), there exists a null set $N_0 \subset \Omega$ such that for all $x \in \Omega \setminus N_0$ it holds that: for almost all $y \in \Omega$ the functions K_x and K_y are linearly dependent. Since we assumed that $K \neq 0$ in $L^2(\Omega \times \Omega)$, we can choose $x_0 \in \Omega \setminus N_0$ such that $K_{x_0} \neq 0$ in $L^2(\Omega)$. Then there exists a null set $N \subset \Omega$ such that for $y \in \Omega \setminus N$ the function K_y is a multiple of K_{x_0}, i.e. there exists a function $\alpha : \Omega \setminus N \to \mathbb{C}$ such that for $y \in \Omega \setminus N$

$$K(z,y) = \alpha(y)K(z,x_0) \qquad \text{for almost all } z \in \Omega.$$

Setting $K_1(z) := K(z,x_0)$ and $K_2(y) := \alpha(y)$, it follows that

$$K(z,y) = K_1(z)K_2(y) \qquad \text{for almost all } (z,y) \in \Omega \times \Omega.$$

Fubini's theorem then yields that $K_1, K_2 \in L^2(\Omega)$. $\qquad\qquad \square$

A10 Calderón-Zygmund inequality

We present a proof of the L^p-estimate in 10.20. To this end, we begin with the following

A10.1 Definition. Let $D \subset \mathbb{C}$ be open and let $f : D \to Y$ be (real) continuously differentiable, where Y is a Banach space over $\mathbb{C}$. Then we define

$$\partial_{\bar{z}} f := \frac{1}{2}(\partial_x f + i \partial_y f) \quad \text{and} \quad \partial_z f := \frac{1}{2}(\partial_x f - i \partial_y f),$$

where we denote complex numbers by $z = x + iy$, $x, y \in \mathbb{R}$.

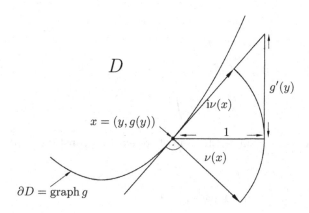

Fig. 10.2. *Outer normal and oriented tangent in* $\mathbb{C}$

Now let $D \subset \mathbb{C}$ be open and bounded with Lipschitz boundary (see A8.2). For functions $f \in C^0(\overline{D}; Y)$ we define the ***oriented boundary integral***

$$\int_{\partial D} f(z)\, dz := i \int_{\partial D} \nu(x) f(x)\, dH^1(x),$$

where $\nu : \partial D \to \mathbb{C}$ is the outer normal to D (see A8.5(3) and Fig. 10.2) and $\nu(x)f(x)$ denotes the complex product of $\nu(x)$ and $f(x)$. Then ***Cauchy's integral theorem*** states that for $f \in C^1(\overline{D}; Y)$

$$\int_{\partial D} f(z)\, dz = 2i \int_D \partial_{\bar{z}} f(z)\, dL^2(z).$$

In the special case where $\partial_{\bar{z}} f = 0$ in D, the function f is called ***holomorphic*** in D.

Proof. Let $y' \in Y'$ and set $g(z) := \langle f(z),\, y' \rangle_Y$. Then (see 5.11)

$$\left\langle \int_{\partial D} f(z) \, dz \, , \, y' \right\rangle_Y = \int_{\partial D} g(z) \, dz = i \int_{\partial D} (\operatorname{Re} g) \nu \, dH^1 - \int_{\partial D} (\operatorname{Im} g) \nu \, dH^1 .$$

It follows from Gauß's theorem (see A8.8) that this is

$$= \int_D \left(i\nabla(\operatorname{Re} g) - \nabla(\operatorname{Im} g) \right) dL^2$$

$$= \int_D \left(i(\partial_x + i\partial_y)\frac{g + \overline{g}}{2} - (\partial_x + i\partial_y)\frac{g - \overline{g}}{2i} \right) dL^2$$

$$= 2i \int_D \partial_{\overline{z}} g \, dL^2 = \left\langle 2i \int_D \partial_{\overline{z}} f \, dL^2 \, , \, y' \right\rangle_Y .$$

$\square$

First we consider the case $n = 1$ in Theorem 10.20.

A10.2 Theorem. If $f \in C_0^\infty(\mathbb{R})$ and $1 < p < \infty$, then

$$T_1 f(x) := \int_{\mathbb{R} \setminus B_1(x)} \frac{f(s)}{x - s} \, ds$$

defines a function $T_1 f$ in $L^p(\mathbb{R})$ and there exists a constant $C(p)$ such that for all f

$$\|T_1 f\|_{L^p(\mathbb{R})} \le C(p) \|f\|_{L^p(\mathbb{R})} .$$

Therefore 10.20 holds in the case $n = 1$.

Proof. As $f \in C_0^0(\mathbb{R})$ we have that $|T_1 f(x)| \le \frac{C}{|x|}$ for large x, and so $T_1 f \in L^p(\mathbb{R})$. In addition, the representation

$$T_1 f(x) = \int_{\mathbb{R} \setminus B_1(0)} \frac{f(x - s)}{s} \, ds$$

shows that $T_1 f \in C^0(\mathbb{R})$. For the proof of the bound we may assume without loss of generality that $f \ge 0$, otherwise consider $\max(f, 0)$ and $\max(-f, 0)$. We extend $T_1 f$ to the upper half-plane

$$D := \{z \in \mathbb{C} ; \operatorname{Im} z > 0\} .$$

To this end we define

$$\varphi(z) := \frac{1}{z} \left(\log(1 + z) - \log(1 - z) \right) \quad \text{for } z \in D,$$

where

$$\log(z) := \log(|z|) + i \arg(z) \quad \text{for } z \in \mathbb{C} \setminus \,] - \infty, 0] ,$$
$$\arg(re^{i\theta}) := \theta \quad \text{for } r > 0, \, |\theta| < \pi.$$

Consider the function

$$F(z) := \int_{\mathbb{R}} \varphi(z - s) f(s) \, ds \quad \text{for } z \in D.$$

Let $x \neq 0, \pm 1$ and $y \searrow 0$. Then

$$\operatorname{Re} \varphi(x + iy) \longrightarrow \frac{1}{x} \big(\log|1 + x| - \log|1 - x| \big) =: \psi(x) \geq 0 \,,$$

and, examining how $1 \pm (x + iy)$ approaches the positive and negative real axis, respectively,

$$\operatorname{Im} \varphi(x + iy) \longrightarrow \begin{cases} \dfrac{\pi}{x} & \text{if } |x| > 1, \\[2mm] 0 & \text{if } |x| < 1. \end{cases}$$

On noting in addition that $|\varphi(x + iy)| \leq C \cdot \log|x \pm 1|$ for $|x \pm 1| \leq \frac{1}{2}$, and that otherwise φ is a bounded function, it follows from Lebesgue's convergence theorem that

$$F(x + iy) \to (\psi * f)(x) + i\pi T_1 f(x) \quad \text{as } y \searrow 0,$$

locally uniformly in x, i.e. $\operatorname{Im}(F)$ is a continuous extension of $\pi T_1 f$ to D. Since $\psi \in L^1(\mathbb{R})$ (observe that $0 \leq \psi(x) \leq \frac{C}{x^2}$ for large $|x|$), it holds that $\psi * f \in L^p(\mathbb{R})$ with the convolution estimate

$$\|\psi * f\|_{L^p(\mathbb{R})} \leq \|\psi\|_{L^1(\mathbb{R})} \cdot \|f\|_{L^p(\mathbb{R})} \,.$$

In addition we have that $\operatorname{Re} F(z) \geq 0$ for all $z \in D$, because for $z = x + iy$

$$\operatorname{Re} \varphi(z) = \frac{1}{|z|^2} \big(x(\log|1 + z| - \log|1 - z|) + y(\arg(1 + z) - \arg(1 - z)) \big)$$

is nonnegative, and f is assumed to be nonnegative. Hence $z \mapsto F(z)^p$ is a well-defined function that is continuous in $\overline{D}$, where

$$z^p := e^{p \log z} \quad \text{for } z \in \mathbb{C} \setminus]-\infty, 0].$$

As φ is holomorphic in D, and hence so is F, and then also F^p, it follows from Cauchy's integral theorem for $R > 0$ that

$$0 = \int_{\partial(D \cap B_R(0))} F(z)^p \, dz = \int_{-R}^{R} F(x)^p \, dx + \int_{D \cap \partial B_R(0)} F(z)^p \, dz \,.$$

Since f has compact support, we have that $|F(z)| \leq C \frac{\log|z|}{|z|}$ for large $|z|$, and so as $R \to \infty$

$$\left| \int_{D \cap \partial B_R(0)} F(z)^p \, dz \right| \leq CR \Big(\frac{\log R}{R} \Big)^p \longrightarrow 0 \,.$$

This shows that

$$\int_{\mathbb{R}} F(x)^p \, dx = 0 \, .$$

Writing $F(x) = F_1(x) + iF_2(x)$, it follows from the identity

$$F(x)^p - \big(iF_2(x)\big)^p = p \int_0^1 \big(tF_1(x) + iF_2(x)\big)^{p-1} \, dt \cdot F_1(x)$$

that

$$\left| \int_{\mathbb{R}} \big(iF_2(x)\big)^p \, dx \right| \le C(p) \int_{\mathbb{R}} \big(|F_1(x)|^{p-1} + |F_2(x)|^{p-1}\big) |F_1(x)| \, dx \, .$$

From the generalized Young's inequality it follows for $0 < \delta \le 1$ that this is

$$\le \delta \int_{\mathbb{R}} |F_2(x)|^p \, dx + \frac{C(p)}{\delta^{p-1}} \int_{\mathbb{R}} |F_1(x)|^p \, dx \, .$$

Since $\mathrm{Re}(iF_2(x))^p = \cos(p\frac{\pi}{2})|F_2(x)|^p$, we have

$$\left| \cos(p\frac{\pi}{2}) \right| \int_{\mathbb{R}} |F_2(x)|^p \, dx = \left| \mathrm{Re} \int_{\mathbb{R}} (iF_2(x))^p \, dx \right|$$

$$\le \left| \int_{\mathbb{R}} (iF_2(x))^p \, dx \right| \le \delta \int_{\mathbb{R}} |F_2(x)|^p \, dx + \frac{C(p)}{\delta^{p-1}} \int_{\mathbb{R}} |F_1(x)|^p \, dx \, .$$

In the case $\cos(p\frac{\pi}{2}) \ne 0$, on choosing $\delta = \frac{1}{2}|\cos(p\frac{\pi}{2})|$, it then follows (employing the usual convention on constants) that

$$\int_{\mathbb{R}} |F_2|^p \, d\mathrm{L}^1 \le C(p) \int_{\mathbb{R}} |F_1|^p \, d\mathrm{L}^1 = C(p) \|\psi * f\|_{L^p(\mathbb{R})}^p \le C(p) \|f\|_{L^p(\mathbb{R})}^p \, .$$

This is the desired result when $\cos(p\frac{\pi}{2}) \ne 0$, which for example is satisfied for $1 < p \le 2$. For $2 \le p < \infty$ the claim follows with a duality argument. In particular, it then holds that $1 < p' \le 2$, and so for all $g \in C_0^0(\mathbb{R})$ we have that

$$\left| \int_{\mathbb{R}} g T_1 f \, d\mathrm{L}^1 \right| = \left| \int_{\mathbb{R}} f T_1 g \, d\mathrm{L}^1 \right|$$

$$\le \|f\|_{L^p(\mathbb{R})} \|T_1 g\|_{L^{p'}(\mathbb{R})} \le C(p') \|f\|_{L^p(\mathbb{R})} \|g\|_{L^{p'}(\mathbb{R})} \, ,$$

which together with 6.13 implies that

$$\|T_1 f\|_{L^p(\mathbb{R})} \le C(p') \|f\|_{L^p(\mathbb{R})} \, .$$

$\square$

In conjunction with the following lemma, we obtain 10.20 in the case $n = 1$.

A10.3 Lemma. The result in 10.20 holds true if there exists a constant $C(n, p)$ such that

$$\|T_1 f\|_{L^p(\mathbb{R}^n)} \le C(n, p)\|f\|_{L^p(\mathbb{R}^n)} \quad \text{for all } f \in C_0^\infty(\mathbb{R}^n).$$

Remark: For $f \in C_0^\infty(\mathbb{R}^n)$ it holds that $T_1 f \in L^\infty(\mathbb{R}^n)$. Moreover $|T_1 f(x)| \le C\|f\|_{\sup} \cdot |x|^{-n}$ for large $|x|$, and so $T_1 f \in L^p(\mathbb{R}^n)$.

Proof. Let $f \in L^p(\mathbb{R}^n)$ and $f_k \in C_0^\infty(\mathbb{R}^n)$ with $\|f - f_k\|_{L^p} \to 0$ as $k \to \infty$. It follows from the Hölder inequality that for $x \in \mathbb{R}^n$

$$|T_1 f(x) - T_1 f_k(x)| \le C \cdot \int_{\mathbb{R}^n \setminus B_1(x)} \frac{|f(y) - f_k(y)|}{|x - y|^n} \, dy$$

$$\le C \cdot \|f - f_k\|_{L^p} \left(\int_{\mathbb{R}^n \setminus B_1(0)} \frac{dy}{|y|^{np'}} \right)^{\frac{1}{p'}} \longrightarrow 0 \quad \text{as } k \to \infty,$$

and, in addition, if C_0 denotes the constant $C(n, p)$ from the assumptions, that

$$\|T_1 f_k - T_1 f_l\|_{L^p} = \|T_1(f_k - f_l)\|_{L^p} \le C_0 \|f_k - f_l\|_{L^p} \longrightarrow 0 \quad \text{as } k, l \to \infty.$$

Hence $(T_1 f_k)_{k \in \mathbb{N}}$ is a Cauchy sequence in $L^p(\mathbb{R}^n)$ with limit $T_1 f$, and so the assumed L^p-estimate also holds for f, i.e.

$$\|T_1 f\|_{L^p} \le C_0 \|f\|_{L^p}.$$

Now let $\varepsilon > 0$ and set $f_\varepsilon(y) := f(\varepsilon y)$. Then

$$T_\varepsilon f(x) = \int_{\mathbb{R}^n \setminus B_\varepsilon(x)} \frac{\omega(x - y)}{|x - y|^n} f(y) \, dy$$

$$= \int_{\mathbb{R}^n \setminus B_1(\frac{x}{\varepsilon})} \frac{\omega(\frac{x}{\varepsilon} - y)}{|\frac{x}{\varepsilon} - y|^n} f_\varepsilon(y) \, dy = T_1 f_\varepsilon(\frac{x}{\varepsilon}).$$

This yields that $T_\varepsilon f \in L^p(\mathbb{R}^n)$, with

$$\|T_\varepsilon f\|_{L^p} = \left(\int_{\mathbb{R}^n} \left| T_1 f_\varepsilon(\frac{x}{\varepsilon}) \right|^p dx \right)^{\frac{1}{p}} = \left(\varepsilon^n \int_{\mathbb{R}^n} |T_1 f_\varepsilon(x)|^p \, dx \right)^{\frac{1}{p}}$$

$$\le C_0 \left(\varepsilon^n \int_{\mathbb{R}^n} |f_\varepsilon(x)|^p \, dx \right)^{\frac{1}{p}} = C_0 \|f\|_{L^p}.$$

It follows for $0 < \varepsilon_1 < \varepsilon_2$ that

$$\|T_{\varepsilon_1} f - T_{\varepsilon_2} f\|_{L^p}$$

$$\le \|T_{\varepsilon_1}(f - f_k)\|_{L^p} + \|T_{\varepsilon_2}(f - f_k)\|_{L^p} + \|T_{\varepsilon_1} f_k - T_{\varepsilon_2} f_k\|_{L^p}$$

$$\le \underbrace{2C_0 \|f - f_k\|_{L^p}}_{\to 0 \text{ as } k \to \infty} + \|T_{\varepsilon_1} f_k - T_{\varepsilon_2} f_k\|_{L^p}.$$

Since the mean of ω vanishes, for all x

$$|T_{\varepsilon_1} f_k(x) - T_{\varepsilon_2} f_k(x)|$$

$$= \left| \int_{B_{\varepsilon_2}(x) \setminus B_{\varepsilon_1}(x)} \frac{\omega(x-y)}{|x-y|^n} f_k(y) \, dy \right|$$

$$= \left| \int_{B_{\varepsilon_2}(x) \setminus B_{\varepsilon_1}(x)} \frac{\omega(x-y)}{|x-y|^n} (f_k(y) - f_k(x)) \, dy \right|$$

$$\leq C \cdot \int_{B_{\varepsilon_2}(x) \setminus B_{\varepsilon_1}(x)} \frac{dy}{|x-y|^{n-1}} \cdot \|\nabla f_k\|_{\sup}$$

$$\leq C(n) \varepsilon_2 \|\nabla f_k\|_{\sup}.$$

Since in addition $T_{\varepsilon_1} f_k(x) = T_{\varepsilon_2} f_k(x)$ for $x \in \mathbb{R}^n \setminus B_{\varepsilon_2}(\operatorname{supp} f_k)$, we obtain for every k that

$$\|T_{\varepsilon_1} f_k - T_{\varepsilon_2} f_k\|_{L^p}$$

$$\leq C(n) \varepsilon_2 \|\nabla f_k\|_{\sup} L^n \big(B_{\varepsilon_2}(\operatorname{supp} f_k)\big)^{\frac{1}{p}} \longrightarrow 0 \qquad \text{as } \varepsilon_2 \to 0.$$

This proves that the functions $T_\varepsilon f$ for $\varepsilon \to 0$ form a Cauchy sequence in $L^p(\mathbb{R}^n)$. Hence it also holds that

$$\left\| \lim_{\varepsilon \searrow 0} T_\varepsilon f \right\|_{L^p} \leq C_0 \|f\|_{L^p}.$$

$\square$

A10.4 Theorem. Theorem 10.20 also holds in the case $n > 1$.

Proof. We need to prove a bound for T_1 similarly to A10.3. Since we can decompose ω as

$$\omega(\xi) = \frac{\omega(\xi) + \omega(-\xi)}{2} + \frac{\omega(\xi) - \omega(-\xi)}{2},$$

it is sufficient to consider separately the two cases: ω is an even function, i.e. $\omega(-\xi) = \omega(\xi)$, or an odd function, i.e. $\omega(-\xi) = -\omega(\xi)$.

We begin with the case when ω is odd. (Observe that odd kernels always satisfy the vanishing mean value property (10-33).) For $f \in C_0^\infty(\Omega)$ it then holds, upon using polar coordinates, that

$$T_1 f(x) = \int_{\mathbb{R}^n \setminus B_1(0)} \frac{\omega(y)}{|y|^n} f(x-y) \, dy$$

$$= \int_{\partial B_1(0)} \omega(\xi) \int_1^\infty \frac{f(x - r\xi)}{r} \, dr \, d\mathrm{H}^{n-1}(\xi).$$

As ω is odd, this is

$$= \frac{1}{2} \int_{\partial B_1(0)} \omega(\xi) \int_1^\infty \frac{f(x - r\xi) - f(x + r\xi)}{r} \, dr \, dH^{n-1}(\xi)$$

$$= \frac{1}{2} \int_{\partial B_1(0)} \omega(\xi) \left(\int_{\{|t| \geq 1\}} \frac{f(x - t\xi)}{t} \, dt \right) dH^{n-1}(\xi),$$

and so the Hölder inequality yields that

$$|T_1 f(x)|^p \leq 2^{-p} \left(\int_{\partial B_1(0)} |\omega(\xi)|^{\frac{1}{p'} + \frac{1}{p}} \left| \int_{\{|t| \geq 1\}} \frac{f(x - t\xi)}{t} \, dt \right| dH^{n-1}(\xi) \right)^p$$

$$\leq 2^{-p} \|\omega\|_{L^1(\partial B_1(0))}^{\frac{p}{p'}} \int_{\partial B_1(0)} |\omega(\xi)| \left| \int_{\{|t| \geq 1\}} \frac{f(x - t\xi)}{t} \, dt \right|^p dH^{n-1}(\xi).$$

For every $\xi \in \partial B_1(0)$ we decompose the space $\mathbb{R}^n$ as

$$\mathbb{R}^n = Z_\xi \perp \mathrm{span}\{\xi\}.$$

For $z \in Z_\xi$ it then follows from A10.2 that

$$\Phi_\xi(z) := \int_{\mathbb{R}} \left| \int_{\{|t| \geq 1\}} \frac{f(z + (s - t)\xi)}{t} \, dt \right|^p ds \leq C(p) \int_{\mathbb{R}} |f(z + s\xi)|^p \, ds,$$

and so, setting $M_\omega := \|\omega\|_{L^1(\partial B_1(0))}$, that

$$\int_{\mathbb{R}^n} |T_1 f(x)|^p \, dx$$

$$\leq 2^{-p} M_\omega^{p-1} \int_{\partial B_1(0)} |\omega(\xi)| \left(\int_{Z_\xi} \Phi_\xi(z) \, dL^{n-1}(z) \right) dH^{n-1}(\xi)$$

$$\leq C(p) M_\omega^{p-1} \int_{\partial B_1(0)} |\omega(\xi)| \int_{Z_\xi} \int_{\mathbb{R}} |f(z + s\xi)|^p \, ds \, dL^{n-1}(z) \, dH^{n-1}(\xi).$$

This shows that

$$\|T_1 f\|_{L^p} \leq C(p) \|\omega\|_{L^1(\partial B_1(0))} \cdot \|f\|_{L^p},$$

which proves 10.20 for odd ω. Observe that the proof did not use the boundedness of ω: it suffices to assume that ω is integrable over $\partial B_1(0)$.

We now assume that ω is even and reduce this case to the odd case. To this end we define the convolution operator

$$S_\varepsilon g(x) := \int_{\mathbb{R}^n \setminus B_\varepsilon(x)} g(y) \frac{x - y}{|x - y|^{n+1}} \, dy \quad \text{and} \quad Sg(x) := \lim_{\varepsilon \searrow 0} S_\varepsilon g(x).$$

As the vector-valued integral kernel of S_ε is odd, what was shown above implies that for $g \in L^q(\mathbb{R}^n)$ with $1 < q < \infty$,

$$S_\varepsilon g \longrightarrow Sg \quad \text{in } L^q(\mathbb{R}^n; \mathbb{R}^n)$$

$$\text{with} \quad \|S_\varepsilon g\|_{L^q} \le C(n,q)\|g\|_{L^q}. \tag{A10-1}$$

We begin by establishing that there exists a $c_0 > 0$ such that

$$\sum_{i=1}^n S_{i\varepsilon} S_{i\varepsilon} g \longrightarrow -c_0 g \quad \text{in } L^q(\mathbb{R}^n) \text{ for } g \in C_0^\infty(\mathbb{R}^n), \tag{A10-2}$$

where $S_{i\varepsilon}$ denotes the i-th coordinate of the operator S_ε. We will use this property to bound $T_1 f$ in a first step in terms of $ST_1 f$. In the last part of the proof we then show that ST_1 is also a singular integral operator with an odd kernel.

In order to prove (A10-2) we write

$$\sum_{i=1}^n S_{i\varepsilon} S_{i\varepsilon} g(x) = \int_{\mathbb{R}^n} \left(\int_{\mathbb{R}^n \setminus B_\varepsilon(x) \setminus B_\varepsilon(y)} \frac{(x-z)\cdot(z-y)}{|x-z|^{n+1}|z-y|^{n+1}} \, dz \right) g(y) \, dy.$$

With the change of variables $z = -z' + \frac{x+y}{2}$ this becomes

$$= -\int_{\mathbb{R}^n} \varphi_\varepsilon\left(\frac{x-y}{2}\right) g(y) \, dy,$$

where

$$\varphi_\varepsilon(x) := \int_{\{|z \pm x| \ge \varepsilon\}} \frac{(z+x)\cdot(z-x)}{|z+x|^{n+1}|z-x|^{n+1}} \, dz.$$

With the change of variables $z = \varepsilon z'$ we obtain that $\varphi_\varepsilon(x) = \varepsilon^{-n}\varphi_1\left(\frac{x}{\varepsilon}\right)$. Hence assertion (A10-2) follows from 4.15(2), if we show that φ_1 is a nonnegative integrable function. If $D \subset \mathbb{R}^n$ is open and invariant under the reflection in $\partial B_{|x|}(0)$, i.e. $\frac{|x|^2}{|z|^2} z \in D$ for $z \in D$, then the change of variables $z = \frac{|x|^2}{|z'|^2} z'$ yields, on noting that

$$dz = \left(\frac{|x|}{|z'|}\right)^{2n} dz' \quad \text{and} \quad |z \pm x| = \frac{|x|}{|z'|}|z' \pm x|,$$

that

$$\int_D \frac{|z|^2 - |x|^2}{|z+x|^{n+1}|z-x|^{n+1}} \, dz = \int_D \frac{|x|^2 - |z'|^2}{|z'+x|^{n+1}|z'-x|^{n+1}} \, dz',$$

i.e. this integral vanishes. Applying this result to the domain $D = \{z; |z \pm x| > 1, |z' \pm x| > 1\}$, we obtain that

$$\varphi_1(x) = \int_E \frac{|z|^2 - |x|^2}{|z+x|^{n+1}|z-x|^{n+1}} \, dz$$

with $E := \{|z \pm x| \geq 1\} \cap (\{|z + x| \leq \frac{|z|}{|x|}\} \cup \{|z - x| \leq \frac{|z|}{|x|}\})$, and so $|z| \geq |x|$ for $z \in E$, which implies that $\varphi_1 \geq 0$. Clearly φ_1 is continuous on $\mathbb{R}^n \setminus \{0\}$, and for $|x| \leq \frac{1}{2}$

$$\varphi_1(x) \leq C \int_{\{|z| \geq \frac{1}{2}\}} \frac{dz}{|z|^{2n}} < \infty,$$

while for $|x| \geq 2$

$$\varphi_1(x) \leq \int_E \frac{dz}{|z + x|^n |z - x|^n}.$$

We partition E into $\{z \in E; \; z \bullet x \geq 0\}$ and $\{z \in E; \; z \bullet x \leq 0\}$. For z in the first set it holds that $|z + x| \geq |x|$ and with $z' := z - x$ we have that $1 \leq |z'| \leq \frac{|x|}{|x|-1}$. An analogous result holds for the second set. Overall we obtain that

$$\varphi_1(x) \leq \frac{2}{|x|^n} \int_{\{1 \leq |z'| \leq \frac{|x|}{|x|-1}\}} \frac{dz'}{|z'|^n} \leq \frac{C}{|x|^{n+1}}.$$

The last inequality follows from the fact that we integrate over an annular region of width $\frac{1}{|x|-1}$. This shows that φ_1 is integrable and the result (A10-2) is shown.

Now let $f \in C_0^\infty(\mathbb{R}^n)$ as before. It follows from (A10-2) and the L^p-bound for S_ε that for $\zeta \in C_0^\infty(\mathbb{R}^n)$ and as $\varepsilon \searrow 0$

$$c_0 \left| \int_{\mathbb{R}^n} \zeta T_1 f \, dL^n \right| \longleftarrow \left| \int_{\mathbb{R}^n} \left(\sum_{i=1}^n S_{i\varepsilon} S_{i\varepsilon} \zeta \right) T_1 f \, dL^n \right|$$

$$= \left| \int_{\mathbb{R}^n} \sum_{i=1}^n S_{i\varepsilon} \zeta \cdot S_{i\varepsilon} T_1 f \, dL^n \right| \leq \|S_\varepsilon \zeta\|_{L^{p'}} \|S_\varepsilon T_1 f\|_{L^p},$$

with

$$\|S_\varepsilon \zeta\|_{L^{p'}} \leq C(n, p') \|\zeta\|_{L^{p'}}.$$

As $T_1 f \in L^p(\mathbb{R}^n)$ (see the remark in A10.3), it holds in addition that

$$\|S_\varepsilon T_1 f\|_{L^p} \longrightarrow \|S T_1 f\|_{L^p} \quad \text{as } \varepsilon \searrow 0.$$

Hence, on recalling 6.13, we obtain the bound

$$\|T_1 f\|_{L^p} \leq C(n, p) \|S T_1 f\|_{L^p}.$$

Now we show that $S T_1$, too, is essentially a singular integral operator with an odd kernel. It holds that

$$S_\varepsilon T_1 f(x) = \int_{\mathbb{R}^n} \left(\int_{\{|z-x| \geq \varepsilon, \; |z-y| \geq 1\}} \frac{x - z}{|x - z|^{n+1}} \frac{\omega(z - y)}{|z - y|^n} \, dz \right) f(y) \, dy$$

$$= \int_{\mathbb{R}^n} \Phi_\varepsilon(x - y) f(y) \, dy,$$

where

$$\Phi_\varepsilon(x) := \int_{\{|z-x|\geq\varepsilon,\,|z|\geq1\}} \frac{x-z}{|x-z|^{n+1}} \frac{\omega(z)}{|z|^n}\, dz\,.$$

Since

$$\Phi_\varepsilon(x) = S_\varepsilon h(x) \quad \text{with} \quad h(z) := \mathcal{X}_{\mathbb{R}^n\setminus B_1(0)}(z)\frac{\omega(z)}{|z|^n}\,,$$

and since $h \in L^q(\mathbb{R}^n)$ for every $1 < q < \infty$ (not for $q = 1$!), it follows from the previously shown convergence in (A10-1) that

$$\Phi_\varepsilon = S_\varepsilon h \to S h =: \Phi \quad \text{in } L^q(\mathbb{R}^n;\mathbb{R}^n)\,,$$
$$\text{with} \quad \|\Phi\|_{L^q} \leq C(n,q)\|h\|_{L^q}\,.$$

Here we have that

$$\|h\|_{L^q} = \left(\int_1^\infty r^{n-1-nq} \int_{\partial B_1(0)} |\omega(\xi)|^q\, d\mathcal{H}^{n-1}(\xi)\, dr\right)^{\frac{1}{q}}$$
$$= c_1(n,q)\|\omega\|_{L^q(\partial B_1(0))}\,,$$

with

$$c_\varrho(n,q) := \left(\int_\varrho^\infty r^{-1-n(q-1)}\, dr\right)^{\frac{1}{q}}\,.$$

In addition,

$$ST_1 f(x) = \int_{\mathbb{R}^n} \Phi(x-y)f(y)\, dy\,.$$

Similarly to Φ_ε, for every $\delta > 0$

$$\psi_\delta(x) := \int_{\{\delta\leq|z|\leq1\}} \frac{x-z}{|x-z|^{n+1}} \frac{\omega(z)}{|z|^n}\, dz$$

defines a function $\psi_\delta \in L^p(\mathbb{R}^n;\mathbb{R}^n)$. Moreover, the limit

$$\psi(x) := \lim_{\delta\searrow0} \psi_\delta(x) = \int_{\{|z|\leq1\}} \frac{x-z}{|x-z|^{n+1}} \frac{\omega(z)}{|z|^n}\, dz$$

exists pointwise for $x \neq 0$. In order to prove this, choose for $\varrho > 0$ an $\eta \in C_0^\infty(B_\varrho(0))$ with $\eta = 1$ in $B_{\frac{\varrho}{2}}(0)$ and decompose $\psi_\delta(x)$ for $|x| > \varrho$ as induced by the decomposition

$$\frac{x-z}{|x-z|^{n+1}} = \eta(z)\frac{x-z}{|x-z|^{n+1}} + (1-\eta(z))\frac{x-z}{|x-z|^{n+1}}\,.$$

The first term is a Lipschitz continuous (in fact smooth) function of z. Hence the corresponding integral converges as $\delta \searrow 0$ (see the first part of the proof of the Hölder-Korn-Lichtenstein inequality 10.19). The integral over the second term is independent of δ for $\delta < \frac{\varrho}{2}$.

On employing the change of variables $z = |x|z'$ we now see that there exists a measurable function $\omega_0 : \partial B_1(0) \to \mathbb{R}^n$ such that

$$\Phi(x) + \psi(x) = \frac{\omega_0(x)}{|x|^n}, \tag{A10-3}$$

where $\omega_0(x) := \omega_0\left(\frac{x}{|x|}\right)$ and

$$\omega_0(\xi) := \int_{\mathbb{R}^n} \frac{\xi - z}{|\xi - z|^{n+1}} \frac{\omega(z)}{|z|^n} \, dz \quad \text{for almost all } \xi \in \partial B_1(0).$$

As ω is an even function, ω_0 must be odd. Moreover, for $|x| \geq 2$ and $|z| \leq 1$ (cf. the proof of 10.19)

$$\left| \frac{x - z}{|x - z|^{n+1}} - \frac{x}{|x|^{n+1}} \right| \leq C|z| \left(\frac{1}{|z - x||x|^n} + \frac{1}{|x||z - x|^n} \right) \leq \frac{C|z|}{|x|^{n+1}},$$

which in view of the mean value property of ω implies that

$$|\psi(x)| = \left| \int_{\{|z| \leq 1\}} \left(\frac{x - z}{|x - z|^{n+1}} - \frac{x}{|x|^{n+1}} \right) \frac{\omega(z)}{|z|^n} \, dz \right|$$

$$\leq \frac{C}{|x|^{n+1}} \int_{\{|z| \leq 1\}} \frac{|\omega(z)|}{|z|^{n-1}} \, dz \leq \frac{C}{|x|^{n+1}} \|\omega\|_{L^1(\partial B_1(0))}.$$

Hence $\psi \in L^q(\mathbb{R}^n \setminus B_2(0); \mathbb{R}^n)$ for $1 \leq q < \infty$ (here the case $q = 1$ is included (!)), with

$$\|\psi\|_{L^q(\mathbb{R}^n \setminus B_2(0))} \leq C(n, q) \|\omega\|_{L^1(\partial B_1(0))}.$$

Therefore, on recalling (A10-3), we obtain for $1 < q < \infty$ that

$$c_2(n, q) \|\omega_0\|_{L^q(\partial B_1(0))} = \left\| \frac{\omega_0}{|\cdot|^n} \right\|_{L^q(\mathbb{R}^n \setminus B_2(0))}$$

$$\leq \|\Phi\|_{L^q(\mathbb{R}^n)} + \|\psi\|_{L^q(\mathbb{R}^n \setminus B_2(0))}$$

$$\leq C(n, q) \left(c_1(n, q) \|\omega\|_{L^q(\partial B_1(0))} + \|\omega\|_{L^1(\partial B_1(0))} \right),$$

and so

$$\|\omega_0\|_{L^1(\partial B_1(0))} \leq C(n) \|\omega\|_{L^\infty(\partial B_1(0))} < \infty.$$

Hence the previously shown L^p-bound for kernels induced by odd ω can be applied to the kernel induced by ω_0. We note from (A10-3) that

$$\Phi(x) = \mathcal{X}_{\mathbb{R}^n \setminus B_2(0)}(x) \frac{\omega_0(x)}{|x|^n} - \widetilde{\Phi}(x),$$

where

$$\widetilde{\Phi}(x) := \mathcal{X}_{\mathbb{R}^n \setminus B_2(0)}(x)\psi(x) - \mathcal{X}_{B_2(0)}(x)\Phi(x),$$

and we note that for every $1 < q < \infty$

$$\left\|\widetilde{\Phi}\right\|_{L^1(\mathbb{R}^n)} \leq \|\psi\|_{L^1(\mathbb{R}^n \setminus B_2(0))} + C(n,q)\|\Phi\|_{L^q(\mathbb{R}^n)} < \infty.$$

We obtain using the L^p-bound for the kernel induced by ω_0 and the convolution estimate that

$$\|ST_1 f\|_{L^p(\mathbb{R}^n)}$$

$$\leq \left(\int_{\mathbb{R}^n} \left|\int_{\{|y|\geq 2\}} \frac{\omega_0(x-y)}{|x-y|^n} f(y)\,dy\right|^p dx\right)^{\frac{1}{p}} + \left\|\widetilde{\Phi} * f\right\|_{L^p(\mathbb{R}^n)}$$

$$\leq \left(C(p)\|\omega_0\|_{L^1(\partial B_1(0))} + \left\|\widetilde{\Phi}\right\|_{L^1(\mathbb{R}^n)}\right)\|f\|_{L^p(\mathbb{R}^n)}.$$

This proves 10.20 also for even kernels. $\square$

11 Spectrum of compact operators

We begin with some general results on the spectrum of continuous operators (11.1–11.5), where we always assume that X is a Banach space over $\mathbb{C}$ (!), i.e. $\mathbb{K} = \mathbb{C}$, and that $T \in \mathscr{L}(X)$ (for the real case see 11.14). The main topic of this chapter is the Riesz-Schauder theory on the spectrum of compact operators (theorem 11.9).

11.1 Spectrum. We define the *resolvent set* of T by

$$\varrho(T) := \left\{ \lambda \in \mathbb{C} ; \ \mathscr{N}(\lambda\mathrm{Id} - T) = \{0\} \ \text{ and } \ \mathscr{R}(\lambda\mathrm{Id} - T) = X \right\}$$

and the *spectrum* of T by

$$\sigma(T) := \mathbb{C} \setminus \varrho(T) .$$

The spectrum can be decomposed into the *point spectrum*

$$\sigma_p(T) := \left\{ \lambda \in \sigma(T) ; \ \mathscr{N}(\lambda\mathrm{Id} - T) \neq \{0\} \right\} ,$$

the *continuous spectrum*

$$\sigma_c(T) := \left\{ \lambda \in \sigma(T) ; \ \mathscr{N}(\lambda\mathrm{Id} - T) = \{0\} \ \text{ and } \right.$$
$$\left. \mathscr{R}(\lambda\mathrm{Id} - T) \neq X, \ \text{ but } \ \overline{\mathscr{R}(\lambda\mathrm{Id} - T)} = X \right\}$$

and the *residual spectrum*

$$\sigma_r(T) := \left\{ \lambda \in \sigma(T) ; \ \mathscr{N}(\lambda\mathrm{Id} - T) = \{0\} \ \text{ and } \ \overline{\mathscr{R}(\lambda\mathrm{Id} - T)} \neq X \right\} .$$

11.2 Remarks.

(1) It holds that $\lambda \in \varrho(T)$ if and only if $\lambda\mathrm{Id} - T : X \to X$ is bijective. The inverse mapping theorem 7.8 yields that this is equivalent to the existence of

$$R(\lambda; T) := (\lambda\mathrm{Id} - T)^{-1} \in \mathscr{L}(X) .$$

The inverse $R(\lambda; T)$ is called the *resolvent* of T in λ and as a function of λ is called the *resolvent function*.

(2) $\lambda \in \sigma_p(T)$ is equivalent to:

> There exists an $x \neq 0$ with $Tx = \lambda x$.

λ is then called an **eigenvalue** and x an **eigenvector** of T. If X is a function space, then x is also called an **eigenfunction**. The subspace $\mathcal{N}(\lambda \mathrm{Id} - T)$ is the **eigenspace** of T corresponding to the eigenvalue λ. The eigenspace is a T-invariant subspace. (A subspace $Y \subset X$ is called **T-invariant** if $T(Y) \subset Y$.)

11.3 Theorem. The resolvent set $\varrho(T)$ is open and the resolvent function $\lambda \mapsto R(\lambda; T)$ is a complex analytic map from $\varrho(T)$ to $\mathscr{L}(X)$. It holds that

$$\|R(\lambda; T)\|^{-1} \leq \mathrm{dist}(\lambda, \sigma(T)) \quad \text{for } \lambda \in \varrho(T).$$

Remark: A map $F : D \to Y$, with $D \subset \mathbb{C}$ open and Y a Banach space, is called **complex analytic** if for every $\lambda_0 \in D$ there exists a ball $\mathrm{B}_{r_0}(\lambda_0) \subset D$ such that $F(\lambda)$ for $\lambda \in \mathrm{B}_{r_0}(\lambda_0)$ can be written as a power series in $\lambda - \lambda_0$ with coefficients in Y. Complex analytic maps are holomorphic (see A10.1).

Proof. Let $\lambda \in \varrho(T)$. Then we have for $\mu \in \mathbb{C}$ that

$$(\lambda - \mu)\mathrm{Id} - T = (\lambda \mathrm{Id} - T) \underbrace{\left(\mathrm{Id} - \mu R(\lambda; T)\right)}_{=:\, S(\mu)}.$$

It follows from 5.7 that $S(\mu)$ is invertible if

$$|\mu| \cdot \|R(\lambda; T)\| < 1,$$

and then $\lambda - \mu \in \varrho(T)$, with

$$R(\lambda - \mu; T) = S(\mu)^{-1} R(\lambda; T) = \sum_{k=0}^{\infty} \mu^k R(\lambda; T)^{k+1}.$$

Setting $d := \|R(\lambda; T)\|^{-1}$ yields that $\mathrm{B}_d(\lambda) \subset \varrho(T)$, i.e. $\mathrm{dist}(\lambda, \sigma(T)) \geq d$. □

11.4 Theorem. The spectrum $\sigma(T)$ is compact and nonempty (if $X \neq \{0\}$), with

$$\sup_{\lambda \in \sigma(T)} |\lambda| = \lim_{m \to \infty} \|T^m\|^{\frac{1}{m}} \leq \|T\|.$$

This value is called the **spectral radius** of T.

Proof. Let $\lambda \neq 0$. We have from 5.7 that $\mathrm{Id} - \frac{T}{\lambda}$ is invertible if $\left\|\frac{T}{\lambda}\right\| < 1$, i.e. if $|\lambda| > \|T\|$, and then

$$R(\lambda; T) = \frac{1}{\lambda}\left(\mathrm{Id} - \frac{T}{\lambda}\right)^{-1} = \sum_{k=0}^{\infty} \frac{T^k}{\lambda^{k+1}}.$$

This shows that

$$r := \sup_{\lambda \in \sigma(T)} |\lambda| \le \|T\| .$$

Since

$$\lambda^m \mathrm{Id} - T^m = (\lambda \mathrm{Id} - T)p_m(T) = p_m(T)(\lambda \mathrm{Id} - T)$$

with

$$p_m(T) := \sum_{i=0}^{m-1} \lambda^{m-1-i} T^i ,$$

we conclude that

$$
\begin{aligned}
\lambda \in \sigma(T) \quad &\Longrightarrow \quad \lambda^m \in \sigma(T^m) \\
&\Longrightarrow \quad |\lambda^m| \le \|T^m\| \quad \text{(recall the bound established above)} \\
&\Longrightarrow \quad |\lambda| \le \|T^m\|^{\frac{1}{m}} .
\end{aligned}
$$

This proves that

$$r \le \liminf_{m \to \infty} \|T^m\|^{\frac{1}{m}} .$$

Next we show that

$$r \ge \limsup_{m \to \infty} \|T^m\|^{\frac{1}{m}} .$$

We recall from 11.3 that $R(\cdot; T)$ is a complex analytic map in $\mathbb{C} \setminus \overline{\mathrm{B}_r(0)}$ (if $\sigma(T)$ is empty, in $\mathbb{C}$). Hence, by Cauchy's integral theorem (see A10.1),

$$\frac{1}{2\pi \mathrm{i}} \int_{\partial \mathrm{B}_s(0)} \lambda^j R(\lambda; T) \, \mathrm{d}\lambda$$

is independent of s for $j \ge 0$ and $s > r$. However, if we choose $s > \|T\|$, then we obtain with the help of the representation of $R(\lambda; T)$ at the beginning of the proof that this integral is equal to

$$
\begin{aligned}
&= \frac{1}{2\pi \mathrm{i}} \int_{\partial \mathrm{B}_s(0)} \sum_{k=0}^{\infty} \lambda^{j-k-1} T^k \, \mathrm{d}\lambda = \frac{1}{2\pi} \sum_{k=0}^{\infty} s^{j-k} \left(\int_0^{2\pi} e^{\mathrm{i}\theta(j-k)} \, \mathrm{d}\theta \right) T^k \\
&\qquad\qquad\qquad = \sum_{k=0}^{\infty} s^{j-k} \delta_{j,k} T^k = T^j .
\end{aligned}
$$

Hence, for $j \ge 0$ and $s > r$,

$$\|T^j\| = \frac{1}{2\pi} \left\| \int_{\partial \mathrm{B}_s(0)} \lambda^j R(\lambda; T) \, \mathrm{d}\lambda \right\| \le s^{j+1} \sup_{|\lambda|=s} \|R(\lambda; T)\| .$$

Consequently we obtain for $s > r$ and every subsequence $j \to \infty$ that

$$\|T^j\|^{\frac{1}{j}} \le s \cdot \left(s \sup_{|\lambda|=s} \|R(\lambda; T)\| \right)^{\frac{1}{j}} \longrightarrow s \text{ or } 0,$$

and hence

$$\limsup_{j \to \infty} \left\| T^j \right\|^{\frac{1}{j}} \le s \,.$$

As this holds for all $s > r$, we obtain the desired result on the spectral radius. In addition, if $\sigma(T)$ was empty, we would obtain for $j = 0$ and as $s \searrow 0$ that

$$\|\mathrm{Id}\| \le s \cdot \sup_{|\lambda| \le 1} \|R(\lambda; T)\| \longrightarrow 0 \,,$$

i.e. $\mathrm{Id} = 0$, and so $X = \{0\}$. $\qquad\qquad\qquad\qquad\qquad\qquad\qquad\square$

11.5 Remarks.

(1) If $\dim X < \infty$, then $\sigma(T) = \sigma_p(T)$.
(2) If $\dim X = \infty$ and $T \in \mathcal{K}(X)$, then $0 \in \sigma(T)$. But in general 0 is not an eigenvalue.

Proof (1). If $\lambda \in \sigma(T)$, then $\lambda\mathrm{Id} - T$ is not bijective, and so, as $\dim X < \infty$, it is also not injective, i.e. $\lambda \in \sigma_p(T)$. $\qquad\qquad\qquad\qquad\qquad\square$

Proof (2). Let $T \in \mathcal{K}(X)$ and assume that $0 \in \varrho(T)$. Then (see 11.2(1)) $T^{-1} \in \mathcal{L}(X)$, and (see 10.3) so $\mathrm{Id} = T^{-1}T \in \mathcal{K}(X)$, which on recalling 4.10 implies that X is finite-dimensional.
Example without eigenvalue 0: The operator $T : C^0([0,1]) \to C^1([0,1])$ in 5.6(3) is injective. As an operator in $\mathcal{L}(C^0([0,1]))$ it is a compact operator in $\mathcal{K}(C^0([0,1]))$, by theorem 10.6. $\qquad\qquad\qquad\qquad\square$

   In the following we are interested in the point spectrum $\sigma_p(T)$ of an operator $T \in \mathcal{L}(X)$, i.e. we consider the **eigenvalue problem** corresponding to T: For a given $y \in X$ we look for all solutions $\lambda \in \mathbb{K}$ and $x \in X$ to

$$Tx - \lambda x = y \,.$$

If $\lambda \in \varrho(T)$, then there exists a uniquely determined solution x to this equation. If $\lambda \in \sigma_p(T)$, then the solution, if one exists, is not unique, i.e. on setting $A_\lambda := \lambda\mathrm{Id} - T$ we see that adding an element from $\mathcal{N}(A_\lambda)$ to a solution yields another solution (for $T \in \mathcal{K}(X)$ see also 11.11). On the other hand, the condition $y \in \mathcal{R}(A_\lambda)$ needs to be satisfied for a solution to the eigenvalue problem to exist at all. An important class of operators A_λ are those operators for which both the number of degrees of freedom for the solution x and the number of side conditions on y are finite:

11.6 Fredholm operators. A map $A \in \mathcal{L}(X; Y)$ is called a **Fredholm operator** if:

(1) $\dim \mathcal{N}(A) < \infty$,
(2) $\mathcal{R}(A)$ is closed,
(3) $\mathrm{codim}\, \mathcal{R}(A) < \infty$.

The *index* of a Fredholm operator is defined by

$$\operatorname{ind}(A) := \dim \mathcal{N}(A) - \operatorname{codim} \mathcal{R}(A).$$

Remark: The **codimension** of the image of A being finite means that $Y = \mathcal{R}(A) \oplus Y_0$ for a finite-dimensional subspace $Y_0 \subset Y$. Then $\operatorname{codim} \mathcal{R}(A) := \dim Y_0$ is independent of the choice of Y_0: indeed, if $Y_1 \subset Y$ is a subspace with $\mathcal{R}(A) \cap Y_1 = \{0\}$, then Y_1 is finite-dimensional with $\dim Y_1 \leq \dim Y_0$, with equality if and only if $Y = \mathcal{R}(A) \oplus Y_1$.

Proof. We have from (2) and 4.9, respectively, that $Z := \mathcal{R}(A)$ and Y_0 are closed subspaces. Now let $P \in \mathcal{P}(Y)$ be the projection onto Y_0 with $Z = \mathcal{N}(P)$, as in 9.15. Then

$$S := P\big|_{Y_1} : Y_1 \to Y_0 \text{ is linear and injective,}$$

because if $y \in Y_1$ with $P(y) = 0$, then $y \in Z \cap Y_1 = \{0\}$. As Y_0 is finite-dimensional, it follows that Y_1 is also finite-dimensional, with $\dim Y_1 \leq \dim Y_0$.

If $Y = Z \oplus Y_1$, then it follows as above (interchange Y_0 and Y_1) that $\dim Y_0 \leq \dim Y_1$, and so $\dim Y_1 = \dim Y_0$. Conversely, if this holds, then S is bijective. For $x \in Y$ we then have that $y := S^{-1}Px \in Y_1$ with $Py = SS^{-1}Px = Px$, and so $x - y \in \mathcal{N}(P) = Z$, which proves that $Y = Z \oplus Y_1$. $\square$

11.7 Example. Let $X = W^{1,2}(\Omega)$ and $Y = W^{1,2}(\Omega)'$. Then $A : W^{1,2}(\Omega) \to W^{1,2}(\Omega)'$, defined by

$$\langle v, Au \rangle_{W^{1,2}} := \int_\Omega \sum_{i,j} \partial_i v \cdot a_{ij} \partial_j u \, d\mathrm{L}^n \quad \text{for } u, v \in W^{1,2}(\Omega),$$

is a weak elliptic differential operator with Neumann boundary conditions. (We consider the homogeneous case in 6.5(2) with $h_i = 0$ and $b = 0$.) We have from 8.18(2) (where the symmetry $a_{ij} = a_{ji}$ was assumed) that:

The null space $\mathcal{N}(A)$ consists of the constant functions, and therefore $\dim \mathcal{N}(A) = 1$. The image of A is $\mathcal{R}(A) = \{F \in Y ; \langle 1, F \rangle_{W^{1,2}} = 0\}$, and so it is closed, with $\operatorname{codim} \mathcal{R}(A) = 1$. It holds that

$$Y = \mathcal{R}(A) \oplus \operatorname{span}\{F_0\}, \quad \text{where} \quad \langle v, F_0 \rangle_{W^{1,2}} := \int_\Omega v \, d\mathrm{L}^n.$$

Hence A is a Fredholm operator with index 0.

Observe: For the homogeneous Dirichlet problem (see 10.14(2)) the operator $A : W_0^{1,2}(\Omega) \to W_0^{1,2}(\Omega)'$ is an isomorphism.

A large class of Fredholm operators with $Y = X$ is given by compact perturbations of the identity (see also 12.8):

11.8 Theorem. Let $T \in \mathcal{K}(X)$. Then $A := \mathrm{Id} - T$ is a Fredholm operator with index 0. We prove this in several steps:

(1) $\dim \mathcal{N}(A) < \infty$,

(2) $\mathcal{R}(A)$ is closed,

(3) $\mathcal{N}(A) = \{0\} \implies \mathcal{R}(A) = X$,

(4) $\operatorname{codim} \mathcal{R}(A) \le \dim \mathcal{N}(A)$,

(5) $\dim \mathcal{N}(A) \le \operatorname{codim} \mathcal{R}(A)$.

Proof (1). On noting that $Ax = 0$ is equivalent to $x = Tx$, we have that $\mathrm{B}_1(0) \cap \mathcal{N}(A) \subset T\big(\mathrm{B}_1(0)\big)$, i.e. the unit ball in $\mathcal{N}(A)$ is precompact, and so 4.10 yields that $\mathcal{N}(A)$ is finite-dimensional. $\qquad\square$

Proof (2). Let $x \in \overline{\mathcal{R}(A)}$ and let $Ax_n \to x$ as $n \to \infty$. We may assume without loss of generality that

$$\|x_n\| \le 2\,d_n \quad \text{with } d_n := \operatorname{dist}(x_n, \mathcal{N}(A)),$$

because otherwise we choose $a_n \in \mathcal{N}(A)$ with $\|x_n - a_n\| \le 2\operatorname{dist}(x_n, \mathcal{N}(A))$ and then proceed with $\widetilde{x}_n := x_n - a_n$, where

$$\operatorname{dist}(\widetilde{x}_n, \mathcal{N}(A)) = \operatorname{dist}(x_n, \mathcal{N}(A)).$$

First we assume that $d_n \to \infty$ for a subsequence $n \to \infty$. Setting

$$y_n := \frac{x_n}{d_n} \quad \text{it holds that} \quad Ay_n = \frac{Ax_n}{d_n} \to 0$$

as $n \to \infty$. Noting that the y_n are bounded and recalling that T is compact yields that there exists a subsequence such that $Ty_n \to y$ as $n \to \infty$. It follows that

$$y_n = Ay_n + Ty_n \to y,$$

and so, by the continuity of A,

$$Ay = \lim_{n \to \infty} Ay_n = 0.$$

Hence $y \in \mathcal{N}(A)$, which implies that

$$\|y_n - y\| \ge \operatorname{dist}(y_n, \mathcal{N}(A)) = \operatorname{dist}\Big(\frac{x_n}{d_n}, \mathcal{N}(A)\Big) = \frac{\operatorname{dist}(x_n, \mathcal{N}(A))}{d_n} = 1,$$

a contradiction. This shows that the d_n are bounded, and so are the x_n. For a subsequence we then have that $Tx_n \to z$ as $n \to \infty$, and so

$$x \longleftarrow Ax_n = A(Ax_n + Tx_n) \longrightarrow A(x + z),$$

which means that $x \in \mathcal{R}(A)$. $\qquad\square$

Proof (3). Assume that there exists an $x \in X \setminus \mathscr{R}(A)$. Then

$$A^n x \in \mathscr{R}(A^n) \setminus \mathscr{R}(A^{n+1}) \quad \text{for all } n \geq 0,$$

because otherwise $A^n x = A^{n+1} y$ for some y, then $A^n(x - Ay) = 0$, and from $\mathscr{N}(A) = \{0\}$ it then would follow (inductively) that $x - Ay = 0$, i.e $x \in \mathscr{R}(A)$, a contradiction. In addition $\mathscr{R}(A^{n+1})$ is closed, on noting that

$$A^{n+1} = (\mathrm{Id} - T)^{n+1} = \mathrm{Id} + \underbrace{\sum_{k=1}^{n+1} \binom{n+1}{k} (-T)^k}_{\in \mathscr{K}(X) \text{ on recalling } 10.3},$$

and so (2) yields that $\mathscr{R}(A^{n+1})$ is closed. Hence there exists an $a_{n+1} \in \mathscr{R}(A^{n+1})$ with

$$0 < \|A^n x - a_{n+1}\| \leq 2 \operatorname{dist}(A^n x, \mathscr{R}(A^{n+1})).$$

Now consider

$$x_n := \frac{A^n x - a_{n+1}}{\|A^n x - a_{n+1}\|} \in \mathscr{R}(A^n).$$

We have that

$$\operatorname{dist}(x_n, \mathscr{R}(A^{n+1})) \geq \tfrac{1}{2}, \tag{11-4}$$

because for $y \in \mathscr{R}(A^{n+1})$

$$\|x_n - y\| = \frac{\|A^n x - (a_{n+1} + \|A^n x - a_{n+1}\| y)\|}{\|A^n x - a_{n+1}\|}$$

$$\geq \frac{\operatorname{dist}(A^n x, \mathscr{R}(A^{n+1}))}{\|A^n x - a_{n+1}\|} \geq \frac{1}{2}.$$

For $m > n$, we have $Ax_n + x_m - Ax_m \in \mathscr{R}(A^{n+1})$, and hence (11-4) implies that

$$\|Tx_n - Tx_m\| = \|x_n - (Ax_n + x_m - Ax_m)\| \geq \frac{1}{2}.$$

Hence $(Tx_n)_{n \in \mathbb{N}}$ contains no convergent subsequence, even though $(x_n)_{n \in \mathbb{N}}$ is a bounded sequence. This is a contradiction to the compactness of T. $\quad\square$

Proof (4). By (1), the number $n := \dim \mathscr{N}(A)$ is finite. Let $\{x_1, \ldots, x_n\}$ be an arbitrary basis of $\mathscr{N}(A)$. If we assume that the claimed inequality is false, then there exist linearly independent vectors $y_1, \ldots, y_n$, such that $\operatorname{span}\{y_1, \ldots, y_n\} \oplus \mathscr{R}(A)$ is a proper subspace of X. Moreover, 9.16(1) yields the existence of $x'_1, \ldots, x'_n \in X'$ with

$$\langle x_l, x'_k \rangle = \delta_{k,l} \quad \text{for } k, l = 1, \ldots, n.$$

Setting

$$\tilde{T}x := Tx + \sum_{k=1}^{n} \langle x\,,\, x_k' \rangle\, y_k$$

then defines an operator $\tilde{T} \in \mathcal{K}(X)$, indeed, T is compact and $\tilde{T} - T$ has a finite-dimensional image. In addition, $\mathcal{N}(\tilde{A}) = \{0\}$, where $\tilde{A} := \mathrm{Id} - \tilde{T}$, because $\tilde{A}x = 0$ implies, on recalling the choice of the y_k, that $Ax = 0$ and $\langle x\,,\, x_k' \rangle = 0$ for $k = 1, \ldots, n$. Therefore $x \in \mathcal{N}(A)$, and hence there exists a representation

$$x = \sum_{k=1}^{n} \alpha_k x_k\,, \quad \text{and so} \quad 0 = \langle x\,,\, x_l' \rangle = \sum_{k=1}^{n} \alpha_k \langle x_k\,,\, x_l' \rangle = \alpha_l$$

for $l = 1, \ldots, n$, which yields that $x = 0$. On applying (3) to the operator $\tilde{A}$, it follows that $\mathcal{R}(\tilde{A}) = X$. On noting that $\tilde{A}x_l = -y_l$ for $l = 1, \ldots, n$ and that

$$\tilde{A}\Big(x - \sum_{l=1}^{n} \langle x\,,\, x_l' \rangle\, x_l\Big) = Ax \quad \text{for all } x \in X\,,$$

we conclude that $X = \mathcal{R}(\tilde{A}) \subset \mathrm{span}\{y_1, \ldots, y_n\} \oplus \mathcal{R}(A)$, a contradiction to the above property. $\qquad\square$

Proof (5). We have from (4) that $m := \mathrm{codim}\,\mathcal{R}(A) \leq n := \dim \mathcal{N}(A)$.

First we reduce the claim to the case $m = 0$. To this end, choose $x_1, \ldots, x_n$ and $x_1', \ldots, x_n'$ as in the proof of (4) and $y_1, \ldots, y_m$ with

$$X = \mathrm{span}\{y_1, \ldots, y_m\} \oplus \mathcal{R}(A)\,.$$

As in the proof of (4), the operator

$$x \longmapsto \tilde{T}x := Tx + \sum_{k=1}^{m} \langle x\,,\, x_k' \rangle\, y_k$$

is compact and $\tilde{A} := \mathrm{Id} - \tilde{T}$ is surjective with $\mathcal{N}(\tilde{A}) = \mathrm{span}\{x_i\,;\ m < i \leq n\}$. We need to show that $\mathcal{N}(\tilde{A}) = \{0\}$. Hence the claim is reduced to the case $m = 0$.

In the case $m = 0$ it holds that $\mathcal{R}(A) = X$. We assume that there exists an $x_1 \in \mathcal{N}(A) \backslash \{0\}$. The surjectivity of A then yields that we can inductively choose $x_k \in X$, $k \geq 2$, with $Ax_k = x_{k-1}$. Then $x_k \in \mathcal{N}(A^k) \backslash \mathcal{N}(A^{k-1})$. It follows from the theorem on the almost orthogonal element that there exists a $z_k \in \mathcal{N}(A^k)$ with $\|z_k\| = 1$ and $\mathrm{dist}(z_k, \mathcal{N}(A^{k-1})) \geq \frac{1}{2}$. For $l < k$ this implies that $Az_k + z_l - Az_l \in \mathcal{N}(A^{k-1})$, and so the choice of z_k yields that

$$\|Tz_k - Tz_l\| = \|z_k - (Az_k + z_l - Az_l)\| \geq \tfrac{1}{2}\,.$$

This shows that $\{Tz_k \; ; \; k \in \mathbb{N}\}$ contains no convergent subsequence. This is a contradiction to the sequence $\{z_k \; ; \; k \in \mathbb{N}\}$ being bounded and the operator T being compact.

A second possible proof for $m = 0$ is as follows: We start with a decomposition $X = \widetilde{X} \oplus \mathscr{N}(A)$ with a closed subspace $\widetilde{X}$ (this follows from (1) and 9.16(2) for $Y = \{0\}$). Then $A : \widetilde{X} \to X$ is bijective, and so 7.8 yields that $\widetilde{A} := (A|_{\widetilde{X}})^{-1} : X \to \widetilde{X}$ is continuous. Now consider $\widetilde{A}$ as an element in $\mathscr{L}(X)$. Then $\widetilde{T} := \mathrm{Id} - \widetilde{A} \in \mathscr{K}(X)$, because if $\{x_k \; ; \; k \in \mathbb{N}\}$ is bounded in X, then so is $\{\widetilde{A}x_k \; ; \; k \in \mathbb{N}\}$, and hence there exists a subsequence with $T\widetilde{A}x_k \to x$ as $k \to \infty$. On the other hand,

$$T\widetilde{A}x_k = (\mathrm{Id} - A)\widetilde{A}x_k = \widetilde{A}x_k - x_k = -\widetilde{T}x_k \, .$$

Now (3) implies that $\mathscr{R}(\widetilde{A}) = X$, i.e. $\mathscr{N}(A) = \{0\}$.

A further possible proof of (5) will be given in 12.7. □

The fundamental theorem of this chapter is the

11.9 Spectral theorem for compact operators (Riesz-Schauder). For every operator $T \in \mathscr{K}(X)$ it holds that:

(1) The set $\sigma(T) \setminus \{0\}$ consists of countably (finitely or infinitely) many eigenvalues with 0 as the only possible cluster point. So if $\sigma(T)$ contains infinitely many elements, then $\overline{\sigma(T)} = \sigma_p(T) \cup \{0\}$, hence 0 is a cluster point of $\sigma(T)$.

(2) For $\lambda \in \sigma(T) \setminus \{0\}$

$$1 \le n_\lambda := \max\left\{ n \in \mathbb{N} \; ; \; \mathscr{N}\big((\lambda\mathrm{Id} - T)^{n-1}\big) \ne \mathscr{N}\big((\lambda\mathrm{Id} - T)^n\big) \right\} < \infty \, .$$

The number $n_\lambda \in \mathbb{N}$ is called the **order** (or **index**) of λ and $\dim \mathscr{N}(\lambda\mathrm{Id} - T)$ is called the **multiplicity** of λ.

(3) *Riesz decomposition.* For $\lambda \in \sigma(T) \setminus \{0\}$

$$X = \mathscr{N}\big((\lambda\mathrm{Id} - T)^{n_\lambda}\big) \oplus \mathscr{R}\big((\lambda\mathrm{Id} - T)^{n_\lambda}\big) \, .$$

Both subspaces are closed and T-invariant, and the **characteristic subspace** $\mathscr{N}\big((\lambda\mathrm{Id} - T)^{n_\lambda}\big)$ is finite-dimensional.

(4) For $\lambda \in \sigma(T) \setminus \{0\}$ it holds that $\sigma(T|_{\mathscr{R}((\lambda\mathrm{Id}-T)^{n_\lambda})}) = \sigma(T) \setminus \{\lambda\}$.

(5) If E_λ for $\lambda \in \sigma(T) \setminus \{0\}$ denotes the projection onto $\mathscr{N}\big((\lambda\mathrm{Id} - T)^{n_\lambda}\big)$ corresponding to the decomposition in (3), then

$$E_\lambda E_\mu = \delta_{\lambda,\mu} E_\lambda \quad \text{for } \lambda, \mu \in \sigma(T) \setminus \{0\}.$$

Proof (1). Let $0 \ne \lambda \notin \sigma_p(T)$. Then $\mathscr{N}(\mathrm{Id} - \frac{T}{\lambda}) = \{0\}$, and so $\mathscr{R}(\mathrm{Id} - \frac{T}{\lambda}) = X$ (recall 11.8(3)), i.e. $\lambda \in \varrho(T)$. This shows that

$$\sigma(T) \setminus \{0\} \subset \sigma_p(T).$$

If $\sigma(T) \setminus \{0\}$ is not finite, then we choose $\lambda_n \in \sigma(T) \setminus \{0\}$, $n \in \mathbb{N}$, pairwise distinct and eigenvectors $e_n \neq 0$ to λ_n and define

$$X_n := \mathrm{span}\{e_1, \ldots, e_n\}.$$

The eigenvectors e_k, $k = 1, \ldots, n$, are linearly independent, because if there exists (this is an inductive proof) $1 < k \leq n$ with

$$e_k = \sum_{i=1}^{k-1} \alpha_i e_i$$

with already linearly independent vectors $e_1, \ldots, e_{k-1}$, then it follows that

$$0 = T e_k - \lambda_k e_k = \sum_{i=1}^{k-1} \alpha_i (T e_i - \lambda_k e_i) = \sum_{i=1}^{k-1} \alpha_i \underbrace{(\lambda_i - \lambda_k)}_{\neq 0} e_i,$$

and so $\alpha_i = 0$ for $i = 1, \ldots, k-1$, i.e. $e_k = 0$, a contradiction. This shows that X_{n-1} is a proper subspace of X_n. Hence the theorem on the almost orthogonal element (see 4.5) yields the existence of an $x_n \in X_n$ with

$$\|x_n\| = 1 \quad \text{and} \quad \mathrm{dist}(x_n, X_{n-1}) \geq \tfrac{1}{2}. \tag{11-5}$$

On noting that $x_n = \alpha_n e_n + \tilde{x}_n$ with certain $\alpha_n \in \mathbb{C}$ and $\tilde{x}_n \in X_{n-1}$, it follows from the T-invariance of the subspace X_{n-1} that $T x_n - \lambda_n x_n = T \tilde{x}_n - \lambda_n \tilde{x}_n \in X_{n-1}$, and so it holds for $m < n$ that

$$\frac{1}{\lambda_n}(T x_n - \lambda_n x_n) - \frac{1}{\lambda_m} T x_m \in X_{n-1}.$$

Hence it follows from (11-5) that

$$\left\| T\Big(\frac{x_n}{\lambda_n}\Big) - T\Big(\frac{x_m}{\lambda_m}\Big) \right\| = \left\| x_n + \frac{1}{\lambda_n}(T x_n - \lambda_n x_n) - \frac{1}{\lambda_m} T x_m \right\| \geq \frac{1}{2}.$$

This shows that the sequence $\left(T\big(\frac{x_n}{\lambda_n}\big) \right)_{n \in \mathbb{N}}$ has no cluster point. As T is compact, this implies that $\left(\frac{x_n}{\lambda_n} \right)_{n \in \mathbb{N}}$ contains no bounded subsequences, which yields that

$$\frac{1}{|\lambda_n|} = \left\| \frac{x_n}{\lambda_n} \right\| \longrightarrow \infty \quad \text{as } n \to \infty,$$

i.e. $\lambda_n \to 0$ as $n \to \infty$. Hence we have shown that 0 is the only cluster point of $\sigma(T) \setminus \{0\}$. In particular, it then holds that $\sigma(T) \setminus B_r(0)$ is finite for every $r > 0$, and so $\sigma(T) \setminus \{0\}$ is countable. $\square$

Proof (2). Let $A := \lambda \text{Id} - T$. Then $\mathcal{N}(A^{n-1}) \subset \mathcal{N}(A^n)$ for all n. First we assume that:

$\mathcal{N}(A^{n-1})$ is a proper subset of $\mathcal{N}(A^n)$ for all $n \geq 1$.

Similarly to the proof of (1), and on recalling the theorem on the almost orthogonal element, we choose an $x_n \in \mathcal{N}(A^n)$ with

$$\|x_n\| = 1 \quad \text{and} \quad \text{dist}(x_n, \mathcal{N}(A^{n-1})) \geq \tfrac{1}{2}. \tag{11-6}$$

Then it follows for $m < n$ that

$$A x_n + \lambda x_m - A x_m \in \mathcal{N}(A^{n-1}),$$

and so with (11-6) that

$$\|T x_n - T x_m\| = \|\lambda x_n - (A x_n + \lambda x_m - A x_m)\| \geq \frac{|\lambda|}{2} > 0.$$

On the other hand, $\{x_n; \, n \in \mathbb{N}\}$ is a bounded sequence. This contradicts the compactness of T. Hence we can find an $n \in \mathbb{N}$ with $\mathcal{N}(A^{n-1}) = \mathcal{N}(A^n)$. This implies for $m > n$ that

$$x \in \mathcal{N}(A^m) \quad \Longrightarrow \quad A^{m-n} x \in \mathcal{N}(A^n) = \mathcal{N}(A^{n-1})$$
$$\Longrightarrow \quad A^{n-1+m-n} x = 0$$
$$\Longrightarrow \quad x \in \mathcal{N}(A^{m-1}),$$

and so $\mathcal{N}(A^m) = \mathcal{N}(A^{m-1})$, and it follows inductively that $\mathcal{N}(A^m) = \mathcal{N}(A^n)$ for all $m \geq n$. Hence we have shown that $n_\lambda < \infty$. Since $\mathcal{N}(A) \neq \{0\}$ it holds that $n_\lambda \geq 1$. □

Proof (3). Let $A := \lambda \text{Id} - T$ as before. Then

$$\mathcal{N}(A^{n_\lambda}) \oplus \mathcal{R}(A^{n_\lambda}) \subset X,$$

because if $x \in \mathcal{N}(A^{n_\lambda}) \cap \mathcal{R}(A^{n_\lambda})$, then $A^{n_\lambda} x = 0$ and $x = A^{n_\lambda} y$ for a $y \in X$. Then $A^{2n_\lambda} y = 0$, and so $y \in \mathcal{N}(A^{2n_\lambda}) = \mathcal{N}(A^{n_\lambda})$ and hence $x = A^{n_\lambda} y = 0$. Now A^{n_λ} can be written as

$$A^{n_\lambda} = \lambda^{n_\lambda} \text{Id} + \underbrace{\sum_{k=1}^{n_\lambda} \binom{n_\lambda}{k} \lambda^{n_\lambda - k} (-T)^k}_{\in \mathcal{K}(X) \text{ by } 10.3}. \tag{11-7}$$

Hence $\text{codim} \, \mathcal{R}(A^{n_\lambda}) \leq \dim \mathcal{N}(A^{n_\lambda}) < \infty$ (recall 11.8(4) and 11.8(1)), which yields that

$$X = \mathcal{N}(A^{n_\lambda}) \oplus \mathcal{R}(A^{n_\lambda}).$$

As T commutes with A, i.e. $TA = AT$, T also commutes with A^{n_λ}, and so both subspaces are T-invariant. □

Proof (4). We denote by T_λ the restriction of T to $\mathscr{R}(A^{n_\lambda})$, where A^{n_λ} has been computed in (11-7). Then $T_\lambda \in \mathscr{K}(\mathscr{R}(A^{n_\lambda}))$, where $\mathscr{R}(A^{n_\lambda})$ is a closed subspace (recall 11.8(2)), and so a Banach space. Here we have used the fact that T and A^{n_λ} commute. Moreover, we have that

$$\mathscr{N}(\lambda\mathrm{Id} - T_\lambda) = \mathscr{N}(A) \cap \mathscr{R}(A^{n_\lambda}) = \{0\},$$

and hence $\mathscr{R}(\lambda\mathrm{Id} - T_\lambda) = \mathscr{R}(A^{n_\lambda})$ (apply 11.8(3) to T_λ), which shows that $\lambda \in \varrho(T_\lambda)$. It remains to show that

$$\sigma(T_\lambda) \setminus \{\lambda\} = \sigma(T) \setminus \{\lambda\}.$$

Let $\mu \in \mathbb{C} \setminus \{\lambda\}$. We recall from above that $\mathscr{N}(A^{n_\lambda})$ is invariant under $\mu\mathrm{Id} - T$. Moreover, $\mu\mathrm{Id} - T$ is injective on this subspace. To see this, note that $x \in \mathscr{N}(\mu\mathrm{Id} - T)$ implies that $(\lambda - \mu)x = Ax$. If in addition $A^m x = 0$ for some $m \geq 1$, it follows that

$$(\lambda - \mu)A^{m-1}x = A^{m-1}((\lambda - \mu)x) = A^m x = 0,$$

and since $\lambda \neq \mu$ this means that $A^{m-1}x = 0$. Inductively (for decreasing m) this yields that $x = A^0 x = 0$. Hence we have shown that

$$\mathscr{N}(\mu\mathrm{Id} - T) \cap \mathscr{N}(A^m) = \{0\} \quad \text{for all } m \geq 1.$$

Setting $m = n_\lambda$ yields the injectivity of $\mu\mathrm{Id} - T$ on $\mathscr{N}(A^{n_\lambda})$. As this space is finite-dimensional, we have that $\mu\mathrm{Id} - T$ is also bijective on $\mathscr{N}(A^{n_\lambda})$. But this means that $\mu \in \varrho(T)$ if and only if $\mu \in \varrho(T_\lambda)$. This shows that by removing the (finite-dimensional) characteristic subspace corresponding to the eigenvalue λ we obtain a remaining operator T_λ for which $\sigma(T_\lambda) = \sigma(T) \setminus \{\lambda\}$. □

Proof (5). Let $\lambda, \mu \in \sigma(T) \setminus \{0\}$ be distinct, and let $A_\lambda := \lambda\mathrm{Id} - T$ and $A_\mu := \mu\mathrm{Id} - T$. Now every $x \in \mathscr{N}(A_\mu^{n_\mu})$, corresponding to the Riesz decomposition of X into $\mathscr{N}(A_\lambda^{n_\lambda}) \oplus \mathscr{R}(A_\lambda^{n_\lambda})$, has a representation $x = z + y$. As both subspaces are invariant under T, and hence also under A_μ, it follows that

$$0 = A_\mu^{n_\mu}x = \underbrace{A_\mu^{n_\mu}z}_{\in \mathscr{N}(A_\lambda^{n_\lambda})} + \underbrace{A_\mu^{n_\mu}y}_{\in \mathscr{R}(A_\lambda^{n_\lambda})}$$

and so $0 = A_\mu^{n_\mu}z$. On recalling from the above proof that A_μ is bijective on $\mathscr{N}(A_\lambda^{n_\lambda})$, and hence also $A_\mu^{n_\mu}$, it follows that $z = 0$, i.e. $x \in \mathscr{R}(A_\lambda^{n_\lambda})$. Therefore we have shown that

$$\mathscr{N}(A_\mu^{n_\mu}) \subset \mathscr{R}(A_\lambda^{n_\lambda}),$$

in other words

$$\mathscr{R}(E_\mu) \subset \mathscr{N}(E_\lambda),$$

and hence $E_\lambda E_\mu = 0$. □

11.10 Corollary. If $T \in \mathcal{K}(X)$ and $\lambda \in \sigma(T) \setminus \{0\}$, then the resolvent function $\mu \mapsto R(\mu; T)$ has an **(isolated) pole** of **order** n_λ in λ, i.e. the function $\mu \mapsto (\mu - \lambda)^{n_\lambda} R(\mu; T)$ can be complex analytically extended to the point λ, and the value at the point λ is different from the null operator.

Proof. Consider the decomposition

$$X = \underbrace{\mathcal{N}\big((\lambda\mathrm{Id} - T)^{n_\lambda}\big)}_{=\mathcal{R}(E_\lambda)} \oplus \underbrace{\mathcal{R}\big((\lambda\mathrm{Id} - T)^{n_\lambda}\big)}_{=\mathcal{N}(E_\lambda)}$$

and the restrictions

$$T_0 := T \quad \text{to } \mathcal{R}(E_\lambda), \qquad T_1 := T \quad \text{to } \mathcal{N}(E_\lambda).$$

Since λ is an isolated point of $\sigma(T)$, there exists an $r > 0$ with $\mathrm{B}_r(\lambda) \setminus \{\lambda\} \subset \varrho(T)$. Then $\mathrm{B}_r(\lambda) \setminus \{\lambda\} \subset \varrho(T_0)$ and we have from 11.9(4) that $\mathrm{B}_r(\lambda) \subset \varrho(T_1)$, and it holds for $0 < |\mu| < r$ that

$$R(\lambda + \mu; T) = R(\lambda + \mu; T_0)E_\lambda + R(\lambda + \mu; T_1)(\mathrm{Id} - E_\lambda).$$

It follows from 11.3 that $R(\lambda + \cdot; T_1)$ is complex analytic in $\mathrm{B}_r(0)$, and so it remains to show that $R(\lambda + \cdot; T_0)$ has a pole of order n_λ in 0. Consider

$$S(\mu) := \sum_{k=1}^{n_\lambda} \mu^{-k}(T_0 - \lambda\mathrm{Id})^{k-1} \quad \text{for } \mu \neq 0.$$

It holds that

$$S(\mu)\big((\lambda + \mu)\mathrm{Id} - T_0\big) = \sum_{k=1}^{n_\lambda} \mu^{1-k}(T_0 - \lambda\mathrm{Id})^{k-1} - \sum_{k=1}^{n_\lambda} \mu^{-k}(T_0 - \lambda\mathrm{Id})^k$$

$$= \mathrm{Id} - \mu^{-n_\lambda}(T_0 - \lambda\mathrm{Id})^{n_\lambda} = \mathrm{Id}$$

and similarly $\big((\lambda + \mu)\mathrm{Id} - T_0\big)S(\mu) = \mathrm{Id}$, i.e. $R(\lambda + \mu; T_0) = S(\mu)$. $\qquad\square$

The assertion $\sigma(T) \setminus \{0\} \subset \sigma_p(T)$ in 11.9(1) can also be formulated as follows:

11.11 Fredholm alternative. If $T \in \mathcal{K}(X)$ and $\lambda \neq 0$, then it holds that:

Either the equation $Tx - \lambda x = y$ is uniquely solvable for every $y \in X$,

or the equation $Tx - \lambda x = 0$ has nontrivial solutions.

Note: See also theorem 12.8.

11.12 Finite-dimensional case. Let X be a finite-dimensional vector space over $\mathbb{C}$ and let $T : X \to X$ be linear. Then there exist pairwise distinct $\lambda_1, \ldots, \lambda_m \in \mathbb{C}$, where $1 \leq m \leq \dim X$, such that

$$\sigma(T) = \sigma_p(T) = \{\lambda_1, \ldots, \lambda_m\},$$

and orders n_{λ_j} with the properties in 11.9(2) – 11.9(5), so that

$$X = \mathcal{N}\big((\lambda_1\mathrm{Id} - T)^{n_{\lambda_1}}\big) \oplus \cdots \oplus \mathcal{N}\big((\lambda_m\mathrm{Id} - T)^{n_{\lambda_m}}\big).$$

Proof. We equip X with an arbitrary norm. Then $T \in \mathcal{K}(X)$ (see 10.2(3)), and similarly $T_\mu := T - \mu \mathrm{Id}$ for $\mu \in \mathbb{C}$. Now apply 11.9 to e.g. T_0 and T_1.   □

11.13 Jordan normal form. Let $T \in \mathcal{K}(X)$ and let $\lambda \in \sigma_p(T)$ be as in 11.9 or 11.12, respectively. Set $A := \lambda \mathrm{Id} - T$. Then:

(1) For $n = 1, \ldots, n_\lambda$ there exist subspaces E_n with $\mathcal{N}(A^{n-1}) \oplus E_n \subset \mathcal{N}(A^n)$ such that

$$\mathcal{N}(A^{n_\lambda}) = \bigoplus_{k=1}^{n_\lambda} N_k , \quad \text{where} \quad N_k := \bigoplus_{l=0}^{k-1} A^l(E_k) .$$

(2) The subspaces N_k, $k = 1, \ldots, n_\lambda$, are T-invariant and the dimensions $d_k := \dim A^l(E_k)$ are independent of $l \in \{0, \ldots, k-1\}$.

(3) If $\{e_{k,j} ; \, j = 1, \ldots, d_k\}$ are bases of E_k, then

$$\{A^l e_{k,j} ; \, 0 \le l < k \le n_j, \, 1 \le j \le d_k\}$$

is a basis of $\mathcal{N}(A^{n_\lambda})$ and with

$$x = \sum_{k,j,l} \alpha_{k,j,l} \, A^l e_{k,j} \quad \text{and} \quad y = \sum_{k,j,l} \beta_{k,j,l} \, A^l e_{k,j}$$

it holds that $Tx = y$ is equivalent to

$$\begin{bmatrix} \beta_{k,j,0} \\ \vdots \\ \vdots \\ \beta_{k,j,k-1} \end{bmatrix} = \begin{bmatrix} \lambda & -1 & & 0 \\ 0 & \ddots & \ddots & \\ & \ddots & \ddots & -1 \\ 0 & & 0 & \lambda \end{bmatrix} \begin{bmatrix} \alpha_{k,j,0} \\ \vdots \\ \vdots \\ \alpha_{k,j,k-1} \end{bmatrix} ,$$

i.e. the matrix representing T with respect to this basis has a **Jordan normal form**.

Proof. If E is a subspace with $\mathcal{N}(A^{n-1}) \oplus E \subset \mathcal{N}(A^n)$, then

$$\mathcal{N}(A^{n-l-1}) \oplus A^l(E) \subset \mathcal{N}(A^{n-l}) \quad \text{for } 0 \le l < n,$$

and A^l is injective on E. To see this, note that if $x \in E$ with $A^l x = 0$, then also $A^{n-1}x = 0$ because $l \le n - 1$, and so $x \in \mathcal{N}(A^{n-1}) \cap E = \{0\}$. Based on this observation we inductively choose E_n for $n = n_\lambda, \ldots, 1$ such that

$$\mathcal{N}(A^n) = \mathcal{N}(A^{n-1}) \oplus \bigoplus_{l=0}^{n_\lambda - n} A^l(E_{n+l}) .$$

This yields the desired results.   □

11.14 Real case. If X is a Banach space over $\mathbb{R}$ and if $T \in \mathcal{K}(X)$, then the spectral theorem can be applied to their **complexification**, i.e. let

$$\widetilde{X} := X \times X$$

and for $x = (x_1, x_2) \in \widetilde{X}$, $\alpha = a + ib$ with $a, b \in \mathbb{R}$, let

$$\alpha x := (ax_1 - bx_2, ax_2 + bx_1), \quad \overline{x} := (x_1, -x_2).$$

With the above $\widetilde{X}$ becomes a vector space over $\mathbb{C}$. On setting

$$\|x\|_{\widetilde{X}} := \sup_{\theta' \in \mathbb{R}} \left(\|\cos(\theta')x_1 - \sin(\theta')x_2\|_X^2 + \|\sin(\theta')x_1 + \cos(\theta')x_2\|_X^2 \right)^{\frac{1}{2}}$$

it holds that $\left\| e^{i\theta} x \right\|_{\widetilde{X}} = \|x\|_{\widetilde{X}}$ for $x \in \widetilde{X}$ and $\theta \in \mathbb{R}$, and equipped with this norm $\widetilde{X}$ becomes a Banach space over $\mathbb{C}$. Then

$$\widetilde{T}x := (Tx_1, Tx_2)$$

defines the corresponding operator $\widetilde{T} \in \mathcal{K}(\widetilde{X})$, so that theorem 11.9 can now be applied.

Now if $\lambda \in \sigma_p(\widetilde{T})$ with eigenvector e, then

$$\widetilde{T}\overline{e} = \overline{\widetilde{T}e} = \overline{\lambda e} = \overline{\lambda}\, \overline{e},$$

and so $\overline{\lambda} \in \sigma_p(\widetilde{T})$ with eigenvector $\overline{e}$. If $\lambda \in \mathbb{R}$, then the vectors $e_{k,j}$ in 11.13(3) can be chosen to satisfy $\overline{e_{k,j}} = e_{k,j}$. If $\lambda \notin \mathbb{R}$ and $e_{k,j}$ as in 11.13(3), then the vectors $\overline{e_{k,j}}$ have the properties in 11.13(3) with respect to $\overline{\lambda}$.

Remark: In the case when X is a Hilbert space, the above norm satisfies

$$\|x\|_{\widetilde{X}} = \left(\|x_1\|_X^2 + \|x_2\|_X^2 \right)^{\frac{1}{2}}.$$

12 Self-adjoint operators

First we prove some fundamental results for the adjoint map (see 12.1–12.6) and then present a version of the spectral theorem 11.9 for compact normal operators (theorem 12.12). Here we employ the notation $\langle x\,, x'\rangle = \langle x\,, x'\rangle_X = x'(x)$ from 7.4. We remark that the adjoint map of an operator has already been defined in 5.5(8).

12.1 Adjoint operator. Let X, Y be normed spaces. Then

$$\langle x\,, T'y'\rangle_X := \langle Tx\,, y'\rangle_Y \quad \text{for } x \in X,\ y' \in Y'$$

defines an isometric embedding $T \mapsto T'$ from $\mathscr{L}(X;Y)$ to $\mathscr{L}(Y';X')$. We call T' the **adjoint operator** (or the **dual operator**, or the **adjoint**) of T.

Proof. For $T \in \mathscr{L}(X;Y)$ and $y' \in Y'$ we have that $\langle Tx\,, y'\rangle_Y$ is linear in x, with

$$|\langle Tx\,, y'\rangle_Y| \le \|Tx\|_Y \cdot \|y'\|_{Y'} \le \|T\| \cdot \|x\|_X \cdot \|y'\|_{Y'}\,.$$

Hence $\langle x\,, T'y'\rangle_X := \langle Tx\,, y'\rangle_Y$ defines an element $T'y' \in X'$ with

$$\|T'y'\|_{X'} \le \|T\| \cdot \|y'\|_{Y'}\,.$$

In addition, $T'y'$ is linear in y', and so $T' \in \mathscr{L}(Y';X')$ with $\|T'\| \le \|T\|$. Now it holds for $\|y'\|_{Y'} \le 1$ and $\|x\|_X \le 1$ that

$$\|T'\| \ge \|T'y'\|_{X'} \ge |\langle x\,, T'y'\rangle_X| = |\langle Tx\,, y'\rangle_Y|\,.$$

If $Tx \ne 0$, then it follows from 6.17(1) that there exists a $y' \in Y'$ with $\|y'\|_{Y'} = 1$ and $\langle Tx\,, y'\rangle_Y = \|Tx\|_Y$, and hence $\|T'\| \ge \|Tx\|_Y$. Therefore we have shown that

$$\|T'\| \ge \sup_{\|x\|_X \le 1} \|Tx\|_Y = \|T\|\,.$$

$\square$

12.2 Hilbert adjoint. If X and Y are Hilbert spaces and if $R_X : X \to X'$ and $R_Y : Y \to Y'$ denote the isometries from the Riesz representation theorem 6.1, then for $T \in \mathscr{L}(X;Y)$ let

$$T^* := R_X^{-1} T' R_Y \,.$$

Then we have that $T^* \in \mathscr{L}(Y; X)$ and it is characterized by the relationship

$$(x\,, T^*y)_X = (Tx\,, y)_Y \quad \text{for all } x \in X,\ y \in Y.$$

In the special case $Y = X$ we call $T \in \mathscr{L}(X)$ **self-adjoint** if

$$T^* = T \,.$$

12.3 Algebraic properties.

(1) $(\alpha T_1 + T_2)' = \alpha T_1' + T_2'$ for $T_1, T_2 \in \mathscr{L}(X; Y)$, $\alpha \in \mathbb{K}$.

(2) If X and Y are Hilbert spaces, then

$$(\alpha T_1 + T_2)^* = \overline{\alpha} T_1^* + T_2^* \quad \text{for } T_1, T_2 \in \mathscr{L}(X; Y),\ \alpha \in \mathbb{K}.$$

(3) $\mathrm{Id}' = \mathrm{Id}$.

(4) $(T_2 T_1)' = T_1' T_2'$ for $T_1 \in \mathscr{L}(X; Y)$, $T_2 \in \mathscr{L}(Y; Z)$.

(5) $T'' J_X = J_Y T$ for $T \in \mathscr{L}(X; Y)$, where $J_X : X \to X''$ and $J_Y : Y \to Y''$ are the canonical embeddings from 8.2.

(6) If X and Y are Hilbert spaces, then $T^{**} = T$ for $T \in \mathscr{L}(X; Y)$.

Proof (4),(5),(6). We have

$$\langle x\,, (T_2 T_1)' z' \rangle_X = \langle T_2 T_1 x\,, z' \rangle_Z = \langle T_1 x\,, T_2' z' \rangle_Y = \langle x\,, T_1' T_2' z' \rangle_X \,,$$
$$\langle y'\,, T'' J_X x \rangle_{Y'} = \langle T' y'\,, J_X x \rangle_{X'} = \langle x\,, T' y' \rangle_X = \langle Tx\,, y' \rangle_Y$$
$$= \langle y'\,, J_Y T x \rangle_{Y'} \,.$$

In the Hilbert space case we have that $(Tx\,, y)_Y = (x\,, T^*y)_X = (T^{**}x\,, y)_Y.$ □

The following result on the null space of T' is an immediate consequence from the definition of the adjoint map.

12.4 Annihilator. For subspaces $Z \subset X$ the **annihilator** Z^0 is defined by

$$Z^0 := \{x' \in X';\ \langle x\,, x' \rangle_X = 0 \text{ for all } x \in Z\},$$

in words: Z^0 consists of all the functionals that vanish on Z. It holds that:

(1) If X is a Hilbert space and if R_X is as in 12.2, then $Z^0 = R_X(Z^\perp)$.

(2) For $T \in \mathscr{L}(X; Y)$ it holds that $\mathscr{N}(T') = \mathscr{R}(T)^0$.

(3) If Z is closed with codim $Z < \infty$, then dim $Z^0 = $ codim Z.

Proof (2). We have that $y' \in \mathscr{N}(T')$ if and only if for all $x \in X$ it holds that $0 = \langle x\,, T'y' \rangle_X = \langle Tx\,, y' \rangle_Y.$ □

Proof (3). Let $x_1, \ldots, x_n \in X$ be linearly independent with

$$X = Z \oplus \operatorname{span}\{x_1, \dots, x_n\}.$$

By 9.16(1), there exist functionals $x_1', \dots, x_n' \in X'$ with $x_j' = 0$ on Z and $\langle x_i, x_j' \rangle = \delta_{i,j}$ for $i, j = 1, \dots, n$. Then $x_j' \in Z^0$ and the x_j' are linearly independent. If x' is an arbitrary functional from Z^0 and

$$x = z + \sum_{i=1}^n \alpha_i x_i \in X \quad \text{with } z \in Z, \ \alpha_i \in \mathbb{K},$$

then

$$\langle x, x' \rangle = \sum_{i=1}^n \alpha_i \langle x_i, x' \rangle = \sum_{i,j=1}^n \alpha_i \langle x_j, x' \rangle \langle x_i, x_j' \rangle$$
$$= \left\langle x, \sum_{j=1}^n \langle x_j, x' \rangle x_j' \right\rangle,$$

i.e. x' is a linear combination of $x_1', \dots, x_n'$. Hence we have shown that $Z^0 = \operatorname{span}\{x_1', \dots, x_n'\}$ and so $\dim Z^0 = n$. $\qquad\square$

12.5 Theorem. Let X and Y be Banach spaces and let $T \in \mathscr{L}(X; Y)$. Then $T^{-1} \in \mathscr{L}(Y; X)$ exists if and only if $(T')^{-1} \in \mathscr{L}(X'; Y')$ exists, and

$$(T^{-1})' = (T')^{-1}.$$

Proof. If T is invertible, then it follows from 12.3(3) and 12.3(4) that

$$\operatorname{Id} = (T^{-1}T)' = T'(T^{-1})' \quad \text{and similarly} \quad \operatorname{Id} = (T^{-1})'T',$$

i.e. $(T')^{-1} = (T^{-1})' \in \mathscr{L}(X'; Y')$. Conversely, if T' is invertible, then the above result yields that T'' is invertible, and hence it maps closed sets into closed sets. It follows from 12.3(5), since J_X and J_Y are isometries, that

$$\mathscr{R}(J_Y T) = \mathscr{R}(T'' J_X) = T''(\mathscr{R}(J_X))$$

is closed in Y''. Hence,

$$\mathscr{R}(T) = J_Y^{-1}(\mathscr{R}(J_Y T))$$

is also closed. As T' is injective, it follows from 12.4 that $\{0\} = \mathscr{N}(T') = \mathscr{R}(T)^0$. Recalling theorem 6.16 then yields that $Y = \overline{\mathscr{R}(T)} = \mathscr{R}(T)$. This shows that T is surjective. The injectivity of T'' yields that $\{0\} = \mathscr{N}(T'' J_X) = \mathscr{N}(J_Y T) = \mathscr{N}(T)$, which shows that T is also injective. The continuity of T^{-1} then follows from the inverse mapping theorem 7.8. $\qquad\square$

We now investigate the adjoint of compact operators and then in particular normal compact operators.

12.6 Theorem (Schauder). Let X and Y be Banach spaces and let $T \in \mathscr{L}(X; Y)$. Then:

$$T \in \mathscr{K}(X; Y) \quad \Longleftrightarrow \quad T' \in \mathscr{K}(Y'; X').$$

Proof $\Rightarrow$. We have that $S := \overline{T(B_1(0))}$ is compact in Y, and for $y' \in Y'$

$$\|T'y'\|_{X'} = \sup_{\|x\|_X < 1} |\langle x, T'y'\rangle| = \sup_{\|x\|_X < 1} |\langle Tx, y'\rangle|$$
$$= \sup_{y \in S} |\langle y, y'\rangle| = \|y'\|_{C^0(S)}.$$

By 10.1, we have to show the precompactness of $T'(B_1(0)) \subset X'$. This follows from the precompactness of the set

$$A := \{ y'|_S \in C^0(S) ; \ y' \in Y', \ \|y'\|_{Y'} < 1 \} \text{ in } C^0(S).$$

Indeed, if A is precompact, then every sequence $(y'_n|_S)_{n \in \mathbb{N}}$ in A contains a subsequence that converges in $C^0(S)$, so that the above identity yields that $(T'y'_n)_{n \in \mathbb{N}}$ is a Cauchy sequence in X', and so convergent in X'.

Now we prove the precompactness of A. By the Arzelà-Ascoli theorem (see 4.12 in the version for compact metric spaces S), the precompactness of A follows if A is a bounded and equicontinuous subset of $C^0(S)$. But for $y'|_S \in A$ and $y = \lim_{n \to \infty} Tx_n \in S$ with $x_n \in B_1(0) \subset X$ it holds that

$$|\langle y, y'\rangle| = \lim_{n \to \infty} |\langle Tx_n, y'\rangle| \le \|T\|,$$

and for $y'|_S \in A$ and $y_1, y_2 \in S$ we have that

$$|\langle y_1, y'\rangle - \langle y_2, y'\rangle| = |\langle y_1 - y_2, y'\rangle| \le \|y_1 - y_2\|_Y,$$

which yields the desired result.

Remark: The claim can also be shown with the help of the sequential compactness, where a diagonal sequence has to be selected.   □

Proof $\Leftarrow$. The previously shown result yields that $T'' \in \mathcal{K}(X''; Y'')$. But by 12.3(5), we have $T = J_Y^{-1} T'' J_X$, because the right-hand side is well-defined, since by 12.3(5) the range of $T'' J_X$ is contained in the closed subspace $\mathcal{R}(J_Y T)$. The compactness of T then follows from 10.3.   □

12.7 Remark. With the help of 12.6 we can give an alternative proof of 11.8(5). We had to show that $\dim \mathcal{N}(A) \le \operatorname{codim} \mathcal{R}(A)$. Now it follows from 12.6 and 11.8(4), applied to $A' = \operatorname{Id} - T'$, that

$$\operatorname{codim} \mathcal{R}(A') \le \dim \mathcal{N}(A').$$

Then we obtain the desired result, if we can prove the two inequalities

$$\dim \mathcal{N}(A) \le \dim \mathcal{N}(A'') \quad \text{and} \quad \dim \mathcal{N}(A') \le \operatorname{codim} \mathcal{R}(A).$$

This is because we obtain $\dim \mathcal{N}(A'') \le \operatorname{codim} \mathcal{R}(A')$, on applying the latter inequality to A', and all four inequalities together then yield the claim. Subsequently we even obtain equality in all inequalities, and so in particular

$$\dim \mathcal{N}(A) = \dim \mathcal{N}(A') = \operatorname{codim} \mathcal{R}(A).$$

Now by 12.3(5) (see the proof of 12.5), $\mathcal{N}(A''J_X) = \mathcal{N}(J_Y A) = \mathcal{N}(A)$, which implies that $\dim \mathcal{N}(A) \leq \dim \mathcal{N}(A'')$ using the injectivity of J_X. Moreover, it follows from 12.4 that

$$\dim \mathcal{N}(A') = \dim \mathcal{R}(A)^0 = \operatorname{codim} \mathcal{R}(A).$$

This also implies an improvement of the result in 11.11.

12.8 Theorem (Fredholm). Let X be a Banach space, let $T \in \mathcal{K}(X)$ and let $\lambda \neq 0$. Then: For $y \in X$ the system of equations

$$Tx - \lambda x = y$$

has a solution $x \in X$, if and only if $\langle y, x' \rangle = 0$ for all solutions $x' \in X'$ of the homogeneous adjoint equation

$$T'x' - \lambda x' = 0.$$

The corresponding (finite) number of side conditions on y is equal to the number of linearly independent solutions z of the homogeneous equation

$$Tz - \lambda z = 0.$$

Proof. On setting $A := \lambda \mathrm{Id} - T$ the last claim follows from $\dim \mathcal{N}(A') = \dim \mathcal{N}(A)$, which was shown in 12.7. The condition on y is, recall 12.4(2),

$$\langle y, x' \rangle = 0 \quad \text{for all } x' \in \mathcal{N}(A') = \mathcal{R}(A)^0.$$

Combining 11.8(2) and theorem 6.16 yields that this is equivalent to $y \in \mathcal{R}(A)$, i.e. to the solvability of the equation $Ax = y$. □

12.9 Normal operators. Let X be a Hilbert space (over $\mathbb{K}$). Then the operator $T \in \mathcal{L}(X)$ is called *normal* if

$$T^*T - TT^* = 0,$$

i.e. if T and T^* commute. Every self-adjoint operator is normal. If T is normal, then so is $\lambda \mathrm{Id} - T$ for all $\lambda \in \mathbb{K}$. Moreover:

$$T \text{ normal} \quad \Longleftrightarrow \quad \|Tx\|_X = \|T^*x\|_X \quad \text{for all } x \in X.$$

In particular, this implies that for normal operators and all $\lambda \in \mathbb{K}$ it holds that

$$\mathcal{N}(\lambda \mathrm{Id} - T) = \mathcal{N}(\bar{\lambda} \mathrm{Id} - T^*).$$

Proof $\Rightarrow$. $(Tx, Tx)_X = (x, T^*Tx)_X = (x, TT^*x)_X = (T^*x, T^*x)_X.$ □

Proof $\Leftarrow$. The identity

$$\tfrac{1}{4}\big(\|a + b\|_X^2 - \|a - b\|_X^2\big) = \operatorname{Re}(a, b)_X \quad \text{for } a, b \in X$$

implies that

$$\operatorname{Re}(Tx\,,\,Ty)_X = \operatorname{Re}(T^*x\,,\,T^*y)_X \quad \text{for } x, y \in X.$$

Replacing y with iy in the case $\mathbb{K} = \mathbb{C}$ then yields that

$$0 = (Tx\,,\,Ty)_X - (T^*x\,,\,T^*y)_X = (T^*Tx - TT^*x\,,\,y)_X \quad \text{for } x, y \in X,$$

and so $T^*T - TT^* = 0$. $\square$

Next we investigate the spectrum of normal operators in Hilbert spaces. We begin with a strengthening of the result 11.4 for the spectral radius.

12.10 Lemma. If X is a Hilbert space over $\mathbb{C}$, $X \neq \{0\}$ and if $T \in \mathscr{L}(X)$ is normal, then

$$\sup_{\lambda \in \sigma(T)} |\lambda| = \|T\|.$$

Proof. Let $T \neq 0$. By 11.4, we obtain the desired result if we can show that

$$\|T^m\| \geq \|T\|^m \quad \text{for } m \geq 0 \quad \text{(then equality holds as well)}.$$

For $m = 0, 1$ this holds trivially. For $m \geq 1$ and $x \in X$

$$
\begin{aligned}
\|T^m x\|_X^2 &= (T^*T^m x\,,\,T^{m-1}x)_X \leq \|T^*T^m x\|_X \, \|T^{m-1}x\|_X \\
&= \|T^{m+1}x\|_X \, \|T^{m-1}x\|_X \quad \text{(recall 12.9)} \\
&\leq \|T^{m+1}\| \, \|T\|^{m-1} \|x\|_X^2\,,
\end{aligned}
$$

and so

$$\|T^m\|^2 \leq \|T^{m+1}\| \, \|T\|^{m-1}.$$

Hence on assuming that $\|T^m\| \geq \|T\|^m$ already holds, it follows that

$$\|T^{m+1}\| \geq \frac{\|T^m\|^2}{\|T\|^{m-1}} \geq \|T\|^{2m-(m-1)} = \|T\|^{m+1}.$$

$\square$

12.11 Example. Let $(e_k)_{k \in N}$, $N \subset \mathbb{N}$, be an orthonormal system in the Hilbert space X and let $\lambda_k \in \mathbb{K}$ with $|\lambda_k| \leq r < \infty$ for $k \in N$. Then

$$Tx := \sum_{k \in N} \lambda_k \, (x\,,\,e_k)_X \, e_k$$

defines an operator $T \in \mathscr{L}(X)$, since (recall 9.6 and 9.7)

$$\|Tx\|_X^2 = \sum_{k \in N} |\lambda_k|^2 |(x\,,\,e_k)_X|^2 \leq r^2 \sum_{k \in N} |(x\,,\,e_k)_X|^2 \leq r^2 \|x\|_X^2.$$

On noting that

$$(Tx\,,\,y)_X = \sum_{k \in N} \lambda_k \, (x\,,\,e_k)_X \, (e_k\,,\,y)_X = \left(x\,,\,\sum_{k \in N} \overline{\lambda_k} \, \overline{(e_k\,,\,y)_X} \, e_k\right)_X$$

it follows that

$$T^*x = \sum_{k \in N} \overline{\lambda}_k \, (x \, , \, e_k)_X \, e_k \, ,$$

and so

$$T^*Tx \text{ (as well as } TT^*x) \; = \sum_{k \in N} |\lambda_k|^2 \, (x \, , \, e_k)_X \, e_k \, .$$

Hence T is normal. If N is finite, then T is also compact (recall 10.2(3)).
Claim: In the case $N = \mathbb{N}$ we have that

$$T \text{ is compact} \quad \Longleftrightarrow \quad \lim_{k \to \infty} \lambda_k = 0 \, .$$

Proof $\Leftarrow$. Setting

$$T_n x := \sum_{k \leq n} \lambda_k \, (x \, , \, e_k)_X \, e_k$$

defines $T_n \in \mathscr{K}(X)$ and similarly to the above estimete we have

$$\|Tx - T_n x\|_X^2 \leq \sup_{k > n} |\lambda_k|^2 \, \|x\|_X^2 \, ,$$

and so

$$\|T - T_n\| \leq \sup_{k > n} |\lambda_k| \longrightarrow 0 \quad \text{as } n \to \infty \, .$$

This shows that $T \in \mathscr{K}(X)$ on recalling 10.2(2).   □

Proof $\Rightarrow$. If we assume that there exists a subsequence $k_j \to \infty$ with $|\lambda_{k_j}| \geq c > 0$, then the e_{k_j} are bounded and for $i \neq j$ it follows that

$$\left\| Te_{k_i} - Te_{k_j} \right\|^2 = \left\| \lambda_{k_i} e_{k_i} - \lambda_{k_j} e_{k_j} \right\|^2 = \sqrt{|\lambda_{k_i}|^2 + |\lambda_{k_j}|^2} \geq c \, ,$$

which contradicts the compactness of T.   □

Next we want to show that every compact normal operator can be written in the form of the previous example.

12.12 Spectral theorem for compact normal operators. If X is a Hilbert space over $\mathbb{C}$ and if $T \in \mathscr{K}(X)$ is normal, $T \neq 0$, then:

(1) There exist an orthonormal system $(e_k)_{k \in N}$ in X and a sequence $(\lambda_k)_{k \in N}$ in $\mathbb{C}$ with $N \subset \mathbb{N}$ such that $\lambda_k \neq 0$ and

$$Te_k = \lambda_k e_k \text{ for } k \in N \, , \quad \sigma(T) \setminus \{0\} = \{\lambda_k \, ; \, k \in N\} \, ,$$

i.e. the numbers λ_k are the nonzero eigenvalues of T with eigenvectors e_k. If N is infinite, then $\lambda_k \to 0$ as $k \to \infty$.

(2) The orders satisfy: $n_{\lambda_k} = 1$ for all k.

(3) $X = \mathcal{N}(T) \perp \text{clos}\,(\text{span}\{e_k\,;\; k \in N\})$.

(4) $Tx = \sum_{k \in N} \lambda_k\,(x\,,\,e_k)_X\,e_k$ for all $x \in X$.

Observe: In this notation the values λ_k need not be distinct for different k.

Proof. By the spectral theorem for compact operators (theorem 11.9), we know that $\sigma(T) \setminus \{0\}$ consists of eigenvalues λ_k, $k \in N \subset \mathbb{N}$, with $\lambda_k \to 0$ for $k \to \infty$, if N is infinite. In this numeration we assume that all the values λ_k are pairwise distinct. In addition, $E_k := \mathcal{N}(\lambda_k \text{Id} - T)$ are finite-dimensional. Moreover, we define $E_0 := \mathcal{N}(T)$ and $\lambda_0 := 0$. We have from 12.9 that

$$E_k = \mathcal{N}(\overline{\lambda}_k \text{Id} - T^*) \quad \text{for } k \in N \cup \{0\}. \tag{12-8}$$

This implies that the eigenspaces are perpendicular to each other, i.e. that

$$E_k \perp E_l \quad \text{for } k, l \in N \cup \{0\} \text{ with } k \neq l.$$

Indeed, if $x_k \in E_k$ and $x_l \in E_l$, then

$$\lambda_k\,(x_k\,,\,x_l)_X = (Tx_k\,,\,x_l)_X = (x_k\,,\,T^*x_l)_X = \left(x_k\,,\,\overline{\lambda}_l x_l\right)_X = \lambda_l\,(x_k\,,\,x_l)_X\,.$$

Since $\lambda_k \neq \lambda_l$, we conclude that $(x_k\,,\,x_l)_X = 0$. We claim that

$$X = \text{clos}\left(\underset{k \in N \cup \{0\}}{\perp} E_k\right). \tag{12-9}$$

To show this, let

$$y \in Y := \left(\underset{k \in N \cup \{0\}}{\perp} E_k\right)^{\perp}.$$

If $x \in E_k$, $k \in N \cup \{0\}$, then (12-8) yields

$$(Ty\,,\,x)_X = (y\,,\,T^*x)_X = \left(y\,,\,\overline{\lambda}_k x\right)_X = \lambda_k\,(y\,,\,x)_X = 0\,,$$

since $Y \subset E_k^{\perp}$. Hence $Ty \in Y$, i.e. Y is a T-invariant, closed subspace. Now consider

$$T_0 := T\big|_Y\,, \quad \text{and so } T_0 \in \mathcal{K}(X) \text{ and } T_0 \text{ is normal.}$$

As T_0 is normal, it follows from 12.10, on assuming that $Y \neq \{0\}$, that there exists a $\lambda \in \sigma(T_0)$ such that $|\lambda| = \|T_0\|$. If we assume that $T_0 \neq 0$, then by the spectral theorem 11.9(1), λ is an eigenvalue of T_0, and hence also of T, i.e. $E_k \cap Y \neq \{0\}$ for some $k \in N$, which contradicts the definition of Y.

Hence, $T_0 = 0$, i.e. $Y \subset \mathcal{N}(T) = E_0$, and so $Y \subset E_0 \cap E_0^{\perp} = \{0\}$. This proves the decomposition of the space X (see 9.17(3)). Now if Q_k for $k \in N \cup \{0\}$ denotes the orthogonal projection onto E_k, then it follows that (see 9.20(2))

$$x = \sum_{k \in N \cup \{0\}} Q_k x \quad \text{for } x \in X,$$

and so

$$Tx = \sum_{k \in \mathbb{N} \cup \{0\}} TQ_k x = \sum_{k \in \mathbb{N}} \lambda_k Q_k x \,. \tag{12-10}$$

This now implies the desired representation of T: With $d_k := \dim E_k$, choose orthonormal bases $(e_{k1}, \ldots, e_{kd_k})$ of E_k. Then (see 9.6)

$$Q_k x = \sum_{j=1}^{d_k} (Q_k x \,, e_{kj})_X \, e_{kj} = \sum_{j=1}^{d_k} (x \,, e_{kj})_X \, e_{kj} \,.$$

From (12-10) it follows in particular that $E_k = \mathscr{N}\big((\lambda_k \mathrm{Id} - T)^2\big)$, and so $n_{\lambda_k} = 1$. To see this, note that for $x \in \mathscr{N}\big((\lambda_k \mathrm{Id} - T)^2\big)$ we have that

$$0 = (\lambda_k \mathrm{Id} - T)^2 x = \sum_{j \in \mathbb{N} \cup \{0\}} (\lambda_k - \lambda_j)^2 Q_j x \,,$$

and so $Q_j x = 0$ for $j \neq k$, i.e. $x = Q_k x \in E_k$.   □

If the operator T in 12.12 is self-adjoint, then the following holds true:

12.13 Remark. Let X be a Hilbert space over $\mathbb{C}$ and let $T \in \mathscr{L}(X)$.

(1) If T is self-adjoint, i.e. if $T^* = T$, then $\sigma_p(T) \subset [-\|T\|, \|T\|] \subset \mathbb{R}$, and if T is compact, then $\|T\|$ or $-\|T\|$ is an eigenvalue.

(2) If T is self-adjoint and **positive semidefinite**, i.e. if $(Tx \,, x)_X \geq 0$ for all $x \in X$, then $\sigma_p(T) \subset [0, \|T\|]$, and if T is compact, then $\|T\|$ is an eigenvalue.

Proof. If λ is an eigenvalue and x a corresponding eigenvector, then

$$\lambda \|x\|_X^2 = (\lambda x \,, x)_X = (Tx \,, x)_X = (x \,, T^* x)_X$$
$$= (x \,, Tx)_X = (x \,, \lambda x)_X = \overline{\lambda} \|x\|_X^2 \,,$$

and so $\lambda = \overline{\lambda}$, since $x \neq 0$. The second claim in (1) then follows from 11.9(1) and 12.10. In (2) we have in addition that

$$\lambda \|x\|_X^2 = (Tx \,, x)_X \geq 0 \,,$$

and so $\lambda \geq 0$.   □

12.14 Eigenvalue problem as a variational problem. Let X be a Hilbert space over $\mathbb{C}$ and let $T \in \mathscr{K}(X)$, $T \neq 0$, be self-adjoint and positive semidefinite. Since $(Tx \,, x)_X \in \mathbb{R}$ for all $x \in X$, we can consider the following variational problem:

Maximize $(Tx \,, x)_X$ under the constraint $\|x\| = 1$.

We want to show that this variational problem has a solution, and that

$$\sup_{\lambda \in \sigma(T)} \lambda = \sup_{\|x\| \leq 1} (Tx, x)_X .$$

(On recalling from 12.13(2) that $\sigma(T) \subset [0, \|T\|]$, we note that the supremum on the left-hand side is well defined and finite.) For the proof we denote the supremum on the right-hand side by s. For $\lambda \in \sigma(T) \setminus \{0\}$ it holds that $\lambda = (Tx, x)_X \leq s$ if x is an eigenvector of T corresponding to λ with $\|x\| = 1$, and hence 12.10 yields that

$$0 < \|T\| = \sup_{\lambda \in \sigma(T)} \lambda \leq s .$$

We now choose points $x_k \in X$ with $\|x_k\| = 1$ and

$$(Tx_k, x_k)_X \nearrow s \quad \text{as } k \to \infty.$$

It follows from theorem 8.10 (see 8.11(1)) that there exist a subsequence and an $x \in X$ such that

$$x_k \to x \text{ weakly in } X \text{ as } k \to \infty,$$

and, by 8.3(4), it holds that $\|x\| \leq 1$. Being compact, it follows from 10.2(1) that T is also completely continuous, and so $Tx_k \to Tx$ (strongly) in X, which on recalling 8.3(6) yields that

$$(Tx_k, x_k)_X \to (Tx, x)_X \quad \text{as } k \to \infty,$$

i.e. $(Tx, x)_X = s$. Moreover, $s > 0$ implies that $x \neq 0$. In fact, $\|x\| = 1$, because otherwise

$$\left(T\left(\frac{x}{\|x\|} \right), \frac{x}{\|x\|} \right)_X = \frac{s}{\|x\|^2} > s .$$

Hence we have shown that the supremum is attained for x. In addition, it follows for all $y \in X$ and $\varepsilon \in \mathbb{R}$ that

$$s \geq \frac{(T(x + \varepsilon y), x + \varepsilon y)_X}{\|x + \varepsilon y\|^2} = \frac{(Tx, x)_X + 2\varepsilon \operatorname{Re}(Tx, y)_X + \varepsilon^2 (Ty, y)_X}{\|x\|^2 + 2\varepsilon \operatorname{Re}(x, y)_X + \varepsilon^2 \|y\|^2} ,$$

where we used the fact that T is self-adjoint, and hence that

$$0 = \frac{\mathrm{d}}{\mathrm{d}\varepsilon} \frac{(T(x + \varepsilon y), x + \varepsilon y)_X}{\|x + \varepsilon y\|^2} \bigg|_{\varepsilon = 0}$$

$$= 2 \operatorname{Re}(Tx, y)_X - s\, 2 \operatorname{Re}(x, y)_X = 2 \operatorname{Re}(Tx - sx, y)_X .$$

As this holds for all y, it follows that $Tx = sx$, and so

$$s \leq \sup_{\lambda \in \sigma(T)} \lambda .$$

For the determination of all eigenvectors consider the space $\widetilde{X} := \operatorname{span}\{x\}^\perp$. Since for $\tilde{x} \in \widetilde{X}$

$$(T\widetilde{x},\, x)_X = (\widetilde{x},\, Tx)_X = s\,(\widetilde{x},\, x)_X = 0\,,$$

we have that $\widetilde{X}$ is a T-invariant subspace. Hence we can repeat the above procedure on $\widetilde{X}$ in order to obtain all nonzero eigenvalues of T and the decomposition of the space X according to 12.12.

Next let us apply the spectral theorem 12.12 for self-adjoint operators to integral operators (12.15) and differential operators (12.17).

12.15 Self-adjoint integral operator. Let $\Omega \subset \mathbb{R}^n$ be open, let $K \in L^2(\Omega \times \Omega)$ with $K(x,y) = \overline{K(y,x)}$, and let T be the corresponding Hilbert–Schmidt operator from 5.12 (see 10.15). Then there exist a finite or countably infinite orthonormal system $(f_k)_{k\in N}$ in $L^2(\Omega)$ with $N \subset \mathbb{N}$, and real numbers $\lambda_k \neq 0$ for $k \in N$ with (if N infinite) $\lambda_k \to 0$ as $k \to \infty$, such that

$$Tf_k = \lambda_k f_k\,, \quad \sum_{k\in N} \lambda_k^2 \leq \|K\|^2_{L^2(\Omega\times\Omega)}$$

and

$$K(x,y) = \sum_{k\in N} \lambda_k f_k(x)\overline{f_k(y)} \quad \text{for almost all } (x,y),$$

where (if N is infinite) this series converges in $L^2(\Omega \times \Omega)$.

Proof. We recall from 10.15 that $T \in \mathscr{K}\left(L^2(\Omega)\right)$. For $f, g \in L^2(\Omega)$

$$(g\,,\, Tf)_{L^2} = \int_\Omega \int_\Omega \overline{K(x,y)}\, g(x)\,\overline{f(y)}\,\mathrm{d}y\,\mathrm{d}x\,.$$

It follows from $\overline{K(x,y)} = K(y,x)$ that this is equal to $\overline{(f\,,\, Tg)_{L^2}}$, and so T is self-adjoint and we can apply 12.12. Let $f_k,\, k \in N \subset \mathbb{N}$, be the eigenfunctions of T from 12.12 and let λ_k be the corresponding eigenvalues, which by 12.13 are real. It follows from the proof of 10.15 that

$$\sum_{k\in N} \lambda_k^2 = \sum_{k\in N} \|Tf_k\|^2_{L^2} \leq \|K\|^2_{L^2}\,.$$

Consider the series

$$H(x,y) := \sum_{k\in N} \lambda_k f_k(x)\overline{f_k(y)}\,.$$

In the case that N is infinite, and so $N = \mathbb{N}$ without loss of generality, it holds for $m \leq l$, since $\|f_k\|_{L^2} = 1$, that

$$\int_\Omega \int_\Omega \left|\sum_{j=m}^{l} \lambda_j f_j(x)\overline{f_j(y)}\right|^2 \mathrm{d}x\,\mathrm{d}y$$

$$= \sum_{j,k=m}^{l} \lambda_j \lambda_k \int_\Omega \int_\Omega f_j(x)\overline{f_j(y)}\,\overline{f_k(x)}f_k(y)\,\mathrm{d}x\,\mathrm{d}y = \sum_{j=m}^{l} \lambda_j^2 \longrightarrow 0$$

as $m \to \infty$, i.e. the series converges in $L^2(\Omega \times \Omega)$.

We now prove the representation of K. For $g \in L^2(\Omega)$ it follows from 12.12(4) that

$$Tg = \sum_{k \in N} \lambda_k \, (g \, , \, f_k)_{L^2} \, f_k \, ,$$

which means that for almost all x

$$\int_\Omega K(x,y)g(y) \, \mathrm{d}y = \sum_{k \in N} \lambda_k f_k(x) \int_\Omega g(y) \overline{f_k(y)} \, \mathrm{d}y \, , \tag{12-11}$$

with the set where this need not hold initially depending on g. However, on noting that Fubini's theorem yields for almost all x that

$$K(x, \cdot) \quad \text{and} \quad H(x, \cdot) = \sum_{k \in N} \lambda_k f_k(x) \overline{f_k(\cdot)} \quad \text{in } L^2(\Omega),$$

where the series, similarly to the above, converges in $L^2(\Omega)$, it follows that

$$\int_\Omega K(x,y)g(y) \, \mathrm{d}y = \int_\Omega H(x,y)g(y) \, \mathrm{d}y$$

for almost all x. For such x both sides of this identity depend continuously on $g \in L^2(\Omega)$. As $L^2(\Omega)$ is separable, it follows that for almost all such x this identity holds for all $g \in L^2(\Omega)$, which on recalling 4.22 implies the representation of K. □

12.16 Eigenvalue problem for the Laplace operator. Let $\Omega \subset \mathbb{R}^n$ be open and bounded. Consider functions $v : \,]0, \infty[\, \times \overline{\Omega} \to \mathbb{R}$ that satisfy the *linear wave equation*

$$\partial_t^2 v - \Delta v = 0 \quad \text{in }]0, \infty[\, \times \Omega, \tag{12-12}$$

with the boundary condition

$$v = 0 \quad \text{on }]0, \infty[\, \times \partial\Omega. \tag{12-13}$$

Here t denotes the time variable and $\Delta v := \sum_{i=1}^n \partial_{x_i}^2 v$ is the **Laplace operator**. The equation describes an approximation of the oscillation of an idealized membrane (in the case $n = 2$) in the linear case that is clamped along $\partial\Omega$, where $x \mapsto (x, v(t,x))$ denotes the membrane at time t, with the membrane at rest being described by the set $\overline{\Omega} \times \{0\}$ (the map $x \mapsto (x,0)$). Hence v denotes the displacement of the membrane, and the differential equation is valid for small displacements of a thin membrane. The *separation of variables*

$$v(t,x) = w(t)u(x)$$

yields (if w and u are twice continuously differentiable) that

$$w''u = w\Delta u,$$

which in the case $v \neq 0$ is only possible if there exists a $\lambda \in \mathbb{R}$ with

$$\Delta u + \lambda u = 0 \quad \text{in } \Omega,$$
$$w'' + \lambda w = 0 \quad \text{in }]0, \infty[.$$

The boundary condition (12-13) implies that $u(x) = 0$ for $x \in \partial\Omega$. Then we formally obtain, on multiplying the equation for u by u and integrating over Ω, that

$$\lambda \int_\Omega u^2 \, \mathrm{dL}^n = -\int_\Omega \Delta u \cdot u \, \mathrm{dL}^n$$

$$= -\underbrace{\int_{\partial\Omega} u\nabla u \bullet \nu_\Omega \, \mathrm{dH}^{n-1}}_{=0} + \int_\Omega |\nabla u|^2 \, \mathrm{dL}^n \geq 0,$$

and so $\lambda > 0$, because $u \neq 0$.

For every $\lambda > 0$ the ordinary differential equation for w has the general solution

$$w(t) = \mathrm{Re}(\bar{a}e^{i\mu t}) = a_1 \cos(\mu t) + a_2 \sin(\mu t) \quad \text{with } a \in \mathbb{C} \text{ and } \mu = \sqrt{\lambda}.$$

To see this, let $w \in C^2([t_0, t_1[; \mathbb{R})$ be an arbitrary solution on the time interval $[t_0, t_1[$. Setting

$$a_1 := w(t_0), \quad a_2 := \tfrac{1}{\mu}w'(t_0), \quad a := a_1 + ia_2$$

we have that

$$w_*(t) := \mathrm{Re}(\bar{a}\,e^{i\mu(t-t_0)}) = a_1 \cos(\mu(t - t_0)) + a_2 \sin(\mu(t - t_0))$$

is a particular solution. Then the vector-valued function

$$W := \begin{bmatrix} w - w_* \\ w' - w'_* \end{bmatrix}$$

satisfies

$$W' = \begin{bmatrix} w' - w'_* \\ -\lambda(w - w_*) \end{bmatrix} = \begin{bmatrix} 0 & 1 \\ -\lambda & 0 \end{bmatrix} W,$$

and so $|W'| \leq C|W|$. Since $W(t_0) = 0$, it follows for $t_0 < t < t_1$ that

$$|W(t)| = \left| \int_{t_0}^t W'(s) \, \mathrm{d}s \right| \leq C \cdot \int_{t_0}^t |W(s)| \, \mathrm{d}s \leq C|t - t_0| \sup_{[t_0, t]} |W|.$$

Taking the supremum over all $t \in [t_0, t_0 + \delta]$ yields that

$$\sup_{[t_0, t_0+\delta]} |W| \leq C\,\delta \sup_{[t_0, t_0+\delta]} |W|,$$

and so, say for $\delta = \frac{1}{2C}$, that $W = 0$ on $[t_0, t_0 + \delta]$. Repeating the argument now for $t_0 + \delta$ in place of t_0 yields, on noting that the choice of δ only depended on λ, after finitely many steps that $W = 0$ on $[t_0, t_1[$, which we wanted to show.

Now we consider the eigenvalue problem for u. The boundary condition $u = 0$ on $\partial\Omega$ means that the weak formulation of the eigenvalue problem, analogously to 6.5, is given by:

Find $\lambda \in \mathbb{R}$ and $u \in W_0^{1,2}(\Omega; \mathbb{R})$ with $u \neq 0$ and

$$\int_\Omega (\nabla\zeta \bullet \nabla u - \lambda\zeta u)\, d\mathrm{L}^n = 0 \quad \text{for all } \zeta \in W_0^{1,2}(\Omega; \mathbb{R}). \tag{12-14}$$

12.17 Theorem. The eigenvalue problem (12-14) has the following solution: There exist pairwise distinct $\lambda_k > 0$ (the eigenvalues) for $k \in \mathbb{N}$ and finite-dimensional subspaces $E_k \subset W_0^{1,2}(\Omega; \mathbb{R})$ (the eigenspaces) with the following properties:

(1) It holds that $\lambda_k \to \infty$ as $k \to \infty$ and

$$L^2(\Omega; \mathbb{R}) = \mathrm{clos}\left(\underset{k \in \mathbb{N}}{\perp} E_k\right).$$

(2) It holds that $(\lambda, u) \in \mathbb{R} \times W_0^{1,2}(\Omega; \mathbb{R})$ with $u \neq 0$ is a solution to the eigenvalue problem in (12-14) if and only if

$$\lambda = \lambda_k\,, \quad u \in E_k \quad \text{for some } k \in \mathbb{N}.$$

(3) It holds that $E_k \subset C^\infty(\Omega; \mathbb{R})$ for $k \in \mathbb{N}$, and hence $u \in E_k$ are classical solutions of

$$\Delta u + \lambda_k u = 0 \quad \text{in } \Omega.$$

(4) The functions in E_k are real analytic. The analyticity of the eigenfunctions $u \in E_k$ implies that $\Omega \cap \{u = 0\}$ is an analytic set. For $n = 2$ these sets are called the **nodal lines** of the membrane corresponding to the eigensolution u.

Proof (1), (2). In the following let $\mathbb{K} = \mathbb{R}$. It follows from theorem 6.8 that for $f \in L^2(\Omega)$ there exists a unique $u_f \in W_0^{1,2}(\Omega)$ such that

$$\int_\Omega (\nabla\zeta \bullet \nabla u_f - \zeta f)\, d\mathrm{L}^n = 0 \quad \text{for all } \zeta \in W_0^{1,2}(\Omega),$$

and with a constant C that is independent of f we have

$$\|u_f\|_{W^{1,2}(\Omega)} \leq C \cdot \|f\|_{L^2(\Omega)} \quad \text{for all } f \in L^2(\Omega).$$

On noting that u_f depends linearly on f, the rule $Tf := u_f$ defines an operator in $\mathscr{L}(L^2(\Omega); W_0^{1,2}(\Omega))$. Now the embedding of $W_0^{1,2}(\Omega)$ into $L^2(\Omega)$ is compact (see 10.9 or Rellich's embedding theorem A8.1). Hence if we consider Tf as an element in $L^2(\Omega)$, then $T \in \mathscr{K}(L^2(\Omega))$. (Note that $T = A^{-1}J$ in 10.14(3).)

In addition, T is self-adjoint and positive semidefinite, because upon choosing $\zeta = Tg$ with an arbitrary $g \in L^2(\Omega)$ in the weak differential equation for Tf, we obtain, due to symmetry, that

$$(Tg\,,\,f)_{L^2(\Omega)} = \int_\Omega \nabla Tg \bullet \nabla Tf \,\mathrm{dL}^n = (g\,,\,Tf)_{L^2(\Omega)}\,.$$

Setting here $g = f$ we find that

$$(Tf\,,\,f)_{L^2(\Omega)} = \int_\Omega |\nabla Tf|^2 \,\mathrm{dL}^n \geq 0\,.$$

Moreover, T is injective, since $Tf = 0$ implies that

$$\int_\Omega \zeta f \,\mathrm{dL}^n = 0 \quad \text{for all } \zeta \in C_0^\infty(\Omega)\,,$$

and hence $f = 0$ almost everywhere in Ω, recall 4.22. With the help of the complexification in 11.14 it then follows that: The spectral theorem 12.12 for compact normal operators and 12.13 yield the existence of pairwise distinct (as in 11.9, we group equal eigenvalues together) values $0 < \lambda_k \in \mathbb{R}$, $k \in \mathbb{N}$, with $\lambda_k \to \infty$ (!) as $k \to \infty$, such that the subspaces $E_k := \mathscr{N}(\lambda_k^{-1}\mathrm{Id} - T) \neq \{0\}$ are finite-dimensional and

$$L^2(\Omega) = \overline{\perp_{k\in\mathbb{N}} E_k}\,.$$

Here we have used that T is injective. We observe that there must exist infinitely many eigenvalues because $L^2(\Omega)$ is infinite-dimensional, while each of the null spaces is finite-dimensional. Thus we have included all the eigenvalues of T.

Since T maps into $W_0^{1,2}(\Omega)$, we have that $E_k \subset W_0^{1,2}(\Omega)$. Hence $u \in E_k$ means that $u \in W_0^{1,2}(\Omega)$ with $T(\lambda_k u) = u$, i.e.

$$\int_\Omega (\nabla\zeta \bullet \nabla u - \lambda_k\zeta u)\,\mathrm{dL}^n = 0 \quad \text{for all } \zeta \in W_0^{1,2}(\Omega)\,. \tag{12-15}$$

$\square$

Proof (3). If $u \in E_k$, then (12-15) holds for $\zeta \in C_0^\infty(\Omega)$. Since $\lambda_k u \in L^2_{\mathrm{loc}}(\Omega)$, it follows from Friedrichs' theorem (see A12.2) that $u \in W^{2,2}_{\mathrm{loc}}(\Omega)$. Here the space $W^{2,2}_{\mathrm{loc}}(\Omega)$ is defined as in 5.13(4). Then for $i = 1,\dots,n$ we replace the test function ζ in (12-15) by $-\partial_i\zeta$ and obtain from the definition of the Sobolev spaces that

$$\int_\Omega (\nabla\zeta \bullet \nabla\partial_i u - \lambda_k\zeta\partial_i u)\,\mathrm{dL}^n = 0\,,$$

where now $\lambda_k \partial_i u \in L^2_{\mathrm{loc}}(\Omega)$, so that Friedrichs' theorem yields that $\partial_i u \in W^{2,2}_{\mathrm{loc}}(\Omega)$ for $i = 1, \ldots, n$, and hence $u \in W^{3,2}_{\mathrm{loc}}(\Omega)$. Repeatedly applying this argumentation we can successfully increase the regularity of u and conclude, thanks to Sobolev's embedding theorem 10.13, that

$$u \in \bigcap_{m \in \mathbb{N}} W^{m,2}_{\mathrm{loc}}(\Omega) \subset C^\infty(\Omega).$$

□

Proof (4). (For $n = 1$ we obtain the desired result on writing u, similarly to the solution of the eigenvalue problem for w, as an exponential function.) We need to quantify the bounds on the derivatives of u. For all multi-indices α

$$\Delta \partial^\alpha u = -\lambda_k \partial^\alpha u.$$

Now if $B_{3R}(x_0) \subset \Omega$ and $0 < \kappa_0 < \kappa_1 \le 3$, then it follows from A12.1, on setting $B_r := B_r(x_0)$, that

$$\|\nabla \partial^\alpha u\|_{L^2(B_{\kappa_0 R})} \le C\left(\kappa_1 R \lambda_k + \frac{1}{(\kappa_1 - \kappa_0)R}\right) \|\partial^\alpha u\|_{L^2(B_{\kappa_1 R})}.$$

Iterating this inequality we obtain for $0 < \kappa_0 < \kappa_1 < \ldots < \kappa_l \le 3$ and β with $|\beta| = l$ that

$$\|\partial^\beta u\|_{L^2(B_{\kappa_0 R})} \le \left(\frac{C}{R}\right)^l \prod_{j=1}^l \left(3R^2 \lambda_k + \frac{1}{(\kappa_j - \kappa_{j-1})}\right) \cdot \|u\|_{L^2(B_{\kappa_l R})}.$$

Set $\kappa_j := 2 + \frac{j}{l}$. Then for large l

$$\|\partial^\beta u\|_{L^2(B_{2R})} \le (Cl)^l \|u\|_{L^2(B_{3R})},$$

where C and all of the following constants only depend on n and R, and where here and below we use the usual convention on constants. Now on choosing l for a given large $m \in \mathbb{N}$ such that $l - \frac{n}{2} - 1 \le m < l - \frac{n}{2}$, we obtain with Sobolev's embedding theorem 10.8 for all $|\alpha| = m$ the bound

$$\|\partial^\alpha u\|_{C^0(B_R)} \le C \sum_{|\beta| \le l} \|\partial^\beta u\|_{L^2(B_{2R})}$$

$$\le C\left(1 + \sum_{0 < |\beta| \le l} (C|\beta|)^{|\beta|}\right) \|u\|_{L^2(B_{3R})}$$

$$\le C\left(1 + (Cl)^l (l+1)^n\right) \|u\|_{L^2(B_{3R})}$$

$$\le (Cm)^{m+2n} \|u\|_{L^2(B_{3R})} \quad \text{for large } m.$$

This implies that the power series (let $x_0 = 0$ without loss of generality)

$$u_*(x) := \sum_{|\alpha| \geq 0} \frac{\partial^\alpha u(0)}{\alpha!} x^\alpha$$

has a positive radius of convergence. To see this, note that with $e := \frac{x}{|x|}$ we have

$$\sum_{|\alpha| \geq 1} \frac{|\partial^\alpha u(0)|}{\alpha!} |x^\alpha| = \sum_{m=1}^{\infty} \Big(\sum_{|\alpha|=m} \frac{|\partial^\alpha u(0)| \, |e^\alpha|}{\alpha!} \Big) |x|^m$$

$$\leq \sum_{m=1}^{\infty} \Big(\sum_{|\alpha|=m} \frac{|e^\alpha|}{\alpha!} \Big) (Cm)^{m+2n} |x|^m \|u\|_{L^2(B_{3R})} \,.$$

Setting

$$p(x) := \frac{(x \bullet \widetilde{e})^m}{m!} \quad \text{with } \widetilde{e}_i := \operatorname{sign} e_i \text{ for all } i$$

yields, if ∂_e denotes the partial derivative in the direction e, that

$$\sum_{|\alpha|=m} \frac{|e^\alpha|}{\alpha!} = \sum_{|\alpha|=m} \frac{\partial^\alpha p(0)}{\alpha!} e^\alpha = \frac{1}{m!} \partial_e^m p(0) = \frac{(e \bullet \widetilde{e})^m}{m!} \leq \frac{n^m}{m!} \,.$$

Hence the radius of convergence is positive, if

$$\sqrt[m]{\frac{n^m m^{m+2n}}{m!}} = n \Big(\sqrt[m]{m} \Big)^{2n} \frac{m}{\sqrt[m]{m!}}$$

remains bounded as $m \to \infty$. But this is easily seen, since

$$\log \sqrt[m]{m!} = \frac{1}{m} \sum_{j=1}^{m} \log j \geq \sum_{j=1}^{m} \int_{1+\frac{j-1}{m}}^{1+\frac{j}{m}} \log(m(s-1)) \, ds$$

$$= \int_1^2 \log(m(s-1)) \, ds = \log m - \int_0^1 |\log s| \, ds \,.$$

Similarly, we obtain that $u = u_*$ in a neighbourhood of 0, because the Taylor expansion yields for $|x| \leq R$ that

$$\left| u(x) - \sum_{|\alpha| \leq m-1} \frac{\partial^\alpha u(0)}{\alpha!} x^\alpha \right| = \left| m \int_0^1 (1-s)^{m-1} \sum_{|\alpha|=m} \frac{\partial^\alpha u(sx)}{\alpha!} x^\alpha \, ds \right|$$

$$\leq \sum_{|\alpha|=m} \frac{|e^\alpha|}{\alpha!} \|\partial^\alpha u\|_{C^0(B_R)} |x|^m \,,$$

which, in much the same way as above, converges to 0 as $m \to \infty$, if $|x|$ is sufficiently small. $\qquad \square$

It remains to discuss under what conditions the eigenfunctions $u \in E_k$ satisfy the boundary condition $u = 0$ in the classical sense. This can be answered, for example, with the regularity theory up to the boundary. With the help of theorem A12.3 it follows, as in the proof of 12.17(3), that $u \in W^{m,2}(\Omega)$, provided Ω has a $C^{m-1,1}$-boundary. Then it follows from 10.13 that $u \in C^0(\overline{\Omega})$ (upon modification on a null set), if $m - \frac{n}{2} > 0$ (for $n = 2$ this is satisfied for $m = 2$, i.e. for domains Ω with $C^{1,1}$-boundary).

The eigenvalue problem for the Laplace operator can also be treated as a minimum problem. To this end, consider

$$\lambda_* := \inf \left\{ \int_\Omega |\nabla u|^2 \, \mathrm{dL}^n \; ; \; u \in W_0^{1,2}(\Omega) \text{ and } \int_\Omega |u|^2 \, \mathrm{dL}^n = 1 \right\} .$$

The infimum is attained (cf. 8.17), because if $(u_j)_{j \in \mathbb{N}}$ is a minimizing sequence, i.e. if

$$\int_\Omega |\nabla u_j|^2 \, \mathrm{dL}^n \longrightarrow \lambda_* \quad \text{as } j \to \infty \quad \text{and} \quad \int_\Omega |u_j|^2 \, \mathrm{dL}^n = 1 ,$$

then $(u_j)_{j \in \mathbb{N}}$ is bounded in the Hilbert space $W_0^{1,2}(\Omega)$, and so there exists a subsequence such that $u_j \to u_*$ weakly in $W_0^{1,2}(\Omega)$. Rellich's embedding theorem A8.1 then yields that $u_j \to u_*$ converges (strongly) in $L^2(\Omega)$, and hence (use 8.3(3))

$$\int_\Omega |\nabla u_*|^2 \, \mathrm{dL}^n \leq \liminf_{k \to \infty} \int_\Omega |\nabla u_k|^2 \, \mathrm{dL}^n = \lambda_* \quad \text{and} \quad \int_\Omega |u_*|^2 \, \mathrm{dL}^n = 1 ,$$

i.e. u_* solves the minimum problem. It then holds for all $v \in W_0^{1,2}(\Omega)$ and $\varepsilon \in \mathbb{R}$ that

$$\frac{\int_\Omega |\nabla(u_* + \varepsilon v)|^2 \, \mathrm{dL}^n}{\int_\Omega |u_* + \varepsilon v|^2 \, \mathrm{dL}^n} \geq \frac{\int_\Omega |\nabla u_*|^2 \, \mathrm{dL}^n}{\int_\Omega |u_*|^2 \, \mathrm{dL}^n} ,$$

which implies, as in 12.14, that

$$\int_\Omega (\nabla u_* \bullet \nabla v - \lambda_* u_* v) \, \mathrm{dL}^n = 0 \quad \text{for all } v \in W_0^{1,2}(\Omega) ,$$

and so u_* is an eigenfunction, hence by 12.17(2) there exists $k \in \mathbb{N}$ with $\lambda_* = \lambda_k$. In addition, observing that for every eigenvalue λ_k, $k \in \mathbb{N}$, and for every eigenfunction u_k corresponding to the eigenvalue λ_k it holds that (set $\zeta = u_k$ in the weak differential equation)

$$\lambda_* = \frac{\int_\Omega |\nabla u_*|^2 \, \mathrm{dL}^n}{\int_\Omega |u_*|^2 \, \mathrm{dL}^n} \leq \frac{\int_\Omega |\nabla u_k|^2 \, \mathrm{dL}^n}{\int_\Omega |u_k|^2 \, \mathrm{dL}^n} = \lambda_k ,$$

we conclude that λ_* must be the smallest eigenvalue. Splitting off the eigenfunction u_*, i.e. considering the minimum problem now on the space

$$\left\{ u \in W_0^{1,2}(\Omega) \ ; \ \int_\Omega u \, u_* \, \mathrm{dL}^n = 0 \right\},$$

we inductively obtain, similarly to 12.14, the remaining eigenfunctions and eigenvalues.

The variational problem in 12.14 may be interpreted as the dual problem to the one considered here. To this end, note that it follows from 12.14 and the relation for $(Tf, f)_{L^2(\Omega)}$ in the proof of 12.17 that

$$\inf_{\substack{u \in W_0^{1,2}(\Omega) \\ \|u\|_{L^2} = 1}} \int_\Omega |\nabla u|^2 \, \mathrm{dL}^n = \inf_{k \in \mathbb{N}} \lambda_k$$

$$= (\sup_{k \in \mathbb{N}} \lambda_k^{-1})^{-1} = \left(\sup_{\substack{f \in L^2(\Omega) \\ \|f\|_{L^2} = 1}} \int_\Omega |\nabla (Tf)|^2 \, \mathrm{dL}^n \right)^{-1}.$$

E12 Exercises

E12.1 Adjoint map on C^0. Let $f \in X := C^0([0,1]; \mathbb{R})$ and let $T \in \mathscr{L}(X)$ be defined by

$$Tg := f \cdot g.$$

Calculate T' with the help of the isomorphism between X' and $rca([0,1]; \mathbb{R})$ and show that:

(1) T is injective $\Longleftrightarrow f^{-1}(\mathbb{R} \setminus \{0\})$ is dense in $[0,1]$.

(2) T' is surjective $\Longleftrightarrow f$ has no roots.

Solution. Let $Y := rca([0,1]; \mathbb{R})$, let $J : Y \to X'$ denote the isometric isomorphism from 6.23, and let

$$T^* := J^{-1} T' J \in \mathscr{L}(Y).$$

For $g \in X$ and $\nu \in Y$ we then have that

$$\int_0^1 gf \, \mathrm{d}\nu = \langle Tg, J\nu \rangle_{C^0} = \langle g, T'J\nu \rangle_{C^0} = \int_0^1 g \, \mathrm{d}(T^*\nu).$$

Hence the measure $T^*\nu$ is given by

$$(T^*\nu)(E) = \int_E f \, \mathrm{d}\nu \quad \text{for Borel sets } E \subset [0,1].$$

If f has no roots, then $\frac{1}{f} \in X$. Defining for $\mu \in Y$

$$\nu(E) := \int_E \frac{1}{f} \, \mathrm{d}\mu,$$

it follows that

$$(T^*\nu)(E) = \int_E f \cdot \frac{1}{f}\,d\mu = \mu(E)\,,$$

i.e. T^* is surjective. Conversely, if $x_0 \in [0,1]$ is a root of f, then we have for all $\nu \in Y$ that

$$(T^*\nu)(\{x_0\}) = \int_{\{x_0\}} f\,d\nu = 0\,,$$

i.e. $\mathscr{R}(T^*)$ does not contain the Dirac measure at the point x_0. This proves (2).

If $f^{-1}(\mathbb{R}\setminus\{0\})$ is dense in $[0,1]$, then $Tg = 0$ implies that $g = 0$ in the set $f^{-1}(\mathbb{R}\setminus\{0\})$. The continuity of g then yields that $g = 0$ also in $\overline{f^{-1}(\mathbb{R}\setminus\{0\})} = [0,1]$. Hence T is injective. Conversely, if $f = 0$ on an open interval $I \subset [0,1]$, choose $g \in X$ with $g \neq 0$ and $g = 0$ on $[0,1]\setminus I$. Then $Tg = 0$, and so T is not injective. This proves (1). $\qquad\square$

E12.2. If X, Y are Banach spaces and if $T \in \mathscr{L}(X;Y)$, then:

(1) T is surjective $\Longrightarrow$ T' is injective.

(2) T' is surjective $\Longrightarrow$ T injective.

Remark: In (2) the converse in general does not (!) hold, as can be seen from the example $f(x) := x$ in E12.1.

Solution (1). See 12.4(2). $\qquad\square$

Solution (2). We have from (1) that T'' is injective, and then from 12.3(5) that so is T. $\qquad\square$

E12.3. Consider the Hilbert-Schmidt operator in 12.15: Let Ω be bounded, let $K \in C^0(\overline{\Omega} \times \overline{\Omega})$ and let $\lambda_k > 0$ for all k (i.e. T is positive semidefinite). Then the series in the representation of K converges uniformly in $\overline{\Omega} \times \overline{\Omega}$.

Solution. Without loss of generality let $N = \mathbb{N}$. It follows that $x \mapsto (Tf)(x)$ is continuous in $\overline{\Omega}$ for $f \in L^2(\Omega)$, and so $\mathscr{R}(T) \subset C^0(\overline{\Omega})$. Hence the eigenfunctions f_k are continuous on $\overline{\Omega}$. In the proof of 12.15 we established that for $g \in L^2(\Omega)$ it holds for almost all x that

$$\int_\Omega K(x,y)g(y)\,dy = \sum_{k\in\mathbb{N}} \lambda_k f_k(x) \int_\Omega g(y)\overline{f_k(y)}\,dy\,. \tag{E12-1}$$

We now show that

$$\sum_{k\in\mathbb{N}} \lambda_k |f_k(x)|^2 = K(x,x) \quad \text{for all } x \in \overline{\Omega}, \tag{E12-2}$$

where $K(x,x) = \overline{K(x,x)}$ is real, by our assumptions. To this end, for $g \in L^2(\Omega)$ we define

$$\widetilde{g} := g - \sum_{k \leq n} (g \,, f_k)_{L^2} \, f_k \,.$$

Noting that $(Tg \,, f_k)_{L^2} = (g \,, T f_k)_{L^2} = \lambda_k \, (g \,, f_k)_{L^2}$, we have

$$0 \leq (T\widetilde{g} \,, \widetilde{g})_{L^2} = (Tg \,, g)_{L^2} - \sum_{k \leq n} \lambda_k |(g \,, f_k)_{L^2}|^2 \,.$$

Set $g(y) = g_\varepsilon(y) := \varphi_\varepsilon(z - y)$ with a standard Dirac sequence $(\varphi_\varepsilon)_{\varepsilon > 0}$ and $z \in \Omega$. Then

$$(Tg_\varepsilon \,, g_\varepsilon)_{L^2} \to K(z, z) \quad \text{and} \quad (g_\varepsilon \,, f_k)_{L^2} \to \overline{f_k(z)} \quad \text{as } \varepsilon \to 0,$$

and consequently

$$0 \leq K(z, z) - \sum_{k \leq n} \lambda_k |f_k(z)|^2 \,.$$

The continuity of f_k and K then yields for all $x \in \overline{\Omega}$ that

$$\sum_{k \leq n} \lambda_k |f_k(x)|^2 \leq K(x, x) \quad \text{and hence} \quad \sum_{k \in \mathbb{N}} \lambda_k |f_k(x)|^2 \leq K(x, x) \,.$$

In order to prove the reverse inequality we observe that (E12-1) holds for all $x \in \overline{\Omega}$. To see this, note that

$$\left(\sum_{k=m}^{l} |\lambda_k f_k(x) \, (g \,, f_k)_{L^2}| \right)^2 \leq \sum_{k \in \mathbb{N}} \lambda_k^2 |f_k(x)|^2 \cdot \sum_{k=m}^{l} |(g \,, f_k)_{L^2}|^2$$

$$\leq \sup_{k \in \mathbb{N}} \lambda_k \cdot K(x, x) \cdot \sum_{k=m}^{l} |(g \,, f_k)_{L^2}|^2$$

$$\longrightarrow 0 \quad \text{as } m \to \infty \text{ uniformly in } x,$$

and so the series on the right-hand side of (E12-1), as a function of x, converges uniformly in $\overline{\Omega}$. Since the left-hand side of (E12-1) is also a continuous function of x, we see that (E12-1) holds for all $x \in \overline{\Omega}$. On the other hand,

$$Tg(x) = \sum_k \lambda_k f_k(x) \, (g \,, f_k)_{L^2}$$

$$\leq \left(\sum_k \lambda_k |f_k(x)|^2 \right)^{\frac{1}{2}} \left(\sum_k \lambda_k |(g \,, f_k)_{L^2}|^2 \right)^{\frac{1}{2}},$$

with

$$\sum_k \lambda_k |(g \,, f_k)_{L^2}|^2 = \left(\sum_k \lambda_k \, (g \,, f_k)_{L^2} \, f_k \,, g \right)_{L^2} = (Tg \,, g)_{L^2} \,.$$

Hence for $z \in \Omega$ and $\varepsilon \to 0$

$$K(x,z) \longleftarrow \int_\Omega K(x,y)g_\varepsilon(y)\,\mathrm{d}y = Tg_\varepsilon(x)$$

$$\leq \left(\sum_k \lambda_k |f_k(x)|^2\right)^{\frac{1}{2}} \cdot \left((Tg_\varepsilon\,,\,g_\varepsilon)_{L^2}\right)^{\frac{1}{2}}$$

$$\longrightarrow \left(\sum_k \lambda_k |f_k(x)|^2\right)^{\frac{1}{2}} \cdot K(z,z)^{\frac{1}{2}}\,,$$

and so letting $z \to x$ we obtain the desired inequality, which proves (E12-2). It follows from Dini's theorem that the series in (E12-2) converges uniformly for $x \in \overline{\Omega}$. Noting that then also

$$\left|\sum_{j=m}^l \lambda_k f_k(x)\overline{f_k(y)}\right| \leq \left(\sum_{j=m}^l \lambda_k |f_k(x)|^2\right)^{\frac{1}{2}} \left(\sum_{j=m}^l \lambda_k |f_k(y)|^2\right)^{\frac{1}{2}} \longrightarrow 0$$

as $m \to \infty$ uniformly in $(x,y) \in \overline{\Omega} \times \overline{\Omega}$, yields the desired result.   □

E12.4. Calculate the eigenvalues λ and eigenfunctions u of

$$u'' + \lambda u = 0 \quad \text{in } \Omega, \qquad u = 0 \quad \text{on } \partial\Omega$$

where $\Omega = \,]0, R[$, $R > 0$.

Solution. By 12.17, we know $\lambda > 0$. Similarly to w in 12.16 it then holds that

$$u(x) = \alpha_1 \cos(\sqrt{\lambda}\, x) + \alpha_2 \sin(\sqrt{\lambda}\, x)$$

with $\alpha_1, \alpha_2 \in \mathbb{R}$. The boundary conditions yield $\alpha_1 = 0$ and $\alpha_2 \sin(\sqrt{\lambda}\, R) = 0$. Since $u \neq 0$, it follows that $\alpha_2 \neq 0$ and $\sqrt{\lambda}\, R = k\pi$ for a $k \in \mathbb{N}$. Hence the eigensolutions are given by

$$\lambda_k := \left(\frac{k\pi}{R}\right)^2 \qquad \text{for } k \in \mathbb{N},$$

$$u_k(x) := a_k \sin(\sqrt{\lambda_k}\, x) \qquad \text{with } a_k \neq 0.$$

It follows that the eigenspaces E_k in 12.17 are one-dimensional and that

$$L^2(]0, R[) = \overline{\text{span}\{u_k\,;\, k \in \mathbb{N}\}}$$

(compare with 9.9 (!)).   □

E12.5. Consider the weak eigenvalue problem

$$\Delta u + \lambda u = 0 \text{ in } \Omega, \qquad u = 0 \text{ on } \partial\Omega$$

on the cuboid

$$\Omega = \underset{i=1}{\overset{n}{\times}} \,]0, R_i[\, \subset \mathbb{R}^n , \quad \text{where } R_1, \ldots, R_n > 0.$$

Show: Define for $k = (k_1, \ldots, k_n) \in \mathbb{N}^n$

$$\lambda_k := \sum_{i=1}^{n} \lambda_{k_i}^i \qquad \text{with } \lambda_j^i := \left(\frac{j\pi}{R_i}\right)^2,$$

$$e_k(x) := \prod_{i=1}^{n} e_{k_i}^i(x_i) \quad \text{with } e_j^i(z) := \sqrt{\frac{2}{R_i}} \sin\left(\frac{j\pi}{R_i} z\right).$$

Then $(\lambda, u) \in \mathbb{R} \times W_0^{1,2}(\Omega)$ is an eigensolution if and only if there exists a $k \in \mathbb{N}^n$ such that

$$\lambda = \lambda_k, \quad u \in \text{span}\{e_l \, ; \, \lambda_l = \lambda_k\}.$$

Solution. We will prove this via induction on n. For $n = 1$ this was shown in E12.4. Let $n \geq 2$ and let (λ, u) be an eigensolution. Define for $j \in \mathbb{N}$

$$u_j(y) := \int_0^{R_n} u(y, z) \, e_j^n(z) \, dz \quad \text{for } y \in \widetilde{\Omega} := \underset{i=1}{\overset{n-1}{\times}} \,]0, R_i[.$$

Then $u_j \in W_0^{1,2}(\widetilde{\Omega})$ and for $\zeta \in W_0^{1,2}(\widetilde{\Omega})$

$$\int_{\widetilde{\Omega}} \nabla_y \zeta \bullet \nabla_y u_j \, dL^{n-1} = \int_\Omega \nabla_y \big(\zeta(y) e_j^n(z)\big) \bullet \nabla_y u(y, z) \, d(y, z)$$

$$= \int_\Omega \zeta(y) e_j^n(z) \, \lambda u(y, z) \, d(y, z) - \int_\Omega \zeta(y) \int_0^{R_i} (e_j^n)'(z) \, \partial_n u(y, z) \, dz \, dy$$

$$= \int_{\widetilde{\Omega}} \zeta(y) (\lambda - \lambda_j^n) u_j(y) \, dy,$$

since $(e_j^n)'' + \lambda_j^n \, e_j^n = 0$. If $u_j \neq 0$, then the inductive hypothesis implies that $\lambda = \lambda_l$ for an $l \in \mathbb{N}^n$ with $l_n = j$ and u_j is a finite linear combination of functions

$$y \longmapsto \prod_{i=1}^{n-1} e_{l_i}^i(y_i) \quad \text{with } l \in \mathbb{N}^n, \, l_n = j, \, \lambda_l = \lambda.$$

Since this is only possible for finitely many j, and since $(e_j^n)_{j \in \mathbb{N}}$ is an orthonormal basis of $L^2(]0, R_n[)$, it follows that with a finite sum

$$u(y, z) = \sum_j u_j(y) \, e_j^n(z).$$

This yields the desired result.   □

A12 Elliptic regularity theory

We prove the regularity theorems which were announced in 6.9, and we will do this on the basis of Friedrichs' theorem. Together with the Sobolev embedding theorems one then obtains that under suitable conditions on the data the weak solutions in 6.6 and 6.8 are indeed classical solutions. This has already been used in the proof of 12.17. The regularity theory hinges on the following local estimate.

A12.1 Lemma. Let $u \in W^{1,2}(B_R)$ with $B_R := B_R(x_0) \subset \mathbb{R}^n$ be a weak solution of the differential equation

$$\sum_i \partial_i \Big(\sum_j a_{ij} \partial_j u + q_i \Big) = f \quad \text{in } B_R \,,$$

i.e. we assume that $q_i, f \in L^2(B_R)$ and that

$$\int_{B_R} \Big(\sum_i \partial_i \zeta \Big(\sum_j a_{ij} \partial_j u + q_i \Big) + \zeta f \Big) \, \mathrm{d}L^n = 0 \tag{A12-1}$$

for all $\zeta \in C_0^\infty(B_R)$, where $(a_{ij})_{i,j}$ is measurable, elliptic (see 6.4) and bounded, that is, there are constants $C_0 \geq c_0 > 0$ such that for almost all $x \in B_R$,

$$\Big| (a_{ij}(x))_{i,j} \Big| := \Big(\sum_{i,j=1}^n |a_{ij}(x)|^2 \Big)^{\frac{1}{2}} \leq C_0 \quad \text{and}$$

$$\sum_{i,j=1}^n a_{ij}(x) \xi_i \xi_j \geq c_0 |\xi|^2 \quad \text{for } \xi \in \mathbb{R}^n.$$

Then (with the notation $B_r := B_r(x_0)$ for $r > 0$) for all $0 < \kappa < 1$ we have

$$\|\nabla u\|_{L^2(B_{(1-\kappa)R})} \leq C \left(\frac{\kappa R}{C_0} \|f\|_{L^2(B_R)} + \frac{1}{c_0} \|q\|_{L^2(B_R)} + \frac{1}{\kappa R} \frac{C_0}{c_0} \|u\|_{L^2(B_R)} \right),$$

with a universal constant C.

Remark: Since for every $c \in \mathbb{R}$ the function $u - c$ is also a solution, we can replace u by $u - c$ in the last term on the right-hand side of the estimate. The best choice for c is the mean value of u over B_R.

Proof. It follows that (A12-1) holds for all $\zeta \in W_0^{1,2}(B_R)$. Choose a cut-off function $\eta \in C_0^\infty(B_R)$ with $0 \leq \eta \leq 1$, $\eta = 1$ on $B_{(1-\kappa)R}$, and $|\nabla \eta| \leq \frac{2}{\kappa R}$. Now set $\zeta = \eta^2 u$ and obtain

$$0 = \int_{B_R} \eta^2 \sum_{i,j} \partial_i u \, a_{ij} \partial_j u \, \mathrm{d}L^n + \int_{B_R} \sum_i \partial_i u \Big(\sum_j 2 a_{ji} \eta u \partial_j \eta + q_i \eta^2 \Big) \, \mathrm{d}L^n$$

$$+ \int_{B_R} u \Big(\sum_i 2 q_i \eta \partial_i \eta + f \eta^2 \Big) \, \mathrm{d}L^n \,,$$

which using the ellipticity yields that

$$c_0 \int_{B_R} \eta^2 |\nabla u|^2 \, d\mathrm{L}^n \le \int_{B_R} \eta |\nabla u| (2C_0 |u| \, |\nabla \eta| + |q| \eta) \, d\mathrm{L}^n$$
$$+ \int_{B_R} \eta |u| (2|q| \, |\nabla \eta| + |f| \eta) \, d\mathrm{L}^n \, .$$

Note that for all $\delta > 0$ (use Young's inequality (8-12))

$$\eta |\nabla u| (2C_0 |u| \, |\nabla \eta| + |q| \eta) \le \delta \eta^2 |\nabla u|^2 + \frac{1}{\delta} \Big(C_0^2 \, u^2 |\nabla \eta|^2 + \frac{1}{4} |q|^2 \eta^2 \Big) \, .$$

Then choosing $\delta = \frac{c_0}{2}$ we have

$$\int_{B_R} \eta^2 |\nabla u|^2 \, d\mathrm{L}^n \le \frac{4C_0^2}{c_0^2} \int_{B_R} u^2 |\nabla \eta|^2 \, d\mathrm{L}^n$$
$$+ \int_{B_R} \Big(\frac{1}{c_0^2} \eta^2 |q|^2 + \frac{2}{c_0} \eta |u| (2|q| \, |\nabla \eta| + |f| \eta) \Big) \, d\mathrm{L}^n \, .$$

For the last two terms it holds for all $\delta > 0$ and $\varepsilon > 0$ that

$$\frac{4}{c_0} \eta |u| \cdot |q| |\nabla \eta| \le \frac{1}{\delta} u^2 |\nabla \eta|^2 + \frac{4\delta}{c_0^2} \eta^2 |q|^2 \, ,$$
$$\frac{2}{c_0} \eta |u| \cdot |f| \eta \le \frac{1}{\varepsilon} \eta^2 u^2 + \frac{\varepsilon}{c_0^2} \eta^2 |f|^2 \, .$$

Overall we obtain the estimate

$$\int_{B_R} \eta^2 |\nabla u|^2 \, d\mathrm{L}^n \le \int_{B_R} \Big(\big(\frac{4C_0^2}{c_0^2} + \frac{1}{\delta} \big) u^2 |\nabla \eta|^2 + \frac{1}{\varepsilon} \eta^2 u^2 \Big) \, d\mathrm{L}^n$$
$$+ \int_{B_R} \Big(\frac{1+4\delta}{c_0^2} \eta^2 |q|^2 + \frac{\varepsilon}{c_0^2} \eta^2 |f|^2 \Big) \, d\mathrm{L}^n \, .$$

Now we exploit the properties of η and $\nabla \eta$ and observe that the optimal choice of δ and ε is given by

$$\varepsilon = (\kappa R)^2 \cdot \delta \, , \quad \delta = \Big(\frac{c_0}{C_0} \Big)^2 \le 1 \, .$$

This yields the desired result, for instance with $C = 5$. $\qquad\square$

A12.2 Friedrichs' theorem. Let $\Omega \subset \mathbb{R}^n$ be open and let $u \in W^{1,2}(\Omega)$ be a weak solution of the differential equation in A12.1 in Ω, and let $(a_{ij})_{i,j}$ satisfy the assumptions in A12.1 in Ω. Moreover, let $m \ge 0$ and let

$$f \in W^{m,2}(\Omega) \quad (\text{i.e. } f \in L^2(\Omega) \text{ for } m = 0),$$
$$q_i \in W^{m+1,2}(\Omega) \, , \quad a_{ij} \in C^{m,1}(\Omega) \, .$$

Then $u \in W^{m+2,2}_{\mathrm{loc}}(\Omega)$, i.e. $u \in W^{m+2,2}(D)$ for every open set $D \subset \Omega$, for which $\overline{D} \subset \Omega$ is compact, i.e. $D \subset\subset \Omega$. The solution u can be bounded in the $W^{m+2,2}(D)$-norm by the data.

Proof. First let $m = 0$. Taking for $h < \text{dist}(D, \partial\Omega)$ the difference between the differential equation in coordinate direction k with step size h, we obtain with the difference quotients

$$\partial_k^h v(x) := \frac{1}{h}\big(v(x + h\mathbf{e}_k) - v(x)\big)$$

that for all $\zeta \in C_0^\infty(D)$

$$\int_\Omega \left(\sum_i \partial_i \zeta \Big(\sum_j a_{ij}\partial_j\partial_k^h u + \sum_j (\partial_k^h a_{ij})\partial_j u(\cdot + h\mathbf{e}_k) + \partial_k^h q_i \Big) + \zeta \partial_k^h f \right) \mathrm{dL}^n = 0.$$

Upon setting

$$f_h(x) := \frac{1}{h}\int_0^h f(x + s\mathbf{e}_k)\,\mathrm{d}s$$

the last term satisfies

$$\int_\Omega \zeta \partial_k^h f\,\mathrm{dL}^n = -\int_\Omega \partial_k\zeta \cdot f_h\,\mathrm{dL}^n\,,$$

and consequently

$$\int_D \sum_i \partial_i\zeta\Big(\sum_j a_{ij}\partial_j(\partial_k^h u) + q_{hi}\Big)\,\mathrm{dL}^n = 0\,, \qquad\qquad \text{(A12-2)}$$

where

$$q_{hi} := \sum_j (\partial_k^h a_{ij})\partial_j u(\cdot + h\mathbf{e}_k) + \partial_k^h q_i - f_h\delta_{i,k}\,. \qquad\qquad \text{(A12-3)}$$

We observe that the weak differential equation (A12-2) for $\partial_k^h u$ is of the same type as (A12-1) for u, and so A12.1 yields for every $D' \subset\subset D$ the estimate

$$\left\|\partial_k^h u\right\|_{W^{1,2}(D')} \le C\big(\|q_h\|_{L^2(D)} + \left\|\partial_k^h u\right\|_{L^2(D)}\big)\,. \qquad\qquad \text{(A12-4)}$$

Here C depends on a_{ij}, D and D'. (More precisely, we apply A12.1 to balls $B_{(1-\kappa)R}(x_l)$, which form a finite cover of D' such that $B_R(x_l) \subset D$.) Now it holds

$$\int_D \left|\partial_k^h u(x)\right|^2\,\mathrm{d}x = \int_D \left|\frac{1}{h}\int_0^h \partial_k u(x + s\mathbf{e}_k)\,\mathrm{d}s\right|^2\,\mathrm{d}x$$

$$\le \int_D \frac{1}{h}\int_0^h |\partial_k u(x + s\mathbf{e}_k)|^2\,\mathrm{d}s\,\mathrm{d}x \le \int_\Omega |\partial_k u|^2\,\mathrm{dL}^n$$

and correspondingly

$$\|q_h\|_{L^2(D)} \le C\big(\|\nabla u\|_{L^2(\Omega)} + \|\partial_k q\|_{L^2(\Omega)} + \|f\|_{L^2(\Omega)}\big)\,,$$

which shows that $\partial_k^h u$ is bounded in $W^{1,2}(D')$. It follows from theorem 8.10 (see 8.11(1)) that there exists a $v_k \in W^{1,2}(D')$ such that for a subsequence $h \to 0$

$$\partial_k^h u \to v_k \quad \text{weakly in } W^{1,2}(D').$$

Therefore, if $\zeta \in C_0^\infty(D')$ then as $h \to 0$

$$\int_{D'} \zeta \cdot \partial_l v_k \, dL^n \longleftarrow \int_{D'} \zeta \cdot \partial_l \partial_k^h u \, dL^n = -\int_{D'} \partial_k^{-h} \zeta \cdot \partial_l u \, dL^n$$

$$= \int_{D'} \partial_l \partial_k^{-h} \zeta \cdot u \, dL^n \longrightarrow \int_{D'} \partial_{lk} \zeta \cdot u \, dL^n,$$

and so it follows from the definition of the Sobolev spaces that $u \in W^{2,2}(D')$, with $\partial_{kl} u = \partial_k v_l = \partial_l v_k$.

The same argumentation shows that $\partial_k^h q_i \to \partial_k q_i$ converges weakly in $L^2(D)$, and similarly $\partial_k^h a_{ij} \to \partial_k a_{ij}$ (recall from 10.5(2) that $a_{ij} \in W^{1,\infty}(D)$) weakly in $L^2(D)$. In addition, $\partial_j u(\cdot + h e_k) \to \partial_j u$ and $f_h \to f$ converge strongly in $L^2(D)$. Overall this yields (use 8.3(6) for Hilbert spaces) the weak convergence in $L^2(D)$ of the functions q_{hi} defined in (A12-3). Hence in the weak differential equation (A12-2) we can pass to the limit and obtain for $\zeta \in C_0^\infty(D')$ that

$$\int_{D'} \sum_i \partial_i \zeta \left(\sum_j a_{ij} \partial_j (\partial_k u) + \sum_j \partial_k a_{ij} \partial_j u + \partial_k q_i + f \delta_{i,k} \right) dL^n = 0.$$

An $W^{1,2}$-bound of $\partial_k u$ locally in D' then follows from A12.1 (since this equation is of the same type as (A12-1)), or by passing to the limit in (A12-4).

In the case $m \geq 1$ we can apply the proof presented above for u on Ω to this differential equation for $\partial_k u$ on D' and hence iteratively obtain the desired result. $\square$

We now prove the regularity of weak solutions to elliptic boundary value problems up to the boundary.

A12.3 Theorem. Let $\Omega \subset \mathbb{R}^n$ be open and bounded with Lipschitz boundary, and let $u \in W^{1,2}(\Omega)$ be the weak solution of the homogeneous Dirichlet problem

$$\sum_i \partial_i \left(\sum_j a_{ij} \partial_j u + q_i \right) = f \quad \text{in } \Omega,$$

$$u = 0 \quad \text{on } \partial\Omega,$$

from theorem 6.8 with the assumptions as in A12.2. If we assume in addition that $\partial\Omega$ can locally be represented as the graph of a $C^{m+1,1}$-function (as in A8.2), and that $a_{ij} \in C^{m,1}(\overline{\Omega})$, then

$$u \in W^{m+2,2}(\Omega).$$

The solution u can be bounded in the $W^{m+2,2}(\Omega)$-norm by the data.

Proof. It follows from theorem A12.2 that we need to prove the result on u only locally at the boundary. Hence let $g \in C^{m+1,1}(\mathbb{R}^{n-1})$ and $r, s > 0$, with

$$\{(y, g(y)) \in \mathbb{R}^n ; |y| < r\} \subset \partial\Omega,$$
$$\{(y, h) \in \mathbb{R}^n ; 0 < h - g(y) < s\} \subset \Omega.$$

Then (see A8.11)

$$\widetilde{u}(y, h) := u(y, h + g(y))$$

satisfies on $\widetilde{\Omega} := B_r(0) \times {]0, s[}$ the weak differential equation

$$\int_{\widetilde{\Omega}} \left(\sum_i \partial_i \zeta \left(\sum_j \widetilde{a}_{ij} \partial_j \widetilde{u} + \widetilde{q}_i \right) + \zeta \widetilde{f} \right) d\mathrm{L}^n = 0 \quad \text{for all } \zeta \in C_0^\infty(\widetilde{\Omega}),$$

where, we denote $\tau(y, h) := (y, h + g(y))$, the coefficients are given by

$$\widetilde{a} := D\tau^{-1} a \circ \tau (D\tau^{-1})^T, \quad \widetilde{q} := D\tau^{-1} q \circ \tau, \quad \widetilde{f} := f \circ \tau$$

and satisfy the assumptions of A12.2 (apply 10.5(2) for g).

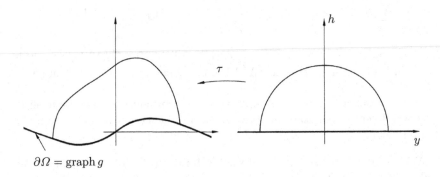

$\partial\Omega = \text{graph } g$

Fig. 12.1. *Straightening of the boundary*

We need to show that $\widetilde{u} \in W^{m+2,2}(\widetilde{D})$ for all $\widetilde{D} = B_{r'}(0) \times {]0, s'[}$ with $r' < r$ and $s' < s$. Once again we first consider the case $m = 0$. As in the proof of A12.2, we can form the difference of this differential equation in the coordinate directions $k < n$ and then obtain analogously to A12.2 (where now A12.1 also holds for $B_R(x_0) \cap \widetilde{\Omega}$ with $x_0 \in \partial\widetilde{\Omega}$ in place of $B_R(x_0) \subset \widetilde{\Omega}$) that for all domains $\widetilde{D}$ as above it holds that $\partial_k \widetilde{u} \in W^{1,2}(\widetilde{D})$ for $k < n$.

In order to establish that $\widetilde{u} \in W^{2,2}(\widetilde{D})$ it remains to show that $\partial_{nn}\widetilde{u} \in L^2(\widetilde{D})$, which now is a consequence of the differential equation for $\widetilde{u}$. This is because we have for $\zeta \in W_0^{1,2}(\widetilde{\Omega})$

$$\int_{\widetilde{\Omega}} \left(\partial_n \zeta \widetilde{a}_{nn} \partial_n \widetilde{u} + \zeta F \right) d\mathrm{L}^n = 0 \,,$$

where it follows from the previously shown that

$$F := - \sum_{\substack{i,j \le n \\ i+j < 2n}} \partial_i (\widetilde{a}_{ij} \partial_j \widetilde{u}) - \sum_i \partial_i \widetilde{q}_i + \widetilde{f} \in L^2(\widetilde{D}) \,.$$

Recalling that $\widetilde{a}_{nn} \in C^{0,1}(\overline{\widetilde{\Omega}})$ with $\widetilde{a}_{nn} \ge c_0 > 0$, we have that $\zeta \in W_0^{1,2}(\widetilde{\Omega})$ implies that also $\frac{\zeta}{\widetilde{a}_{nn}} \in W_0^{1,2}(\widetilde{\Omega})$. Since

$$\partial_n \left(\frac{\zeta}{\widetilde{a}_{nn}} \right) = \frac{\partial_n \zeta}{\widetilde{a}_{nn}} - \frac{\partial_n \widetilde{a}_{nn}}{\widetilde{a}_{nn}^2} \zeta \,,$$

we have

$$\int_{\widetilde{\Omega}} \partial_n \zeta \partial_n \widetilde{u} \, d\mathrm{L}^n = - \int_{\widetilde{\Omega}} \zeta \underbrace{\left(\frac{F}{\widetilde{a}_{nn}} - \frac{\partial_n \widetilde{a}_{nn}}{\widetilde{a}_{nn}} \partial_n \widetilde{u} \right)}_{\in L^2(\widetilde{D})} d\mathrm{L}^n \,,$$

which means that $\partial_{nn} \widetilde{u} \in L^2(\widetilde{D})$. This proves the regularity up to the boundary for $m = 0$. The case $m \ge 1$ can be shown inductively, by applying the above proof to derivatives of the weak differential equation in directions $i < n$.

□

The theorem also implies the corresponding regularity of the solutions in 6.8 with nonzero b-term, if $b \in W^{m,\infty}(\Omega)$. The proof is analogous to the proof of 12.17(3). In addition, the theorem and the proof carry over to the solutions of the Neumann problem in 6.6, since then in A12.1 the employed test functions of the type $\eta^2 \zeta$ with $\zeta \in W^{1,2}(\Omega)$ are admissible.

For weak solutions with mixed Dirichlet and Neumann boundary conditions (see 8.18(5)) the regularity theorem A12.3 holds, but in general it does not (!) hold for all of Ω. The standard counterexample is the following:

$$\Omega = \{ x \in \mathrm{IR}^2 \,;\, |x| < 1,\ x_2 > 0 \} \,,$$
$$u(x) = \mathrm{Im}\sqrt{z} \quad \text{with } z = x_1 + ix_2 \,.$$

Then $u \in W^{1,2}(\Omega)$ is a weak solution of the mixed boundary value problem

$$\begin{aligned}
\Delta u &= 0 & &\text{in } \Omega, \\
u(e^{i\theta}) &= \sin \tfrac{\theta}{2} & &\text{for } 0 \le \theta \le \pi, \\
u(0,r) &= 0 & &\text{for } 0 \le r \le 1, \\
\nu_\Omega \bullet \nabla u(0,-r) &= 0 & &\text{for } 0 < r < 1.
\end{aligned}$$

However, it is easily seen that u is not (!) an element of $W^{1,4}(\Omega)$. It follows from the embedding theorem 10.9 that neither is u an element of $W^{2,p}(\Omega)$ for $p \ge \frac{4}{3}$, something which can also be shown directly without great difficulty.

Nevertheless, $u \in C^{0,\frac{1}{2}}(\overline{\Omega})$, where we recall from 10.13 that the Sobolev space $W^{1,4}(\Omega)$ is continuously embedded into this space. However, the regularity theorem A12.2 is applicable.

References

Books on linear functional analysis:

[Aubin] *J.P. Aubin:* Applied Functional Analysis.
   2nd ed. Wiley 2000

[BachmanNarici] *G. Bachman, L. Narici:* Functional Analysis.
   Dover Publications 2000

[Baggett] *L.W. Baggett:* Functional Analysis. Marcel-Dekker 1991,
   or: [spot.colorado.edu/~baggett/functional.html]

[Berezansky] *Y.M. Berezansky, Z.G. Sheftel, G.F. Us:* Functional Analysis.
   Operator Theory: Advances and Applications Vol. 85, Birkhäuser 1996

[Boccara] *N. Boccara:* Functional Analysis. An Introduction for Physicists.
   Academic Press 1990

[Brezis] *H. Brezis:* Functional Analysis, Sobolev Spaces and
   Partial Differential Equations. Springer 2010

[Conway] *J.B. Conway:* A Course in Functional Analysis. 4th ed. Springer 1997

[Dobrowolski] *M. Dobrowolski:* Angewandte Funktionalanalysis.
   2nd ed. Springer 2010

[DunfordSchwartz] *N. Dunford, J.T. Schwartz:* Linear Operators I, II.
   2nd ed. Interscience Publishers 1988

[EidelmanMilmanTsolomitis] *Y. Eidelman, V. Milman, A. Tsolomitis:* Functional
   Analysis. An Introduction. Graduate Studies in Math. Vol. 66, AMS 2004

[Heuser] *H. Heuser:* Funktionalanalysis. Theorie und Anwendung.
   4th ed. Teubner 2006

[HirzebruchScharlau] *F. Hirzebruch, W. Scharlau:* Einführung in die Funktionalanalysis. Spektrum Akademischer Verlag 1996

[KurdilaZabarankin] *A.J. Kurdila, M. Zabarankin:* Convex Functional Analysis.
   Birkhäuser 2005

[Lax] *P.D. Lax:* Functional Analysis.
   Pure and Applied Mathematics Vol. 55. Wiley 2002

[LebedevVorovich] *L.P. Lebedev, I.I. Vorovich:* Functional Analysis in Mechanics.
   Springer 2002 Reprint: 2012

[MeiseVogt] *R. Meise, D. Vogt:* Introduction to Functional Analysis.
   Clarendon Press 1997

[PugachevSinitsyn] *V.S. Pugachev, I.N. Sinitsyn:* Lectures on Functional
   Analysis and Applications. World Scientific 1999

[ReedSimon] *M. Reed, B. Simon:* Methods of Modern Math. Physics I – IV.
   I: Functional Analysis. 2nd ed. Academic Press 1980

[Reddy] *B.D. Reddy:* Introductory Functional Analysis. Springer 1998

[Rudin1] *W. Rudin:* Functional Analysis. 2nd ed. 1991,
   12th Reprint: McGraw–Hill 2006

[Rudin2] *W. Rudin:* Real and Complex Analysis.
   3rd ed. McGraw–Hill 1987

[Schechter] *M. Schechter:* Principles of Functional Analysis.
Academic Press 1971, 2nd ed. AMS 2002

[Schroeder] *H. Schröder:* Funktionalanalysis. 2nd ed. Verlag Harri Deutsch 2000

[TaylorLay] *A.E. Taylor and D.C. Lay:* Introduction to Functional Analysis.
2nd ed. Wiley 1980

[Werner] *D. Werner:* Funktionalanalysis. 7th ed. Springer 2011

[Wloka] *J. Wloka:* Funktionalanalysis und ihre Anwendungen.
De Gruyter 1971

[Yosida] *K. Yosida:* Functional Analysis. 6th ed. Springer 1980, Reprint: 1995

Further literature:

[Adams] *R.A. Adams:* Sobolev Spaces. Pure and Applied Mathematics 65.
New edition: *R.A. Adams, J.J.F. Fournier:* Pure and Applied Mathematics 140,
2nd ed. Academic Press 2003, Reprint: 2009

[Braess] *D. Braess:* Finite Elements. Theory, Fast Solvers, and Application
in Solid Mechanics. 3rd ed. Cambridge Univ. Press 2007

[GilbargTrudinger] *D. Gilbarg, N.S. Trudinger:* Elliptic Partial Differential
Equations of Second Order. 3rd ed. Springer 1998, Reprint: 2001

[Halmos] *P.R. Halmos:* Measure Theory. 2nd ed. Springer 1978

[Kelley] *J.L. Kelley:* General Topology. Springer 1975, Reprint: 2012

[Mazya] *V.G. Maz'ya:* Sobolev Spaces. 2nd ed. Springer 2011

[Ruzicka] *M. Ružička:* Nichtlineare Funktionalanalysis. Eine Einführung.
Springer 2004. See also: [portal.uni-freiburg.de/aam/abtlg/ls/lsru/lehrbrz/]

[Sauvigny1] *F. Sauvigny:* Partielle Differentialgleichungen der Geometrie und
der Physik. Grundlagen und Integraldarstellungen. Springer 2004

[Sauvigny2] *F. Sauvigny:* Partielle Differentialgleichungen der Geometrie und
der Physik 2. Funktionalanalytische Lösungsmethoden. Springer 2005

[Schubert] *H. Schubert:* Topologie. Eine Einführung. Teubner 1975

[Schwartz] *L. Schwartz:* Théorie des Distributions. Hermann, Paris 1966

[Schweizer] *B. Schweizer:* Partielle Differentialgleichungen. Springer Spektrum 2013

[Simon] *L. Simon:* Lectures on Geometric Measure Theory. Proc. of
the Centre for Math. Analysis, Vol. 3, ANU Canberra 1984

Cited original articles:

[Murray] *F. J. Murray:* On complementary manifolds and projections in spaces
L_p and ℓ_p. Trans. Amer. Math. Soc., **41**, New York 1937, 138–152

[MRiesz] *M. Riesz:* Sur les ensembles compacts de fonctions summables.
Acta Sci. Math. Szeged, **6**, 1933, 136–142

Symbols

$\mathscr{L}(X) : \mathscr{L}(X;X)$, 142
$\mathscr{L}(X;Y)$: linear operators from X to Y, 142

(M1) : axiom, 16
(M2) : axiom, 16
(M3) : axiom, 16
$M(\mu;Y)$: μ-measurable functions with values in Y, 48

ν_Ω : outer normal at $\partial\Omega$, 263
(N1) : axiom, 13
(N2) : axiom, 13
(N3) : axiom, 13
$NBV(S)$: normalized functions of bounded variation, 204
$\mathscr{N}(T)$: null space of the operator T, 144

(P1) : axiom, 303
(P2) : axiom, 303
$\mathscr{P}(X)$: projections in X, 144

(Q1) : axiom, 304
(Q2) : axiom, 304

$\varrho(\cdot)$: notation for Fréchet metric, 16
$rba(\cdot)$, $rca(\cdot)$: space of regular measures, 186
$\mathscr{R}(T)$: image of the operator T, 144

R_X : Riesz's isomorphism from X to X', 163

(S1) : axiom, 9
(S2) : axiom, 9
(S3) : axiom, 9
(S4') : axiom, 9
(S4) : axiom, 9
$\mathrm{span}(E)$: linear hull of the set E, 115
$\mathrm{supp}(f)$: support of the function f, 41

(T1) : axiom, 20
(T2) : axiom, 20
(T3) : axiom, 20
(T4) : axiom, 20
T' : adjoint map of T, 145
$T(\mu;Y)$: step functions with respect to μ with values in Y, 74

$\mathrm{var}(f,\cdot)$: variation of a function f, 201

$W^{m,p}(\Omega)$: Sobolev space of order m for the power p, 64
$W_0^{m,p}(\Omega)$: Sobolev space with zero boundary values, 66
$W_{\mathrm{loc}}^{m,p}(\Omega)$: local Sobolev functions, 150

X' : dual space to X, 144

Z^0 : annihilator of Z, 390

Index

Printed in the United States
By Bookmasters